City of Westminster College

217305

KT-499-308

Introduction to Ergonomics

R.S. Bridger

Routledge
Taylor & Francis Group

LONDON AND NEW YORK

17-10-07

No. 217305 BRI

620.82

Published in 2003 by
CRC Press
Taylor & Francis Group
6000 Broken Sound Parkway NW, Suite 300
Boca Raton, FL 33487-2742

© 2003 by Robert Bridger
CRC Press is an imprint of Taylor & Francis Group

No claim to original U.S. Government works
Printed in the United States of America on acid-free paper
10 9 8 7 6 5 4

International Standard Book Number-10: 0-415-27378-1 (Softcover)
International Standard Book Number-13: 978-0-415-27378-7 (Softcover)

This book contains information obtained from authentic and highly regarded sources. Reprinted material is quoted with permission, and sources are indicated. A wide variety of references are listed. Reasonable efforts have been made to publish reliable data and information, but the author and the publisher cannot assume responsibility for the validity of all materials or for the consequences of their use.

No part of this book may be reprinted, reproduced, transmitted, or utilized in any form by any electronic, mechanical, or other means, now known or hereafter invented, including photocopying, microfilming, and recording, or in any information storage or retrieval system, without written permission from the publishers.

Trademark Notice: Product or corporate names may be trademarks or registered trademarks, and are used only for identification and explanation without intent to infringe.

Library of Congress Cataloging-in-Publication Data

Catalog record is available from the Library of Congress

informa

Taylor & Francis Group
is the Academic Division of Informa plc.

Visit the Taylor & Francis Web site at
http://www.taylorandfrancis.com

and the CRC Press Web site at
http://www.crcpress.com

Para Barbara, Daniella y Angelo

Contents

Preface to the second edition

Can it work, does it work, is it worth it?[1] These three questions have been foremost in my mind throughout the revision of '*Introduction to Ergonomics*'. In revising and updating the text, I have tried to attain three goals. First, to update the scientific content of the book to reflect the state of our knowledge at the beginning of the twenty-first century. Second, to maintain the book's essential character as a general introductory text that teaches the basic science that ergonomists use at work. Third, to add new material at the end of every chapter to answer the three questions above.

There is a great deal of evidence that ergonomics does work. It really does improve the interactions between people and machines and it really can make systems work better. How to demonstrate this has been one of the challenges in the process of revision. Several criteria have influenced the selection of supporting evidence. It would be ideal if evidence for the benefits of all of the diverse practices and subdisciplines of ergonomics came in the form of randomised controlled trials with double-blind application of treatments to satisfy even the most sceptical reviewer. This is not the case, and it never can be, so I have tried to present a variety of evidence that best exemplifies what each particular area has to offer.

The evidence comes in the form of field trials, field experiments, longitudinal studies and even a few laboratory experiments. Such a variety of research methods will never please everyone – what satisfies the university academic may seem dry and other-worldly to the production manager. Uncontrolled trials in real factories may not impress the academic, but practitioners may find therein much useful ammunition for arguing their cause.

Another theme that is pursued throughout the book is that engineering and design are increasingly driven by standards. Probably the best evidence that the value of ergonomics is now recognised is the publication of international standards for ergonomics. These standards are paving the way for a new, quantitative and much more precise form of practice. With this in mind, I have tried to inform the reader about these standards, wherever possible (with the rider that this is a textbook, not a design manual). The reader is encouraged to use these standards in practice and, with due deference to national bodies, the International Organization for Standardization (IOS) is recommended as the first port of call.

In keeping with these modern trends, some new essays and exercises have been added to encourage the learning of quantitative skills. Some of the older, perhaps

---

[1] Haynes. 1999. Can it work? Does it work? Is it worth it? *British Medical Journal*, **319** (Sep. 11): 652–653.

more frivolous, illustrations have been replaced by new drawings illustrating modern research in ergonomics. These have been rendered, in her customary 'estilo cuadradito', by Rina Araya, my graphic artist in Chile. Restrictions on the length of the book have led to some fairly ruthless editing to make way for new material. It is hoped that lecturers will continue to find much useful material to structure their courses and to illustrate the concepts of ergonomics and, at the end of their courses, that students will be able to give the following correct answers to the three opening questions:

'Yes, it can, Yes, it does, Yes it is!'

R. S. Bridger
Lee-on-the-Solent
Hants, UK
2002

1   Introduction

In the past, the man has been first; in the future, the system must be first.
(Frederick Winslow Taylor, *The Principles of Scientific Management*, 1911, p. 7)

Ergonomics is the study of the interaction between people and machines and the factors that affect the interaction. Its purpose is to improve the performance of systems by improving human machine interaction. This can be done by 'designing-in' a better interface or by 'designing-out' factors in the work environment, in the task or in the organisation of work that degrade human–machine performance.

Systems can be improved by

- Designing the user-interface to make it more compatible with the task and the user. This makes it easier to use and more resistant to errors that people are known to make.
- Changing the work environment to make it safer and more appropriate for the task.
- Changing the task to make it more compatible with user characteristics.
- Changing the way work is organised to accommodate people's psychological, and social needs.

In an information processing task, we might redesign the interface so as to reduce the load on the user's memory (e.g. shift more of the memory load of the task onto the computer system or redesign the information to make it more distinctive and easier to recall). In a manual handling task, we might redesign the interface by adding handles or using lighter or smaller containers to reduce the load on the musculoskeletal system. Work environments can be improved by eliminating vibration and noise and providing better seating, desking, ventilation or lighting, for example. New tasks can be made easier to learn and to perform by designing them so that they resemble tasks or procedures that people are already familiar with. Work organisation can be improved by enabling workers to work at their own pace, so as to reduce the psychophysical stresses of being 'tied to the machine' or by introducing subsidiary tasks to increase the range of physical activity at work and provide contact with others.

The implementation of ergonomics in system design should make the system work better by eliminating aspects of system functioning that are undesireable, uncontrolled or unaccounted for, such as

- Inefficiency – when worker effort produces sub-optimal output.
- Fatigue – in badly designed jobs people tire unnecessarily.
- Accidents, injuries and errors – due to badly designed interfaces and/or excess stress either mental or physical.
- User difficulties – due to inappropriate combinations of subtasks making the dialogue/interaction cumbersome and unnatural.
- Low morale and apathy.

In ergonomics, absenteeism, injury, poor quality and unacceptably high levels of human error are seen as system problems rather than 'people' problems, and their solution is seen to lie in designing a better system of work rather than in better 'man management' or incentives, by 'motivating' workers or by introducing safety slogans and other propaganda.

The name 'ergonomics' comes from the Greek words 'ergon', which means work and 'nomos' which means law.

The focus of ergonomics

The focus is on the interaction between the person and the machine and the design of the interface between the two (Figure 1.1). Every time we use a tool or a machine we interact with it via an interface (a handle, a steering wheel, a computer keyboard and mouse, etc.). We get feedback via an interface (the dashboard instrumentation in a car, the computer screen, etc.) The way this interface is designed dertermines how easily and safely we can use the machine.

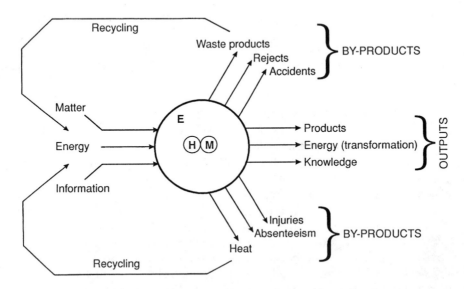

Figure 1.1 A simple work system. People interact with machines to turn inputs into outputs. System capacity refers to amount of input that can be processed over time. Productivity refers to the ratio of outputs to inputs. Efficient systems minimise by-products of all kinds (E = local environment, M = machine, H = human operator).

When faced with productivity problems, engineers might call for better machines, personnel management might call for better-trained people. Ergonomists call for a better interface and better interaction between the user and the machine – better task design.

Human–machine systems

A system is a set of elements, the relations between these elements and the boundary around them. Most systems consist of people and machines and perform a function to produce some form of output. Inputs are received in the form of matter, energy and information. For ergonomics, *the human is part of the system* and must be fully integrated into it at the design stage. Human requirements are therefore system requirements, rather than secondary considerations and can be stated in general terms as requirements for

- Equipment that is usable and safe
- Tasks that are compatible with people's expectations, limitations and training
- An environment that is comfortable and appropriate for the task
- A system of work organisation that recognises people's social and economic needs.

Compatibility – matching demands to capabilities

Compatibility between the user and the rest of the system can be achieved at a number of levels. Throughout this book we will encounter compatibility at the biomechanical, anatomical, physiological, behavioural and cognitive levels. It is a concept that is common to the application of ergonomics across a wide range of settings and disciplines. In order to achieve compatibility, we need to assess the demands placed by the technological and environmental constraints and weigh them against the capabilities of the users. The database of modern ergonomics contains much information on the capabilities and characteristics of people and one of the main purposes of this book is to introduce the reader to this information and show how it can be used in practice.

Ergonomic entropy (Karwowski *et al.*, 1994) is disorder in system functioning that occurs owing to a lack of compatibility in some or all of the interactions involving the human operator. This incompatibility can occur for a variety of reasons, for example:

- Human requirements for optimum system functioning were never considered at the design stage (e.g. there was a failure to consult appropriate standards, guidelines or textbooks).
- Inappropriate task design (e.g. new devices introduce unexpected changes in the way tasks are carried out and these are incompatible with user knowledge, habits or capacity, or they are incompatible with other tasks).
- Lack of prototyping (e.g. modern software development is successful because it is highly iterative; users are consulted from the conceptual stage right through to pre-production protoypes).

Disorder in the way systems function usually costs money and examples of the economic benefits of ergonomics are given throughout the book.

Table 1.1 Basic interactions in a work system and their evaluation[a]

Interaction	Evaluation
H > M: The basic control actions performed by the human on the machine. Application of large forces, 'fine tuning' of controls, stocking raw materials, maintenance, etc.	*Anatomical*: Body and limb posture and movement, size of forces, cycle time and frequency of movement, muscular fatigue. *Physiological*: Work rate (oxygen consumption, heart rate), fitness of workforce, physiological fatigue. *Psychological*: Skill requirements, mental workload, parallel/sequential processing of information, compatibility of action modalities.
H > E: Effects of the human on the local environment. Humans emit heat, noise, carbon dioxide, etc.	*Physical*: Objective measurement of working environment. Implications for compliance with standards.
M > H: Feedback and display of information. Machine may exert forces on the human due to vibration, acceleration, etc. Machine surfaces may be excessively hot or cold and a threat to the health of the human.	*Anatomical*: Design of controls and tools. *Physical*: Objective measurement of vibrations, reaction forces of powered machines, noise and surface temperatures in the workspace. *Physiological*: Does sensory feedback exceed physiological thresholds? *Psychological*: Application of grouping principles to design of faceplates, panels and graphic displays. Information load. Compatibility with user expectations.
M > E: Machine may alter working environment by emitting noise, heat, noxious gases.	Mainly by industrial/site engineers and industrial hygienists
E > H: The environment, in turn, may influence the human's ability to interact with the machine or to remain part of the work system (owing to smoke, noise, heat, etc.).	*Physical/physiological*: Noise, lighting and temperature surveys of entire facility
E > M: The environment may affect the functioning of the machine. It may cause overheating or freezing of components, for example. Many machines require oxygen to operate. Oxygen is usually regarded as unlimited and freely available rather than part of the fuel.	Industrial/site engineers, maintenance personnel, facilities management, etc.

[a] H = human, M = machine, E = environment, > = causal direction

A basic worksystem

Even in a simple system consisting of one person, one machine and an environment, six directional interactions are possible (H > M, H > E, M > H, M > E, E > H, E > M) and four of these involve the person (Table 1.1). Each of the components of a particular work system may interact either directly or indirectly with the others. For example, the machine may change the state of the environment (by emitting noise or heat, for example) and this may affect the user.

All work systems have a physical or functional boundary around them that separates them from adjacent systems. *Systems analysis* is the name of the discipline that studies the structure and function of work systems and provides the means by which simple systems may be combined to form more complex systems. Systems analysis is an integral part of all advanced work in ergonomics.

Application of ergonomics

The purpose of ergonomics is to enable a work system to function better by improving the interactions between users and machines. Better functioning can be defined more closely, for example, as more output from fewer inputs to the system (greater 'productivity') or increased reliability and efficiency (a lower probability of inappropriate interactions between the system components). The precise definition of better functioning depends on the context. Whatever definition is used should, however, be made at the level of the total work system and not just one of the components. Improved machine performance that increased the psychological or physical stress on workers or damaged the local environment would not constitute improved performance of the total work system or better attainment of its goals. Workstation redesign to make workers more 'comfortable' is an incorrect reason for the application of ergonomics if it is done superficially, for its own sake, and not to improve some aspect of the functioning of the total work system (such as reduced absenteeism and fewer accidents due to better working conditions).

There are two ways in which ergonomics impacts upon systems design in practice. Firstly, many ergonomists work in research organisations or universities and carry out basic research to discover the characteristics of people that need to be allowed for in design. This research often leads, directly or indirectly, to the drafting of standards, legislation and design guidelines. Secondly, many ergonomists work in a consultancy capacity either privately or in an organisation. They work as part of a design team and contribute their knowledge to the design of the human–machine interactions in work systems. This often involves the application of standards guidelines and knowledge to specify particular characteristics of the system.

Real work systems are hierarchical. This means that the main task is made up of sub-tasks (the next level down) and is governed by higher-level constraints that manifest as style of supervision, type of work organisation, working hours and shiftwork, etc. If we want to optimise a task in practice, we rarely redesign the task itself. We either change or reorganise the elements of the task (at the next level down) or we change the higher-level variables. For example, to optimise a data entry task we might look at the style of human–computer dialogue that has been chosen. We might find that there are aspects of the dialogue that cause errors to be transmitted to the system (e.g. when the operator mistakenly reverses two numbers in a code, the system recognises it as a different code rather than rejecting it). Alternatively, we might find there are insufficient rest periods or that most errors occur during the night shift.

To optimise a task we first have to identify the level of the task itself (e.g. a repetitive manual handling task), the next level down (the weight and characteristics of the load and the container) and the next level up (the workload and work organisation). To optimise the task we can either redesign it from the bottom up (e.g. use lighter containers and stabilise the load) or from the top down (e.g. introduce job rotation or more rest periods) or both. At the same time we can look at extraneous or environmental factors at the level of the task but external to it, factors that also degrade performance (e.g. slippery floors in the lifting example or bad lighting and stuffy air in the data entry example).

Having redesigned the task and evaluated the improvements to task performance, we then monitor it over time to detect improvements in system performance.

'Set designers' versus 'choreographers'

One of the historical trends that has influenced the development of ergonomics is the wide variety of individuals who enter the field. In the ergonomics literature, there is a wealth of information about the physical structure of systems and standards for noise, lighting, seating, climate, etc. There are also information and methods for designing tasks and interactions between people and machines. To use a theatrical analogy, ergonomists are involved in both designing the set and choreographing the performance in the work system. Often, but not always, the 'set designers' have a background in biological sciences or engineering and the 'choreographers' have a background in psychology or computer science.

This has tended to cause a split between the 'physical ergonomists' and the 'human factors specialists', particularly in academia. However, many practitioners are called upon to do both types of work and there is now a requirement from the Centre for the Registration of European Ergonomists (CREE) for prospective members to demonstrate formal education in all areas of ergonomics, even if they only specialise in one area. This is a positive step as it will give the discipline more intellectual coherence and a stronger identity, bringing it into line with professions such as medicine where people only specialise after a basic education in the whole discipline.

Description of human–machine systems

Figure 1.2 depicts a descriptive human–machine model.

Human components

The human body is part of the physical world and obeys the same physical laws as other animate and inanimate objects. The goal of ergonomics at this level is to optimise the interaction between the body and its physical surroundings. This means ensuring that physical space requirements are met (using data on human 'anthropometry') and that internal and external forces acting on the body are not harmful. Ergonomic problems often arise because, although the operator is able to carry out the task, the effort required overloads the sustaining and supportive processes of the body and causes fatigue, injury or errors.

The effectors   The three primary effectors are the hands, the feet and the voice. More generally, the musculoskeletal system and body weight can be regarded as effectors – no purposeful physical activity of the limbs can be carried out without maintenance of the posture of the body and stabilisation of the joints.

The senses   The senses are the means by which we are made aware of our surroundings. Human beings are often said to have five senses – sight, hearing, touch, taste

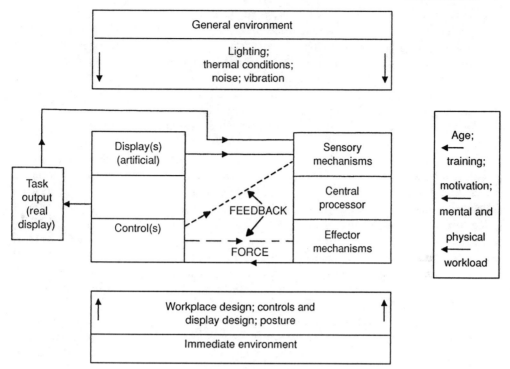

Figure 1.2 A human–machine model. (Adapted from prEN 14386: 2002(E).)

and smell. The existence and nature of a 'sixth sense' remains debatable, although a sense of balance and of the body's position in space and a sense of the passage of time are some of the less controversial candidates. Vision and hearing are the most relevant to ergonomics, although smell is important in detecting leaks, fires, and so on.

Although vision and hearing may be dealt with separately at an introductory level (as they are here), people typically utilise a combination of senses to carry out an activity. Vision is often oriented by hearing: a young baby will move its head to fixate a rattle placed at one side. Vision is often complemented by touch. When encountering a novel object, we may hear someone say, 'Let me see that' while at the same time reaching for it. In this example, the word 'see' really means 'see and touch'.

Central processes In order to carry out work activities, we require energy and information. Physiological processes provide energy to the working muscles and dissipate waste products. The brain can be regarded as an information processing centre, that contains low-level programs to control the basic sensori-motor work activities and higher-level cognitive processes that make possible the planning, decision making and problem solving activities of work. The human operator can be thought of both as a user and as a source of energy. An understanding of these basic processes is essential in work design to determine workers' capacity for physical work and to investigate the factors (such as climate and individual differences) that influence work capacity.

Information is obtained via the senses, as feedback from the joints and muscles and from memory. Modern approaches model the brain as an information processing system like a computer – an analogy that has some value in directing attention to the types of programs that underlie human information processing, the limitations of the system and the circumstances under which it can break down. Of particular interest are the implications of this approach for information design. It raises the question of how best to design the information content of jobs to be compatible with the information acquisition and storage characteristics of the human information processing system.

Although the computer analogy has some value, it is also clear that, in many ways, humans process information quite differently from computers. Apart from the conitive aspect, humans excel at activities such as recognising a face or classifying an event on the basis of incomplete data, whereas computers excel at numerical computation and logical problem solving. From a structural viewpoint, computer and human information processing systems also differ. Memory and processing (the CPU) are separate in the computer system, whereas in humans we can almost think of processing as that part of memory that is active. An ergonomic approach would suggest that the strengths of each of these information processing systems should be used to support the weaknesses of the other in a complementary way.

Finally, energy and information can only lead to purposeful work activity if the human is sufficiently *motivated*. Motivation is the force that directs behaviour and is here regarded as a supportive process. Some would argue that motivation lies in the realm of occupational psychology rather than ergonomics. Although this may be true, the ergonomist cannot ignore such a fundamental determinant of human behaviour – no work system can operate in a goal-directed and purposeful way if the human component itself is not goal-directed and purposeful.

Machine components

A 'machine' can be any man-made device that augments work capacity. The prototypical human–machine system of ergonomics is illustrated by the car driver or machine operator where the machine component and the links between the human and the machine (the displays and controls) are tangible. Recent developments in information technology have shifted much of the attention of ergonomics to information systems in which the work system is abstract and has no unique spatial location. A piece of software on a computer network is an example of such a machine. The local environment may be the network itself and many users, in different locations, may interact with the machine at the same time.

The controlled process   This is the basic operation of the machine on its local environment as controlled by the human. Digging a vegetable garden with a spade is an example of such a process. Nuclear fission is another example of a controlled process that is used to produce electricity in a nuclear power station. In information systems, the controlled process is often more difficult to categorise because of the abstract nature of the concepts used in these systems. Automatically sorting files, sending electronic mail and searching a database for an item of information or a directory for a particular entry are examples of controlled processes in information systems.

Displays In simple work systems, the display is often just the action of the machine on its local environment. The process is its own display (as in chopping wood with an axe). With increasing technological sophistication, the distance between the controlled process and the human component is increased and artificial displays have to be designed. Driving a motor car or operating a lathe are intermediate in the sense that the display comes both directly from the controlled process (the view of the road or the action of a machine tool on a part) and indirectly from gauges, dials, etc. The complexity and hazardous nature of the controlled processes in nuclear power generation or chemical process industries necessitate the use of artificial displays. In these systems, the human operator has no direct access to the controlled process and interacts with the machine entirely by means of artificial displays. The way these displays are designed need have no one-to-one correspondence with the actual controlled processes they represent. In information systems, the controlled process may be so abstract that it cannot be faithfully represented at all using physical displays (for example, compiling of code written in a high-level language). The designer's challenge is then to construct an interface that will provide an appropriate metaphor or way of thinking about the process.

Controls Human interaction with machines depends on the provision of suitable controls, which can be acted on by the effectors. With simple technology, the machine component itself is often the control. The handle of an axe is a lever that enables the cutting edge to be accelerated towards the target via the pivoting action at the hand–handle interface. At this level, the design requirements centre around the interaction of the effectors and the control and the mechanical advantage provided by the design of the control.

Controls are also an important source of feedback during the execution of control actions. For example, the resistance to cutting of a saw tells us something about the sharpness of the blade or the hardness of the wood, the resistance to turning of a motorcar steering wheel provides feedback about the road surface or the tyre pressures.

The immediate environment

This refers to the place and the circumstances in which work is carried out and consists of the physical workspace, the physical environment and the social and technical constraints under which the work is done.

Workspace The workspace is the three-dimensional space in which work is carried out. In simple work systems, the workspace may be just the place in which work is being carried out at any point in time as we move from one location to another. In more complex systems, workspaces usually become fixed and this introduces design issues such as the need to determine the workspace dimensions.

The physical environment Many aspects of the physical environment can affect workers. Ergonomists are most interested in those that have an influence on the way the human and machine components interact. Noise, vibration, lighting and climate are of most concern to the ergonomist. Contamination and pollution of the environment are matters best dealt with by industrial hygienists, because they presumably have direct effects on health irrespective of any other work system factors. However, an awareness of these aspects is also important from an ergonomic perspective because they may have effects on human abilities and motivation as well as on health.

Work organisation   Work organisation at its most basic level refers to the immediate organisation of human–machine interaction – the rate of work, whether the human works at his own pace or whether the machine sets the pace and whether people work alone or are dependent on others. More broadly, it refers to the organisational structure in which the work activity is embedded, the technical system and the social system which supports it.

Ergonomics and its areas of application in the work system

One of the problems facing the ergonomist both in the design of new work systems and in the evaluation of existing ones is to ensure that all aspects are considered in a systematic way. The human–machine approach enables key areas to be identified irrespective of the particular system so that ergonomics can be applied consistently in different systems.

The first step is to describe the work system and its boundaries. This enables the content and scope of the application of ergonomics to be specified. Next, the human and machine components and the local environment are defined and described in terms of their main components as depicted in Figure 1.2. Following this, the interactions between the various components can be analysed to identify the points of application of basic knowledge to the design/evaluation process. Examples of interactions are the interaction between the displays and the workspace – this directs attention to the positioning of the displays in the workspace so that the operator can see them when working. The interaction between the effectors and the workspace introduces considerations about the space requirements for body movements required by the task. Table 1.2 summarises the basic ergonomic interactions between the components of a work system.

This book has been written from the human outwards to provide a general introduction to ergonomics. It contains several chapters that describe the basic processes that support behaviour at the physical/musculoskeletal, physiological and psychological levels. The remaining chapters discuss various aspects of human–machine

Table 1.2 Common interactions between the components of a work system

Interactions	Design issues
Display:Workspace	Location of displays
Display:Environment	Effects of lighting, vibration, noise on legibility
Senses:Workspace	Sensory access to task
Senses:Environment	Environmental requirements for operation of the senses
Processing:Environment	Effects on perception and cognition
Processing:Organisation	Skill levels, training, fatigue, motivation
Effectors:Workspace	Determination of workspace envelope
Effectors:Environment	Effects of vibration and climate on effectors
Controls:Workspace	Task description needed to optimise control layout
Controls:Environment	Effects of environment on usability of controls

systems in terms of basic human characteristics and their interactions with other parts of the work system. For example, vision and lighting are discussed together in relation to the workspace and the requirements for seeing the displays. In addition, light is part of the local environment and has psychological and biological effects on workers that should also be taken into account in any design of work system. Sound and hearing are discussed together both fundamentally and in terms of the effects of noise on task performance and people's satisfaction with the environment. The design of symbolic information is discussed in relation to the characteristics of the human memory system and the processing of language. The basic scientist might find this approach unsatisfactory or even confusing because issues from different areas of science are treated under the same heading. From the perspective of ergonomics the approach can be defended: in real work systems, conceptually different issues do overlap and interact, sometimes in complex ways, and investigations that are restricted to one core science and one level of analysis are almost always unsatisfactory. An analysis of noise or lighting, for example, that took only numerical measurements and ignored questions of environmental quality and people's attitudes would clearly be incomplete. Similarly, an investigation of workspace design and musculoskeletal pain that took only physical factors into account and ignored work organisation and mental workload would also be lacking. In practice, the ergonomist must take a holistic view of work systems and be able to apply knowledge from different areas of science to the analysis of problems and the specification of designs.

A brief history of ergonomics

Ergonomics came about as a consequence of the design and operational problems presented by technological advances in the last century. It owes its development to the same historical processes that gave rise to other disciplines such as industrial engineering and occupational medicine.

Scientific management and work study

Scientific management, developed by F. W. Taylor, and workstudy, developed by the Gilbreths, are precursors of ergonomics. Both were developed at the beginning of the twentieth century and were based on the realisation that productivity could be improved by redesigning the way work was done and not just by using better machines. Taylor was a mechanical engineer who is famous for his book, *The Principles of Scientific Management* (although he also wrote a book about concrete, for which he is not famous). Scientific management was a reaction against the then prevalent management methods inherited from the Victorians. Factory owners supplied premises, power and raw materials, etc. and hired foremen to organise the work. These foremen acted rather like subcontractors and were left to themselves to organise the basic industrial tasks as best they could. Management was concerned only with output and had only a global notion of 'productivity', regarding the work itself with disdain. Incentives were provided for employees to suggest improvements and profits depended on getting a 'good man' in to organise the workers.

Taylor realised that there were many drawbacks to this 'incentive and initiative' style of mangement. Nobody was directly responsible for productivity and the system was open to the corruption and exploitation of workers. Weekly 'kickbacks' to supervisors were common as was the sexual harassment of females (Stagner, 1982). There were few formal ways of generating better designs of systems or work procedures or of evaluating current practice on a day-to-day basis. Workers, by being too narrowly focused on carrying out daily work activities might be unaware of the scope for improvement through the implementation of new methods, or even unaware of the methods themselves. They might be unwilling to suggest changes that might be in the best interests of the company but not in their own best interests. Management, on the other hand, through its failure to focus on way basic tasks were carried out, was incapable of maximising productivity. As Taylor, (1911) himself put it (p. 7)

> . . . The remedy for this inefficiency lies in systematic management, rather than in searching for some unusual or extraordinary man.

Taylor emphasised that every job, no matter how small, was worthy of study and improvement. Furthermore, he emphasised that it was management's responsibility to see that this was done in the interests of maximising returns for the financial benefit of the company, for its shareholders and for its employees. In practice, tasks were to be broken down into their simplest possible form and, with the introduction of an appropriate bonus scheme, it was assumed that workers would maximise their output.

One of the main lessons to be learned from the writings of Taylor is that it is management's job to manage and that this means having formal systems in place to monitor working practices and their relationship to productivity. However, there are many criticisms of Taylor's approach – the social aspects of work and the intrinsic satisfaction that could be gleaned from carrying out work tasks were ignored. A more fundamental criticism, though, is that throughout Taylor's writings there is an assumption that all parties in an industry can and do have the same interests. It is assumed that all will cooperate to maximise their own economic returns:

> Scientific management, on the contrary, has for its very foundation the firm conviction that they (i.e. the interests of employer and employee) are one and the same (p. 10).

Clearly, the interests of management and workers are not the same. One of management's main goals is to maximise profits for shareholders and one of the best ways to do this is to minimise the cost of labour. This conflict of interest had implications for the way Taylor's ideas were implemented. It also provided the impetus for the infamous 'Hawthorne experiments'.

Work study was developed by the Gilbreths at around the same time. They developed methods for analysing and evaluating the way tasks were performed. A task would be broken down into 'elements' – the basic movements and procedures required to perform it. Inefficient or redundant movements would be eliminated. By redesigning and reconfiguring the remaining elements, productivity was enhanced.

Work study and scientific management were the forerunners of time and motion study and human engineering. Companies employed a new breed of specialists to investigate human–machine interaction and to design tasks. Working practices were no longer assumed to be at the worker's discretion or determined by tradition or technology, but were regarded as something to be bought under management control. This change in outlook was an essential requirement for the introduction of mass assembly and production line techniques.

Time and motion study (methods engineering) can be criticised on many grounds: for example, that it looks only at the superficial features of task performance, that it makes unwarranted assumptions about people and that it is little more than common sense. Many have argued that Taylorism succeeded in 'de-skilling' craftsman and created mundane, repetitive jobs. This may be true, but one might counter-argue that it made industrial work available to unskilled people and mass-produced products more affordable. Taylorism had many advantages for management:

1. It allowed greater flexibility in allocating operators to easily learnt tasks.
2. Fewer skilled workers were needed. Skill shortages were avoided and training costs and wages could be more easily contained.
3. The introduction of paced work enabled production schedules to be more rigorously quantified. Better predictions of output could be made.
4. If everyone worked at the same pace, the result was always a finished product.

Human relations and occupational psychology

Occupational psychology developed in the 1920s and 1930s. The essence of Taylorism had been to regard the worker as an isolated individual whose output was determined by physical factors such as fatigue or poor job design and by economic incentives. A job would be redesigned to make it as simple as possible to learn and to perform. A production standard and rate of pay would be set and a bonus scheme introduced as an incentive for workers to produce more than the standard. It was assumed that 'rational economic men' would maximise their productivity to maximise the bonus. The social context in which work took place was ignored.

Despite its advantages, Taylorism also presented management with a dilemma. Continually increasing productivity had to be met with continual increases in pay. To avoid this, new techniques were employed. New and higher production standards were introduced whenever sustained increases in output were achieved, bonuses only being paid when the new standard was exceeded. Not surprisingly, workers reacted by restricting their output to prevent the standard being raised and placed social pressure on 'rate-busters'.

The Hawthorne experiments

In the 1920s and early 1930s a series of experiments were carried out over 12 years by Elton Mayo and his colleagues at the Hawthorne Works of the Western Electric Company in the USA. The experiments are of historical interest more for their influence, which was to draw attention to social factors at work, than for their findings. The investigators began by examining the effects of illumination, rest pauses and shorter hours on productivity and fatigue but soon ran into difficulties because they were

> trying to maintain a controlled experiment in which they could test for the effects of single variables while holding all other factors constant. . . . By Period XIII it had become evident that in human situations, not only was it practically impossible

to keep all other factors constant but trying to do so in itself introduced the biggest change of all; in other words, the investigators had not been studying an ordinary shop situation but a socially contrived situation of their own making.

(Rothlisberger and Dickson, 1939)

The most often quoted experiment concerned a relay assembly task. The experimenters began by manipulating the lighting levels to observe the effect on output. Unexpectedly, it was found that output increased – even when the illumination was reduced. This result is the basis of belief in the 'Hawthorne effect' and is used as evidence for the importance of social rather than physical factors in determining worker performance. The usual interpretation is that the changes in lighting levels reminded workers that they were in an experiment, being observed, and that this motivated them to work harder.

The Hawthorne experiments earned their place in history, ushering in a new era of research into 'Human Relations' in the workplace and drawing attention to the importance of social and personal determinants of worker behaviour. However, the experiments themselves and their interpretation have always been controversial. Gibson, in 1946, put it this way:

To take years to find out that employee attitudes and the 'total situation' are important and that workers appreciate human and kindly interest in their personal affairs seems absurd to any garden variety of industrial personnel worker.

Gillespie (1991) provides fascinating insights into the historical context of the Hawthorne experiments and the motives of the principal players, particularly Elton Mayo. Mayo had been impressed by the advances in industrial efficiency bought about by engineering and by the introduction of mass production methods. He had also been impressed by the work of Sigmund Freud, in particular Freud's writings on the 'Psychopathology of Everyday Life' and had formed the opinion that uncooperative workers were suffering from a kind of psychopathology amenable to treatment. These ideas were behind his interest in 'Human Relations' in the workplace and his belief that the next wave of productivity increases would be achieved using psychological techniques, rather than economic incentives, to motivate workers. The possibility that productivity might be increased by manipulating the social environment rather than by paying workers more money was obviously attractive to Western Electric management, although there is scant evidence that the Hawthorne experiments provided any support for these claims.

One of the immediate consequences of Mayo's experiments was the introduction of a personnel counselling service at the factory. In 1950, Western Electric employed 20 000 workers at Hawthorne and had a counselling department of about 40 people who interviewed over half the workforce over a 10-year period. The method used was non-directive, confidential and neutral, similar in some ways to the client-centred psychotherapy of Carl Rogers, which dates from the same era. The costs and benefits of the service do not seem to have been assessed, although Wilensky and Wilensky (1951) concluded that counselling appeared to be effective as a supplementary means of exerting management control over workers, by draining-off resentment and bitterness that might otherwise have gained expression through militant unionism. Some have claimed that the counselling service was really a kind of surreptitious continuous attitude survey that enabled management to pre-empt and deal with industrial relations problems.

Sociotechnical systems theory

Sociotechnical Systems Theory emerged in the UK after the Second World War. Trist and Bamforth (1951) investigated the social and psychological consequences of mechanised coal-mining in the context of a reportedly higher incidence of psychosomatic disorders among miners working under mechanised conditions. They pointed out that the mechanised method owed its origins and scale to the factory floor rather than to coal mines. In coal mines, a different form of social organisation was necessary because of the intrinsically unpredictable nature of the working environment in a mine compared to a factory. The organisation of technology, the social organisation and the local environment had to be seen as interconnected and designed to be mutually compatible if low productivity and pathological psychological stresses were to be avoided. Sociotechnical systems theory is discussed in the final chapter of this book; its emphasis on the importance of the social organisation of work for effective functioning of a work system is of clear relevance to ergonomics. That the theory has survived to the 1990s (e.g. Hitchcock, 1992) attests to the value of its fundamental tenets.

Participation

The idea that workers should participate in decisions about work dates from the same era. Coch and French (1948) carried out an investigation of participation in a US pyjama factory staffed by young, unskilled women. Staff turnover was high and workers were paid on the basis of individual performance. Management collaborated with two sociologists in an attempt to deal with the problems. Unusually for such studies, it was possible to introduce a degree of experimental control with a control group, where change was introduced by management without the involvement of employees, a 'representative group', where change was introduced after discussions with employee representatives and a 'total participation' group where all employees were involved. Production fell by 10–20% in the control group immediately after the change, whereas it was maintained in the representative group, rising 10–15% after a few weeks. In the participation group, production rose immediately and the gains were maintained for several months. It was argued that the participatory approach had produced better group cohesion, which countered any frustrations bought about during the change, and that people will accept change more readily if they feel they have some control over its effects. If this explanation is correct, then participation has to become an established management process if its benefits are to be sustained.

Occupational medicine

Occupational medicine had its origins in the eighteenth century when Ramazzini (1717) wrote his *Treatise on the Diseases of Tradesmen* but became more formalised at the beginning of the twentieth century. Around the period 1914–1918, a number of government institutions were founded in Britain as interest in working conditions spread to scientists and medical doctors. The Health and Munitions Workers

Committee studied conditions in munitions factories and factors influencing product-ivity such as the length of the working day. It subsequently became the Industrial Health Research Board and its area of interest was fairly wide, covering ventilation, the effects of heat and shiftwork and training. Recommendations were made at about this time for a variety of aspects of industrial work including the types of food served in factory canteens, taking into account the likely nutritional deficiencies of the workforce and the demands of work. This was appropriate in Britain in the early years of the twentieth century and is still appropriate in developing countries today. Attention was also directed at fatigue and the problems of repetitive work. Vernon (1924) investigated postural and workspace factors related to fatigue and concluded that 'Any form of physical activity will lead to fatigue if it is unvarying and constant'.

It is from foundations such as these that industrial physiology and occupational health have arisen (it is surprising, given the current high incidence of cumulative trauma disorders, how few industries have acted upon Vernon's valid conclusion).

Human performance psychology

Human performance psychology had its roots in the practical problem of how to reduce the time taken to train a worker to carry out a task. Taylor and the Gilbreths had gone some way to the solution of this problem by simplifying tasks. However, the emergence of new, more complex machines and manufacturing processes created new jobs whose skill requirements could not be met from the existing labour pool. Traditional methods of training (such as apprenticeship to a skilled craftsman and 'sitting by Nellie') were not always appropriate to the industrial system with its fundamental requirement to standardise all aspects of the production process. Psychologists began investigating the variables influencing the learning of work skills – for example, whether tasks should be broken down into elements to be learnt separately and then bought together, the advantages and disadvantages of 'massed' versus 'spaced' practice and the duration of rest periods. After the Second World War, interest in training continued. The cybernetic approach investigated the use of feedback (knowledge of results) and its effects on learning. The theoretical ideas of B. F. Skinner and the behaviourist school of psychology were implemented in the form of 'programmed learning'. Behaviourism saw learning as the chaining together of stimulus–response pairs under the control of reinforcing, or rewarding stimuli from the environment. In programmed learning, the material to be learnt was presented in a stepwise fashion and the order of presentation of information to be learnt was determined by whether the trainees' previous responses were correct or incorrect.

The pressure for the productive and efficient use of machines was amplified by the demands of the Second World War and brought psychologists into direct contact with the problems of human–machine interaction. The famous Cambridge psycho-logist Sir Frederick Bartlett built a simulator of the 'Spitfire' aircraft and investigated the effects of stress and fatigue on pilot behaviour. This led to an increased under-standing of individual differences in response to stress and enabled the breakdown of skilled performance to be described in psychological, rather than machine-based terms. Perceptual narrowing, which occurs as a result of fatigue or as a maladaptive response to severe stress, is an example. Craik studied a class of tasks known as 'tracking tasks' (which involve following a target as in gunnery or steering a vehicle). The beginnings of user-interface (display and control) design emerged from this in the form of recommendations for gear ratios and lever sizes. This became an important area of ergonomics, particularly in post-war military applications but also in civilian vehicle design and the aerospace industry.

Operations research

Operations research attempts to build mathematical models of industrial processes. It was also stimulated to grow by the demands for prediction and control brought about to satisfy the requirements of the military during the Second World War. It had become clear that further advances in system performance would depend on how well technology was *used*, not just how well it was *designed*. This shift of attention from the machine to the man-machine system gave birth to the subject of ergonomics.

FMJ versus FJM

A number of general trends can be identified in the historical review. First, organisations attempted to increase their productivity by introducing new methods and machines. In the era of pure engineering this worked because great improvements in machine design were possible (many existing processes had not been mechanised at all until then). Second, attempts to increase productivity tried to optimise the design of tasks to minimise apparently unproductive effort. After the First World War, a movement arose stimulated by the development of psychometric testing, which tried to develop tools to objectively measure various human characteristics such as intelligence or personality.

Attempts to 'fit the man to the job' (*FMJ*) were based on the idea that productivity or efficiency could be improved by selecting workers with the right 'aptitudes' for a particular job. This approach, which forms one of the roots of modern occupational psychology, is based on the assumption that important aptitudes for any particular job really do exist and that they can be identified and objectively measured. This is certainly true in the sense of selecting people with formal qualifications or skills to fill particular posts and there is plenty of evidence that it is cost-effective (see Anastasi, 1990, for example).

It is also true for some jobs – even today firemen, lifesavers and those working in the armed services are restricted to individuals with certain specific aptitudes and/or physical characteristics. Historically, anthropometric selection criteria were made use of. Stature is generally a good surrogate measure for strength and size. However, owing to changes in labour legislation and the increased emphasis on equal opportunities, static anthropometric criteria have given way to functional tests. Instead of selecting on the basis of size or chest circumference, recruitment depends on the applicant's ability to carry out some generic or job-critical task under defined conditions and to a pre-set standard.

However, selection tests are still a source of controversy. Trade unions, for example, may object to management attempts to select workers with 'strong backs' for a particular job, arguing that the problem lies with the job, not the worker, and that the job should be redesigned to be performable by anyone in the workforce.

An alternative approach, which is the guiding philosophy of ergonomics, is known as 'fitting the job to the man' (FJM). Much of the early human engineering and workspace design attempted to design tasks to suit the characteristics of the worker. The underlying assumptions of the FJM approach are that a suitable set of worker characteristics can be specified around which the job can be designed and that this can be done for any job. A large part of this book is devoted to describing these characteristics at the anatomical, physiological and psychological levels and explaining their design implications.

FMJ came about at a time when the demographics of the populations of many countries were very different from those of today. There were many young people, there were no shortages of the necessary skills and there was a great deal of unskilled work. With few options for alternative employment, it was common for workers to remain in a job for life. Management could afford to be selective. In many countries this is no longer true, owing to demographic ageing of the workforce and no shortage of vacant jobs. FJM seems to be the most practical of the two approaches, wherever it can be implemented, since many organisations now find themselves with a dwindling applicant pool and are loath to reject otherwise suitable people. There are also political pressures on organisations in many countries to make employment as widely available as possible. Equal opportunities legislation prevents discrimination against women and minorities and employers are being forced select on the basis of functional tests of task performance rather than applicants' characteristics.

Only under extreme circumstances is the FMJ approach taken, as when acclimatising workers who have to work in hot conditions that cannot be changed. Stature is still used to select military pilots in some cases, owing to limited leg room in the cockpit and the dangers of knee amputation during ejection. However, in less extreme conditions, there are usually many FJM options such as designing a better work–rest schedule, providing protective clothing or building a 'cool spot'.

Human factors and ergonomics

In 1857, Jastrzebowski produced a philospohical treatise on 'An Outline of Ergonomics or the Science of Work' but it seems to have remained unknown outside Poland, until recently. In Britain, the field of ergonomics was inaugurated after the Second World War. The name was re-invented by Murrell in 1949 despite objections that people would confuse it with economics. The emphasis was on equipment and workspace design and the relevant subjects were held to be anatomy, physiology, industrial medicine, design, architecture and illumination engineering. In Europe, ergonomics was even more strongly grounded in biological sciences. In the USA, a similar discipline emerged (known as 'Human Factors'), but its scientific roots were grounded in psychology (applied experimental psychology, engineering psychology and 'human engineering').

Human factors and ergonomics have always had much in common, but their development has moved along somewhat different lines. Human factors puts much emphasis on the integration of the human considerations into the total system design process. It has achieved remarkable success in the design of large systems in the aerospace industry and notably NASA and the US space programme. European ergonomics is sometimes more piecemeal and has traditionally been more tied to its basic sciences or to a particular topic or application area.

Despite these differences, the reader should not be concerned about the two terms. In the USA, the Human Factors Society has recently changed its name to the Human

Factors and Ergonomics Society. Presumably this was done to indicate that the two areas now have so much in common that one society can represent the interests of people who see themselves as working in only one or the other area. In the author's opinion, few objections would be raised today if the title of the present book were *Introduction to Human Factors*.

Both human factors and ergonomics take the FJM approach and state that jobs should be made appropriate for people rather than the other way around.

Modern work systems and 'Neo-Taylorism'

A major criticism of both Taylorism and of the Hawthorne counselling programme is that both did nothing to ameliorate the alienation of workers from their work and from the products of their work. Taylorism may have been appropriate in the early twentieth century, acting as a catalyst for industrialisation and mass production, but its legacy endures: even in modern organisations, workers have little scope for organising, implementing and completing their daily tasks, as these are firmly under management control. Whether this style of management is still appropriate is far from obvious. Despite calls for job reform in Sweden, for the humanisation of work in Germany in the 1970s and 1980s and for a renewed emphasis on the quality of working life in the UK, there has arguably been little change. Indeed, many modern jobs in the twenty-first century (such as those found in call centres) are intrinsically fragmented, piecemeal or extremely repetitive. Margulies (1981) complained that few designers of the then modern computer systems felt responsible for ergonomic issues, nor did they feel the need to consult users about anything other than minor issues. New designs were justified by claiming that the constraints imposed on users were enforced by the requirements of implementing the technology. Margulies called for a change in outlook based on the fact that computer systems are not intrinsically deterministic and that people have a need for a satisfying work over which they have some control – a need for psychological ownership of the job. This was to be achieved by means of participation between employees and their representatives, computer systems designers and social scientists:

> Why should it not be possible to initiate preparations for a new computer system by informing everybody involved in its use about the qualities, possibilities and limitations of the system and then ask them to consider in small working groups what part of their work they would like to assign to the computer and what other tasks they would like to assume instead? . . . It is for the human being, individually and collectively that that the optimal operating mode must be found, and not for the computer.
>
> (Margulies, 1981)

Attempts to 'humanise' work

In the 1960s, 1970s and 1980s a number of large-scale programmes were initiated in several European countries. These programmes were motivated by a variety of factors. For example, successive generations of school-leavers in the countries in question had increasingly higher levels of education and higher expectations of work. The programs attempted to provide higher-quality jobs through changes in

Table 1.3 General features of a good job[a]

---

1. Optimum work loading
2. Minimum role ambiguities (each worker has a clearly defined set of responsibilities which they can relate to the objectives of the organisation)
3. Minimum role conflict (there is clear relationship between an individual's role, how they are rewarded and they and their co-workers' expectations)
4. Support from colleagues or management
5. Social content: interaction with others is 'built-in' to the design of the job and the execution of basic work functions

---

[a] From Sell (1980).

work organisation. Some general characteristics of a good (psychologically reward-ing) job are given in Table 1.3.

In Sweden, the Volvo motor car company (which was suffering from high absen-teeism and labour turnover in the 1960s) tried to find new ways of assembling cars in an attempt to have a more stable, motivated and productive workforce. Conven-tional production line methods were replaced by 'unit production'. Teams of workers manned electric assembly wagons that moved around the assembly area stopping at centralised stores to collect the various components.

It is frequently difficult to disentangle the effects of such programmes on either productivity or the psychological rewards of work because many different factors are involved. Unit production changes the social relations between people but it also eliminates pacing (where the rate of work is set by machines) and lengthens cycle times. Both of these latter factors are known to influence job satisfaction. The British 'Quality of Working Life' programme (Tynan, 1980) attempted to combine new approaches to job design with technological change under the premise that, since technological change forces job redesign anyway, the opportunities presented by new technologies could best be realised by optimising the work organisation and design of jobs. Management, trade unions and workers were to be included in a participatory approach towards job design.

Success of work humanisation programmes

The modern workplace, according to this view, is characterised by flexibility and individual discretion over work elements. Traditional fragmented, repetitive tasks and rigid organisational hierarchies have been replaced with more decentralised sys-tems. The extent to which this thinking has really penetrated organisations and re-placed traditional styles of management is a question open to empirical investigation. It has been investigated by Boreham (1992), who carried out an international com-parative study of the organisation of work and the amount of discretion available to employees in a variety of organisations in the USA, Australia, Britain, Canada, Germany, Japan and Sweden. The following employee groups were sampled: man-agers, professionals, clerks, skilled workers, semiskilled and unskilled workers. More than half of the employees sampled reported negligible freedom to put their ideas into practice or to introduce new tasks. Autonomy was found to be a property of higher-status individuals. Lower-status individuals were almost totally excluded from participation in decisions about production in their organisations. Citing such findings

Table 1.4 Contribution of modern ergonomics in systems design and management[a]

1. A standard format for describing human–machine systems
2. Identification, classification and resolution of design issues involving the human component
3. Task and human–machine interaction analysis
4. Specification of system design and human behaviour. Implementation of controls
5. Identification of core trends in human and biological science and their implications for system design and management
6. Generation of new concepts for the design and analysis of human–machine systems
7. Evaluation of the sociotechnical implications of design options

[a] Adapted from Van Wyk (1992).

as evidence for the 'Myth of Post-Fordist Management', Boreham concluded that there was scant evidence to suggest that truly participative organisational practices had been implemented in the countries studied. In one sense, then, Taylorism lives on, alive and well in the call centres of modern computerised offices.

Despite these reservations, participative practices have been accepted in some organisations and in some countries (e.g. Germany). According to Bernoux (1994) the requirements for effective participation are that

- Employees have to acknowledge the need for participation.
- Employees have to trust that their participation will not have negative effects and that they will have some control over the final decisions.
- Employees have to perceive that changes are being introduced in a legitimate way.
- Employees have to believe that change is being implemented correctly.
- Employees have to be given a real role to play in the introduction and testing of new ways of working.

Further information on participation can be obtained from the European Foundation for the Improvement of Living and Working Conditions.

Modern ergonomics

Modern ergonomics contributes to the design and evaluation of work systems and products. Unlike in earlier times when an engineer designed a whole machine or product, design is a team effort nowadays. The ergonomist usually has an important role to play both at the conceptual phase and in detailed design as well as in prototyping and the evaluation of existing products and facilities. Modern ergonomics contributes in a number of ways to the design of the work system (Table 1.4). These activities should be seen as an integral part of the design and management of systems rather than 'optional extras'.

Standard format for describing human–machine systems

This book attempts to provide a standard way of describing and understanding work systems, irrespective of the application area or the technology. This may be used to generate checklists and methods for evaluating prototypes or existing systems. A standard format for describing work systems forces the ergonomist to consider all of the issues at least in the early stages.

The two most important first steps when using the human–machine model are to describe the technology and to describe the user or operator. Machines are normally well-described and plenty of information in the form of manuals and textbooks is available. Designers of human–machine systems normally operate with much less detailed and less formal information about people and make assumptions about people in a way that they would never do about machines.

User descriptions can be physical, in terms of dimensions and abilities and including physiological factors such as age and fitness. Psychological descriptions include details of skills, knowledge, experience and motivation and may include detailed considerations such as preferred ways of working, jargon, etc.

Identification and classification of design issues

A major role of ergonomics is to identify design issues that affect people and to classify them in order to render them amenable to further analysis using appropriate knowledge.

All work systems are specific, yet many published design guidelines are general. For this reason, there is normally no body of literature appropriate to all of the needs of a particular application or industry and a standard method is required to enable system-specific design issues to be identified and interpreted in general terms amenable to analysis using available data. For example, many of the issues in office design are the same whether it is an office in a bank, an insurance company or the administrative centre of a heavy industry, but often the people working on the project will only know about the problems of their own particular industry.

Proper interpretation of ergonomic issues and their human implications is a crucial step in the design or redesign process. Most modern research on back pain and upper limb disorders takes account of psychosocial factors as well as the physical environment and the design of jobs. Simmonds and colleagues, for example, investigated the prevalence of back problems in industrial workers and found that psychosocial factors accounted for around 30% of the variance in outcome. Psychosocial variables of interest are typically job satisfaction, mental stress and social support at work. Other personal psychological factors that are sometimes of importance are the presence of fatalistic beliefs about back trouble and the extent to which people perceive themselves to be disabled by the problem. These findings are of interest because they imply that effective prevention and management must take psychosocial as well as physical factors into account.

There is often a tendency to attribute accidents, breakdowns and low productivity to the human component of work systems. Human behaviour at work takes place in the context of a system and is shaped by the way the system is designed. System malfunctions that involve humans must therefore be analysed in context, which requires that the focus of the analysis be shifted from the human to the human–machine system. For example, the design of the machine may invite the human to make a particular type of error, or the machine itself may not be immune to the types of errors that humans are known to make. The pattern of work organisation may cause undue fatigue and increase the likelihood of inappropriate actions by the human. It is the task of the ergonomist to investigate problems of reliability from a systems perspective to determine the relative contributions of inappropriate human behaviour and inappropriate system design. Further analysis can then be undertaken to determine which aspects of the design of the work system degrade performance and how the system can be redesigned to solve the problem.

Task and human–machine interaction analysis

Task analysis is dealt with in a later chapter. Simple flow diagrams can be used to 'choreograph' human–machine interaction and, at each stage, the appropriateness and suitability of the design can be assessed.

Specification of system design

After many years, research in ergonomics and its related disciplines has yielded standards and guidelines for tighter control of working conditions. The International Organization for Standardization in Geneva, Switzerland publishes a growing body of ergonomics standards. Examples are given in Table 1.5.

Standards specify 'how we do things' and are usually adopted by agreement, often between different countries. They do not have the same force as regulations but are normally more specific than guidelines and contain more technical information. Ergonomists in design teams often find that they can communicate their ideas more effectively when they express them with reference to standards. Standards facilitate interchangeability of systems and system components and can facilitate evolution, rather than revolution, in design, ensuring that key features are kept the same as the overall design evolves. Ergonomists are often most effective when they can give focused, quantitative specification to engineers and designers. National and international standards bodies are often the best source of this information and examples of what they offer are given throughout the book.

Identification and analysis of core trends

An important role for ergonomics generally and for the ergonomist working in a large organisation is to act as an interface between developments in basic human and biological sciences and organisational needs. The ergonomist is frequently one of the few, if not the only, member of the design team with formal training in these areas capable of interpreting the latest legislation, findings and reports. The current confusion and superficiality of the debate surrounding 'repetitive strain injuries' exemplifies the need for a more critical application of fundamental principles.

Ergonomics has generated many design guidelines and recommendations, which are frequently made available to engineers and designers. This information is usually of a general nature and cannot always be used in a straightforward 'cookbook fashion' as an 'off-the-shelf' solution. It is the function of the ergonomist to use background knowledge of human sciences to interpret general guidelines so that they are appropriate to the particular system. For example, design manuals often give the dimensions of workstations irrespective of the particular industry. Because ergonomists knows the rationale for a given guideline or specification, they can decide whether it needs to be modified in accordance with local requirements.

Table 1.5 Some ISO standards for ergonomists

Standard	Year published	Title
General		
ISO 7317	1987	Ergonomics: Standards Guidelines for Designers
ISO 6385	1981	Ergonomic Principles in the Design of Worksystems
ISO 18529	2000	Ergonomics – Ergonomics of Human–System Interaction – Human-Centred Lifecycle Process Descriptions
ISO 13407	1999	Human-Centred Design Processes for Interactive Systems
Physical workload		
ISO 1128-1-3	2000	Ergonomics – Manual Handling Parts 1 to 3
ISO 53120	1996	Equipment for Manual Handling
ISO 11226	2000	Ergonomics – Evaluation of Static Work Postures
Physical environment		
ISO 13340-1	1995	Protective Equipment in General
ISO 13340	1996	Protective Gloves
Workspace design		
ISO 9241	1990s	Ergonomic Requirements for Office Work with Visual Display Terminals – Parts 1–9
ISO 11064-1		Ergonomic Design of Control Centres. Control Room Layout
Physical work/stress		
ISO 8996	1990	Ergonomics – Determination of Metabolic Heat Production
Mental work/HCI		
ISO 9241	1990s	Ergonomic Requirements for Office Work with Visual Display Terminals – Parts 10–17
Displays and controls		
ISO 9355	1999	Ergonomic Principles for the Design of Signals
ISO 11429	1996	Ergonomics – System of Auditory and Visual Danger and Information Signals
ISO 7731	1986	Ergonomics. Danger Signals for Public and Work Areas Auditory Danger Signals
Work environment		
ISO 13731	1995	Ergonomics of the Thermal Environment. Vocabulary and Symbols
ISO 11399	2001	Ergonomics of the Thermal Environment – Principles and Application of Relevant International Standards
ISO 9116-1	1989	Lighting in General
ISO 91160-10	1989	Interior Lighting
ISO 8995	1989	Principles of Visual Ergonomics – The Lighting of Indoor Work Systems
ISO 1996-1	1982	Acoustics – Description and Measurement of Environmental Noise
ISO 9921-1	1996	Ergonomic Assessment of Speech Communication
Safety		
ISO 15534-1-3	2000	Ergonomic Design for the Safety of Machinery, Parts 1–3
ISO 14121	1999	Safety of Machinery – Principles of Risk Assessment
ISO 13854	1996	Safety of Machinery – Minimum Gaps to Avoid Crushing Parts of the Human Body

Generation and implementation of new concepts

The design of the human–machine interface is the classical point of departure for the application of ergonomics. A very large literature is available to assist designers with the design of the user–interface. The ergonomist has an important role to play at the concept generation stage in trying to anticipate what the demands of the new system will be and how they will affect operators. Organisations sometimes exhibit a certain inertia that manifests itself as a desire to 'do things in the way we've always done them'. The ergonomist must analyse the reasons for current or proposed designs and suggest improvements and alternative concepts. Technological advances now offer many new concepts for the design of interfaces that are radical departures from traditional methods. Control of machines by voice and the use of synthetic speech as a display are examples. Computer generated 2- and 3-dimensional graphic displays offer designers much more flexibility and capacity for information display than their electromechanical counterparts. They have application in the aerospace and and process industries and in the design of everyday products such as the automobile dashboard.

It is necessary to develop a sensitivity to the cost–benefit implications and practicalities of new design ideas. This is particularly true when an existing system is being improved. Several schemes exist so that recommendations for ergonomic improvements can be prioritised. The category chosen for a recommendation depends on the need for change, as perceived by the ergonomist, and the implementation costs in terms of money, time, expertise, impact on day-to-day running, etc. One way of prioritising recommendations (used, for example, by the University of Vermont Rehabilitation Engineering Center) is as follows:

1. Implement recommendation immediately (e.g. there is a serious design flaw threatening employee health or system reliability or contravening legislation).
2. Implement recommendation soon (e.g. the current way of working is unsatisfactory but there is no immediate danger).
3. Implement when equipment is shut down (e.g. if stoppages are expensive and there is no immediate danger, wait until the system is shut down for regular maintenance or repair and then implement the idea).
4. Implement when cost–benefit is acceptable (e.g. when the financial situation improves, when implementation costs are lower).
5. Implement when equipment is built or purchased (e.g. phase-in new products or items on a replacement basis as old ones are discarded).

Evaluation of sociotechnical implications

The design of new systems and the redesign of existing ones can have serious implications for the organisational 'climate'. Technological and organisational changes can have profound effects on the working lives of individuals and it is part of the ergonomist's function to determine what these effects might be and to anticipate future problems. There is a growing body of literature on the sociotechnical aspects of technological change and the importance of involving employees as part of a participatory approach to system design (e.g. Sell, 1986).

Effectiveness and cost-effectiveness

Three key concepts in any discussion of the economic benefits of ergonomics are

- Efficacy – whether the applications work under ideal circumstances.
- Effectiveness – whether the applications work under normal conditions of use.
- Efficiency – whether the applications save more resources than they consume.

Most of the material presented in this book is of an introductory and explanatory nature and describes the basic concepts used in ergonomics. Throughout each chapter, the results of efficacy studies are included. Additionally, at the end of each chapter is a section devoted to effectiveness and efficiency studies, wherever possible in the form of field trials with data on costs and benefits. It is not the purpose of this book to introduce the reader to techniques of cost–benefit analysis since this is an application of economic theory and is a field of study in its own right. The goal is to illustrate to the reader the kinds of analyses that have been done and, more importantly, to show how the findings demonstrate the efficacy and efficiency of the application of ergonomics.

Cost–benefit models and methods

In general, there is more information on the cost of problems than on the economic benefits of solving them. However, the main approaches are as follows.

Oxenburgh productivity model

The approach is to identify jobs with a high level of absenteeism, injury or staff turnover. Ergonomic modifications are made to the workplace, tools, equipment or work organisation. The costs of these modifications can be calculated in terms of the materials and labour charges needed to implement them, as well as any downtime during redesign. The benefits are expressed in terms of the reductions in absenteeism, expressed as a monetary value made up of direct and indirect costs:

Direct wage costs = wage + obligatory charges to the wage + personnel
+ admin costs

Indirect costs = costs of losing trained employees + costs of hiring and training
new employees + costs of additional overtime
+ overemployment costs

There may be other benefits as well, such as productivity improvements and these then have to be included in the cost–benefit calculations. Oxenburgh (1994) provides a detailed step-by-step procedure for carrying out cost–benefit analyses using the productivity model. In practice, an ergonomics consultant would need the assistance of the client company's finance department to make detailed calculations of the cost savings brought about by reductions in absenteeism. However, once an estimate has been made, the payback period for the investment in ergonomic improvements is simple to calculate by dividing the costs of the improvements by the monthly savings due to reductions in work loss, plus the value of any productivity gains. Oxenburgh gives some examples of cost-effective modifications to single machines or workplaces (similar to many of the examples described in this book) that paid for themselves in less than one year and often in less than 6 months.

The model can be used in a wide variety of industries in different countries. The cost savings will vary between countries owing to differences in employment practices; in particular, the social costs of the wage depend on whether the country has some kind of worker's compensation and social security scheme and how it is funded – by private insurance or by general taxation. As a rule of thumb, Oxenburgh suggests that, in developed-country businesses, one day of absenteeism costs up to 3.5 times one day's pay when all the direct and indirect costs are included. It is easy to see how ergonomic improvements can be cost-effective when the approach is used to justify change in high-risk jobs. Used correctly, the model should not overestimate the costs of absenteeism because it takes into account the costs of employing temporary workers or of paying others overtime. It would seem to be most applicable to blue-collar jobs where output is tightly managed and easily quantified.

Limitations of the approach   The productivity model has a number of limitations. It is only usable in companies or jobs with high absenteeism. It assumes that changes that will increase the scope for productivity gains and/or savings will do so in practice, i.e. that productivity will go up and stay up. Tight control over the life of the exercise is needed to implement the changes, control the costs and demonstrate the benefits. The 'time saved' hypothesis is based on simplistic notions such as 'time is money'. This may work in rigid production line environments but not in others because workers may perceive sick leave as part of the package and take time-off anyway; or, in non-paced jobs, the work may 'expand to use up the time available'. In some occupations, it may be possible to postpone work or there may be flexibility in the remainder of the workforce (everyone else has to work harder). In service industries, the penalties of failure to deliver (e.g. patients may have to wait longer to be admitted to hospital) are social rather than financial.

Although 'time is money', time saved does not always save money. Change needs to be managed to ensure that the extra time made available by ergonomic improvements is used productively. In production line systems, for example, speeding up one part of the line may be ineffective if the rest of the system is unable to keep up with the new pace. Improvements need to be carried out systematically and integrated into the rest of the system.

Prevention is better than cure

One of commonest arguments for the application of ergonomics is to prevent accidents, illnesses and low productivity. Stamper (1987) reported that, at that time, the Boeing Company spent $3500 per year on employee health care ($350 million annually), more than it spent on aluminium to build aeroplanes. In a similar vein, in the 1980s, the Chrysler car manufacturer found that its health care obligations amounted to 10% of the cost of one of its basic models of car. An insurance carrier, not a steelmaker was the company's biggest supplier. Shelton and Mann-Janosi (1992) demonstrated that between 1985 and 1989 the cost of employer-provided employee health care rose by an annual compound rate of 13%, much higher than inflation at the time. This led many companies to implement cost-containment programmes, including ergonomics programmes. The results of effectiveness and efficiency studies of these programmes are reported in later chapters.

Limitations of the approach   Despite the moral and ethical arguments for prevention, it is not always cost-effective, according to some studies. An investigation into the costs and benefits of preventive health care was carried out by the US Office of Technology Assessment (Leutwyler, 1995). Historically, public health measures such as sanitation and immunisation have had a massive impact on life expectancy and quality of life. However, of all modern preventive health measures only prenatal care for poor women, testing of neonates for congenital diseases and most childhood immunisations ultimately paid for themselves. Screening for high blood pressure was found to cost more than treating stroke victims and preventing disease was often more expensive than treating it. These are salutory lessons for ergonomists using cost–benefit arguments to justify their recommendations. Preventive efforts can be cost effective, but only when

- The measure is inexpensive and only has to be applied once
- The incidence of the problem is high
- The cost of dealing with it is high
- The measure is effective in preventing the problem and has few side effects.

In practice, cost-effective interventions should be targeted at high-risk groups and should be similar to 'primary prevention' used in medicine (e.g. redesign the task to reduce stresses on the back) rather than 'secondary' prevention (don't allow people with a history of back problems to do the job because even those who get through the screening process may still injure themselves later on).

In some countries, reductions in injury may have few immediate benefits for the employer if workers' compensation and health care are funded indirectly via taxation rather than directly through employers' insurance premiums.

Examples of industrial ergonomics programmes

Ergonomics programmes are often introduced via the existing safety infrastructure. A survey may be carried out to determine the scope and cost of any existing ergonomic problems. In the Goodyear study by Geras and his colleagues, an ergonomics section was added to the safety programme audit manual. Goodyear plants were required to establish ergonomics committees, give general awareness training to personnel and carry out audits to indentify ergonomic problems in the plant. In this way, an 'internal market' was created for ergonomics and the programme proceeded with a reactive phase in which existing problem areas were identified and problems were rectified.

An ergonomics training programme was set up and aimed at plant nurses, industrial engineers, safety manangers, training specialists, industrial hygienists industrial relations personnel, production specialists, workers' compensation personnel, union safety personnel and engineering maintenance personnel. Participants were trained to conduct ergonomics surveys and use techniques such as video to present the results of their findings. They were also trained in problem solving and general awareness training skills. In practice, trainees were encouraged to 'get their hands dirty' by tackling small, simple problems first (positive feedback gives confidence and encourages

Table 1.6 Goodyear accident rates before and after an ergonomics programme was implemented[a]

	1984	1985	1986	1987	1988
Plant 1	4.9	4.4	4.5[b]	0.8	0.9
Plant 2	9.7	5.6	8.9	7.4/0.8	2.6[c]

[a] From Geras *et al.* (1989).
[b] Start of ergonomics programme.
[c] Ergonomics programme implemented mid-year (figures show incident rate before/after programme).

people to tackle further problems), to combine training with practical intervention, to develop and maintain an operational ergonomics group, to review existing and new equipment, to communicate and to look at problems as opportunities. The success of the programme was quantified overall, in terms of a reduction in lost-time incidents (Table 1.6).

Plant 2 went from being the plant with the highest incident rate to one of the lowest after introduction of the programme. The programme was also evaluated using productivity data from sections of plants in which redesign efforts has been carried out, In one section, a 60% increase in productivity was achieved by reducing fatigue caused by uneccesary operator movement and movements requiring extended hand reach.

Car assembly Eklund (1995) investigated the relationship between ergonomics and quality in a Swedish car assembly plant and found clear relationships between ergonomic demands and quality problems. One of his suggestions was that the quality was degraded by the sociotechnical design, which separated inspection from quality, and that better quality could be obtained by making production line workers responsible for their own quality control and inspection.

Three categories of ergonomic problem were identified across a range of jobs: musculoskeletal loading due to the task or the posture; 'difficult to do' tasks (owing to the design of task objects); and psychologically demanding tasks. Next, quality control data from the assembly of 2000 cars were obtained. Quality data from 12 finished cars were also obtained. The cars were randomly sampled and disassembled in a special department for quality control purposes. Finally, quality inspectors were given a list of tasks and asked to identify those that were associated with quality control problems. Only 25% of all tasks were ergonomically demanding, but these generated 50% of all quality defects. Almost 40% of the faults found in the disassembly exercise arose from the ergonomically demanding tasks. Eklund estimated that the risk of quality deficiencies was almost 3 times as large for the ergonomically demanding tasks as for the other tasks. 'Difficult to do' tasks accounted for most of the problems and psychologically demanding tasks the least. However, 66% of tasks that had been identified as ergonomically demanding had some kind of quality deficiency. There was evidence from interviews with workers that they would settle for imperfect results in order to avoid physical discomfort or 'pass-on' uncompleted work or problems when under time pressure. Quality could be improved, it was concluded, by redesigning problem tasks to reduce the ergonomic demands.

Kochan (1988) reviewed the implementation of advanced technology by car manufacturers in the USA, Europe and Japan. Kochan argued that the implementation of

Table 1.7 Quality and productivity comparison of car assembly plants[a]

	Productivity (man hours/car)	Quality (defects/100 cars in 1st 6 months)	Automation (normalized to 100)[b]
Honda, Ohio	19.2	72.0	77
Nissan, Tennessee	24.5	70.0	89.2
NUMMI, California	19.0	69.0	62.8
Toyota, Japan	15.6	63.0	79.6
GM, Michigan	33.7	137.4	100.0
GM, Massachusetts	34.2	116.5	7.3

[a] From Kochan (1988).
[b] 100 = the normalised highest level of automation.

new technologies works best when it is integrated with the human resource function. In a graphic example, he described how, on a visit to the most 'high-tech' US car assembly plant, he was shown a room in which car door panels, sourced from external suppliers, were inspected for quality by lasers. The suppliers used the same system, which enabled both parties to ensure high quality both when the products left the supplier and when they arrived at the assembly plant. While visiting the US assembly plant of a Japanese car manufacturer, Kochan asked to see the area where incoming door panels were checked and was told that there was no such room. The company regarded it as the supplier's responsibility to deliver door panels of the required quality and, having worked with the supplier from the early stages, was now able to assume that the quality met the required standard. The cost (in 1988 dollars) for the inspection room in the first car plant was $5 million.

Kochan presents data on the productivity of several different car assembly plants (Table 1.7). The two General Motors plants differed strongly in level of automation (a value of 100 is the normalised highest level of automation) but had the same 'Fordist' management style (traditional hierarchical management with high job specialisation and close supervision of workers). The Japanese plants' management style, in contrast, was marked by few job classifications, flexible work organisation and extensive communication. The NUMMI plant, a joint venture between GM and Nissan, had a moderate level of automation but was managed by Nissan. Kochan concluded the following:

- Although GM *invested $650 million* in automating the Michigan plant, the quality and productivity were barely superior to those of the old, almost completely non-automated Massachusetts plant.
- The NUMMI plant, with moderate automation but revamped management style and human resource strategy, took 60% of the time to produce cars with 60% of the defects.
- NUMMI was unionised whereas Nissan and Honda were not. Unionisation does not seem to be a barrier to productivity and quality.

Organisational performance thus depends not just on technology but on how it is used, in other words on sociotechnical considerations. In modern organisations, companies have access to the same technology, so management and human factors have the main role to play in determining company competitiveness. Kochan advocates the 'humanware' approach to management in which a high degree of skill, motivation and adaptability is expected from workers who, in turn, are encouraged to suggest improvements continuously and are made responsible for quality control.

Economics of participation   Many of the arguments for participation are strategic or humanistic rather than economic. Gold (1994) argued that the burden of proof should be reversed and that there is little evidence for negative effects of participation on performance. N. Wilson of the University of Bradford Management Centre reported a 5–10% productivity difference in favour of companies adopting a participatory approach. O'Brien (1994) reported that the implementation of autonomous working groups in an Irish food company resulted in a 36% increase in throughput and increases in maintenance efficiency with savings of 25%. Carter (1994) reports that the introduction of changes centred around a partnership scheme in a UK environmental engineering company led to massive reductions in manufacturing cycle times, an 18-month cut in new product development times and a reduction in manufacturing time for boilers from 6 hours to 52 minutes.

Future directions for ergonomics

'Technology push' is one of the main factors influencing the direction and growth of ergonomics. Rapid development of usable systems is a priority in many organisations. Demographic change in industrially developed countries is imposing new constraints – an ageing workforce and shortages of skilled people. At the same time, equal opportunities legislation demands that employers make work available to all. These trends place pressure on traditional 'FMJ' approaches involving personnel selection in favour of 'FJM' – redesign the work so that anyone can do it. System designers are responding by making more use of automation in new 'lean-manned' systems. Paradoxically, having fewer people around increases the need for ergonomics as the role of those remaining becomes more critical. Physical ergonomic issues are taking on a new importance as workforces age and as more women take on jobs previously done by men. In developing countries, there is still a need for basic ergonomic design of factories and offices.

Summary

Ergonomics occupies the 'no man's land' between engineering and medicine, architecture and health and safety, computer science and consumer product design. It is the only scientific subject that focuses specifically on the interaction between people and machines.

Historically, ergonomics can be seen to have arisen as a response to the need for rapid design of complex systems. The modern ergonomist has an important role to play as a member of the design team, providing scientific information about humans (a scarce commodity in many organisations), and ensuring that all aspects of the system are evaluated from the users' or operators' point of view. The participatory approach seems to be the best way to ensure that the implementation of ergonomics will be effective.

Essays and exercises

1. Was ergonomics invented or discovered, or is it just a buzzword for what happens anyway? Discuss.
2. Search the web for sources of information about ergonomics. Use the categories in the human–machine model to guide your search. (Hint: start with the web sites of the International Organization for Standardization, the American National Standards Institute and the British Standards Institution. Cross reference each published standard or guideline with the corresponding chapter in this book.)
3. Contact five different manufacturers across a range of industries. Find out from them whether they use ergonomic principles in the design of their products. For those companies that do use ergonomics, find out who is responsible and how ergonomics is applied. (Hint: choose a white goods manufacturer, a car manufacturer, computer hardware and software manufacturers and one other such as a manufacturer of medical equipment or of garden tools.)
4. Describe the present scope and concerns of ergonomics. Obtain back issues for the last 5 years of two or more of the journals *Human Factors, Ergonomics* and *International Journal of Industrial Ergonomics and Applied Ergonomics.* Summarise, using keywords of your choice, the topics and research questions being dealt with in each of the papers in these journals.

2   Anatomy, posture and body mechanics

Posture is an active process and is the result of a great number of reflexes, many of which have a tonic character. The attitudinal, as well as the righting reactions, are involuntary.
(H. D. Denniston, 1935)

It is easy to overlook the fact that the human body is a mechanical system that obeys physical laws. Many of our postural and balance control mechanisms, essential for even the most basic activities, operate outside of conscious awareness. Only when these mechanisms break down – as in slipping or losing balance – are we reminded of our physical limitations. An understanding of these limitations is fundamental to practically all applications of ergonomics.

The skeleton plays the major supportive role in the body. It can be likened to the scaffolding to which all other parts are attached. The functions of the skeletal and muscular systems are summarised in Table 2.1.

Like any mechanical system, the body may be stable or unstable and is able to withstand a limited range of physical stresses. Stresses may be imposed both internally or externally and may be acute or chronic. A useful starting point in the discussion of mechanical loading of the body is to distinguish between postural stress and task-induced stress. According to Grieve and Pheasant (1982), postural stress is the term used to denote the mechanical load on the body by virtue of its posture. Posture is defined as the average orientation of the average orientation of the body parts over time. Task-stress depends on the mechanical effort needed to perform the

Table 2.1 Functions of the skeletal and muscular systems

Skeletal system	Muscular system
1. Support 2. Protection (the skull protects the brain and the rib cage protects the heart and lungs 3. Movement (muscles are attached to bone; when they contract, movement is produced by lever action of bones and joints) 4. Homopoiesis (certain bones produce red blood cells in their marrow)	1. To produce movement of the body or body parts 2. To maintain posture 3. Heat production (muscle cells produce heat as a by-product and are an important mechanism for maintaining body temperature)

Table 2.2 Task stress and postural stress

Postural stress	Task stress	
	High	*Low*
High	Digging a trench	Painting a ceiling
Low	Competitive Weightlifting	Reading a book

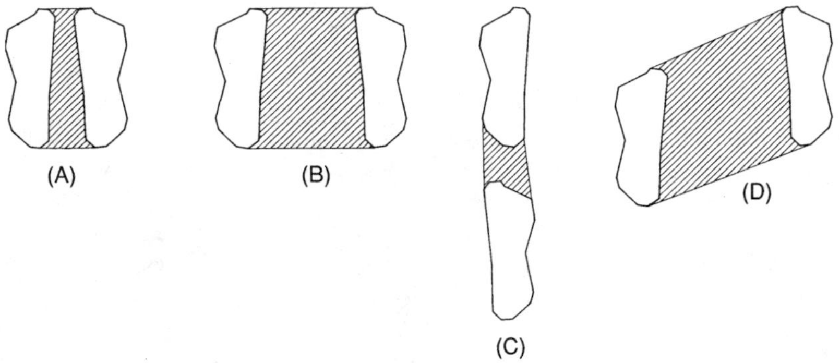

Figure 2.1 The stability of the body parts depends on the shape of the base of support described by the position of the feet. (A) is unstable, (B) is fairly stable in all directions, (C) is stable antero-posteriorly, and (D) is laterally stable.

task. Task and postural stress can vary independently of each other (Table 2.2). Some tasks, such as lifting a barbell are high in task stress but can be performed in non-stressful postures. Painting a ceiling requires little effort to apply the paint but much effort to maintain the posture. Much biomechanical stress in unnecessary because it is postural and can be reduced by redesigning the task to improve the posture.

Postural stability

In order for the body to be stable, the combined centre of gravity (COG) of the various body parts must fall within a base of support (the contact area between the body and the supporting surface). In standing, the weight of the body must be transmitted to the floor through the base of support described by the position of the feet (Figure 2.1). The alignment of the body parts must be maintained to ensure continuing stability and it is in the maintenance of posture that much stress arises.

Some basic body mechanics

The basic limiting condition for postural stability in standing is that the combined COG of the various body parts is within the base of support described by the position of the feet (assuming no other external means of support). Ideally, the lines of action of the masses of the body parts should pass through or close to the relatively incompressible bones of the skeleton (Figure 2.2). The jointed skeleton thus supports the

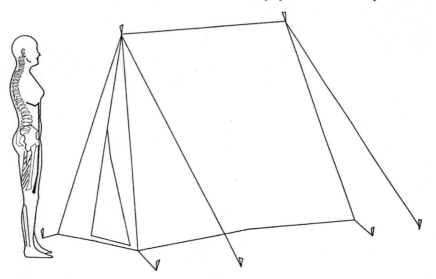

Figure 2.2 The tent analogy. The skeleton is the tent pole, the muscles are the guy ropes and the soft tissues are the canvas.

body parts and is itself stabilised by the action of muscles and ligaments, which serve merely to correct momentary displacements of the mass centres from above their bony supports. Using a rather crude analogy, the skeleton can be likened to an articulated tent pole with guy ropes (postural muscles) on every side. The fabric of the tent corresponds to the soft tissues of the body. Any displacement of the COG of the structure in a given direction leads to tension in the guy ropes on the opposite side. Ligaments can be likened to the springs and rubber fittings that stabilise the articulations of the tent pole, and tendons to the ends of the guy ropes where they insert into the poles.

Postural stress can cause pain. Workers who have to work with the spine flexed forwards (by 60 degrees for more than 5% of the day or 30 degrees for more than 10% of the working day) or rotated (more than 30 degrees) suffer back pain (Hoogendoorn *et al.*, 2000).

Demonstration

To demonstrate the 'tent' analogy, stand upright and relaxed with body weight equally distributed between the feet and neither on the heels or the balls of the feet. Place one hand on the low back muscles (you should feel a muscular ridge in the centre of your back). Place the other hand on your abdominal muscles. Palpate both sets of muscles, which should feel soft. Next let your weight move to the balls of your feet and lean forwards slightly. As you do this, tension should appear in the low back muscles as they act to maintain equilibrium. Repeat by swaying backwards, slightly, with the weight on your heels. Your abdominal muscles will begin to tense and your back muscles will relax as the weight moves onto the heels. In the middle, there will be a neutral position, where your upright stance can be maintained with minimal muscular load. **This neutral position is one of low postural load.**

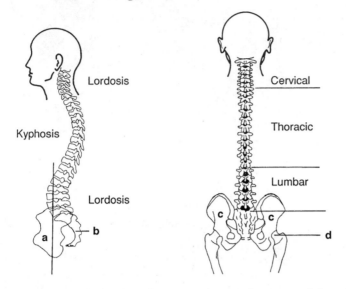

Figure 2.3 The lumbar, thoracic and cervical spines and the pelvis (**a**) and sacrum (**b**). The weight of the upper body is transmitted through the lumbar spine, the iliac bones of the pelvis (**c**) to the hip joints (**d**) and legs.

Without its associated trunk muscles, the human spine is very weak – it buckles under a compressive load of only 90 N. Cholewicki *et al.* (1997) have shown that the function of the trunk muscles is critical in giving the spine its compressive strength. They demonstrated that, although a neutral position of minimal postural stress does exist, it depends on low-level antagonistic co-contraction of the trunk flexors and extensors. This activity increases when the person carries a load and is an example of true postural muscle activity – the muscles act like guy ropes to stiffen the intervertebral joints. The main cost of the co-contraction is increased spinal loading. In upright postures, the benefits of increased spinal stability outweigh these costs (Granata and Marras, 2000).

Anatomy of the spine and pelvis related to posture

The spine and pelvis support the weight of the body parts above them and transmit the load to the legs via the hip joints. They are also involved in movement. Almost all movements of the torso and head involve the spine and pelvis in varying degree. The posture of the trunk may be analysed in terms of the average orientation and alignment of the spinal segments and pelvis. Figure 2.3 depicts the spine and pelvis viewed frontally and sagittally.

The spine

Quadrupedal animals and human babies have a single spinal curve running dorsally from pelvis to head. The thorax and abdomen hang from the spine and exert tension that is resisted by the spinal ligaments, the apophyseal (facet) joints and the back muscles. In adult humans, the spine is shaped such that it is close to or below the

COG of the superincumbent body parts that are supported *axially* – that is, the effect of weightbearing in the standing posture is to compress the spine (Adams and Hutton, 1980). This compression is resisted by the vertebral bodies and the intervertebral discs. The *cervical* and *lumbar* spines are convex anteriorly – a spinal posture known as *lordosis*. It is the presence of these lordotic curves that positions the spine close to or directly below the line of gravity of the superincumbent body parts. The effect is to reduce the energy requirements for the maintenance of the erect posture and place the lumbar motion segments in an advantageous posture for resisting compression (Klausen and Rasmussen, 1968; Adams and Hutton, 1980, 1983; Corlett and Eklund, 1986). The thoracic spine is concave anteriorly and is strengthened and supported by the ribs and associated muscles.

The term 'spinal column', although universally accepted, is something of a misnomer and 'spinal spring' might be more appropriate: the 'S' shape of the spine of a person standing erect gives the entire structure a 'spring-like' quality such that it is better able to absorb sudden impacts, such as the mechanical shock when the heel strikes the ground when walking (Schultz, 1969), than if it were a straight column. The loss of the 'S' shape in sitting may be one of the reasons why drivers of trucks and farm vehicles who are exposed to vibration in the vertical plane are so prone to back trouble.

The cervical and lumbar spines are not fixed in lordosis. Each vertebral body is joined to its superior and inferior counterpart by muscles, ligaments and joints. The spine takes part in functional movements of the body – part of the postural adaptation required to carry out many activities takes place in the lumbar and cervical spines.

The spine can be considered simplistically to consist of three anatomically distinct but functionally interrelated columns (Figure 2.4). The anterior column, consisting of the vertebral bodies, intervertebral discs and anterior and posterior longitudinal ligaments, is the main support structure of the axial skeleton. It resists the compressive stress of the superincumbent body parts. The two identical posterior columns are positioned astride the neural arch (which forms a bony cavity through which passes the spinal cord), and consist of the apophyseal (or 'facet') joints and the associated bony projections, ligaments and muscles. The posterior elements of the spine act as jointed columns that control the movement of the complete spine and provide

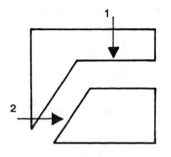
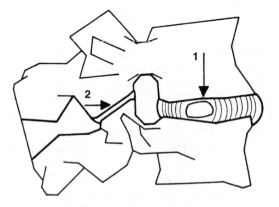

Figure 2.4 Function of (1) intervertebral disc and (2) facet joints. The disc resists the compressive load and the facets resist the intervertebral shear force. (From Kapandji, 1974, with permission.)

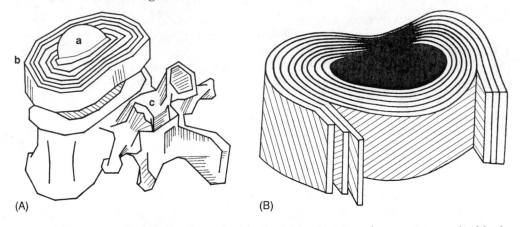

(A)   (B)

Figure 2.5 Intervertebral disc and vertebral body. (a) In this view, the superior vertebral body
has been removed to reveal the intervertebral disc below. a = the nucleus pulposus,
b = the annulus fibrosus. At the rear can be seen the inferior facet joints (c). (From
Kapandji, 1974, with permission.) (B) Detail of the structure of the annulus fibrosus.
The annulus consists of a number of layers of cartilage. The fibres in the layers run
obliquely and in different directions somewhat like the layers of a cross-ply tyre.
The outer layers run perpendicularly to each other. (From Vernon Roberts, 1989,
with permission of Churchill Livingstone.)

attachment points for the back muscles. The vertebral bodies and their related structures increase in size from the top to the bottom of the spine in accordance with the
increased load that they must bear.

The intervertebral discs act as shock absorbers and limit and stabilise the articula-
tion of the vertebral bodies. Each disc consists of concentric layers of cartilage whose
fibres are arranged obliquely in a manner similar to a cross-ply tyre (Figure 2.5). The
layers of cartilage enclose a central cavity that contains a protein–mineral solution
('proteoglycans'). Positive osmotic pressure ensures that water is always tending to
enter the disc. Thus, the discs are *pre-stressed* to withstand loading (in a manner
analogous to reinforced concrete beams used in the construction of modern buildings).
According to Kapandji (1974), the nucleus pulposus functions as a swivel joint.

Intervertebral discs exhibit viscoelastic behaviour. Forces of rapid onset are resisted
in an elastic manner: the disc deforms initially then returns rapidly to its original
shape when the force is removed. Under continuous loading, however, the disc exhibits
a type of viscous deformation known as 'creep'. Creep occurs as a result of loading
above or below a threshold level. Under compressive loading, the disc narrows as
fluid is expelled and the superior and inferior vertebral bodies move closer together
(Eklund and Corlett, 1984). Under traction ('stretching' or 'pulling forces'), fluid
moves into the disc and the disc space becomes wider (Bridger *et al.*, 1990).

The narrowing and expansion of the disc spaces is natural and occurs as a result of
the forces exerted on the spine during daily living activities. Since there are 24 vertebral bodies, all with discs between them, the shrinkage and expansion of the disc
spaces results in measurable changes in stature: most people are about 1% taller
when they wake up in the morning than when they go to bed at night for this reason
(dePuky, 1935). Stature change varies exponentially with loading time: almost 50%
of the stature gained after a night's sleep is lost in the first half-hour after rising.

Grieco (1986) suggests that, since the discs have no direct blood supply, the daily ingress and egress of fluid due to variations in loading is the mechanism whereby nutritional exchange with the surrounding tissues takes place. Postures that exert static loads on the body will interfere with this mechanism and are hypothesised by Grieco to accelerate the degeneration of the discs. Static compression of cells in the discs has been linked to an increase in the rate of cell death (Lotz and Chin, 2000). Although it is too early to specify what the tolerance limits would be for safe exposure to static compression, there is some empirical support for the view that such loading should be avoided. *Stressful postures adopted for 8 hours per day would be regarded as a health hazard according to this view.*

Stature change also occurs with age; after about 30 years of age, the intervertebral discs degenerate, developing micro-tears and scar tissue, fluid is lost more readily and the disc space narrows permanently. At this stage, the spinal motion segments lose stability. It is not surprising then, that most occupationally induced low back pain occurs in middle-aged people. In the elderly, disc degeneration reaches a stage where, together with other degenerative processes, the spine is restabilised but with a corresponding loss of mobility.

The pelvis

The pelvis is a ring-shaped structure made up of three bones: the sacrum and the two innominate bones. The sacrum extends from the lumbar spine and consists of a number of fused vertebrae. The three bones are held together in a ring shape by ligaments (Figure 2.6). The innominate bones are themselves made from the fusion of

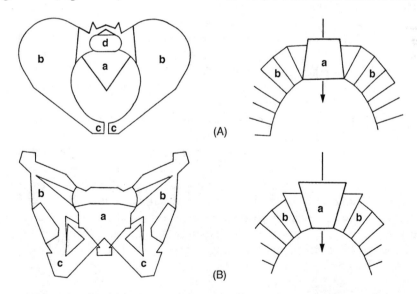

Figure 2.6 The pelvis as an arch: (A) The pelvis viewed from above: a = the sacrum, b = ilium, c = the pubis, d = position of the intervertebral disc between the first sacral and fifth lumbar vertebrae. Under load, the sacrum tends to move forwards, like an inverted keystone in an arch. It has to be held in place by strong ligaments. (B) The pelvis viewed from the rear: a = sacrum, b = ilium, c = ischium. The sacrum acts like a true keystone in this plane. (Redrawn from Tile, 1984.)

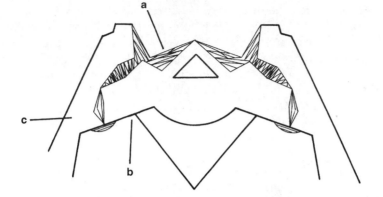

Figure 2.7 View of the sacroIliac joint from above: **a**, ligaments; **b**, sacrum; **c**, pelvis. According to Tile (1984) the ligaments act like the cables of a suspension bridge preventing the sacrum from slipping forwards. If the joint is deformed by loading, the ligaments can be pinched by bone, causing pain in the very low back, usually on one side. (Adapted from Tile, 1984.)

three other bones, the ilium, the ischium and the pubis. The pubis lies at the anterior part of the pelvis. It joins the other bones together, completing the ring shape and acting like a strut to prevent the pelvis from collapsing under weightbearing (Tile, 1984). The posterior structures of the pelvis, the sacrum and the ilia carry out the actual weightbearing function.

The pelvis can be likened to an arch that transfers the load of superincumbent body parts to the femoral heads in standing and to the ischial tuberosities (part of the two ischia) in sitting (Figure 2.6).

When it is viewed from the rear (Figure 2.6), it can be seen that the sacrum resembles the keystone of the arch. The load from above is transmitted through the innominates to the femoral heads. However, when viewed from above, the sacrum has the wrong shape for a keystone – it tends to slide forwards, out of the arch (Figure 2.6). Under weightbearing, the tendency for the sacrum to slide forwards anteriorly is resisted by the strong ligaments between the sacrum and the ilia. It is these posterior sacroiliac ligaments that stabilise the joint between the sacrum and the ilia. DonTigny (1985) has pointed out that standing postures in which the person has to bend forwards slightly from the hip (such as when washing dishes at a sink) increase the tendency for the sacrum to be anteriorly displaced, thereby increasing the tension in the sacroiliac ligaments. Small displacements of the sacrum can occur, causing soft tissues to be 'pinched' and causing pain (Figure 2.7). This pain can be mistaken for low back pain.

The lumbo-pelvic mechanism

The lumbar spine arises from the sacrum and the degree of lumbar lordosis depends on the sacral angle, which, in turn, depends on the tilt of the pelvis (Figure 2.8). The relationship between the posture of the pelvis and that of the lumbar spine was eloquently described by Forrester-Brown (1930) as follows:

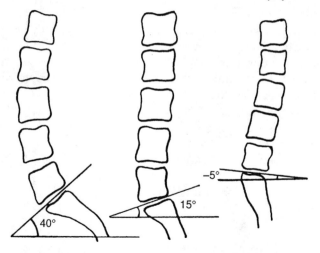

Figure 2.8 Relationship between sacral angle and lumbar angle.

The simplest countryman understands that one cannot put on the top story of a house until one has built the ground floor and foundations; yet medical men are constantly trying to alter the position of the upper bricks of the spinal column without adjusting the base on which they stand.

Some ergonomists could be accused of making the same mistake when trying to design seats to prevent excessive lumbar flexion. Parents' admonitions to young children to 'sit up straight and don't slouch' may be equally mistaken.

When looking at the posture of the spine at work, it is necessary to consider the factors that determine the position of the pelvis, such as the slope of the seat or the position of the feet on the floor. The pelvis can be represented as in Figure 2.9, as

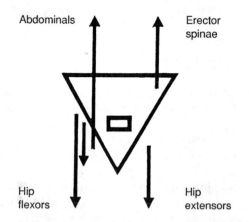

Figure 2.9 Schematic representation of the muscular system of the pelvis (sagittal view). When the abdominal or hip extensor muscles shorten, the pelvis tilts backwards. The result is a flattening of the lumbar spine to maintain the trunk erect. When the hip flexors or erector spinae muscles shorten, the pelvis tilts forwards. This is accompanied by a compensatory increase in the lumbar lordosis.

part of a lever system, with the hip joint regarded as a fulcrum. Many muscles attached to the pelvis can be considered as guy ropes or stays that fixate the pelvis onto the heads of the femora. These muscles can exert torques on the pelvis, which cause change in pelvic tilt (even though this is not their main function). The hamstring, gluteal, iliopsoas, erectores spinae muscles and other muscles together with the ligaments of the hip joint are part of the lumbo-pelvic system. The tilt of the pelvis in the anterior/posterior plane depends on equilibrium of the torques exerted by the antagonistic muscles in the system.

In standing, the line of gravity falls slightly behind the centre of the hip joint, causing the pelvis to *tend* to tilt backwards. This relieves the abdominal muscles of a postural role and explains why these muscles are relaxed in standing (this applies to normal standing, when one is carrying a load on the back or walking down a steep hill, the abdominals do play a role).

Postural stability and postural adaptation

A stable posture can only be maintained if the various body parts are supported and maintained in an appropriate relation to the base of support such as the feet (Figure 2.1). The line of gravity of the body parts must fall within the base of support and postural reflexes exist to ensure that, as one body part moves, the total body mass remains balanced over the feet

For example, when a standing person leans forward, as if to touch the toes, the pelvis moves rearwards to compensate for the forward displacement of the COG of the upper body (Figure 2.10). It is impossible to carry out this movement without falling over if there is insufficient space at the rear to allow for compensatory projection of the buttocks, unless one foot is placed in front of the other. A simple demonstration of the principle can be achieved by standing with one's back and heels against a wall and attempting to pick up an object from the floor in front without moving the feet.

It is important to provide *sufficient space* around standing operators and plenty of *room for the feet* if losses of balance are to be avoided.

For the body to be in a condition of **static equilibrium:**

Upward forces (from floor)	= Downward forces (body weight plus any objects held)
Forward forces (e.g. bending forwards)	= Backward forces (extension of back muscles)
Clockwise torques (e.g. from assymetric load)	= Anticlockwise torques (back and hip muscles)

Ideally, the skeleton should play the major role in supporting the various body parts since this is its function. However, muscles, ligaments and soft tissues can also play a role, but at a cost of increased energy expenditure, discomfort or risk of soft-tissue injury.

When leaning forwards in the manner shown in Figure 2.10, a stable posture can be maintained indefinitely although the posture itself is uncomfortable. This is because strain is placed on the posterior spinal ligaments and lumbar intervertebral discs: the

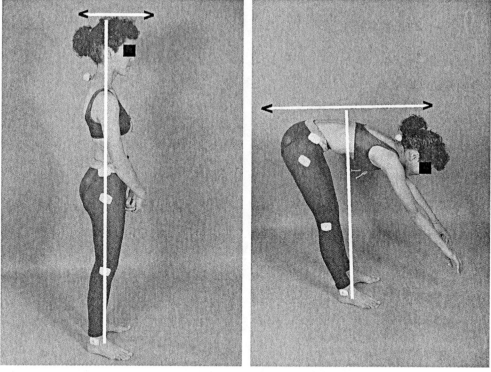

(A)   (B)

Figure 2.10 When the base of support is constrained, compensatory movements occur auto-
        matically to maintain postural stability.

upper-body load is no longer supported by axial compression of spinal structures but
by tension in ligaments and asymmetric compression ('wedging') of the intervertebral
discs (Figure 2.11).

Postures can be stable but stressful if support of body mass depends on soft tissues
rather than bone. Ligaments are able to resist high tensile forces, particularly if these
forces are exerted in the direction of their constituent fibres. They play a major role
in protecting joints by limiting the range of joint movement and by resisting sudden
displacements that might damage the joint. However, injuries can occur if ligaments
are exposed to sudden forces when pre-stressed by extreme joint positions or by
complex movements. This is one of the reasons why ergonomists stress the importance of the posture of the hands, wrists, elbows and trunk when tools or controls are
operated or when loads are lifted. *Poor equipment design that forces the adoption of
extreme joint positions when holding an object predisposes the joint to injury.* Good
design enables equipment to be used with the joints in the middle of their range of
movement.

When a person leans forward in the manner described above but 'arches' the back
in an effort to prevent the spine from losing its shape, the posture that is produced is
still stable but a different cost is incurred. In this position, the muscles of the back
must carry out static work to maintain the shape of the spine against the pull of
gravity which is causing it to flex. *Static muscle contractions rapidly lead to fatigue.*

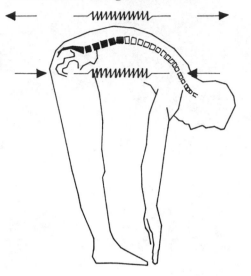

Figure 2.11 In this position, postural stress occurs in the form of compression of abdominal contents and intervertebral discs and stretching of the posterior spinal ligaments.

Low back pain

Low back pain is not a disease, although it can be a symptom of one. It can also be the natural response of healthy tissue to biomechanical loading, in which case it disappears rapidly after the loading ceases.

Epidemiology of low back pain

A survey by Papageorgiou *et al.* (1995) estimated the one-month prevalence of low back pain to be between 35% and 37% in the adult population, with peak prevalence among the 45–59-year age group. Miedema *et al.* (1998) followed up 444 patients who had consulted their doctor because of an episode of back pain. After 7 years, 28% had developed chronic problems. Chronic sufferers were more likely, subsequently, to consult a doctor, had poorer general health and poorer physical and social functioning. McFarlane *et al.* (1997) found an increased risk of pain among those whose jobs involved lifting, pulling or pushing objects over 11 kg or whose jobs required long periods of standing or walking, but there was no clear relation with years of exposure, suggesting that a simple 'dose-response' model does not apply. Richardson *et al.* (1997) investigated the prevalence of discogenic back pain in the relatives of patients who had received surgery for lumbar intervertebral disc lesions and in a comparison group of patients with upper-limb problems. Relatives of 'disc' patients were over 16 times more likely to have discogenic pain than controls, suggesting a genetic predisposition towards lumbar pain and intervertebral disc lesions.

Volinn (1997) investigated low back pain prevalence in low- and middle-income countries. Contrary to his hypothesis, that rates would be lower in more affluent societies because of the unavoidable demands for hard physical work in poorer countries, he found that rates were 2–4 times higher among the populations of Germany and Belgium than those of Nigeria, Southern China, Indonesia and the Philippines. There are many possible explanations for this finding, perhaps subsistence farmers in low-income countries have a higher pain threshold or exercise more. Within low-income countries, rates were higher in urban than in rural areas, particularly among groups such as sewing machine operators in small factories. This draws attention to work organisation factors and task variety – industrial jobs being far more restricted, physically, than small-scale farming actitivities that vary throughout the day and with the season.

Mechanical risk factors for low back pain

Kumar (1990) has shown that mechanical load is a risk factor for low back pain. He used a two-dimensional, static mathematical model of spinal loading to estimate the shear and compression forces at the lumbosacral and thoracolumbar joints. These forces were found to be higher in workers with self-reported pain than in workers without pain.

Klingenstierna and Pope (1987) have shown that vibration increases the rate of creep in the spinal motion segments. Kasra *et al.* (1992) have shown that loss of fluid from the nucleus reduces its incompressibility. This manifests itself as a decrease in the resonant frequency and an increase in the response amplitude of the motion segments when they are exposed to compressive forces. An already degenerated disc will respond in this way when exposed to vibration and will experience greater stresses and strains than a healthy disc; the likelihood of further disc degeneration will be greater and the facet joints will be exposed to greater loading (Shirazi-Adl, 1992). Boshuizen *et al.* (1992) report a very high prevalence of back pain among fork lift truck drivers exposed to vibrations of $0.8 \, \text{m s}^{-2}$ (68% compared to 25% in comparable control subjects). Of particular interest is that the drivers were relatively young (<35 years of age) and the risk of getting pain was greatest in the first five years of driving. Older drivers had a lower risk of pain (presumably older drivers with pain had found alternative employment). Apart from the vibration, the drivers often sat with the trunk twisted while looking backwards (this posture increases the load on the facet joints). Taken together all these findings point to a pattern of loading that will cause chronic pain and hasten the onset of degenerative diseases of the spine.

Clearly, driving tasks that expose the driver to vertical vibrations should not be combined with lifting tasks. Cabs should be fitted with mirrors to obviate the need to look behind, and vibration should be reduced. Drivers of goods delivery vehicles should not have to load and unload the goods or should be provided with lifting and carrying aids.

Constrained sitting for 8 hours per day is a risk factor for low back pain. Burdorf *et al.* (1993) investigated low back pain among crane drivers, straddle-carriers and a control group of office workers. Twelve-month pain prevalence was 50% in crane drivers, 44% in straddle operators and 34% in office workers. Non-neutral postures of the trunk were present in all groups (>20% forward and lateral flexion and >20% rotation). Crane drivers and office workers had more forward flexion and the straddle carriers had more lateral flexion and rotation. In contrast to the office workers, who could stand or walk for 17% of the day, the other groups had to sit for hours in a space that prevented standing or other efforts to change posture. Thus, forced non-neutral postures that have to be sustained are a risk factor for low back pain. This finding supports the ergonomics rule of thumb that as far as possible postural constraint should be designed out of jobs and the goal should be to permit workers to adopt a number of different working and resting postures at will.

Causes of low back pain

Although the anatomy of the spine is well understood, finding the source and cause of low back pain can be a much more elusive problem for clinicians. Pain is unlikely to arise from the intervertebral discs themselves since only the outer parts contain nerve endings in the adult.

Similar reasoning rules out pain from the capsules of the apophyseal joints. Likely sources of pain are the posterior ligaments and the back muscles. These may be irritated by mechanical trauma due to damage to or degeneration of bony structures, or the pain may be due to fatigue. Nerve root compression can also be a source of pain. Pain from the sacroiliac joint can sometimes be mistaken for low back pain (DonTigny, 1985) and evidence is accumulating that the sacroiliac joint is the source of pain below L5–S1 (Schwarzer *et al.*, 1995).

Back pain can occur for non-work-related reasons and from unrelated structures such as the kidneys. Colds and flu may cause complaints of pain in the back. Back pain is a complex problem (Waddell, 1982) and detailed investigation of back problems is best left to expert clinicians. If a worker complains of back pain, the ergonomist's natural inclination is to search for causes in the workplace. In the case of strenuous jobs, this may be appropriate but, in less obvious cases, non-occupational and potentially more serious causes should also be considered.

There is some evidence that chronic low back pain sufferers exhibit inefficient co-contraction of the transversus abdominus muscle, a muscle thought to play a role in stiffening the trunk during postural movements (Hodges and Richardson, 1996). Soft tissues in the spine are then exposed to greater load and, over time, this sets up a cascade of maladaptation as one problem causes another. Jayson (1997) suggested that vascular damage due to disc degeneration causes nerve root damage leading to a chronic pain syndrome.

Back pain and muscular fatigue

It has been shown that the lumbar muscles of chronic low back pain sufferers fatigue more rapidly than those of non-sufferers. Presumably, pain occurs both directly, as a result of stimulation of pain receptors in the muscles due to the biochemical changes that accompany fatigue, and indirectly due to the increased load on soft tissues in the lumbar spine itself. Roy *et al.* (1989) used median frequency analysis of the EMG signal to investigate lumbar muscle fatigue in chronic low back pain sufferers and in normals. They found that the median frequency parameters could be used to reliably classify 91% of the low back pain sufferers and 84% of the healthy subjects. In a second study, Roy *et al.* (1990) showed that the median frequency analysis could distinguish between low back pain sufferers and pain-free individuals in a group of elite rowers. Finally, Klein *et al.* (1991) compared the median frequency technique with conventional clinical diagnostic tests for low back pain (range of motion of the lumbar spine and maximum voluntary contraction). The conventional tests correctly identified 57% of the pain sufferers and 63% of pain-free individuals, whereas median frequency analysis correctly identified 88% of pain sufferers and 100% of pain-free individuals. Other researchers (e.g. Biedermann *et al.*, (1991) have obtained similar findings using median frequency analysis to investigate lumbar muscle fatigue. It seems that a lack of back muscle endurance, rather than a lack of strength, is the key characteristic of the type of chronic low back pain sufferers who suffer no other obvious physical abnormalities or pathological conditions. Van Dieen *et al.* (1993) have shown that the greatest shift in erectores spinae median frequency is at the L5 level.

These findings help explain why chronic low back pain sufferers are at risk in tasks that involve repetitive lifting, carrying of weights in front of the body, leaning forward or with trunk extended: all these activities require sustained activity of the back extensors.

Low back and shoulder girdle pain are major problems in the industrialised world. Frequently, the pain is of an acute form and is due to muscular fatigue. This type of pain usually subsides within hours or days if the sufferer rests. However, for some individuals the pain is chronic and may be indicative of an underlying pathology. In these cases, it may be difficult to construct a complete causal explanation and medical advice may be required.

Many researchers now acknowledge that health problems are often exacerbated, if not caused, by habitual daily activities. It is known that low back problems have a higher incidence among certain groups of individuals such as professional drivers. Magora (1972) carried out an epidemiological survey to investigate the incidence of low back pain in relation to the occupational requirements for sitting, standing and lifting. Low back symptoms were higher among those with uniform occupational requirements than in those whose daily activities were more varied (who were able to alternate between standing and sitting). Gilad and Kirschenbaum (1988) investigated back pain across a broad spectrum of jobs. More back pain was found in groups who worked in unusual body positions or with the trunk flexed laterally or forward in standing or sitting. Keyserling *et al.* (1988) came to similar conclusions after investigating back problems in an automobile assembly plant. Persistent back pain was associated with forward and lateral flexion and twisting (axial rotation) of the spine. Pain prevalence increased substantially if a non-neutral posture was held for more than 10% of the work cycle, suggesting that such postures be designed out of the work cycle or be minimised. Back pain sufferers were 5 times more likely than matched controls to work with the trunk in mild flexion and almost 6 times more likely to work with the trunk in severe flexion.

Although the aetiology of musculoskeletal problems involves several factors, it is known that pain can be caused or exacerbated by excessive loading of joints and muscles. This can occur not only as a result of traumatic events but also owing to sustained exposure to particular working postures. Nachemson (1966) used a needle transducer to measure the hydrostatic pressure in the third lumbar intervertebral disc. Disc pressure was found to be higher in sitting than in lying down but was reduced when the sitter reclined against a backrest. Other researchers (e.g. Adams and Hutton, 1985; Andersson, 1986; Keegan, 1953; Schierhout *et al.*, 1992) have also presented data that support the notion that it is not *whether* we stand or sit that causes undue postural stress but *how*.

Can low back pain be prevented?

Case studies demonstrating the effectiveness of specific ergonomic interventions are given in later chapters. In general, the evidence that low back pain can be prevented in the general population is not promising, as is evident in the following review papers. Linton and van Tulder (2001) reviewed controlled trials of prevention programmes and found that exercise had a mild protective effect. Van Tulder *et al.* (2000) found that exercise was ineffective as a form of therapy for acute low back pain but that there was some evidence that it was effective as therapy for chronic low back pain in facilitating the return to normal daily activities, including work. On balance, there seems to be some evidence that exercise is beneficial, particularly if it strengthens the trunk or improves endurance of the trunk muscles. The mechanisms by which exercise may help are unknown. It is of interest that Stevenson *et al.* (2001) found that personal fitness is an important defence against low back pain. Their prospective study of manual workers handling more than 5000 kg/day showed that those who did not get back problems had stronger static leg strength and endurance and could move their upper bodies faster than those who went on to develop problems.

Risk factors for musculoskeletal disorders in the workplace

The main risk factors for musculoskeletal disorders can be categorised under one of four headings:

- Force
- Posture
- Repetition
- Duration of task

Posture is a key consideration in any assessment of risk. Tasks that require the adoption or repetition of postures at the extremes of the range of repetitive motions can lead to imbalances in antagonistic muscle tendon units, resulting in degradation of joint function.

In addition to tissue strength and loading, the dynamic aspects of task performance are of interest in relation to the prevalence of musculoskeletal disorders. Marras and Schoenmarklin (1993) investigated the wrist motions of workers engaged in jobs entailing high or low risk of contracting carpal tunnel syndrome. Wrist position was characterised in terms of flexion/extension, pronation/supination and ulnar/radial deviation. The angular velocity and accelerations of the wrist when moving in the corresponding spatial planes were also recorded. No significant differences were found between the high- and low-risk groups in wrist posture. However, wrist movements in the high-risk group had greater velocity and acceleration. The findings were interpreted by the authors in terms of Newton's second law of motion (force = mass × acceleration). In order to produce greater accelerations of the wrist, larger muscle forces are required, which are transmitted to the bones via the tendons. The tendons will also be exposed to greater friction by contact with surrounding structures.

In a similar vein, Marras *et al.* (1993) investigated trunk motion in relation to the risk of low back disorder in a variety of jobs. As might be expected, high-risk jobs were associated with high load moments and lifting frequencies and large trunk flexion angles. In addition, lateral trunk velocity and twisting trunk velocity were also associated with high-risk jobs – the faster these movements were, the greater was the risk of injury. This finding can be explained by recalling that a major function of muscles is to stabilise joints. When a body part has to be moved rapidly, in a controlled way, greater coactivation of synergistic and antagonistic muscles is needed to stabilise the joints involved. Since many of the muscles involved will be working against one another, the result is a magnification of the joint loading (Marras and Mirka, 1992).

It appears that many bodily structures will be at risk of injury when exposed to repetitive movements that require rapid acceleration of limbs or fast responses. A general principle for the prevention of disorders would seem to be to reduce the repetition rate and not just the number of repetitions or work cycles in a job. One way to do this is by combining the microelements of tasks into larger units. Further details of the mechanisms of tissue response to mechanical loading resulting in injury can be found in Chapters 5 and 6. A method for characterising postural stress and repetitive motion can be found in Radwin and Lin (1993). A checklist for evaluating risk factors for upper extremity disorders can be found in Keyserling *et al.* (1993).

Any factors that reduce the strength of body parts will increase the risk of injury. McGill has argued that fatigue is such a factor and that attempts to specify maximum 'safe' loads are limited. Rather, the injury threshold varies throughout the day, or over the work shift depending on the level of fatigue. Figure 2.12 illustrates this concept diagrammatically.

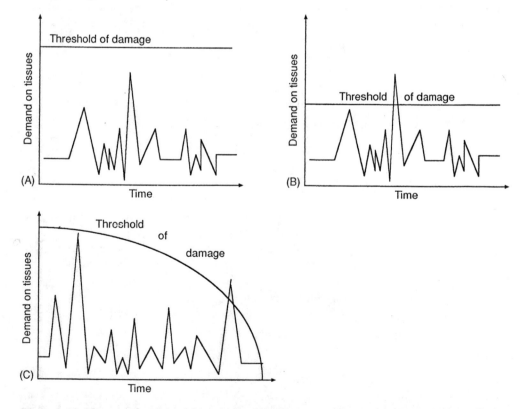

Figure 2.12 The specification of force limits for safety is complicated by factors such as fatigue that can lower the threshold for injury of the tissues. (From Professor S. McGill.)

In a similar vein, it has been hypothesised that forward bending is more hazardous in the early morning than at other times of the day. On rising, increased fluids in the disc may escape into the outer annulus through radial tears, carrying inflammatory material to nerve endings in the outer annulus and resulting in back pain. Snook *et al.* (1998) report significant reductions in back pain in a group of sufferers trained to avoid early morning flexion of the spine. Examples of avoidance strategies were; techniques of rising from bed without flexion, avoidance of sitting or squatting for two hours after rising, the use of reachers to pick things up, etc.

The main risk factors for musculoskeletal injury in the workplace are force, posture, repetition rate and fatigue; their external (task) counterparts are load, layout, cycle time and work organisation (shifts and rest periods). The specification of safe workloads is limited by the fact that human tissue seems to have no absolute threshold for injury. Time of day and time on shift are important determinants at the organisational level.

Biomechanics of spinal loading

The extensor muscles of the back play a major role in maintaining postural stability. Acting like the guy ropes on one side of a mast, they resist any forces that tend to cause the trunk to flex forwards. However, they are at a great mechanical disadvantage with respect to most of the destabilising forces that they are called on to resist. For an object held 50 cm in front of the body, this disadvantage is 10:1. If the object weighed 30 kg, the back muscles would be required to exert a force equivalent to that generated by hanging a 300 kg weight!

Estimating spinal compression

A simple mechanical model can be used to estimate the spinal compression during static work. By definition, force is the product of mass and acceleration:

$$f = m \times a$$

where m = mass of the object in kilograms, a = the acceleration in metres per second per second and f = the force in newtons.

The acceleration due to gravity at the surface of the Earth is approximately 9.81 m/s^2, so a mass of 1 kg resting on the Earth results in a force of

$$f = 1(\text{kg}) \times 9.81(\text{m/s}^2) = 9.81 \text{ N}$$

For clarity of presentation, the value of a has been rounded-up to 10 m/s^2 in the example below (so that a 1 kg mass is taken to give rise to a force of 10 newtons).

EXAMPLE

A person standing erect holds a 10 kg box at elbow height, at a distance of 50 cm in front of the lumbar spine. The spine can be considered as the fulcrum in a lever system. On one side of the fulcrum is the load, which acts at a distance of 50 cm and tends to cause the trunk to flex forwards. On the other side of the fulcrum are the

217305 | 620.82 BRI

back extensor muscles. To maintain postural stability, they need to exert a force to balance the system. Since the person is standing erect, it is assumed that the centre of gravity of the superincumbent body parts is directly above the fulcrum.

The total compressive force acting on the spine (C_t), is calculated as follows:

C_t = (Compressive force due to upper body weight) + (Compressive force due to load) + (Compressive force due to back muscle contraction needed to maintain posture)

If the mass of the upper body is 40 kg then the compressive force is 400 N. Similarly, the compressive force due to the load is 100 N. To maintain the trunk in an erect posture, the back muscles need to counteract the load moment. The unit of load moment is the Newton-metre (N m) and it is the product of the force (in Newtons) and the distance at which it operates (in metres). In this example, a force of 100 N acts at a distance of 50 cm (0.5 m) from the lumbar spine. Therefore, the load moment about the spine, L_m, in newton-metres is

$$L_m = 100(\text{N}) \times 0.5(\text{m}) = 50 \text{ N m}$$

To maintain the posture, the back muscles have to exert a countermoment of equal size (i.e. 50 N m). Assuming the back muscles have a lever arm of 5 cm (0.05 m), the back muscle force for postural stability is

Back muscle extensor force = 50(N m)/0.05(m) = 1000 N

Therefore the total spinal compression, C_t, is

$$C_t = 400 + 100 + 1000 = 1500 \text{ N}$$

At first glance, it seems surprising that such as high compressive force should result when only a 10 kg load is being held. The reason is that the load is held at a distance from the body – 50 cm from the spine. The back muscles are at a mechanical disadvantage of 10:1 because their lever arm is only 5 cm. This means that the back muscle force needed to balance the load is 10 times greater than the force exerted by the load itself. When it is remembered that most muscle–joint systems are at a similar mechanical disadvantage with respect to the outside world, it is easy to understand why good task design is essential. One of the main goals of the designer should be to minimise the adverse effects of this mechanical disadvantage and so prevent the generation of dangerously large internal forces.

According to Chaffin and Andersson (1984), the model can be applied in the evaluation of non-static tasks as long as the accelerations are not large. If large accelerations are involved (when the task involves running, jumping or other fast movements, the model will underestimate the forces generated by the work activities).

Spinal compression tolerance limits

Much is now known about the strength of spinal motion segments and their component discs and intervertebral bodies, owing to the in-vitro studies of Yamada (1970), Adams and Hutton (1980, 1985 for example) and others. For a comprehensive review, see Adams and Dolan (1995). These studies typically involve the removal of motion segments from cadavers and testing them to failure in a compression testing machine. Variations on the method include first placing the motion segment into a flexed posture before compressing it, or combining compression with axial rotation. The load at which failure occurs is recorded and the specimen is removed from the machine and examined to determine where and how failure has occurred (the vertebral body itself can fail or the bony end plates can fail; the disc can also fail).

The data obtained from mechanical testing of spinal motion segments enable tolerable loads to be identified that will not cause failure of lumbar motion segments. The spinal compression tolerance limit (SCTL) is the maximum compressive load to which a specified motion segment can be exposed without failure. In practical situations, manual handling tasks can be evaluated using biomechanical models to estimate the compression load (see next section). If the estimated load exceeds the SCTL, then the tasks must be redesigned, by reducing either the load or the load moment.

According to Genaidy *et al.* (1993) several factors reduce the SCTL and therefore increase the risk of injury. SCTL is greatest in 20–29-year-olds, declining by 22% in the next 10 years, 26% in the next 10 and 42% in the next 10. At 60 or more years of age the SCTL has declined by 53%. Female SCTLs are approximately 67% of male values. SCTLs are lower when spinal motion segments are loaded in complex ways, as when compression and bending are combined. A hyperflexed lumbar spine has a lower SCTL than a flexed lumbar spine (Adams and Hutton, 1985). In a flexed position, however, the compressive load on the thicker part of the disc (the anterior annulus) is increased and the load on the facet joints is reduced. Physical activity seems to strengthen both the vertebral bodies and the intervertebral discs. Women, older workers and those unaccustomed to lifting should not be expected to carry out forceful exertions at work.

Ayoub and Mital (1997) quote SCTLs of 6700 N for people under 40 years of age and 3400 N for people over 60. The regression equation below (Genaidy, 1993) can be used to calculate SCTLs for a given population percentile:

$$CS = 7222.41 - (1047.71 \times Age) - (1279.18 \times Gender) + (56.73 \times PP)$$

where

> CS = compressive strength
> Age = age group expected to carry out task (use 1 for 20–29 years, 2 for 30–39 years, 3 for 40–49 years and 4 for 50+)
> Gender: use 1 for male and 2 for female
> PP = population percentile (e.g. 5 for 5th percentile tolerance limit, 95 for 95th, and so on).

For risk assessment purposes, a margin of safety is needed. For one-off tasks, the task load should be less than 60% of the SCTL and for repetitive loads less than 30%.

Forward-flexed postures

The simplistic biomechanical model is sometimes called a cantilever model. The lumbar spine is regarded as the fulcrum. Behind the lumbar spine are the erector spinae muscles with their lever arm of 5 cm. In front of the lumbar spine is the load. The total spinal compression is the sum of three compressive forces: the compression due to body weight, the compression due to the load, and the compression due to the back muscle force needed to maintain a static posture. In an erect posture, the upper body weight can be regarded as being directly above the spine, imposing a load but no moment. The load moment is the product of the load and its moment arm (horizontal distance from the lumbar spine). The countermoment produced by the back extensors is the product of the back muscle force and the lever arm (distance of the back extensors from the lumbar spine, usually taken to be 5 cm).

In forward-flexed postures, the centre of gravity of the trunk moves in front of the lumbar spine and an additional load moment is created (the product of the weight of the upper body and its horizontal distance from the lumbar spine). The back muscles have to counter two load moments (due to postural load and to task load). At the level of the lumbar spine, the forward-flexed posture introduces two components of force, one of compression and one of shear. The total compressive force acting on the spine now becomes

$$C_t = \text{(Force due to body weight + Force due to load)} \times \text{cosine angle of flexion} + \text{(Compression due to back muscle force needed to maintain posture)}$$

The shear force (F_s) acting on the spine is given by

$$F_s = [\text{Force due to body weight + Force due to load}] \times \text{sine angle of flexion}$$

Behavioural aspects of posture

Body links

Anyone who has ever attempted to hold an unconscious person upright will readily appreciate that the body is like a floppy chain of interlinked segments that can move in unpredictable ways if not stabilised. Stabilisation can take place internally, via low-level muscular activity, and externally, by the person interacting with objects in the environment. In the absence of external stabilisation, tonic (low-level) muscle activity is required to maintain posture and poise. Although we are not normally aware of this activity and do not perceive it as arduous work, it can lead to discomfort if sustained for long periods. Behaviours such as folding the arms and crossing the legs are postural strategies that turn open chains of body links into approximate closed chains stabilised by friction. Closed chains have fewer degrees of freedom for movement and move in more predictable ways when subjected to destabilising forces. The net effect of such postural control strategies is to reduce the postural load on the muscles involved. This may be perceived as being 'comfortable'. Supporting evidence for these assertions comes from Snijders *et al.* (1995). Electromyographic (EMG) activity of the internal and external oblique abdominals was higher in standing than in sitting and was further reduced when the legs were crossed. The authors hypothesised that leg crossing stabilises the sacroiliac joint by compressing it and therefore relieves the muscles, such as the oblique abdominals. EMG activity in rectus abdominis (which runs parallel to the joint and is not involved in its stabilisation) was not influenced by leg crossing.

Dempster (1955) viewed the body as an open-chain system of links. Each joint of the body has a freedom for angular motion in one or more directions. A complex linkage such as that between the shoulder, arm and hand has many degrees of freedom of movement and power transmission is impossible without accessory stabilisation of the joints by muscle action. For example, supination of the wrist may be required to turn a door handle, but this is only possible if the elbow and shoulder joints are stabilised and can counter the reaction at the hand–handle interface.

We are not normally aware of the postural reflexes that exist to control muscle–joint systems, even though these reflexes are essential for the performance of nearly all our daily activities. Individual muscles almost never work in isolation to produce a movement, rather synergistic recruitment of muscles take place in which contraction of the prime mover is accompanied by contraction of surrounding muscles to position and stabilise the joints. In very few activities do the trunk and limb muscles exert direct forces on the environment, rather they maintain joint postures such that shifts in body weight can be transmitted via the chain of body links to exert a force. The use of body weight when kneading dough is a simple example of the importance of joint stabilisation, which enables shifts in body weight to be used to carry out a task.

The same principle also underlies the performance of almost all highly skilled activities in which very large forces are exerted or where body parts or external objects undergo rapid acceleration through pivoting actions. Throwing a discus and swinging a golf club or a tennis racket are excellent examples of this principle.

Preparation for action

In tennis and golf and when using hand tools such as hammers, very large forces can be exerted by skilled performers with little apparent effort. One of the key features of this behaviour is the efficient stabilisation of joint postures and the use of a back swing involving eccentric contraction of muscles and stretching of tendons. This back swing can be likened to an archer pulling back the bowstring before releasing the arrow. In the case of club and racket sports and hand tools, it is the hands, rather than the arrow, that are released when sufficient tension has built up.

In repetitive movements, and particularly in cyclic movements, the function of tendons and ligaments is to store energy. In walking, the stretching and release of the Achilles tendon throughout the gait cycle stores some of the energy when the foot is planted on the ground and this is released during the push-off phase. The deformation of the arch of the foot at these times serves a similar function. The combined effect is to increase the energy efficiency of walking. This illustrates that tendons and ligaments play an active role in many activities and that their continued ability to withstand tension and compression is essential if these activities are to be carried out efficiently. Excessive tension and compression or a regime of loading cycles that exceeds the body's repair capability can occur in industrial jobs as well as in sports if tasks or tools are badly designed. The outcome is pain and diminished capability. Further discussion of musculoskeletal problems of this nature may be found in Chapter 5.

The body-link concept is often of value in posture analysis because the muscular workload of many tasks can be reduced if the operator stabilises a body part by forming temporary, approximate closed chains with his or her environment. For example, upper body weight can be used to stabilise a forearm and hand resting on a work surface. The hand that is stabilised in this manner can then grasp a component while the other hand uses a tool to work on the component.

Effectiveness and cost-effectiveness

Back injuries often occur at work when people are handling heavy objects manually. They can also occur outside of work. These injuries are often the result of many years of stress both inside and outside of work. We can use an analogy with noise-induced hearing loss: people exposed to loud noises eventually go deaf. It makes no difference whether the exposure occurs at work or at home or both. The way to delay or prevent people going deaf is to lower the total exposure. Similarly, people chronically exposed to back stress eventually develop back problems. The way to delay or prevent the occurrence of these problems is to lower the total exposure to back stress. At work, this means identifying stressful tasks and redesigning them to reduce the stress.

High costs of back injury

Snook (1978) has presented evidence that ergonomic workplace design can reduce back pain by up to one-third. Snook's work has been followed-up over many years, both by Snook himself and by his colleagues at Liberty Mutual in the USA. Using Snook's data on maximum acceptable loads for industrial workers, Ciriello *et al.* (1993) report that if a manual handling task is acceptable to less than 75% of workers, the probability of compensatable back pain is three times greater than if the job is acceptable to more than 75% of the population. So, by redesigning jobs to reduce the weight handled, large reductions can be made in back pain compensation claims with immediate and continuing benefits to the company.

Many writers have presented statistics that indicate that back problems are a major cost to many industries and to the national economies of many countries. In the USA, low back pain accounts for 18% of workers' compensation claims and 30% of the costs (Liberty Mutual Research Center for Safety and Health, 1996). In 1995, there were 900 000 cases of back injury in the USA. Webster and Snook (1990) analysed records of the Liberty Mutual Insurance Company in the USA. In 1986, the mean cost of low back pain claims was $6807 and the median was $391 dollars (suggesting that there are a large number of fairly low-cost claims and a small number of very expensive ones, in other words a few cases account for most of the costs). The total cost of claims handled by Liberty Mutual was $673 884 128. From this, Webster extrapolated that the total cost to the USA as a whole was $11.1 billion in 1986, 241% greater than the cost in 1980.

A study by Safety managers at Goodyear Tire and Rubber Company in 1986, revealed that 62% of the company's total workers' compensation payouts resulted from strains due to lifting, pushing and bending. Ergonomic problems were costing millions of workers' health care dollars. It would seem that there is great scope for cost savings. Webster argued that, on average, $6000 spent on a control measure would pay for itself in one year if it prevented one injury.

A recent report by the UK Health and Safety Executive estimates the total cost to Britain of work-related accidents and ill-health to be between 5% and 10% of all UK industrial companies' trading profits. The total cost to society as a whole, including direct and indirect costs, is estimated to be 2–3% of gross domestic product or the equivalent of one year's economic growth (Davies and Teasdale, 1994).

Approximately 43.4% of the UK sickness and injury absence is due to musculoskeletal conditions (49% if so-called 'repetitive strain injuries' are included).

Similar losses occur in developed countries. Much of this absence is caused, exacerbated or triggered by ergonomic risk factors in the work environment (see Kuorinka and Forcier, 1995; Pheasant, 1991). According to Manstead (1984), six times as many man-days were lost to British industry due to back pain than were lost due to strikes.

Role of occupational factors

How much of the high incidence of back pain is actually caused by work is open to debate: 85% of all low back pain has no identifiable cause (Liberty Mutual, 1996). However, it is likely that much of the pain, if not directly caused by occupational factors, may well be exacerbated or amplified by work activities. Pathological degeneration of the components of the human spine occurs with age in most people, but it is eminently feasible that this process of degeneration can be accelerated by work-imposed stress. Similarly, people with an existing medical condition may find it difficult to cope at work if excessive demands are placed on their musculoskeletal systems. It is likely, therefore, that much of the disabling back pain at work is preventable, can be reduced, or is at least postponable and that significant savings could accrue through well-designed ergonomic intervention programmes. Examples of these programmes are given in later chapters.

Research directions

Basic modelling of spine biomechanics in health and disease continues to be one of the main areas of fundamental research of relevance to ergonomics. Better models that account for the loading patterns of spinal motion segments in different activities and that enable loads to be predicted will be of obvious value in drafting safety standards for industry. Together with a better understanding of the processes of spinal degeneration, these models will support a more scientific approach to the prevention of spinal degeneration and disease.

Summary

Evaluation of the physical workplace requires a basic knowledge of human anatomy and body mechanics. In order to carry out a physical task in a safe and comfortable manner, a number of physical requirements must be met. First, the body must be stable. This depends on the relationship between the body parts and the base of support provided by the feet, the seat and any other surfaces in the workplace that can be used to support body weight. The design of a workspace can determine the range of stable postures that can be adopted and can be evaluated from this point of view. In vehicles, destabilising external forces due to motion may also have to be accounted for. If a task requires a posture to be held for any length of time, posture analysis is necessary. A starting point for such an analysis is to determine the mechanisms by which the posture is maintained, whether static muscular effort is required, whether ligaments are being strained and whether parts of the work surface such as the backrest of a chair or a work surface are providing support. Knowledge of the anatomy of the spine and pelvis is particularly valuable here, as is an understanding of the mechanisms of physical fatigue.

Finally, it is essential to observe a worker's postural behaviour. Skilled workers use postural strategies both to minimise fatigue and to enable them to exert large forces efficiently.

Essays and exercises

1. One of the best ways to develop an understanding of anatomy is to practise drawing anatomical structures. Copy the anatomical drawings in this book and label the various structures using the appropriate terms.

2. Observe people carrying out the following or similar activities:

   - Students sitting an exam
   - A bus driver behind the wheel
   - A person washing dishes
   - A computer programmer at work
   - An archer aiming at a target
   - A gardener digging a hole
   - A mother holding a young child
   - A person operating a lathe

   Draw stick figures to indicate the position of the main body parts and the base of support at the feet. Indicate where you think the main areas of static and dynamic body load are by considering how the person maintains the posture. Try to visualise the load on underlying structures and indicate whether the task load is greater than the postural load.

3. A worker has to carry 20 kg sacks of feathers from a packing machine to a despatch zone. He is able to walk in an erect posture. The centre of gravity of the sack is 70 cm from his lumbosacral joint. Given that his upper body weighs 40 kg, what is the total spinal compression force. Comment on the difference between the task load and the postural load and suggest ways of reducing the task load.

4. You have decided to redesign a shovel for shovelling snow. In normal shovelling, people typically flex their trunks by 90 degrees and the distance from the lumbosacral (LS) joint to the blade of the shovel is about the same as the distance of the upper body centre of gravity (COG) from the LS joint. In your redesign, the handle is lengthened so that the person leans forwards by only 10 degrees. However, the loaded blade is now 1.2 metres from the LS joint. Given the following body dimensions, would you say that your redesign is really an improvement, given that when the blade of the shovel is laden with snow it weighs 15 kg?

   - Upper body weight = 45 kg
   - Distance of upper body COG from LS joint = 45 cm

3 Anthropometric principles in workspace and equipment design

He who sees things grow from the beginning will have the best view of them.
(Aristotle)
Humans are, perhaps, the most plastic of all species, and hence, the most variable.
(C. G. N. Mascie-Taylor and B. Bogin, 1995)

The word 'anthropometry' means measurement of the human body. It is derived from the Greek words 'anthropos' (man) and 'metron' (measure). Anthropometric data are used in ergonomics to specify the physical dimensions of workspaces, equipment, furniture and clothing to ensure that physical mismatches between the dimensions of equipment and products and the corresponding user dimensions are avoided.

Designing for a population of users

The first step in designing is to specify the user population and then to design to accommodate as wide a range of users as possible – normally 90% of them. Well-designed products acknowledge and allow for the inherent variability of the user population.

In ergonomics, the word 'population' is used in a statistical sense and can refer to a group of people sharing common ancestors, common occupations, common geographical locations or age groups. A user population may consist of people from different races (i.e. groups differing in their ancestry) or different ethnic groups (different cultures, customs, language, and so on). For design purposes, the criteria for deciding what constitutes a 'population' are functional and are related directly to the problem at hand. If we want to design a cab for bus drivers in Chile, we require data on the anthropometry of Chilean bus drivers. If we want to design workspaces in private hospitals in Saudi Arabia, we need data about the European and Australian nurses who usually work in them.

Sources of human variability

Biological anthropologists distinguish four types of human adaptation. Over many lifetimes, *genetic* changes may occur as a result of natural selection. Over the course of a lifetime, organisms exhibit *plasticity* (literally the capability of being moulded). Over the short term, organisms can exhibit *acclimatisation*, and over the very short term *behavioural adaptation*. Only the last two of these forms of adaptation are reversible.

Plasticity is an intermediate form of adaptation that takes place over the course of a lifetime. Roberts (1995) cites evidence that the heads of people who were habitually placed in a supine position in the first years of life grow to be broader than the heads of those who were more often placed on their sides. Much of the evidence for plasticity comes from studies of human migration, beginning with the work of Boas early in the twentieth century. Boas (1910) found that children born to immigrant parents after arrival in the USA were larger and had differently shaped heads to those born before the parents migrated. This challenged prevailing views about the fixity of racial or ethnic 'types'. More recently, a study of children born of Punjabi parents showed that those born in Canada were taller and weighed more than those born in India, for all age groups. Similar findings have been obtained with studies of Maya children living in Guatamala and Maya living in the USA. The latter were found to be taller and heavier, with more muscle mass than the former (Bogin, 1995).

Factors influencing the change in body size of populations

Many studies indicate that better living conditions are associated with larger body size. Smallness does not appear to be intrinsic to many groups of people but it is related to development in a biologically stressful environment. Thus, smallness in a population may be a plastic response to deprivation.

There is a great deal of evidence to suggest that improved living conditions are accompanied by an increase in body size. Many countries have witnessed an increase in the size of their inhabitants over the last 150 years since the establishment of industrialised societies. Part of this in undoubtedly due to better diet and living conditions – better sanitation, childhood immunisations, refrigerated transportation making available a year-round supply of fresh food, and supplementation of dairy products and cereals with vitamin D. In the USA and Britain, increases in mean stature of 1 cm every 10 years have been observed, although this trend seems to have played itself out in the USA among those whose families have lived there for generations (McCook, 2001). According to the US Center for Disease Control and Prevention, the mean stature of US males and females (175.25 and 162.56 cm, respectively) has not changed since the 1960s, so data from that era are still usable for ergonomic purposes. In the UK, however, mean male stature increased by 1.7 cm between 1981 and 1995 and mean female stature increased by 12 mm (Peebles and Norris, 1998).

As countries industrialised over the last few hundred years, previously isolated rural communities were scattered owing to improvement in transportation and to urbanisation. This resulted in outbreeding, or *heterosis*, which is thought to be result in a genetically healthier population (with 'hybrid vigour'). Boldsen (1995) investigated the secular trend in stature of Danish males over the last 140 years. He found that there has been an increase in mean male stature of 13 cm over this time. Forty-five per cent of this increase was found to be due to a change in the population structure, i.e. due to heterosis, and the rest a plastic response to improved living conditions.

According to Konig *et al.* (1980), this growth 'acceleration' is a worldwide phenomenon and is not restricted to adult stature – it is even apparent in the fetal stage of development. Neonates have increased 5–6 cm in length and 3–5% in birth weight over the last 100 years. Additionally, the age of onset of puberty has decreased by about 2 and 3 years for boys and girls, respectively, and the age of menopause has increased by about 3 years. Konig also remarks that 'acceleration' has not proceeded uniformly. From about 1830 to 1930, the average height of juveniles increased by about 0.5 cm per decade, whereas from 1930 onwards it increased by up to 5 cm per decade.

Implications for ergonomics

These findings have far-reaching implications for ergonomists attempting to design to accommodate a wide range of people. The concept of 'Mr Average' or an ideal body type will become increasingly invalid and designers must expect users to be different from themselves. The structure of populations and their living conditions are changing in many parts of the world. This means that anthropometric data captured in the past may no longer be representative. When designing for international markets, then, each target country has to be considered separately. Anthropometric data should be used with caution, particularly if it is old.

Anthropometry and its uses in ergonomics

As a rule of thumb, if we take the smallest female and the tallest male in a population, the male will be 30–40% taller, 100% heavier and 500% stronger (Grieve and Pheasant, 1982). Clearly, the natural variation of human populations has implications for the way almost all products and devices are designed. Some obvious examples are clothes, furniture and automobiles. *The approach of ergonomics is to consider product dimensions in human terms in view of the constraints placed on their design by body size variability.* For example, a seat should be no higher than the popliteal height of a short user and no deeper than the distance from the buttocks to the knees.

Body size and proportion vary greatly between different populations, a fact that designers must never lose sight of when designing for an international market. A US manufacturer hoping to export to Central and South America or South-east Asia would need to consider in what ways product dimensions optimised for a large US (and probably male) user-group would suit Mexican or Vietnamese users; the latter belong to one of the smallest populations in the world. Ashby (1979) illustrated the importance of anthropometric considerations in design as follows:

> If a piece of equipment was designed to fit 90% of the male US population, it would fit roughly 90% of Germans, 80% of Frenchmen, 65% of Italians, 45% of Japanese, 25% of Thais and 10% of Vietnamese.

It is usually impracticable and expensive to design products individually to suit the requirements of every user. Most are mass-produced and designed to fit a wide range of users: the bespoke tailor, the dressmaker and the cobbler are perhaps the only remaining examples of truly user-oriented designers in Western industrial societies.

Anthropometric surveys

The anthropometry of military populations is usually well-documented and is used in the design of everything from cockpits to ranges and sizes of boots and clothing. Anthropometric surveys are expensive to carry out, since large numbers of measure-

Figure 3.1 Semi-automatic capture of anthropometric data. Variables are measured using digital caliper and tape measure and automatically entered into a computer according to a pre-planned measurement protocol. (Photograph courtesy of Ergotech Consultants, Pretoria, South Africa.)

ments have to be made on sizeable samples of people representative of the population under study. Traditionally, measurements are made using manually operated instruments such as anthropometers and calipers. Automated or semi-automated systems have been developed. Whole-body scanners (Daanen and Water, 1998), based on the principles of laser-stripe projection, stereophotogrammetry or patterned light projection, are commercially available and cost from $50 000 to $410 000, depending on resolution and speed of operation (1–8 mm). A low-cost semi-automatic system is shown in Figure 3.1. Manually operated instruments (such as the 'digital tape measure') are used to capture the measurements according to a user-programmed protocol. The data are automatically digitised and downloaded to computer. The World Health Organization recommends (WHO, 1995) that, if anthropometric data are to be used as reference standards, a minimum sample size of 200 individuals is needed (this gives a standard deviation around the 5th percentile of about 1.54 percentiles, a 95% probability that the true 5th percentile falls within plus or minus 2.25 percentiles of the estimate, approximately).

Problems with much of the anthropometric data from the USA and Europe are the age of the data and the lack of standardisation across surveys (not all researchers measure the same anthropometric variables). Marras and Kim (1993) note that the first large-scale survey of civilian women in the USA was carried out in 1941 for garment sizing purposes. From 1960 to 1962, the National Center for Health Statistics carried out a survey of 20 anthropometric variables of both men and women. A survey of civilian weights and heights was carried out between 1971 and 1974. In Britain, several civilian surveys have been published since 1950 (see Oborne, 1982) but many of these suffer the same limitations as their US counterparts: the data are often fragmentary, are only relevant to the design problems of a particular industry and are probably out of date. Marras and Kim (1993) present recent measurements of 12 anthropometric variables from 384 males and 124 females. Abeysekera and Shahnavaz (1989) present data from industrially developing and developed countries and discuss some of the problems of designing for a global marketplace. The International Standards Organization has done some work on the standardisation of anthropometric measurements (ISO DIS 7250, Technical Committee 159). The ISO list of variables (Table 3.1) will be of use in the planning of anthropometric surveys.

Table 3.1 ISO list of anthropometric variables (ISO/DIS 7250)

Variable	Method of measurement
Body weight	Subject stands on weighing scale
Stature	Vertical distance to highest part of head. Subject stands erect, with feet together with heels, buttocks, shoulders, back of head touching a vertical surface
Eye height	Vertical distance floor to inner corner of eye. Subject stands as above
Shoulder height	Vertical distance from floor to acromium. Subject stands as above
Elbow height	Vertical distance from floor to lowest bony point of elbow. Upper arm hangs freely and elbow is flexed 90 degrees
Spina iliaca height	Height of anterior superior iliac spine above floor
Tibial height	Height of anterior head of tibia above floor
Sitting height (erect)	Distance of highest point of head to horizontal sitting surface. Subject sits against a vertical surface, thighs fully supported and lower legs hanging freely
Eye height (sitting)	As for standing but in the seated posture above
Shoulder height (sitting)	Vertical distance from horizontal sitting surface to acromion
Elbow height (sitting)	As for standing, but in the sitting position above.
Cervical height (sitting)	Vertical distance from seat surface to skin overlying tip of 7th cervical vertebra
Shoulder breadth	Distance between the acromions
Lower leg length	Vertical distance from floor to lowest part of thigh behind knee (90 degrees of knee flexion)
Knee height	Vertical distance from floor to upper surface of thigh (90 degrees of knee flexion)
Hand length	Distance from the tip of the middle finger to the most distal point of the styloid process of the radius with the hand outstretched
Hand breadth (at metacarpal)	Distance between the radial and ulnar metacarpals
Index finger length	Distance from tip of 2nd finger to the proximal skin furrow between the digits
Index finger breadth	Distance between the medial and lateral surfaces of the 2nd finger in the region of the joint between the proximal and medial phalanges (first knuckle)
Index finger breadth (distal)	As above but medial and distal phalanges
Foot length	Maximum distance from the back of the heel to the tip of the longest toe
Foot breadth	Maximum distance between the medial and lateral surfaces of the foot
Head length	Distance along a straight line from the glabella to the rearmost point of the skull
Head breadth	Maximum breadth of the head above the ears

Table 3.1 (cont'd)

Variable	Method of measurement
Head circumference	Maximum circumference of the head over the glabella and the rearmost point of the skull
Sagittal arc	The arc from the glabella to the inion (protrusion where the back of the head meets the neck in the mid-sagittal plane)
Bitragion arc	Arc from one tragion over the top of the head to the other
Face length	Distance between the sellion and the menton with the mouth closed
Forward reach	Maximum distance from a wall against which the subject presses the shoulder blades to the grip axis of the hand
Forearm–hand length	Distance from the back of the upper arm at the elbow to the grip axis of the hand (90 degrees of elbow flexion)
Fist height	Vertical distance from the floor to the grip axis of the hand, with the arms hanging freely
Crotch height	Distance from the floor to the distal part of the pubis
Hip breadth (standing)	Maximum horizontal distance across the hips
Hip breadth (sitting)	As above
Elbow to elbow breadth	Maximum distance between the lateral surfaces of the ebows
Waist circumference	Trunk circumference in the region of the umbilicus
Body depth (sitting)	Horizontal distance from the rear of the knee to the back of the buttock
Buttock–knee length	Horizontal distance from the front of the kneecap to the rearmost part of the buttock
Wrist circumference	The circumference of the wrist between the styloid process and the hand, with the hand outstretched

Types of anthropometric data

Structural anthropometric data   Structural anthropometric data are measurements of the bodily dimensions of subjects in fixed (static) positions. Measurements are made from one clearly identifiable anatomical landmark to another or to a fixed point in space (e.g. the height of the knuckles above the floor, the height of the popliteal fossa (back of the knee) above the floor, and so on). Some examples of the use of structural anthropometric data are to specify furniture dimensions and ranges of adjustment and to determine ranges of clothing sizes. Figure 3.2 shows structural variables that are known to be important in the design of vehicles, products, workspaces and clothing. Figure 3.3 shows examples of vehicle dimensions that would require user anthropometry to be specified. Tables 3.2 to 3.11 present selected anthropometric data from various parts of the world (collated from Ashby, 1979; Pheasant, 1986; Woodson, 1981; and figure of the Ministry of Science and Technology, Brazil, 1988).

The reader is advised to study the tables, looking in particular for differences in body *proportion* between different groups. For example, there is an approximately 100 mm stature difference between US and Japanese males when standing. This decreases to between 5 and 25 mm in the seated position. When considering possible modifications to a vehicle to be exported from one country to the other, the designer might, therefore, first pay attention to the rake of the seat and position of the foot pedals, rather than to the height of the roof.

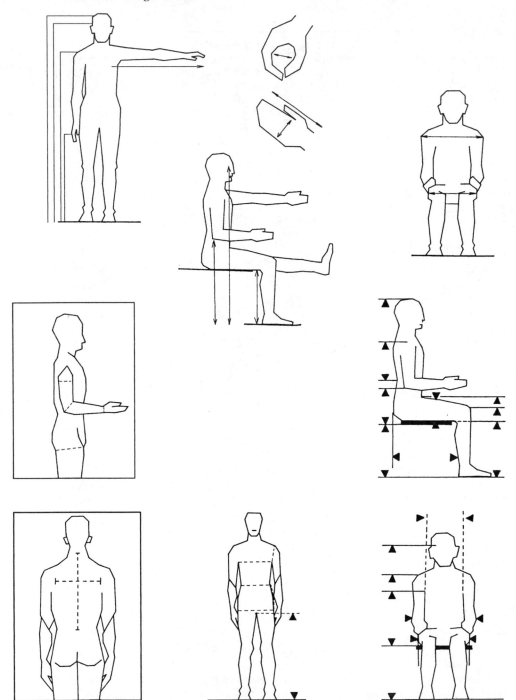

Figure 3.2 Some common structural anthropometric variables.

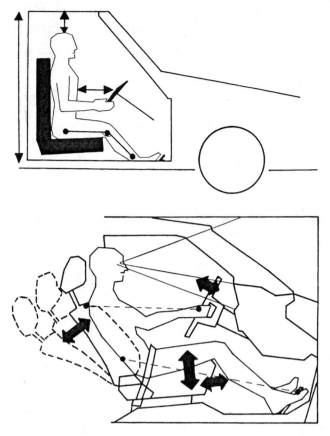

Figure 3.3 Some product dimensions that are determined using anthropometric considerations.

Table 3.2 Stature of selected adult populations (5th and 95th percentiles)

Population	Males		Females	
	5th	95th	5th	95th
US	1640	1870	1520	1730
Northern European	1645	1855	1510	1720
Japanese	1560	1750	1450	1610
Brazilian	1595	1810		
African	1565	1790		

Limitations on the use of structural data Structural data may be used for design in situations where people are adopting static postures. Caution should be used when applying these data to design problems that involve movement, particularly skilled movement. Bilzon *et al.* (2000) investigated the associations between user anthropometry and the ability to escape through vertical hatches and bulkhead doors on ships. It was hypothesised that subjects with large bi-deltoid breadth and abdominal depth would find it more difficult to move through the small spaces.

Table 3.3 Eye height (mm) of selected adult populations (5th and 95th percentiles)

Population	Males		Females	
	5th	95th	5th	95th
US	1595	1825	1420	1630
Northern European	1540	1740	1410	1610
Japanese	1445	1635	1350	1500
Brazilian	1490	1700		
African	1445	1670		

Table 3.4 Elbow height (mm) of selected adult populations (5th and 95th percentiles)

Population	Males		Females	
	5th	95th	5th	95th
US.	1020	1190	945	1095
Northern European	1030	1180	910	1050
Japanese	965	1105	895	1015
Brazilians	965	1120		
Africans	975	1145		

Table 3.5 Fingertip height (mm) of selected adult populations (5th and 95th percentiles)

Population	Males		Females	
	5th	95th	5th	95th
US	565	695	540	660
Northern European	595	700	510	720
Japanese	565	695	540	660
Brazilian	590	685		
African	520	675		

Table 3.6 Sitting height (mm) of selected adult populations (5th and 95th percentiles)

Population	Males		Females	
	5th	95th	5th	95th
US	855	975	800	920
Northern European	865	970	795	895
Japanese	850	950	800	890
Brazilian	825	940		
African	780	910		

Table 3.7 Sitting eye height (mm) of selected adult populations (5th and 95th percentiles)

Population	Males		Females	
	5th	*95th*	*5th*	*95th*
US	740	860	690	810
Northern European	760	845	695	685
Japanese	735	835	690	780
Brazilian	720	830		
African	670	790		

Table 3.8 Elbow rest height (mm) of selected adult populations (5th and 95th percentiles)

Population	Males		Females	
	5th	*95th*	*5th*	*95th*
US	195	295	185	285
Northern European	195	270	165	245
Japanese	220	300	215	285
Brazilian	185	275		
African	175	250		

Table 3.9 Popliteal height (mm) of selected adult populations (5th and 95th percentiles)

Population	Males		Females	
	5th	*95th*	*5th*	*95th*
US	395	495	360	450
Northern European	390	460	370	425
Japanese	360	440	325	395
Brazilian	390	465		
African	380	460		

Table 3.10 Buttock–popliteal length (mm) of selected adult populations (5th and 95th percentiles)

Population	Males		Females	
	5th	*95th*	*5th*	*95th*
US	445	555	440	540
Northern European	455	540	405	470
Japanese	410	510	405	495
Brazilian	435	530		
African	425	515		

Table 3.11 Hip breadth (mm) of selected adult populations (5th and 95th percentiles)

Population	Males		Females	
	5th	95th	5th	95th
US	310	410	310	440
Northern European	320	395	320	440
Japanese	280	330	270	340
Brazilian	306	386		
African	280	345		

Table 3.12 Constants used to estimate population proportions

Required percentile	Number of standard deviations to be subtracted from or added to the mean
50th	0
10th or 90th	1.28
5th or 95th	1.64
2.5th or 97.5th	1.96
1st or 99th	2.32

However, even when subjects were encumbered by bulky protective clothing, it was found that subjects with broad shoulders could escape through the hatch more quickly than those with narrow shoulders. There was a strong positive association between escape time and lean body mass. In dynamic tasks such as these, performance appears to depend more on strength and technique than on body size.

Functional anthropometric data   Functional anthropometric data are collected to describe the movement of a body part with respect to a fixed reference point. For example, data are available concerning the maximum forward reach of standing subjects. The area swept out by the movement of the hand can be used to describe 'workspace envelopes', zones of easy or maximum reach around an operator. These can be used to optimise the layout of controls in panel design. The size and shape of the workspace envelope depends on the degree of bodily constraint imposed on the operator. The size of the workspace envelope increases with the number of unconstrained joints. For example, the area of reach of a seated operator is greater if the spine is unencumbered by a backrest and can flex, extend and rotate. Standing reach is also greater if the spine is unconstrained and greater still if there is adequate foot space to enable one or both feet to be moved. Somewhat counterintuitively, one way to increase a worker's functional hand reach is to provide more space for the feet.

Generally speaking, fewer functional than structural anthropometric data are available. Although clinicians have long been interested in determining normal ranges of joint movement in healthy individuals to assist in assessment of patients, these data are not always quantified or even directly applicable to design problems. However, existing functional anthropometric data are useful for designing workspaces and positioning objects within them, particularly in the design of aircraft cockpits,

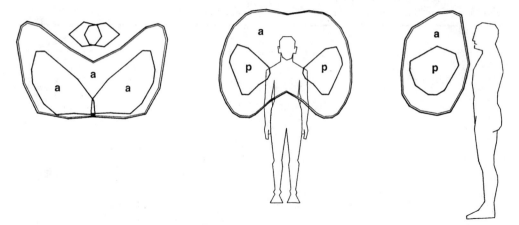

Figure 3.4 Functional anthropometric data. The figure shows the shapes of the reach envelopes and the allowable (a) and preferred (p) zones for the placement of controls
in a workspace.

crane cabs, vehicle interiors and complex control panels in the process industries
(Figure 3.4).

Newtonian anthropometric data Newtonian anthropometric data are used in
mechanical analysis of the loads on the human body. The body is regarded as an
assemblage of linked segments of known length and mass (sometimes expressed as
a percentage of stature and body weight). Ranges of the appropriate angles to be
subtended by adjacent links are also given to enable suitable ranges of working
postures to be defined. This enables designers to specify those regions of the workspace
in which displays and controls may be most optimally positioned. Newtonian data
may be used to compare the load on the spine due to different lifting techniques.
Data on strength are also available (Table 3.13).

Principles of applied anthropometry in ergonomics

Anthropometric variables in the healthy population usually follow a normal distribution as depicted in Figure 3.5.

The normal distribution

For design purposes, two key parameters of the normal distribution are the mean
and the standard deviation. The mean is the sum of all the individual measurements
divided by the number of measurements. It is a measure of central tendency. The
standard deviation is calculated using the difference between each individual measurement and the mean. It is a measure of the degree of dispersion in the normal
distribution. Thus, the value of the mean determines the position of the normal
distribution along the x (horizontal) axis. The value of the standard deviation determines the shape of the normal distribution. A small value of the standard deviation
indicates that most of the measurements are close to the mean value: the distribution
has a high peak that tails off rapidly at both sides. A large value of the standard

Table 3.13 Selected strength data for males (M) and females (F)

		Mean	SD
Pushing (N)[a]			
Handle Height			
1.7 m	M	300	50
	F	181	75
1.3 m	M	337	83
	F	221	103
0.7 m	M	393	134
	F	185	57
Pulling (N)[a]			
Handle Height			
1.7 m	M	263	60
	F	196	56
1.3 m	M	347	55
	F	223	80
0.7 m	M	541	81
	F	292	97
Wrist twisting			
strength (N m)[b]			
(males only)			
Knob diameter (mm)			
9.5		52.75	12.8
12.7		65.21	12.54
19.1		111.65	26.2

[a] Daams (1993).
[b] Swain (1970).

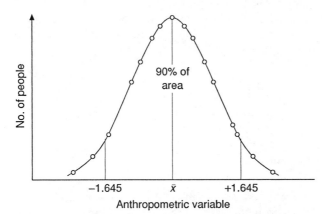

Figure 3.5 The normal distribution: 90% of the measurements made on different people will fall in a range whose width is ~1.64 standard deviations above and below the average.

deviation means that the measurements are scattered more distantly from the mean: the distribution has a flatter shape.

In order to estimate the parameters of stature in a population (the mean and standard deviation) it is necessary to measure a large sample of people who are

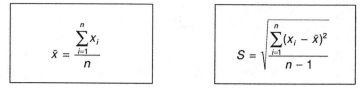

$$\bar{x} = \frac{\sum_{i=1}^{n} x_i}{n}$$

$$S = \sqrt{\frac{\sum_{i=1}^{n} (x_i - \bar{x})^2}{n-1}}$$

Figure 3.6 Equations used to calculate the mean and standard deviation from measurements sampled from a population where x = individual measurement and n = number of measurements.

The mean is the sum of all the measurements divided by the number of measurements. The standard deviation is the square root of the mean squared difference between each score and the mean.

representative of that population. The formulae given in Figure 3.6 can then be used to calculate the estimates of the mean and standard deviation. Estimates of *population parameters* obtained from calculations on data from samples are known as *sample statistics*.

The distribution of stature in a population exemplifies the statistical constraints on design. An important characteristic of the normal distribution is that it is symmetrical – as many observations lie above the mean as below it (or in terms of the figure, as many observations lie to the right of the mean as to the left). If a distribution is normally distributed, 50% of the scores (and thus the individuals from whom the scores were obtained) lie on either side of the mean.

This simple statistical fact applies to very many variables, yet it is often ignored or misunderstood by both specialists and laymen alike. It is common to hear statements such as 'Your child is below average height for his age' or 'You are above average weight for your height', both pronounced and perceived in a negative way. The statistical reality is that half of any normally distributed population is either above or below 'average'. In itself, this has no connotations of either negativity or positivity. What is important is how far from or how different from the mean a particular observation is. Observations that are different from the mean become rarer as the distance increases. The 'distance' of an observation from the mean is usually measured in standard deviations: we can talk about an observation being a certain number of standard deviations greater than or less than the mean.

It is unwise to design for 'Mr Average', a mythical person with the mean stature. As can be seen, very few individuals will be exactly of 'average' height and therefore it is necessary to design to accommodate a range of people – preferably those both above and below the mean.

Estimating the range

The standard deviation contains information about the spread of scores in a sample. It is known, for a normal distribution, that approximately two-thirds of the observations in the population fall within one standard deviation above and below the mean. Thus, for a population with a mean stature of 1.75 metres and standard deviation of 0.10 metres, approximately two-thirds of the population would be between 1.65 and 1.85 metres tall. The remaining one-third would lie beyond these two extremes, at either side.

Using the standard deviation and the mean, estimates of stature can be calculated below which a specified percentage of the population will fall. The area under the normal curve at any point along the x-axis can be expressed in terms of the number of standard deviations from the mean. For example, if the standard deviation is multiplied by the constant 1.64 and *subtracted* from the mean, the height below which 5% of the population falls is obtained. If 1.64 standard deviations are *added* to the mean, the height below which 95% of the population falls is obtained. These are known as the 5th and 95th percentile heights. The 1st and 99th percentile heights are obtained when the constant 2.32 is used (Table 3.12).

Using the mean and standard deviation of an anthropometric measurement and a knowledge of the area under the normal curve expressed as standard deviations from the mean, ranges of body size can be estimated that will encompass a greater and greater proportion of individuals in a population. Thus, given the mean and standard deviation of any anthropometric variable, a range of statures, girths, leg lengths, etc. can be computed within which a known percentage of the population will fall.

Applying statistics to design

Statistical information about body size is not, in itself, directly applicable to a design problem. First, the designer has to analyse in what ways (if any) anthropometric mismatches might occur and then decide which anthropometric data might be appropriate to the problem. In other words, the designer has to develop some clear ideas about what constitutes an appropriate match between user and product dimensions. Next, a suitable percentile has to be chosen. In many design applications, mismatches occur only at one extreme (only very tall or very short people are affected, for example) and the solution is to select either a maximum or a minimum dimension. If the design accommodates people at the appropriate extreme of the anthropometric range, less-extreme people will be accommodated.

Minimum dimensions

A high percentile value of an appropriate anthropometric dimension is chosen. When designing a doorway, for example, sufficient head room for very tall people has to be provided and the 95th or 99th percentile (male) stature could be used to specify a minimum height. The doorway should be no lower than this minimum value and additional allowance would have to be made for the increase in stature caused by items of clothing such as the heels of shoes, protective headgear, etc.

Seat breadth is also determined using a minimum dimension: the width of a seat must be no narrower than the largest hip width in the target population (Figure 3.7). Minimum dimensions are used to specify the placement of controls on machines, door handles, etc. Controls must be sufficiently high off the ground that tall operators can reach them without stooping – i.e. no lower than the 95th percentile standing knuckle height. In the case of door handles, the maximum vertical reach of a small child might also be considered (to prevent young children opening doors when unsupervised).

Circulation space must be provided in offices, factories and storerooms to allow for ingress and egress of personnel and to prevent collisions. In a female or mixed-sex workforce, the body width of a pregnant women would be used to determine the

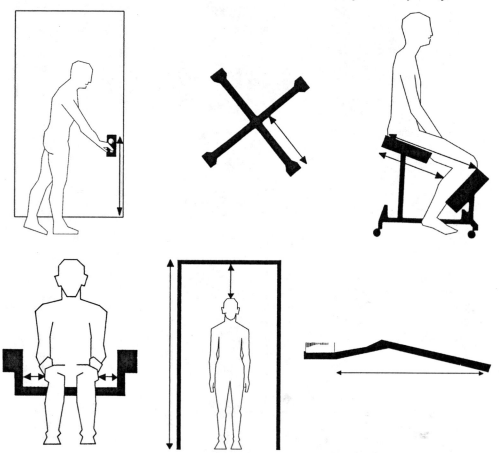

Figure 3.7 Some minimum dimensions. The height of a doorway must be no lower than the stature of a tall man (plus an allowance for clothing and shoes). The width of a chair must be no narrower than the hip breadth of a large woman. A toothbrush must be long enough to reach the back molars of someone with a deep mouth. A door handle must not be lower than the highest standing knuckle height in a population so that all users can open the door without stooping. The distance from the kneepad to the back of the seat of a 'kneeling chair' must exceed the longest buttock–knee lengths in the population of users. The length of a wheel brace must provide sufficient leverage for a weak person to generate sufficient torque to loosen the wheel nuts.

minimum (some anthropometric data of the seated pregnant woman can be found in Culver and Viano, 1990). About 60 cm of clearance space is needed for passages, the separation of machines and the distance of furniture from walls or other objects in a room.

Maximum dimensions

A low percentile is chosen in determining the maximum height of a door latch so that the smallest adult in a population will be able to reach it. The latch must be no higher

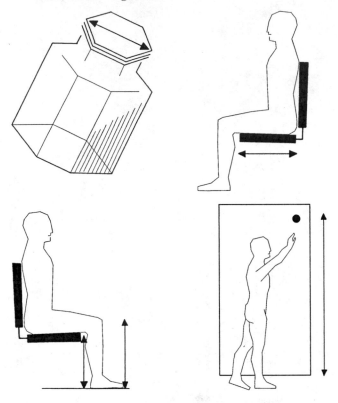

Figure 3.8 Some maximum allowable dimensions. A door lock must be no higher than the maximum vertical reach of a small person. Seat heights and depths must not exceed the popliteal height and buttock–knee lengths of small users. Screw-top lids must be wide enough to provide a large contact area with the skin of the hand to provide adequate friction and so that pressure 'hot spots' are avoided; but they must not exceed the grip diameter of a small person.

than the maximum vertical grip reach of a small person. The height of non-adjustable seats used in public transport systems and auditoria is also determined using this principle: the seat must be low enough so that a short person can rest the feet on the floor when using it. Thus, the seat height must be no higher than the 1st percentile or 5th percentile popliteal height in the population. Figure 3.8 gives examples of some maximum allowable dimensions.

Cost–benefit analysis

Sometimes it is not necessary to use anthropometric data because there may be no costs incurred in designing to suit everyone (a doorway or entrance is a hole and the bigger it is, the less building materials are used). However, there are often trade-offs between the additional costs of designing to suit a wide range of people and the number of people who will ultimately benefit. As can be seen from the normal distribution, the majority of individuals in a population are clustered around the mean and attempts to accommodate more extreme individuals soon incur diminishing returns since fewer and fewer people are accommodated into the range.

EXAMPLE 1

Let us suppose that the minimum height of a car interior has to be specified. Increases in roof height increase wind resistance and construction cost but provide head clearance for tall drivers. Thus, there are both costs and benefits of building cars with plenty of headroom for the occupants. If the mean sitting height for males is 90 cm with a standard deviation of 5 cm, a ceiling height of 91 cm (measured from the seat) will accommodate 50% of drivers with 1 cm clearance to allow for hair or clothing (except hats). An increase in ceiling height of 5 cm will accommodate a further 34% of drivers so that a total of 84% are now catered for. A further increase of 5 cm will accommodate an extra 14% of drivers bringing the total to 98%. However, a further increase of 5 cm will only accommodate an extra 2% of drivers in the population. Because this 2% probably represents a very small number of potential customers, the additional costs of accommodating their anthropometric requirements into the design of every car built become significant. It may be more cost-effective to exclude these individuals in the generic design but retrofit the finished product to accommodate one or two extremely tall buyers (for example, by making it possible to lower the seat).

This example should illustrate why the 5th and 95th percentiles of anthropometric variables are often used to determine the dimensions of products. Ninety per cent of potential users are accommodated using this approach and further sizeable alterations to product dimensions will only accommodate a small number of additional users – the point of diminishing returns has been reached.

EXAMPLE 2

The designers of a new ship for use by a British crew wish to specify the minimum deck height to allow sufficient headroom for the crew. The ship will be built in 2 years' time and will be in service for 30 years. Assume a stature increase due to the secular trend of 1.5 mm/year and recommend an appropriate height, DH.

$$DH = SHx + CA + DA + STA(Yt - Yb) + PA*$$

where

> DH = deck height
> SHx = x percentile stature
> CA = clothing allowance (shoes, helmets, when worn) = 70 mm
> DA = dynamic allowance (accommodates head movements when walking)
>   = 100 mm
> STA = secular trend allowance = 1.5 mm/year
> Yt = target year for specifying deck height = 2035
> Yb = Base year (in which the height data were captured) = 1986
> PA = Psychological allowance (perception of head room) = 50 mm

Use the 95th percentile male stature for the UK population, measured in 1986 at 1870 mm:

$$DH = 1870 + 70 + 100 + 1.5(2035 - 1986) + 50 = 2163.5 \text{ mm}$$

*From an example by Mrs E. Bilzon.

Using this value, 95% of personnel would not be at risk of head injury during normal operation. However, as can be seen from the shape of the normal distribution, even small increases in deck height will provide a large additional margin of safety. This margin of safety can be expected to last the lifetime of the ship because we have designed to accommodate the 95th percentile stature at the end of the product life cycle.

Anthropometric data must always be used in a cautious manner and with a sound appreciation of the design requirements and the practical considerations. In particular, the designer should try to predict the consequences of a mismatch – how serious they would be and who would be affected. The height of the handle of a door leading to a fire escape in an apartment block dramatises the seriousness of such mismatches – it is essential that a very wide range of users, including children, be able to reach and operate the handle in an emergency. The design of passenger seats for urban transportation systems is also important, although somewhat more mundane. Because the seats are used regularly by a very wide range of users, even small imperfections will affect the comfort of a very large number of people every day. When using anthropometric data, the selection of a suitable cut-off point depends on the consequences of an anthropometric mismatch and the cost of designing for a wide range of people. One of the ergonomist's most important tasks is to predict and evaluate what any mismatches are going to be like. It is not normally sufficient only to specify the required dimensions without considering other aspects such as usability and misuse.

Use of mannequins

Design aids such as mannequins (jointed representations of the human form used in conjunction with appropriately scaled drawings or models) can help potential problems of anthropometric fit to be visualised.

Mannequins representing 5th, 50th and 95th percentile users may be used but with caution because of differences in proportion: a person of 5th percentile stature may have 20th percentile reach and 60th percentile girth, which the mannequin will not disclose. It is incumbent on the designer, therefore, to try to anticipate potential anthropometric mismatches rather than using anthropometric data in a 'cook-book' fashion.

Computerised design aids now exist to facilitate visualisation of the physical interaction between users and hardware. SAMMIE (system for aiding man–machine interaction evaluation, SAMMIE CAD Ltd) is an example of such a system (Figure 3.8). The anthropometry of a three-dimensional 'man-model' displayed on the computer screen can be manipulated and the consequences evaluated using a computer-generated representation of the product being designed.

Several CAD systems are available that incorporate model humans whose body dimensions can be manipulated using anthropometric data. The JACK system (Maida, 1993) has 88 articulated joints of the human body and a 17-segment torso, and incorporates data on human body contours and strength limits.

Applications of anthropometry in design

Table 3.14 gives examples of some common anthropometric variables and how they are used in ergonomics. The list is intended to be representative rather than

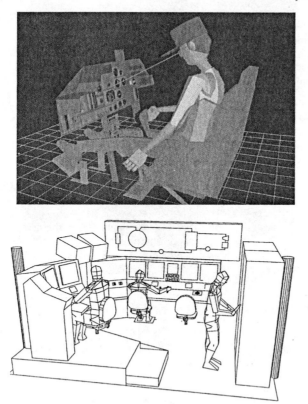

(A)

(B)

Figure 3.9 Computerised anthropometry. (A) JACK (JACK is a registered trademark of the University of Pennsylvania). (B) SAMMIE (SAMMIE CAD Ltd., Loughborough, United Kingdom).

Table 3.14 Description and examples of use of some common anthropometric variables

Standing eye height: Height above the ground of the eye of a person standing erect. Can be used as a maximum allowable dimension to locate visual displays for standing operators. The displays should not be higher than the standing eye height of a short operator. Short operators should not need to extend the neck to look at displays.

Standing shoulder height: Height of the acromion above the ground. Used to estimate the height of the centre of rotation of the arm above the ground and can help specify the maximum allowable height for a control. Short workers should not need to elevate the arms above shoulder height to operate a control.

Standing elbow height: Height above the ground of the elbows of a person standing erect. Used to design the maximum allowable bench height for standing workers. For delicate tasks, bench heights can be slightly higher than elbow height so that the worker can stabilise the forearms by resting them on the bench. For tasks requiring high forces, bench heights should be lower than elbow height to allow workers to apply body weight. For fixed tasks, the height of the work object itself should be included in the calculations (bench height plus work object height in relation to standing elbow height).

Table 3.14 (cont'd)

Standing knuckle height: Height of the knuckles above the ground. Used to determine the minimum height of full grip for a standing operator. Operators with high standing knuckle heights should not have to stoop when grasping objects in the workplace. Could be used to determine the minimum length of the handle of a spade, for example.

Standing fingertip height: Height of the tips of the fingers above the ground. Used to determine the lowest allowable position for controls such as switches.

Sitting height: Distance from the seat to the crown of the head. Can be used to determine ceiling heights in vehicles to provide clearance for users with large sitting heights.

Sitting elbow height: Height of the elbows of a seated person above the chair. Used to determine arm rest heights and work surface heights for seated operators.

Popliteal height: Height of the popliteal fossa (back of the knee) above the ground. The 5th percentile popliteal height may be used to determine the maximum allowable height of non-adjustable seats. The 95th percentile popliteal height may be used to set the highest level of adjustment of height-adjustable seats.

Knee height and thigh depth: Taken together, these variables specify the height above the floor of the upper thigh of a seated person. Can be used to determine the thigh clearance required under a table or console.

Buttock–popliteal length: Distance from the buttocks to the back of the knee. Used to determine the maximum allowable seat depth such that seat depth does not exceed the buttock–popliteal length of short users.

Shoulder width: Widest distance across the shoulders. Used to determine the minimum width of narrow doorways, corridors, etc. to provide clearance for those with wide shoulders.

Hip breadth: Widest distance across the hips. Used to determine the space requirements necessary for clearance and for example, the minimum width of seats to allow clearance for those with wide hips.

Abdominal/chest depth: Widest distance from a wall behind the person to the chest/abdomen in front. Used to determine the minimum clearance required in confined spaces.

Vertical reach (sitting and standing): Highest vertical reach. Used to determine maximum allowable height for overhead controls, door latches, etc. such that they are reachable by the shortest users.

Grip circumference: Internal circumference of grip from the root of the fingers across the tip and to the palm when grasping an object. Used to specify the maximum circumference of tool handles and other objects to be held in the palm of the hand. Handle circumferences should enable those with small hands to grasp the tool with slight overlap of the thumb and fingers. Opening strength and grip strength deteriorate when the circumference of lids and handles exceeds grip circumference.

Reach: The dimensions of the reach envelope around an operator can be used to locate controls so that seated operators can operate them without having to lean forward away from the backrest or twist the trunk and standing operators can operate them without forward, backward or sideways inclination of the trunk. Arm movements should be kept in the normal work area to eliminate reach over 40 cm for repeated actions. These data are applicable to the design of all vehicle cockpits and cabs.

Table 3.15 Self-reported work problems of nurses related to their anthropometry

Reported problem/complaint	Associated anthropometric variables
Reaching for work objects	Low stature and grip reach
Buying clothes	Large hip breadth
Buying shoes	Foot breadth and length
Lower backache	Large stature and abdominal depth
Shoulder/arm pain	Low stature and grip reach
Handles on equipment too small	Large hand and palm length hand breadth
Handles that hurt the hands	Short hand length
Work surfaces too low	High stature and standing elbow height
Work surfaces too high	Low stature and standing elbow height
Inadequate legroom in seated workspaces	High popliteal height and buttock–knee length

exhaustive and further information may be found in Croney (1980), Clark and Corlett (1984) and Pheasant (1986).

Some common approaches to the anthropometric solution of design problems are described below.

Design for everyone

Matching product and user dimensions is important for reasons of safety, health and usability. Botha and Bridger (1998) carried out an anthropometric survey of nurses in a hospital in Cape Town. They also captured data on problems of musculoskeletal pain and equipment usability. The anthropometric variables were divided into quartiles and the frequency of occurrence of problems was counted for each quartile. Many problems were found to be more common in the extreme quartiles. Table 3.15 summarises the findings and demonstrates that many of the reported problems were caused or exacerbated by a work environment that did not fit its occupants.

It is noteworthy that the sample of nurses in the study bore no resemblance to the local population of females in the Western Cape – many of them were not even from South Africa! It is often fallacious to assume that a group of workers in a particular occupational group are representative of the population of the parent country. If they are not representative, then anthropometric data from national databases (if available) cannot be used. An alternative approach (see below) is to use statistical scaling techniques to estimate the required anthropometric data.

The problem of designing to suit a range of users can be approached in several different ways.

Make different sizes

In clothing and school furniture design, a common solution is to design the same product in *several different sizes*. Anthropometric data can be used to determine a minimum number of different sizes (and the dimensions of each size) that will accommodate all users. Mass production or long production runs often bring economies of scale in product design through reduced retooling and stoppages. This usually has economic benefits and demonstrates why it is important to determine the *minimum*

RASH – Rapid Anthropometrics Scaled For Height

Pheasant (1986) recommended the use of the RASH technique to estimate the body dimensions of populations whose anthropometry is unknown. The technique requires data on the stature of the target population. These data are normally available from medical records.

RASH estimates are made by calculating scaling factors for the anthropometric variables of interest using data from a known population and then applying these scaling factors to the height data in the target population. Suppose we want to estimate the sitting height of Chilean bus drivers, but have no data on the anthropometry of the Chilean population – where do we begin?

1. Obtain data on the stature of the drivers from medical records and calculate the mean and standard deviation stature (all data are in cm.):

    $\bar{x} = 166$ s.d. = 5.5

2. Obtain data on the mean and standard deviation stature and sitting heights of a similar group of people, assuming that body proportions will be similar. We will use data for Brazilian males:

    Stature $\bar{x} = 170$ s.d. = 6.6   Sitting height $\bar{x} = 88$ s.d. = 3.5

3. Calculate scaling ratios for the mean and standard deviation stature:

    SR(mean sitting height) = 88/170 = 0.518
    SR(s.d. sitting height) = 3.5/6.6 = 0.53

4. Multiply the mean and s.d. stature of the Chilean bus drivers by the scaling ratios:

    Estimated mean sitting height = 166 × 0.518 = 86
    Estimated s.d. sitting height = 5.5 × 0.53 = 2.92

5. Calculate 5th percentile sitting height by subtracting 1.64 standard deviations from the estimated mean sitting height:

    Estimated 5th percentile sitting height of Chilean bus drivers
    = 86 − (1.64 × 2.92) = 81.2

6. Calculate 95th percentile sitting height by adding 1.64 standard deviations to the estimated mean sitting height:

    Estimated 95th percentile siting height of Chilean bus drivers
    = 86 + (1.64 × 2.92) = 90.8

Use of the technique is best limited to body dimensions that depend on the length of bones – sitting eye height, popliteal height – and is not recommended to estimate circumferential dimensions such as girth or measures, such as hip breadth, that also depend on the distribution of adipose tissue in the body.

---

number of sizes in a product range that will accommodate most of the users in the population in question.

Clothing design is the best example of fitting the product to the user. There are standards, such as ISO 3636 for the designation of clothing sizes. It is left to the manufacturer to decide what sizes to make and this is often done by trial and error or 'knowing the market' rather than by carrying out an anthropometric survey. Head girth, neck girth, chest girth, bust girth, underbust girth, waist girth, hip girth, stature, outside leg and inside leg length are dimensions used to designate the size of a piece of clothing. National and international standards only specify which dimensions the manufacturer must use to designate sizes, they do not specify the number of sizes in a range, nor what actual size of the clothing should be.

Clothing sizing systems divide a variable population into roughly homogeneous subgroups (McCullogh *et al.*, 1998). It is assumed that members of a subgroup are a single size and shape so that a single garment can fit everyone in the group. The simplest approach is to choose one or two dimensions that are thought to be crucial if a good fit is to be achieved, and then to divide the range into an equally spaced series of sizes according to these dimensions. The remaining variables are adjusted across the range in proportion to the criterion sizes. Sizing is difficult because many anthropometric dimensions are not highly correlated – long bone measures and girth measures, for example – and users may struggle to find a shirt in which the collar size and the sleeves both fit. Because anthropometric variables are usually normally distributed, a far greater number of people will fit clothes in the middle of the range and only short production runs are needed to accommodate the extreme sizes, which may be difficult to find in the shops.

An alternative is to expand the range around each size. The best recent example of this approach was the Levi-Strauss mass customisation experiment in which women's jeans were offered in 16 hip sizes. For each hip size, there were 11 waist sizes, 4 crotch depths and 6 lengths, resulting in 4224 different sizes of jeans. This approach is probably only practical for mass-produced items.

More recently, clothing researchers (e.g. Ashdown, 1998) have turned to multivariate statistical methods to tackle the problem. The goal of achieving as good a fit as possible can be re-stated as follows:

- Accommodate as large a percentage of the population as possible with ready-made garments.
- For accommodated individuals, provide as good a fit as possible.
- Use as few sizes as possible.

A statistical technique, known as cluster analysis, is used to search for groups, or clusters, of anthropometric variables in the target market. It is not necessary to specify criterion variables, nor to adjust the remainder proportionally. Rather, any number of variables can be used, so that the combination of dimensions will fit people of a particular size. Each size selected is defined by a set of 'prototype' design values. The cluster analysis approach requires a fair amount of computer power and is an example of an 'optimisation' technique. The optimisation criterion is to select a set of values that will minimise the average distance, where distance corresponds to a lack of fit. Because this applies to all the anthropometric variables selected, an item of clothing that fits the neck will tend to fit the chest and sleeves as well. The result is that clothing sizes do not increase in fixed intervals, but according to the make-up of the target market. Some jumps may be bigger than others between sizes in the range and some measures may change more than others in a single jump. Sizes may sometimes differ in terms of proportion and not in terms of largeness or smallness and this may mean that potential buyers would have to try on more sizes before finding one that fits. However, from the retailer's point of view, the distribution of people fitted by each size is likely to be more uniform, and this may simplify the ordering of stock.

Design adjustable products

An alternative is to manufacture products whose critical dimensions can be *adjusted* by the users themselves. A first step is to determine what the critical dimensions for use are, then to design the mechanism of adjustability, with the emphasis on ease of operation. Finally, some instructions or a training programme may be needed to explain to users the importance of adjusting the product and how to adjust it correctly.

In seated work, for example, the height of the seat and desk are critical dimensions for seated comfort. The seat height should be no higher than the popliteal height of the user so that both feet can be rested firmly on the floor to support the weight of the lower legs (otherwise the soft tissues on the underside of the thigh take the weight and blood circulation is impeded owing to compression of these tissues). Secondly, the desk height (or middle row of keys on a keyboard) should coincide with the user's sitting elbow height. Since popliteal height and elbow height do not correlate strongly in practice (Verbeek, 1991), adjustable seat and desk heights are needed.

Shute and Starr (1984) investigated the effects of adjustable furniture on visual display unit (VDU) users. On-the-job discomfort was reduced when adjustable chairs or adjustable desks were used in place of non-adjustable furniture. The greatest reductions in discomfort were found when the two adjustable items were used in combination.

One problem with adjustability is that users may not use the adjustment facility if they do not expect a product to be adjustable or if they do not understand the reason for incorporating adjustability into the product. Verbeek (1991) investigated the effect of an instruction programme for office workers on the anthropometric fit between users and their chair/desk workstations. Before the programme, a survey of chair/desk settings in an office revealed mean deviations from the ideal of 71 mm for seat height and 70 mm for desk height. A model of 'correct' sitting was used as a criterion to evaluate the chair/desk settings. After the programme, these deviations were reduced by 11 mm and 18 mm, respectively. However, only 7% of users adjusted their seat heights as advised and only 13% adjusted their desk heights as advised. It was concluded that this meagre result was due to practical difficulties, the unaesthetic appearance of adjacent desks having different heights and the suspect validity of the model of 'correct sitting' that had been used to specify the method of adjustment (further discussion of concepts of 'correct sitting' can be found in the following chapter).

'Fit for use' surveys

There are many ways of assessing how well a work environment fits its users. We can measure the critical features of equipment and compare them with user anthropometry: we may unobtrusively observe people at work and note any apparent problems. Alternatively, we may carry out some kind of task analysis and consult the users, asking them to report any problems experienced. A minimal set of items for a questionnaire designed to assess 'fit' is given in Table 3.16.

Table 3.16 Questionnaire items for a basic survey of anthropometric fit

Personal characteristics (self-reported or measured)	Difficulties with equipment items
Age	Equipment difficult to operate
Sex	Requires too much force
Height	Parts that are difficult to grip
Weight	Have to reach too far forward
Chest circumference	Have to reach too far up
Department	Have to reach too far down
Job title	Have to stand in an awkward posture
Length of time in job	Have to sit in an awkward posture
Any work-related injuries?	Unable to see parts of the equipment
	Equipment too heavy
	Equipment too bulky
	Straps do not fit or are uncomfortable
	Cannot get access
	Problems with ingress
	Problems with egress

In practice, the investigation might go through several stages and require all of the methods above.

Anthropometry and personal space

Personal space can be defined as the area immediately around the body. Argyle (1975) describes personal space in the context of territorial behaviour. Many animals regard a certain area of space as their exclusive preserve. The area immediately around an individual's body is usually regarded in this way: two important issues in 'psycho-anthropometry' are the volume of space regarded as personal territory and the consequences of an invasion of this space by others.

Large individual and cultural differences exist. Argyle reports the results of studies that indicate that Arabs stand closer together than Europeans and North Americans, with Latin Americans and Asians intermediate.

Invasion of personal space and crowded conditions appear to be stressful (Dabbs, 1971). In an experiment that is now infamous, Middlemist *et al.* (1976) tested the hypothesis that personal space invasion increases arousal. They measured the time to onset and duration of micturation (urination) of 60 men using a public lavatory each under one of three conditions: subject alone (control condition), subject standing next to a confederate (adjacent invasion), and subject one urinal removed from the confederate (moderate invasion). The confederate was an accomplice of the experimenters and arrived after the subject had unwittingly taken his place. Onset time of micturation increased from 4.9 seconds in the control condition to 6.2 (moderate space invasion) and 8.4 seconds for adjacent invasion. Micturation duration dropped from 24.8 seconds in the control condition to 23.4 and 17.4 seconds in the space invasion conditions, respectively. In the context of public lavatory use, the minimum required personal space would appear to be about the distance of one urinal.

The degree of stress depends on the context. Invasion of personal space in a library, for example, is much more stressful than in a crowded train or lift. Personal space is another important consideration in addition to the purely dimensional ones. Design decisions regarding the size and spacing of seats in public areas, the proximity of desks, and so on need to take account of people's personal space requirements and the particular social context. In the workplace, a minimum separation of desks or benches of approximately 1.2 metres is thought to be necessary.

The reader is referred to Altman (1975) and Argyle (1975) for further discussion of these issues.

Effectiveness and cost-effectiveness

Benefits of protective clothing that fits

Many companies spend large amounts of money on protective clothing and equipment for their employees. Ensuring correct fit is one of the most basic requirements for the item to function correctly. Warktosch (1994) carried out an investigation of leg protectors used by forestry workers in South Africa, a group with considerable ethnic diversity. The existing leg protectors were imported from Brazil and had not succeeded in preventing injury – 5 injuries were occurring per day among the 300 workers. The protectors were modified to suit the anthropometry of the users, together with improvements to the materials and fastenings. The result was complete prevention of leg injuries due to axes and hatchets in the one-year follow-up period. The cost saving was approximately $250 000 and was expected to provide even greater savings when implemented on a larger scale.

Industrial workplace layout

Lim and Hoffman (1997) investigated the performance of a light assembly task when components were laid out within the 'zone of convenient reach' (ZCR). ZCR is defined, for each hand, as the area of the table within the arc swept by the hand with the elbow extended (within this zone is the 'normal working area' defined by a similar sweep of each hand but with 90 degrees of elbow flexion). An approximately 10% improvement in assembly time was achieved when items were placed within the zone of convenient reach and arranged in an ergonomically designed jig. These improvements led to more efficient hand movements (reducing lengthy reaches and contralateral movements and increasing the number of simultaneous hand actions). That the naive operators participating in the experiment did not spontaneously optimise the working arrangements (even though they were permitted to) highlights the need for this to be done when the task is designed and included in the initial training.

Adjustability and adjustment of office furniture

A number of studies have been carried out into productivity in office environments. Some took place in the early 1980s when desktop computers were first being introduced into offices, often without any modification to the rest of the workspace. Modern office work is heavily computerised and requires the adoption of static postures for long periods as people have less reason to leave their desks. It would be expected that a good fit between workers and their equipment is one of the requirements for improved productivity. In 1981, the Merck company (Ruff, 1985) decided to upgrade its office facilities in an attempt to create a more productive and more healthy environment. Two thousand office workers were sent a questionnaire and asked to rate 29 features of workstations regarding job effectiveness and satisfaction. The greatest mismatches between workstations and satisfaction were in air-conditioning, the ability to concentrate and privacy, overall workspace size and work area. Equally important were lighting and adjustability. Merck then selected office furniture to renovate its offices and canvassed the opinions of user-groups at a test facility. Training videos on how to use and adjust the new workstations were also produced. The return on all renovation projects was 25%; for one, an installation for 74 international workers, the return was 50% with an improvement in turnaround of jobs from 4 days to 6 hours.

A limitation of practical case studies, such as that at Merck, is the inability to disentangle the effects of specific interventions in the context of an overall refurbishment project in which many different aspects of the work environment are changed. Springer (1982, quoted from Dainoff and Dainoff (1986)) evaluated four new workstations in a recently computerised office. Two were easily adjustable, a third was less adjustable, and a fourth was new but non-adjustable. One hundred and ten employees compared these with their old office furniture while performing tests lasting less than half an hour. At the beginning of the trial, they adjusted the adjustable workstations to their own requirements. The adjustable furniture yielded 15% faster performance than the non-adjustable furniture, thus demonstrating that the improvement was a result of the adjustability of the furniture rather than a novelty effect.

A study by Cushman (1985), while lacking the 'real world' feel of the Merck study, does provide evidence for a link between productivity and good anthropometric fit. Twenty female employees of Eastman Kodak took part in a trial in which they copy typed for 10 minutes at each of five keyboard heights (70, 74, 78, 82 and 86 cm). Their mean stature was 164.3 cm with a standard deviation of 8.0 cm. Performance was measured by number of words typed per minute and error rate. Keying rate was relatively unaffected by keyboard height. The error rate was about 20% lower when the keyboard was at the 74 cm. height. Posture discomfort ratings were positively associated with the number of errors made. An interesting finding was that errors increased when the difference between the keyboard height and each subject's sitting elbow height was negative (keyboard too low). Subjects made fewer errors when the keyboard was reported to be 'at about the right height'. However, the 'right height' was reported to be 5–10 cm above the subjects' sitting elbow height, contrary to the usual recommendation that it should be at elbow height. Nevertheless, these findings demonstrate the importance of adjustability in workstation design so that workers can achieve a good anthropometric fit, that a good fit feels comfortable and that when people feel comfortable because the furniture fits, they make fewer mistakes.

Effectiveness of office ergonomics interventions

Evanoff *et al.* (2000) evaluated an office ergonomics intervention in a US hospital billing department in which the injury and lost-day rates were higher than in the hospital as a whole. One hundred and fifteen employees were offered a one-hour educational session covering workstation adjustment and arrangement, office ergonomics and prevention of musculoskeletal disorders. Each employees' workstation was then assessed by an ergonomist over the next 6 months. Where needed, adjustments were made and job aids (e.g. armrests, wrist rests and footrests) were provided and leg room was increased. After 18 months, a 15-minute follow-up educational session was given. Data on work-related injuries and illnesses (OSHA 200 log), workers' compensation records and self-assessment were collected.

Total injuries fell from 24 in the 18 months prior to the intervention to 26 in the following 4 years and total lost days as a result of the injury fell from 126 to 31. Workers' compensation costs in the 18 months prior to the intervention amounted to $86 144. In the following 3 years, these costs were $89 331 – an annualised reduction of 57%. These improvements occurred despite an increase in workload of 100% in the follow-up period. As with many such studies, this one is limited by the lack of a control group. Injury rates dropped in the hospital as a whole over the follow-up period, but the authors report that the observed reductions in the billing department were still significant after correcting for this underlying trend.

Research directions

In the light of the increasingly international nature of business and trade, an urgent area of research is to fill in the gaps in the world anthropometric database. Data are lacking on the anthropometry of many populations. Another challenge is to develop inexpensive automated methods of capturing the data – traditional methods requiring the use of manual instruments are prohibitively expensive.

Also of interest is the increase in body size of populations that occurs as countries develop industrially. This is known to have happened in Western industrial countries and Japan and more recently has been happening in southern Europe and parts of South America. In developing countries, there may be differences in the anthropometry of urban versus rural peoples.

Available data are being incorporated into computer-based design aids. Several packages have been developed and they all assist in integrating data with three-dimensional representations of the human form to improve the visualisation and solution of problems of anthropometric fit. Further development of these aids will no doubt take place.

Summary

Anthropometric data provide the designer with quantitative guidelines for dimensioning workspaces. However, a number of precautions are needed if data are to be used correctly. Always:

1. Define the user population and use data obtained from measurements made on that population.
2. Consider factors that might interfere with the assumption of normal distribution of scores. For example, in some countries, stature may be negatively skewed because many individuals do not attain their potential stature owing to disease or malnutrition.
3. Remember that many anthropometric variables are measured using semi-nude subjects. Allowance for clothing is often necessary when designing for real users. Centimetre accuracy is usually appropriate because the effect of clothing on the

estimates of user anthropometry can never be accurately predicted. These considerations are particularly important when using data on stature and leg length: allowances for heel heights of 5 cm or more may be needed depending on the user population and current fashions of dress. The effect of clothing also depends on climate: the colder the climate the more bulky the clothing and the greater the importance of allowing for this in design.

In many parts of the world, biological and social change will cause plastic changes in population parameter values over relatively short time periods. Drift in the anthropometric parameter values should be expected. In practical design situations, the data required for a particular body dimension or population may not be available. However, techniques are available for estimating unknown dimensions and the literature on anthropometry can still be of use to the designer in drawing attention to the various body dimensions that need to be considered in the design and the types of human–machine mismatches that could occur. Anthropometry can provide the designer with a very useful perspective on usability issues at the very early stages of the design process. Later, the designer might then take a more empirical approach and test out prototypes using a small sample of users from the extremes of the anthropometric range.

Essays and exercises

1. Using the data in this chapter, specify seat and work surface dimensions to fit 95% of VDU users in two different populations.
   Comment on any design and equipment procurement implications of this exercise.
2. Carry out an anthropometric evaluation of either a commercial kitchen or a number of domestic kitchens. Measure and comment on the following:

   • The heights of all major work surfaces including the sink
   • The heights of all cupboards and shelves
   • The reach requirements of all major work and storage areas
   • The amount of space for movement and foot position

   Interview users to find out how they spend their time in the kitchen. List the most common to the least common tasks and note any problems and difficulties experienced.
   Use all of the above information to suggest improvements.
3. Write a report that explains to a design engineer why anthropometric data should be used in the design of human–machine systems. Give examples from everyday life to support your arguments.
4. You have been asked to help a large financial institution in Cape Town, South Africa to assist with office chair selection. The workforce is a mixture of different racial and ethnic groups. Because of the cost savings involved in making one large order for 2500 identical chairs, the company wishes to standardise on office chairs. One chair must fit all.
   You decide to get information on the stature of employees from their company medical records (a medical is a requirement for membership of the company

pension scheme). From this, you decide to use the RASH technique to estimate 5th and 95th percentile measurements used in chair design.

(i) List the anthropometric variables you would need to estimate to help the company choose a chair.

(ii) Given that the mean and standard deviation heights of males and females in the company are as tabulated below, estimate 5th and 95th percentile values of the variables of your choice using scaling factors calculated from the tables in this chapter (RASH technique).

	Mean	*SD*
Males	1762.5	72.7
Females	1629.5	69.2

(iii) You decide that seat depth is a critical factor. The company has a shortlist of 14 chairs but none of them has adjustable seat depth. You measure the depth of the seats with the following results:

Seat depth (mm)	*Number of chairs at that depth*
450	3
420	4
400	3
390	1
370	3

Given that the mean female buttock–popliteal length in the company is 490.5 mm (SD 33.2 mm), what percentage of the company's female workers will be able to sit on the seats without having to adopt a posture in which they are 'perched' on the edge of the seat?

4 Static work: Design for standing and seated workers

> In biological terms, posture is constant, continuous adaptation ... Standing is in reality movement upon a stationary base ... From this point of view, normal standing on both legs is almost effortless.
>
> (F. A. Hellebrandt (1938))

Humans are designed to stand on two legs, but they are not designed to stand still. Standing is the position of choice for many tasks in industry but it can lead to discomfort if insufficient rest is provided or if unnecessary postural load is placed on the body. Some advantages of the standing work position are given in Table 4.1.

In everyday life, people *rarely stand still* for any length of time – if not walking or moving, they adopt a variety of resting positions that vary depending on the culture (Hewes, 1957; Bridger *et al.*, 1994). Table 4.2 summarises some behaviours associated with unconstrained standing. In most occupational settings where people engage in 'industrial standing', they are denied the opportunity to practise these behaviours by the design of their workspaces and the design of their jobs.

Short periods of walking and gross body movements are vital to activate the venous pump and assist the return of blood from the lower limbs (Cavanagh *et al.*, 1987; Stranden, 2000), so the idea that workers should stand still is physiologically and mechanically unacceptable. Anecdotal evidence across many cultures and over time tells us that people who *do* have to stand for long periods use standing aids such as the staff of the Nilotic herdsman or the spear of the sentry. An experiment on constrained standing by Whistance (1996) demonstrated that even unpractised users spontaneously make use of such aids when they are provided.

Table 4.1 Some advantages of the standing work position[a]

1. Reach is greater in standing than in sitting.
2. Body weight can be used to exert forces.
3. Standing workers require less leg room than seated workers.
4. The legs are very effective at damping vibration.
5. Lumbar disc pressures are lower.[b]
6. It can be maintained with little muscular activity and requires no attention.[c]
7. Trunk muscle power is twice as large in standing than in semi-standing or sitting.[d]

[a] Singleton (1972).   [c] Hellebrandt (1938).
[b] Nachemson (1966).   [d] Cartas *et al.* (1993).

Table 4.2 Some behaviours characteristic of unconstrained standing

1. Never stand still	10. Hang on elbows
2. Bear weight on one leg	11. Rest body on surface
3. Lean backwards against anything	12. Sit on heels against a wall
4. Lean sideways against a vertical surface	13. Use a footrest
5. Rest pelvis against counter	14. Sit down when tired
6. Maximise contact with fixed objects	15. Hang on to overhead objects
7. Use arms as props resting on a surface	16. Rest head on hand
8. Rest one foot on an object	17. Rest knee on something
9. Use thoracic support (e.g. lean on broom)	18. Rest hands on knees

Prolonged daily standing is known to be associated with low back pain. Where possible, *jobs that require people to stand still for prolonged periods without some external form of aid or support must be redesigned to allow more movement* or to allow the work to be done in a combination of standing and sitting postures.

Fundamental aspects of sitting and standing

Anatomy of standing

The pelvis is held in an anteriorly tilted position by the iliopsoas muscles and the hip joint is free to extend as happens during the stance phase of gait. The trunk and head are rotated until they are vertically above the legs. This is achieved by extension of the lumbar and cervical spines and is why the vertically held spine is 'S' shaped in humans whereas in quadrupeds it is 'C' shaped and held horizontal to the ground.

Bones and joints   In the erect posture, the line of gravity of superincumbent body parts passes through the lumbar, sacral and hip joints and in front of the knee and ankle joints. This places an extension torque around the knee joint, which is resisted because the joint is already fully extended. The flexion torque around the ankle is resisted by the plantar flexors.

Muscles and ligaments   A person standing erect under the influence of gravity is never in a state of passive equilibrium. The body can be conceived of as a pillar of segments stacked one on top of the other and linked by joints. It is momentarily balanced when the resultant of all forces acting on it is zero. The system is designed to minimise any displacement of the line of action beyond the base of support described by the position of the feet. Compensatory mechanisms come into play to maintain balance immediately this happens.

Muscles and ligaments play a stabilising role by means of the active and passive torques they exert around joints to correct small, fleeting displacements of the lines of action away from the joints. A 'good' posture may be defined as one in which destabilising moments are minimised and the posture is maintained by the resistance of the relatively incompressible bones (as well as interleaved soft tissues such as the intervertebral discs).

When the body is pulled 'off-balance' by the requirements of badly designed jobs or workspaces, the anti-gravity muscles come into play and a new equilibrium position is established but with the associated cost of isometric muscle activity.

The erector spinae muscles   These are the main extensors of the trunk. They are also used to control flexion. In relaxed standing, very little muscle activity occurs since the lumbar lordosis minimises the trunk flexion moment. When the trunk is flexed even slightly forwards or when a weight is held in front of the body, the erector spinae muscles come into play. Work situations that set up static loading of these muscles include

- Working with the hands and arms held away from the body
- Holding a tool or a weight
- Standing with the trunk flexed to reach for work objects placed too far away or inaccessible owing to a lack of foot space

The leg muscles   The soleus and gastrocnemius muscles are true postural muscles in the sense that they are always 'switched on' when standing. When a person is leaning forward, the activity of the gastrocnemius muscle increases. Prolonged standing causes significant localised leg muscle fatigue and is one of the causes of leg discomfort.

The abdominal muscles   There is very little abdominal muscle activity in standing and even less in sitting (Burdorf *et al.*, 1993). These muscles may help to maintain a proper relationship between the thorax and pelvis by preventing excessive anterior pelvic tilt and hyperlordosis. The abdominals can prevent trunk extension, caused, for example, by loads placed high on the back (or when putting on a backpack, for example) or when walking down steep hills.

The hamstring and gluteal muscles   The hamstring and gluteal muscles are hip extensors. The gluteal muscles exhibit hypertrophy in humans and their function is to stop the trunk from 'jack-knifing' forwards over the legs, unlike in quadrupeds where the trunk is already 'jack-knifed' and the gluteals are used for locomotion. The gluteals are, however, used for locomotion in climbing ladders or stairs. Activity in the hamstrings is slight in the standing position but increases when the stander leans forward, holds a weight or pulls.

The iliopsoas muscles   Psoas major and iliacus are hip flexors and are constantly active in normal standing as they prevent extension of the hip joint (the trunk 'jack-knifing' backwards over the legs, or loss of lumbar lordosis if the head position is maintained). The iliopsoas muscles act against the hip extensors.

The adductors and abductors of the hip   When a person is standing on two feet, these muscles provide lateral stability, preventing translation of the pelvis in the frontal plane. When a person standing on one foot (and also during the stance phase of gait) the pelvis tends to tilt in the direction of the unsupported side. The hip abductors on the side of the supporting leg contract to keep the pelvis level.

Physiology of standing

The increase in energy expenditure when a person changes from a supine to a standing position is only about 8% (Grandjean, 1980). However, erect standing imposes a hydrostatic handicap that makes humans liable to peripheral circulatory collapse. Peak plantar (foot) pressures of 137 (kilopascals) kPa exceed the normal systolic pressure of 17 kPa, resulting in occlusion of blood flow through the foot (Cavanagh *et al.*, 1987). Walking and fidgeting temporarily reduce the pressure, allowing fresh blood to pervade the tissues. Venous and circulatory insufficiencies in the lower limbs also contribute to the discomfort that results from prolonged standing. It has been shown that venous reflux is more common in symptom-free surgeons (who stand for long periods) than in a comparison group who experience discomfort in standing.

Prolonged standing causes physiological changes including peripheral pooling of blood, a decrease in stroke volume and increases in heart rate, diastolic and mean arterial pressure, peripheral resistance and thoracic impedance. Standing up from a supine position is accompanied by an increase in the dimensions of the nasal passages (Whistance, 1996).

Constrained standing is particularly troublesome for older workers or for those with peripheral vascular disease because the 'venous muscle pump' that returns blood to the heart ceases to function. Fidgeting is a pre-conscious defence against the postural stresses of constrained standing or sitting. Its purpose is to redistribute and relieve loading on bones and soft tissues and to rest muscles.

Varicose veins and standing work

Varicose veins are superficial veins, often in the legs, in which the valves function ineffectively, resulting in pooling of blood and painful swelling. With deep veins, the problem is more serious and can cause blood to return along abnormal pathways, resulting in long-term health problems, including chronic oedema and leg ulcers. Risk factors include obesity, cigarette smoking, high blood pressure and lack of exercise. The disease is one of the 10 leading causes of hospitalisation in Denmark. Occupational standing is thought to be associated with varicose veins in the lower extremities. Tuchsen *et al.* (2000) followed a sample of 1.6 million working Danes for three years from 1991. Men who worked mostly in a standing position were almost twice as likely to experience a first hospitalisation for varicose veins compared to all other men. Women who worked mostly in a standing position, were two and a half times more at risk than all other women. Tomei *et al.* (1999) compared the prevalence of chronic venous diseases in office workers, industrial workers and stoneworkers. The prevalence of the disorders increased with age and number of hours spent standing at work. Controlling for age, the prevalence was higher for workers who stood for *50% or more* of their shift. The findings suggest that standing work should be combined with other types of work in which sitting is possible.

Sitting

When a person flexes the hip and knee joint to sit down, the iliopsoas muscles immediately shorten and the hip extensors lengthen (Link *et al.*, 1990; Bridger *et al.*, 1992). The balance of antagonistic muscle forces, which keep the pelvis in its anteriorly tilted position, is changed and the pelvis tilts posteriorly almost immediately and continues in proportion to the flexion at the hip. In order to maintain the head erect, the lumbar spine flexes to compensate for the tilting pelvis and the lumbar lordosis diminishes and eventually disappears (Figure 4.1).

The spine of a person seated erect exhibits greatly diminished lumbar curvature (mean angle of curvature of 34 degrees compared to 47 degrees in standing, according

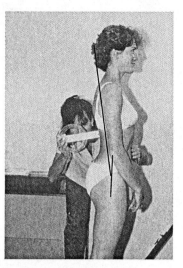

(A)

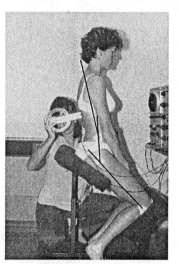

(B)

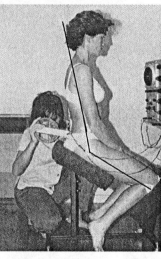

(C)

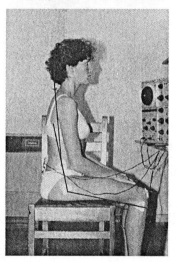

(D)

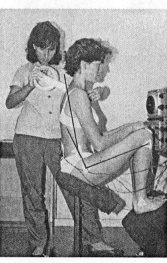

(E)

Figure 4.1 Posterior tilting of the pelvis
and flattenting of the lumbar
curve during the transition
from standing to sitting.
Much of the postural adaptation to sitting takes place
in the back, rather than the
legs.

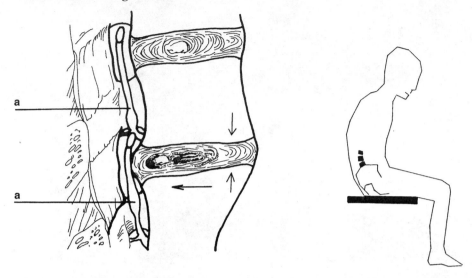

Figure 4.2 Anterior wedging of the intervertebral disc occurs in the slumped sitting position (**a** = posterior ligaments). Soft tissues between the anterior and posterior elements of the spine may be pressurised, resulting in pain. (Adapted from Keegan, 1953, with permission of the *The Journal of Bone and Joint Surgery, Inc.*).

to Lord *et al.*, 1997). The flexion moment about the lumbar spine is increased, putting the posterior spinal ligaments under tension and causing the intervertebral discs to be 'wedged' anteriorly and to protrude into the intervertebral foramen (Keegan, 1953; Figure 4.2). The mechanical effect of the flexion moment is an increase in the pressure in the intervertebral discs.

Disc pressures are lower in standing than in sitting and lower still when lying down. In standing, the load is shared between the facet joint. In sitting, the discs bear more of the load (Adams and Dolan, 1995), whereas when lying the absolute load on all structures is lowered. Rohlmann *et al.* (2001) found that disc pressure was lower in relaxed sitting than in standing, but higher when subjects attempted to extend the spine to sit erect. Both Nachemson (1966) and Rohlmann *et al.* report lower disc pressures when subjects recline against a backrest. This implies that seated workers should be able to adopt relaxed postures.

Brunswic (1984) investigated the relationship between lumbar curvature, hip flexion and knee flexion (Figure 4.3). Brunswic concluded that *if the knees are flexed by 110 degrees*, seat tilts between 5 degrees rearwards to 25 degrees forwards were acceptable. If the knees were flexed by 70 degrees or less, forward tilt of at least 10 degrees is required. Seated workers should not have to fully extend their legs to operate foot pedals, or, in the case of tall workers, workstations should be designed to provide space for the knees to flex.

Working posture

As far as the lumbar spine is concerned, a good working posture is one in which the spine is towards the mid-point of its range of movement and the trunk is unconstrained – free to move anteriorly and posteriorly.

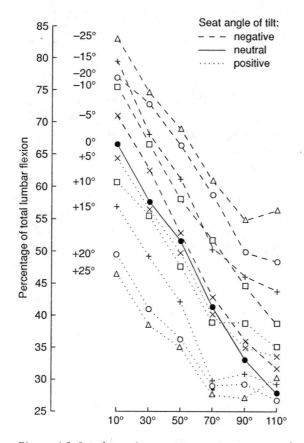

Figure 4.3 Lumbar spine posture as a percentage of maximum lumbar flexion for different hip and knee angles. (From Brunswic, 1984, with permission.)

Spinal problems in standing

Low back pain is common in standing workers and a number of authors have suggested reasons for this. In extended postures (e.g. when standing with a pronounced lumbar lordosis) the facet joints may begin to take on some of the compressive load. If the lumbar intervertebral discs are degenerated, the space between adjacent vertebrae decreases and the load on the facet joints increases even more. Adams and Hutton (1985) suggest that excessive facet joint loading may be a causal factor in the incidence of osteoarthritis. Bough *et al.* (1990) have shown that degeneration of the facet joints is a source of low back and sciatic pain.

Both Adams and Hutton and Yang and King (1984) suggest that excessive loading of the facet joints stresses the soft tissues around the joint and causes low back pain. For these reasons, *excessive lumbar lordosis should be avoided when standing.* Extrapolating from this, any workspace or task factors that require workers to arch the back more than they would normally do should be designed out.

Low back pain can also be caused by *muscular fatigue* if a standing person has to work with the trunk inclined forwards (when doing the washing-up or ironing, for example). In standing, the workspace must be designed to prevent workers from having to stand with excessive lumbar lordosis or having to adopt forward-flexed working positions.

Spinal problems in sitting

Sitting has a number of advantages over standing as a working position. Static, low-level activity of the soleus and tibialis anterior muscles is required in standing and these muscles can fatigue. Because the lower limbs drain blood against gravity, pooling may occur when someone stands still for long periods, causing swelling at the ankles. In extreme circumstances, reduced return of blood to the heart may cause a drop in blood pressure and the person may faint. The hydrostatic head that has to be overcome to return blood to the heart from the lower limbs is less in sitting than in standing, if the seat is correctly designed.

Despite the popular myth that occupational sitting is a risk factor for low back pain, the evidence suggests that sitting at work, in itself, is quite harmless (Hartvigsen *et al.*, 2000). However, prolonged sitting at work (more than 95% of the day) is associated with back pain (Hoggendoorn *et al.*, 2000). Further, many people wth bad backs gravitate towards sedentary work. Poor design of workstations may then exacerbate their problems. In a study at the Eastman Kodak Company in New York, 35% of sedentary workers visited the medical department with complaints of low back pain over a 10-year period. People with existing low back problems often cannot tolerate the sitting position for more than a few hours over the workday.

It is generally accepted that flexed sitting postures should be avoided because of the posterior protrusion of the discs caused by anterior wedging strain. Mandal (1981, 1991) has criticised the '90-degree' concept in seat design, stating that the slumped, flexed posture is inevitable if seats are designed in this way. Mandal has suggested that seats should be higher and should have a forward tilt to reduce the amount of hip flexion needed to use them.

An ergonomic approach to workstation design

Ergonomic workstation design encourages good posture. Figure 4.4 presents a framework for posture. It emphasises the role of three classes of variables (Table 4.3).

Clearly, 'ergonomically designed' furniture cannot be bought off the shelf. Decisions about the appropriateness and relative advantages of different designs can only

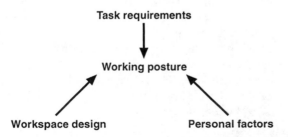

Figure 4.4 The postural triangle. A person's working posture is a result of the requirements of the task, the design of the workspace and personal characteristics such as body size and shape and eyesight. Consideration of all three components is needed in posture analysis and workspace design.

Table 4.3 Examples of factors that influence working posture

Factor	Example
1. User characteristics	Age
	Anthropometry
	Body weight
	Fitness
	Joint mobility (range of movement)
	Existing musculoskeletal problem
	Previous injury/surgery
	Eyesight
	Handedness
	Obesity
2. Task requirements	Visual requirements
	Manual requirements:
	positional
	forces
	Cycle times
	Rest periods
	Paced/unpaced work
3. The design of the workspace	Seat dimensions
	Work surface dimensions
	Seat design
	Workspace dimensions:
	headroom
	legroom
	footroom
	Privacy
	Illumination levels and quality

be made after considering the characteristics of users and the requirements of their jobs (Figure 4.5). Indeed, van Dieen *et al.* (2001) in a study of word processing, CAD work and reading found that trunk loading and trunk kinematics were more affected by the task carried out than the chair that was sat in.

Characteristics of users and workspace/equipment design

One of the most basic considerations in workstation design is the anthropometric fit between users and furniture (Figure 4.6). Designers typically design to ensure that 90% of users will be accommodated. Problems can therefore occur with extremely tall, short or obese individuals and special arrangements may need to be made to accommodate them. Much office furniture is designed around a desk height of approximately 73 cm and assumes provision of a height-adjustable chair. This ensures, within limits, that

1. Short users can raise their chair heights such that the desk height approximates their sitting elbow height.
2. The desk is not so high that the chair height exceeds the popliteal height of a short person in order to achieve (1).

Figure 4.5 Application of the postural triangle to workspace evaluation. The illustration depicts a monk transcribing a text. The slumped sitting posture is a result of the low seat, excessive task distance and the visually demanding task. The right elbow is resting on the thigh to improve the stabilisation of the writing hand. The left hand is steadying the book and the left elbow is resting on the worksurface to close the postural chain and reduce load on the left-hand side of the body. The left foot is resting on a footrest. How would you redesign this workspace to improve the posture?

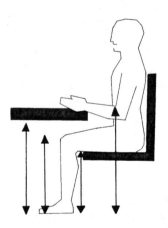

Figure 4.6 Critical user and furniture dimensions

3. The desk is high enough that tall users can approximate desk and elbow heights and still have space under the desk for their legs (i.e. they do not have to lower their chairs too much).

Immediately, it can be seen that with fixed desk heights, there are conflicting requirements in trying to accommodate both tall and short users.

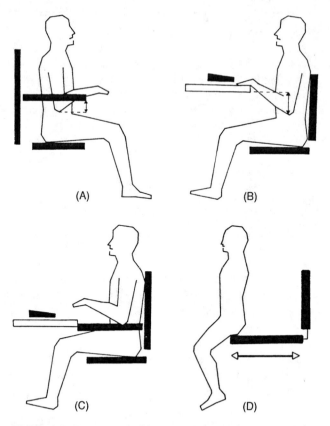

Figure 4.7 Common anthropometric mismatches in seated work: (A) seat too high, elbow rest too high; (B) desk too high (above elbow height); (C) task distance too great (elbow rest prevents access to desk); (D) seat too deep.

Health hazards and anthropometric mismatches

Anthropometric mismatches can have serious consequences for health and efficiency because of the way they increase the postural load on the body (Figure 4.7). Short users, for example, may have to raise *seat heights* beyond popliteal height in order to gain access to the desk. As a consequence, the feet no longer rest firmly on the floor and the floor cannot be used by the legs as a fulcrum for stabilising and shifting the weight of the upper body. The load of stabilisation is now transferred to the muscles of the trunk. The weight of the legs, instead of being borne by the feet, is borne by the underside of the thigh. This can restrict blood flow and is particularly undesirable for those with varicose veins. It can also cause a condition known as 'lipoatrophia semicircularis' (band-like circular depressions and isolated atrophy of subcutaneous fatty tissue on the thighs). Hermans *et al.* (1999) found postural differences between lipoatrophia semicircularis sufferers and non-sufferers – static sitting postures, less use of lumbar support, chair too high and higher pressures at the seat surface.

When seats are too high, the feet can no longer be used to extend the base of support beyond the base of the chair. This makes activities such as reaching and picking up heavy objects more hazardous. Continuous compensatory movements of body parts may be necessary to maintain stability. This hastens fatigue.

From the *Daily Mail*, Monday, 3 September 2001

An accountant stands to receive up to £250 000 in damages after her back condition was aggravated by spending hours on badly-designed office furniture . . . Shortly after her terminal was installed she began suffering back and neck pains, and was forced to take time off work . . . she claimed it was because her seat was too high and her feet did not reach the floor. The seat could not be lowered and her employers failed to provide her with a lower one. 'Subconsciously I was moving forward so that my feet reached the floor', she said, but that meant my back was not supported. When she returned to her office her workload had grown, denying her regular rest breaks.

A spokesman for the Manufacturing, Science and Finance Union predicted it would open the floodgates.

An alternative to raising the chair is for the short user to work with the elbows below desk height, increasing the static loading of the upper body, particularly the shoulder girdle as the elbows are held in an elevated position. Computer and type-writer keyboards increase the effective work surface height, hence the use of sliding keyboard drawers for programmers.

Tall users may find *desk heights* of approximately 70–75 cm too low since, even with the chair at its lowest level, the distance between the eyes and the work surface may be too great for comfortable viewing, causing the user to slump over the desk when writing.

Lack of footspace may also be a problem for tall workers. Mandal (1981, 1991) has called for an increase in desk heights to overcome these problems and to account for the increasing height of the population. Short users can be given *footrests as* standard equipment to ensure anthropometric fit. This is similar to the industrial practice of designing high benches to suit taller standing workers and providing raised platforms for those of lesser height.

Height-adjustable desks have been proposed (Ostberg *et al.*, 1984) for use with height-adjustable chairs in order to increase the range of users accommodated by a workstation. The correct way to adjust one of these workstations is as follows (Starr, 1983):

1.   First adjust the chair height so that the feet are resting firmly on the floor.
2.   Adjust the work surface height for comfortable access to the desktop or keyboard.

Backrest inclination has been found to influence trunk inclination in sitting, as might be expected. However, the location of the lumbar support with respect to the level of the lumbar spine (L1/L2 or L4/L5) does not seem to be of importance (Andersson, 1986). This suggests that a height-adjustable lumbar support is not necessary since the height of a fixed lumbar support can be optimised to contact the lumbar spines of a wide range of users (population differences in stature are largely due to differences in the lengths of the long bones rather than of the lumbar vertebrae).

Avoiding anthropometric mismatches

Anthropometric mismatches are pervasive. In a study of school furniture, Parcells *et al.* (1999) found that less than 20% of the students studied were fitted by the available chair/desk conmbinations; the chairs were typically too high or too deep and the desks were too high.

In the absence of anthropometric data, engineers, designers and facilities personnel can avoid anthropometric mismatches by means of a *participative approach*. A *consumer panel* of workers can work with the design team to evaluate furniture prior to purchase. The panel should consist of managers, dedicated VDU users and secretaries/clerical and other sedentary workers. Fitting trials can be carried to identify mismatches and specify the requirements for accessories such as footrests, document holders, etc.

The basic requirements for adequate workstation design are the anthropometric fit between workers and their furniture and the task. Ergonomically designed workplaces must be *flexible* if *postural fixity,* causing static loading of the musculoskeletal system, is to be avoided. Flexibility implies that the worker can carry out the task, at least some of the time, in more than one working posture with a workspace designed to accommodate both postures. A 'designed-in' resting posture is also desirable. Well-designed seats may enable the user to 'take the weight off of the feet' but will not take the weight off of the spine unless the user reclines against a tilted backrest.

Task requirements

All tasks have three sets of requirements that influence workspace design:

1. Visual requirements
2. Postural (effector) requirements
3. Temporal requirements

Visual requirements   The position of the head is a major determinant of the posture of the body and is very strongly influenced by the visual requirements of tasks. When the head is erect, the 'straight-ahead' line of sight, parallel to the ground, can be comfortably maintained. However, most people will not maintain the head in this position in order to look at objects on a horizontal work surface if this requires the eyes to be rotated downwards by more than 30 degrees. If the main visual area is more than 30 degrees below the 'straight ahead' line of sight, it is viewed by tilting the head forwards. The tilted posture places a static load on the neck muscles and displaces the COG of the body anterior to the lumbar spine, causing the characteristic forward 'slumped' posture (Figure 4.8 in which the backrest or lumbar support of the chair is no longer used.)

If objects are placed above the line of sight, the neck is extended to tilt the head backwards. Frequently used displays should not be above the standing or sitting eye height of a short worker. For VDU users carrying out editing tasks, document holders should be provided and placed orthogonal to the line of sight and adjacent to the VDU screen. Brand and Judd (1993) found that this configuration produced significant reductions in text editing time (of around 15%). According to Woodson (1981), the eye is sensitive to stimuli up to 95 degrees to the left and right, assuming

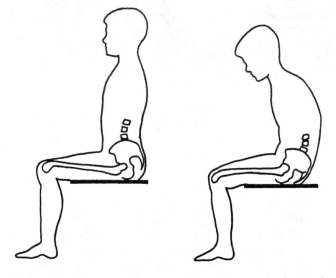

Figure 4.8 Most people are unable to sit erect in the 90 degree posture for long periods and soon adopt a slumped posture.

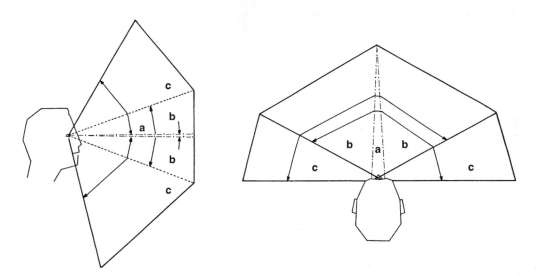

Figure 4.9 Regions of high (a), medium (b) and low (c) resolution in the visual field. a and b are preferred regions for placement of objects. Objects should not, however, be placed above the straight-ahead line of sight, particularly at fixed workstations.

binocular vision. The optimum position for placing objects is 15 degrees either side of the straight-ahead line of sight (Figure 4.9).

These limits describe a visual field in which objects may be placed such that they can be viewed without moving the head from its comfortable erect position. Thus, static loading of neck muscles and other soft tissues in the neck can be avoided if the visual component of the task is kept within a cone from 'straight-ahead' line of sight to 30 degrees below and 15 degrees to the left and right.

The visual requirements of tasks are of major importance in determining postural adaptation. Well-designed furniture can sometimes provide better support for the postures that result from task requirements but only if its design or selection is based on an understanding of what these postures are going to be like.

If the visual requirements exceed worker abilities, poor postures may well result despite careful attention to the furniture design.

Postural requirements The position of the hands, arms and feet is another major determinant of posture and postural load. The working area of a desk or bench should always be close to its front edge and with relatively unobstructed access. In vehicle design, the comfort of the driver's seat depends not just on its particular dimensions but on its positioning in relation to the footpedals and manual controls.

For both standing and sedentary workers, it has been demonstrated that reduction of the magnitude of required forces and redesign of jobs to avoid excessive reach is of value in the prevention of back problems (Rivas *et al.*, 1984). Tasks should be designed to be performable in the zone of optimum reach.

Temporal requirements The temporal requirements of tasks are a major consideration in the design of workspaces and exert a moderating influence on the effects of other factors. For example, in a multiuser computer workstation in which individuals spend only 15–20 minutes at a time interrogating the system, a sit-stand workstation using a high bench and stool might be appropriate. At the other extreme, certain data entry jobs require users to carry out a repetitive task over the whole day. These jobs impose a high degree of postural constraint as the position of the hands and head are fixed by that of the keyboard, documents and screen. Such highly constrained jobs are associated with increased prevalence of health hazards in VDU work (Eason, 1986). It is particularly important when analysing physically constraining jobs to characterise task components in terms of frequency and importance of operation and to ensure that the workspace is arranged optimally for high-ranking elements.

The 'constrained/unconstrained' distinction may be a useful aid in designing or evaluating workstations in an organisation. Jobs may be categorised in terms of the degree of postural constraint they impose on the employee. A manager may be relatively unconstrained, free to arrange the layout of items on the desk, to leave the desk or office at will and to switch from one task to another. Variety may be built into the job (attending meetings elsewhere, for example). A secretary may be relatively constrained inasmuch as work with typewriters and VDUs will be common but there will also be other tasks such as answering the phone or receiving visitors. Workers such as data entry clerks will often be highly constrained.

Jobs falling into the 'highly constrained' category require maximum flexibility to be built into the workspace (in terms of adjustable furniture, etc.) to compensate for the lack of flexibility in the design of the job. In addition, current thinking about VDU work now insists that all VDU jobs must be designed with frequent natural rest breaks (by providing alternative tasks).

Design for standing workers

As a rule of thumb, all objects that are to be used by standing workers should be placed between hip and shoulder height to minimise postural stress caused by

Table 4.4 Recommended worksurface heights (cm) for standing workers[a]

Task requirements	Male	Female
Precision work	109–119	103–113
Light assembly work	99–109	87–98
Heavy work	85–101	78–94

[a] From Ayoub (1973).

stooping or working with the hands and arms elevated. Work surface heights should approximate the standing elbow height of workers, depending on the task – when carrying out fine work, a higher work surface is appropriate to reduce the visual distance and allow the worker to stabilise the forearms by resting them on the work surface. When carrying out heavy work, a lower work surface is needed to permit the worker to apply large vertical forces by transmitting part of the body weight through the arms. Some recommended worksurface heights are given in Table 4.4. (from Ayoub, 1973). These should only be taken as a guide because the actual working height depends on the size of the work objects and the user population.

Some workspace design faults that increase postural stress in standing workers can be summarised as follows:

1. Working with the hands too high and/or too far away: compensatory lumbar lordosis.
2. Work surface too low: trunk flexion and back muscle strain.
3. Constrained foot position due to lack of clearance: worker stands too far away.
4. Working at the corner of the bench: constrained foot position, toes turned out too much.
5. Standing with a twisted spine having to work at the side rather than directly ahead.

Postural constraint in standing workers can be relieved by providing stools to enable workers to rest during quiet periods or to alternate between sitting and standing. Adequate space for the feet should be provided to permit workers to change the position of their feet at will.

Evaluation of standing aids

Several researchers have investigated the effects of standing aids on comfort and fatigue in standing.

Footrests and footrails   Bridger and Orkin (1992) determined the effect of a foot-rest on pelvic angle in standing. The footrest raised the resting foot 250 mm above the level of the floor and resulted in a net posterior rotation of the pelvis of 4–6 degrees. Whistance *et al.* (1995) confirmed this finding: use of a footrail reduced anterior pelvic tilt, straightened the supporting leg and increased the plantar flexion of the supporting foot. The footrest would appear to be a valid way of reducing lumbo-pelvic constraint in standing workers and help prevent discomfort in the lumbo-pelvic region. Rys and Konz (1994) have reviewed the ergonomics of standing. A 100 mm foot platform used by subjects was perceived as more comfortable than normal standing in 9 of 12 body regions, including the neck. Use of either a flat or a 15-degree-tilted platform was percieved to be better than use of a simple footrail, but all standing aids were preferred to standing on a bare floor. During a 2-hour period of standing, subjects placed one foot on the platform 83% of the time, switching their foot from the platform to the floor every 90 seconds on average. *The freedom to stand with one foot forward and elevated seems to be an important feature of a well-designed standing workplace.*

'Anti-fatigue mats' Stuart-Buttle *et al.* (1993) report that prolonged standing causes significant localised leg muscle fatigue, particularly in the gastrocnemius muscles. Footrests seem to relieve some of the load on the resting leg. Mats do not seem to reduce lower leg fatigue although they do reduce discomfort in the lower leg, feet and back (Rys and Konz, 1994) and muscle fatigue in the erector spinae muscles. Resilient rubber mats, being slightly unstable, may stimulate postural muscle activity in the lower legs and activate the venous pump. There are many reasons why standing workers should not have to stand on hard and sometimes cold concrete floors to work. Mats or carpets, or wooden, rubber or plastic platforms provide a more yielding surface and better insulation. They may also offer better friction and therefore aid postural stability and help prevent accidents. In a comparison of three different mats, Konz *et al.* (1990) found that a mat with 5.8% compression was perceived as more comfortable than mats of 7.4% and 18.6% compression. All were preferred over concrete. Further information can be found in Kim *et al.* (1994).

Compression stockings Krijnen *et al.* (1997) evaluated the effects of rubber floor mats and compression stockings on the leg volume of standing workers suffering from *chronic venous insufficiency*. Although there was some evidence for a reduction in complaints of pain and tiredness in the legs among those using rubber floor mats, leg swelling over the workday did not differ from a control group. The wearing of compression stockings bought about a significant reduction in leg swelling and in complaints (from 70% complaining at the beginning of the trial to 27% after 3 months). The wearing of the stockings was found to be acceptable to the male workers in the study.

Toespace Panels or obstructions in front of benches cause users to stand farther away from the worksurface. The postural adaptation is for people to bend forwards. Whistance *et al.* (1995) found that this was achieved by a combination of pelvic tilting and lumbar flexion, placing more stress on the spine. Fox and Jones (1967) observed that, after having to lean forwards to work for many years, dentists did so by arching the back in the thoracic region or by flexing the lumbar spine, using the lumbar spine as a false joint and the pelvis as if it were part of the legs. Toespace (Figure 4.10a) can prevent this from happening (Whistance *et al.*, 1995; DeLaura and Konz, 1990). Figure 4.10b shows a production line designed to provide standing workers with a choice of comfortable standing and sit-standing work positions.

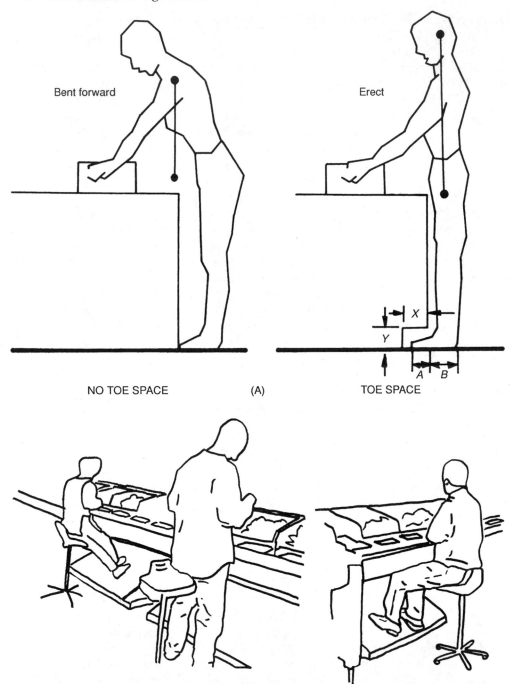

Figure 4.10 (A) Toespace is needed if standing workers are to stand up straight. (B) Sit–stand industrial production line. Note the use of height adjustable sit–stand stools and a large sloping footrest to accommodate a variety of trunk–thigh angles.

Table 4.5 Recommended worksurface heights (cm) for sedentary workers[a]

Task requirements	Male	Female
Fine work	99–105	89–95
Precision work	89–94	82–87
Writing	74–78	70–75
Coarse or medium work	69–72	66–70

[a] From Ayoub (1973).

Design for seated workers

Some recommended worksurface heights for sedentary work are given in Table 4.5. (from Ayoub, 1973). When working with keyboards, the worksurface is often 3–6 cm lower than a writing worksurface to allow for the thickness of the keyboard. In addition, space must be provided for the sitter's legs. An open area approximately 50 cm wide and 65 cm deep (measured from the sitter's ischial tuberosities) is suggested by Ayoub.

Office chairs

Some important features of office chairs are summarised in Table 4.6.

Table 4.6 Key features of chair design

1. Seats should swivel and have heights adjustable between 38 and 54 cm. Footrests should be provided for short users.
2. Free space for the legs must be provided both underneath the seat to allow the user to flex the knees by 90 degrees or more and underneath the work surface to allow knee extension when reclining.
3. A 5-point base is recommended for stability if the chair has castors.
4. The function of the backrest is to stabilise the trunk. A backrest height of approximately 50 cm above the seat is required to provide both lumbar and partial thoracic support.
5. If the backrest reclines, it should do so independently of the seat to provide trunk–thigh angle variation and consequent variation in the distribution of forces acting on the lumbar–pelvic region.
6. Lumbar support can be achieved either by using extra cushioning to form a lumbar pad, or by contouring the backrest. In either case, there must be open space between the lumbar support and the seat pan vertically below it to allow for posterior protrusion of the buttocks.
7. The seat pan must have a slight hollow in the buttock area to prevent the user's pelvis from sliding forwards. This keeps the lower back in contact with the backrest when reclining. The leading edge of the seat should curl downwards to reduce underthigh pressure.
8. Arm rests should be high enough to support the forearms when the user is sitting erect. They should also end well short of the leading edge of the seat so as not to contact the front edge of the desk. If the armrests support the weight of the arms, less load is placed on the lumbar spine.
9. Modern chairs tend to have a thin layer of high-density padding. Layers of thick foam tend to destabilise the sitter. The foam can collapse after constant use.
10. Cloth upholstery provides friction to enhance the stability of the sitter.

A number of researchers have suggested that chairs should be designed with forward-tilted seats (Figure 4.13). These chairs should permit a user to sit with an erect trunk and less posterior pelvic tilting and flattening of the lumbar curve because the tilt of the seat increases the trunk–thigh angle. Comparisons of lumbar angles of people sitting on conventional and forward sloping chairs indicate that this is the case (Mandal, 1981; Bendix and Beiring-Sorensen, 1983; Brunswic, 1984; Frey and Tecklin, 1986; Bridger, 1988). Drury and Francher (1985) found that many users found a particular implementation of this seating concept uncomfortable, probably owing to the lack of postural variation permitted by the particular design. A similar principle has also been incorporated into a form of high stool for use in industry (Gregg and Corlett, 1988). Few extensive field trials of these 'alternative' forms of seating have been reported in the literature, however.

It is sometimes stated in the literature that the lumbar supports of conventional chairs can 'preserve' the lumbar lordosis in sitting (Figure 4.11). A lumbar support acts like a brace that uses the lumbar spine as a lever to tilt the sitter's pelvis forward. Although lumbar supports stabilise the pelvis and reduce posterior pelvic tilting, the amount of lumbar lordosis that is 'preserved' is modest, similar to that of a person sitting in a neutral position such as kneeling (Bridger *et al.*, 1992).

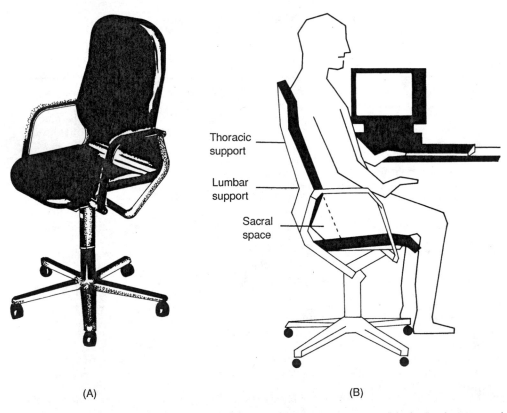

Thoracic
support

Lumbar
support

Sacral
space

(A)   (B)

Figure 4.11 Modern office chair with lumbar and thoracic support suitable for both VDU and non-VDU office work.

Dynamic sitting

Branton (1969) used the body-link concept described in a previous chapter to evaluate the comfort of train seats by observing sitting behaviour. An open-chain system of body links can behave in unpredictable ways when subject to internal or external forces. The prime function of a seat is to support body mass against the forces of gravity. A second function, which was emphasised by Branton, was to stabilise the open-chain system. In the absence of external stabilisation, tonic muscle activity is required, leading to discomfort if sustained. Behaviours such as folding the arms and crossing the legs can be seen as postural strategies to turn open chains into approximate closed chains stabilised by friction. The comfort of a seat depends, in a dynamic sense, on the extent to which it permits muscular relaxation while stabilising the open-chain system of body links.

Bendix *et al.* (1985, 1986) evaluated an office chair with a seat that could tilt from 5 degrees backwards to 10 degrees forwards over a tranverse axis. The purpose of the seat was to prevent postural fixity by providing users with the ability to 'rock' backwards and forwards at will. Subject were found to prefer the tiltable seat and to use the flexibility provided to change posture – tilts exceeding 2 degrees were observed each minute, on average with up to 10 smaller tilts each minute. The seat was moved more frequently and with a greater range when the seat height was 6 cm above the subjects' popliteal height compared to 1 cm below. Foot swelling was found to be greater with the higher seats, whether tiltable or fixed forward sloping, but there were no effects of seat type on lumbar muscle activity. However, van Dieen *et al.* (2001) found greater stature gains (indicative of lower trunk loading) when subjects sat on dynamics chairs compared to a chair with a fixed seat and backrest. Van Deursen *et al.* (1999) designed a dynamic office chair in which rotations about an axis perpendicular to the seat were applied by a motor at an amplitude of 0.6 degrees and a frequency of 0.08 Hz. The application of these gentle twisting motions was found to result in increases in spinal length over a 1-hour period of sitting – significantly more than when subjects sat in a static control chair.

These findings support the hypothesis that rotation applied to the vertebrae during sitting reduces pressure in the nucleus pulposus, allows fluid to enter, increasing disc thickness, and improves the nutritional status of the disc.

When provided with additional workstation flexibility in the form of a tilting seat, subjects do use it and appear to prefer it to similar chairs with fixed seats. Interestingly, the tilt facility was used less frequently and the muscular load was higher in a typing task compared to the subjects' ordinary desk work not involving typing. This indicates that tasks involving typing are constraining and should be combined with other work.

Udo *et al.* (1999) compared a tilting seat with a fixed seat during the performance of a word processing task. The tilting frequencies were much lower than in Bendix's experiments (25 per hour) but a reduction in back discomfort was reported when the tilting seat was used. Greater low back EMG activity was recorded with the tilting seat, suggesting that the reduction in pain was bought about by the reduction of static back muscle activity. As with Bendix's experiments, no differences in lower leg swelling were found. However, Stranden (2000) did find that tilting seats produced a significant reduction in leg swelling (note that the difference was small, of the order 1% of calf volume, which may explain why previous researchers found no difference).

Figure 4.12 Modern school furniture based on ergonomic principles. Note the high desk that slopes towards the user and the seat designed to permit a trunk–thigh angle greater than 90 degrees. (Photograph courtesy of Dr A.C. Manda.)

According to Stranden, the key to venous pump activation is leg movement. Plantar flexion, in particular, is immediately effective and can be elicited by reclining on a tilting seat or by resting the feet on dynamic footrests that are pivoted to allow a 'treadle pumping' kind of action.

Worksurface design

Workstations can be further improved by considering various aspects of desk and bench design. Some important considerations are the provision of tilt in the work surface and/or of document holders and the provision of free space in the working area. Zacharkow (1988) provides many interesting illustrations of Victorian school desks having a 15 degree slope for writing and an integral book holder for reading, angled at 45 degrees to the vertical. Mandal (1981) suggested that chairs with forward-tilting seats be used with desktops that tilt towards the user by about 15 degrees to lessen the visual angle and encourage a more upright posture of the trunk (Figure 4.12).

Several studies have indicated that tilted desktops (of 15 or even 10 degrees) do reduce the trunk and neck flexion of seated persons engaged in reading and writing (Bridger, 1988; de Wall *et al.*, 1991) and thus reduce the load on the corresponding parts of the spine (Figure 4.13). Significant effects have been found on subjects seated on both conventional and forward sloping seats. Porter *et al.* (1992) have reported similar benefits with a sloping computer desk. Burgess and Neal (1989) found that using a document holder when writing on a flat desk significantly reduced the moment of flexion of the head and neck at the C7–T1 level of the spine and was rated by the subjects as more comfortable than not using one. The use of the horizontal desktop for reading and writing would appear to be a rather new and rather inferior development in the history of furniture design.

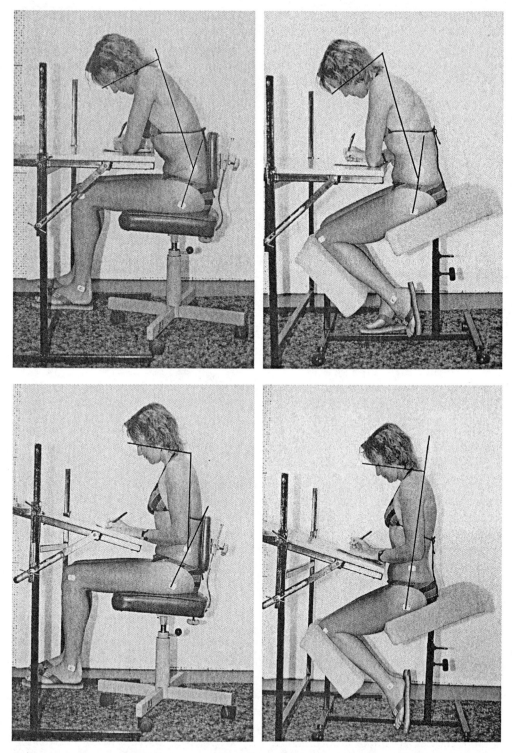

Figure 4.13 Tilted worksurfaces can be effective when used with both conventional and alternative seats.

For both standing and sedentary workers, worksurfaces should be arranged so that the worker does not have to work continually with objects placed at one side or reach excessively to the side. The main working area should be directly in front of the worker's body to minimise any twisting of the trunk when carrying out task-related movements.

Placement of work objects

Pearcy (1993) has shown that the twisting mobility of the human back is increased in sitting compared with standing in either an upright or a forward-flexed position. The increase occurs because the morphology of the lumbar facet joints permits more axial rotation of the superior vertebral body over the inferior body when the spine is flexed. The increase is considerable (38% more twisting in 90 degree sitting and 44% more twisting in long sitting over the whole spine or about 2 degrees more at each intervertebral joint). Because, the posterior fibres of the annulus are already stressed when the spine is flexed, the additional stress of twisting may result in very high annular stresses, predisposing the fibres to rupture. Jobs involving the asymmetric handling of loads from a seated position (such as supermarket checkout personnel) would seem to be particularly hazardous. Seated workers who have to resist sudden, external twisting forces (such as catching an object falling from a supermarket conveyor) have a high risk of injury according to Pearcy. The findings provide further support for the placement of work tasks and principal work objects within the zone of convenient reach, discussed in the previous chapter. For work involving keyboards, there is evidence that shoulder muscle load can best be reduced by placing the keyboard as low as possible (Bendix and Jessen, 1986) or by using a negative-tilt keyboard (with the keys tilting away from the user).

Visual display units

The design of the workspace for the visual display unit/terminal (VDU) operator has received much attention in recent years (e.g. Grandjean, 1987) because of the importance of attaining a proper relationship between user and workspace to support the requirements for interactive computer work. Figure 4.14 summarises the important considerations in the design of the VDU workspace.

Standards for the design of these workspaces have been drafted or proposed (e.g. ANSI Standard No. 100-1988; BS 7179, 1990, BS 5940, 1980; BS 3044, 1990; Health and Safety Commission, 1992). The guidelines proposed in these documents attempt to specify user space requirements and to define appropriate furniture in functional as well as physical terms. Readers are referred to these documents for detailed information.

VDUs support the activities of many different occupations and professions and there is no single, correct workspace arrangement to satisfy the requirements of all workers. Therefore, detachable keyboards and screens with adjustable tilt and swivel are recommended. This permits a copy typist, for example to work with the screen to one side and the document directly ahead and supported in the same visual plane as the screen by means of a document holder. The screen is the main work area for a programmer and thus should occupy the main visual area of the programmer's desk.

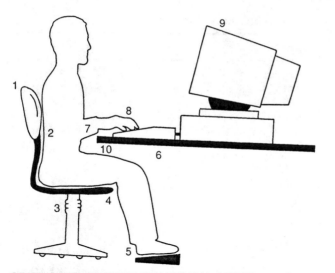

SEATING AND POSTURE FOR TYPICAL OFFICE TASKS
1 SEAT BACK ADJUSTABILTY
2 GOOD LUMBAR SUPPORT
3 SEAT HEIGHT ADJUSTABILITY
4 NO EXCESS PRESSURE ON UNDERSIDE OF THIGHS AND BACKS OF KNEES
5 FOOT SUPPORT IF NEEDED
6 SPACE FOR POSTURAL CHANGE, NO OBSTACLES UNDER DESK
7 FOREARMS APPROXIMATELY HORIZONTAL
8 MINIMAL EXTENSION, FLEXION OR DEVIATION OF WRISTS
9 SCREEN HEIGHT AND ANGLE SHOULD ALLOW COMFORTABLE HEAD POSITION
10 SPACE IN FRONT OF KEYBOARD TO SUPPORT HANDS/WRISTS DURING PAUSES IN KEYING

Figure 4.14 Essential furniture considerations in VDU work. (From Display Screen Equipment – Work Guidance on Regulations, © HMSO, 1992).

Table 4.7 summarises some design solutions for increasing the flexibility of VDU workstations. A participatory approach is often the most effective way to decide which product features and accessories are likely to have the greatest cost-benefits for different groups of workers in an organisation. For example, document holders are often of great benefit to copy typists and data-entry clerks, whereas systems designers may regard increased storage space and shelving as a higher priority. Table 4.8 presents sample workspace dimensions for US workers (from ANSI/HFS 100-1988).

Delleman and Berndsen (2002) found that gaze inclination in VDU work is independent of sitting posture for backrest inclinations of 0–15 degrees recline. A gaze angle of 6–9 degrees below the horizontal appears to be optimal for minimising postural stress.

Guidelines for the design of static work

Regional musculoskeletal pain is associated with non-neutral static work postures (Hunting *et al.*, 1981). A neutral posture of the hands, arms, neck, etc., is simply a posture that one adopts naturally when the muscles are relaxed. Figure 4.15a shows the joint angles of data entry workers observed by Hunting *et al.* and the bony landmarks used to define the angles.

Table 4.7 Product features and accessories for enhancing VDU workstation flexibility

Item	Accessory
Source documents	Document holders
	Bookstands
	Tilted area of work surface
	Shelving
	Noticeboard
VDU screen	Movable screens
	Tilt and swivel adjustment
	Screenholders
Keyboard	Detached from VDU
	Keyboard drawer
Seating	Footrest
	Arm rest
	Lumbar pad
	Narrow backrest (permits trunk lateral flexion and rotation)
	Lumbar and thoracic support
	Recline mechanism
Desks	Height adjustable
	Tilt adjustable
	Extensions can be fitted

Table 4.8 VDU workspace dimensions (cm) for US users

	Male 95th percentile	*Female* 5th percentile
Seat height[a]	49	41
Elbow rest height[a]	29	18
Worksurface height[a]	71	58
Screen height[a]	Adjust screen so that top of screen is no higher than horizontal sight line When reclining by 15 degrees	
Eye height (90 degree sitting)	130	103

[a] From ANSI/HFS 100-1988.

Figure 4.15(a, b and c) shows the prevalence of musculoskelatal pain in relation to the static work posture. Shoulder stiffness was related to increased elbow angle. Neck pain and stiffness were more prevalent in groups with neck angles greater than 55 degrees, and hand pains, stiffness and cramps were increased in prevalence at angles of ulnar deviation greater than 10 degrees. The reader is referred to ISO 1226, which gives maximum joint angles for static work, although even fairly minor deviations from neutral increase the prevalence of discomfort. Work postures should be as close to neutral as possible.

Trunk inclination

ISO 1226 gives time limits for static trunk inclinations (Figure 4.16). Static trunk inclinations greater than 60 degrees are not permitted at all under ISO guidelines,

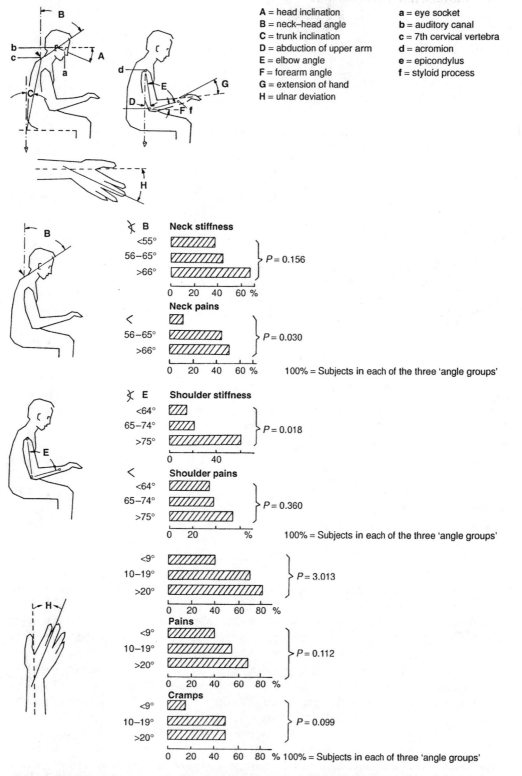

A = head inclination
B = neck–head angle
C = trunk inclination
D = abduction of upper arm
E = elbow angle
F = forearm angle
G = extension of hand
H = ulnar deviation

a = eye socket
b = auditory canal
c = 7th cervical vertebra
d = acromion
e = epicondylus
f = styloid process

B   Neck stiffness
<55°
56–65°
>66°

} P = 0.156

0   20   40   60 %

Neck pains
<
56–65°
>66°

} P = 0.030

0   20   40   60 %   100% = Subjects in each of the three 'angle groups'

E   Shoulder stiffness
<64°
65–74°
>75°

} P = 0.018

0   40

Shoulder pains
<
<64°
65–74°
>75°

} P = 0.360

0   20   %   100% = Subjects in each of the three 'angle groups'

<9°
10–19°
>20°

} P = 3.013

0   20   40   60   80 %

Pains
<9°
10–19°
>20°

} P = 0.112

0   20   40   60   80 %

Cramps
<9°
10–19°
>20°

} P = 0.099

0   20   40   60   80 % 100% = Subjects in each of three 'angle groups'

Figure 4.15 Postural angles associated with joint pain. (Redrawn from Hunting *et al.*, 1980.)

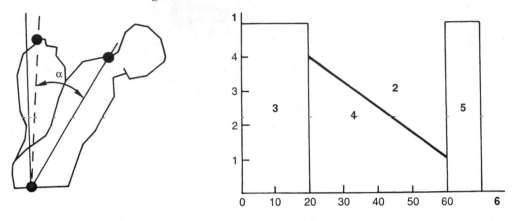

Figure 4.16 Time limits for static forward flexion. (Redrawn from ISO 11226, 2000.)

neither are negative inclinations (leaning backwards) permitted without back support. Angles from 0 to 20 degrees are acceptable for 5 minutes and angles from 20 degrees to 60 degrees can be held from 4 to 1 minutes (sloping line in Figure 4.16).

Effectiveness and cost-effectiveness

Improvement of workstations for computer users

Dressel and Francis (1987) evaluated the effects of systems furniture on the performance and satisfaction of US office workers around the time that desktop microcomputers were introduced. Systems furniture is specifically designed to support work with computers and contains the ergonomic features described in this chapter. The researchers wished to determine whether the furniture would cost-justify itself through improved work performance and satisfaction. A maximum payback period of 8 years was needed.

A group of 198 office workers, ranging from managers to secretaries and clerks, was divided into three groups. The experimental group received systems furniture, a second group received additional conventional furniture, and a third served as a control group. Subjects completed a questionnaire to indicate the types of activities they performed and the percentage of time spent on each task. Workstation designs and layouts were developed on the basis of the questionnaire results and of interviews. The main requirements in the new workstations were for more storage capacity and more privacy. Task lighting and partitions were also introduced. Productivity data 11 months before and 11 months after the introduction of new furniture were collected for all groups. Data were also collected on the level of satisfaction with the office furniture, layout, environment and the building and its features.

Systems furniture users exhibited a 20.6% improvement in productivity after the new furniture was installed compared to a 4% improvement in the group receiving additional conventional furniture (the improvement for this group was not statistically significant). The control group showed a negligible increase of 0.6%. However, both systems furniture and additional furniture users showed an increase in satisfaction with their workstations that persisted for 9 months after the change. Thus, it cannot be assumed that *ergonomic improvements that meet with the approval of users will automatically cost-justify themselves in increased productivity*. Claims for improved performance *need to be demonstrated directly* using appropriate data. The control group exhibited no change in satisfaction over the study period.

Cost-benefit and payback analysis   The systems furniture was more space efficient than the conventional furniture. At the time of the study, office space was leased at 90$/sq ft/yr resulting in saving of US$ 18 245 in 8 years. The increased productivity would pay for the furniture in 11.5 months with a mean 1984 salary of US$ 19 735 since the cost of one workstation was US$ 3894. Taking better space utilisation into consideration, the payback period dropped to 10.6 months.

Improvement of work conditions of data entry clerks

Ong (1990) reports a study carried out on data entry operators who spent more than 85% of the workday at a computer terminal and had more health complaints than other office workers. In particular, complaints of musculoskeletal pain and fatigue were common. Observational and questionnaire data revealed that 60% of the (all female) operators complained that their chairs were uncomfortable, too hard and unsteady. Most did not lean against the adjustable backrests provided. Lighting levels were low (65–250 lux) and the walls were of low reflectance. There was evidence that fatigue built up over the working day. Output was highest in the morning, dropping by 10% in the afternoon and further when overtime was worked. The investigators upgraded the entire working environment, providing new chairs and desks, footrests and document holders. Lighting levels were increased to 450–500 lux and the ceiling walls and floor were rendered with new materials of 100%, 75% and 30% reflectance (see Chapter 10 for more information on lighting). Rest periods, throughout the workday, were increased from 10–15% of total hours to 23%.

As a result of these interventions, the prevalence of musculoskeletal pain and visual fatigue fell and the mean data entry rate increased from 9480 keystrokes/hour in the 12 months prior to the intervention to approximately 13 000/hour in the 12 months immediately after the intervention. The increase in keystroke rate was accompanied by a decrease in keying errors. Following this, 6 years after the ergonomic improvements the productivity gains were still evident with a mean keystroke rate of 11 300 keystrokes/hour and a lower rate of errors.

Training programmes for VDU ergonomics

The introduction of the display screen equipment regulations in Europe has resulted in the provision of training programmes for occupational users of VDUs. These programmes are based on the material presented in the chapter and typically have a knowledge-based component that informs users of the need for good workstation ergonomics and a skill-based component that trains users how to set up their furniture and workstation to obtain a good posture and viewing angle. Training in the use and selection of accessories is frequently provided. The intention is to reduce postural and other stresses associated with VDU work by imparting self-help skills into daily working life.

There is evidence that these programmes work. Brisson *et al.* (1999) evaluated the effects of an ergonomic programme on the workstation layout and the prevalence of musculoskeletal pain in a group of VDU users. A training group was compared to a control group using several criteria. Direct assessment of workstations, a self-administered questionnaire and a physical examination were carried out 2 weeks before the training and 16 months afterwards in both groups. The prevalence of workplace stress decreased significantly in the experimental group after the training. The questionnaire data indicated that musculoskeletal disorder prevalence decreased from 29% to 13% among workers less than 40 years of age. The physical examination data indicated a decrease from 19% to 3%. This study provides further evidence that postural stressors in the workplace cause a higher prevalence of musculoskeletal disorders and that VDU training enables employees to implement their own ergonomic interventions that reduce the prevalence. The effects are greatest in workers under 40 years of age.

Workstation design and viewing angles

Bhatnager *et al.* (1982) investigated performance of a circuitboard inspection task. Circuitboards were projected using 35-mm slides onto a screen. The subjects sat on height-adjustable chairs at height-adjustable desks. The screen was 153 cm in front of the subjects and centred at one of three heights, 0.61 m, 1.53 m or 2.44 m. These different heights gave sight line angles of 23 degrees down (low), 10 degrees up (medium) and 38 degrees up (high).

The results showed that the screen height affected posture. The least comfortable height was with the screen in the high position. The most comfortable was in the low position. Performance (the percentage of errors detected) was best in the low position and worst in the high position. Errors increased from about 4% to 6%. The work sessions lasted 3 hours and during this time errors increased from 2% to 8% overall. Thus, poor posture caused by poor workspace design and inappropriate visual requirements seems to have an effect similar to fatigue on work performance. Alternatively, subjects fatigue more quickly.

Whistance (1996) carried out an investigation of standing VDU work. One of the interesting findings was that screen height did affect the accuracy of a visual checking task. In particular, it was found that accuracy was best when subjects worked with the line of sight approximately 15 degrees below horizontal. Comfort was also greatest at this level and the best postures were achieved.

It seems that poor workspace design with visual targets placed incorrectly in the visual field can affect posture and comfort. It can also degrade the performance of visual checking tasks.

Research directions

About 75% of workers in developed countries are sedentary and the basic requirements for the design of seated workspaces are well understood. However, what is less well understood is the contribution of a sedentary lifestyle to the prevalence of spinal disease and how sedentary jobs should be modified to reduce any risks. Although clear recommendations can be made, more evidence of an epidemiological nature is needed.

A number of alternative seat designs have appeared over the years and new concepts are appearing all the time. Well-designed prospective studies are needed to demonstrate the effectiveness of workstation ergonomics in safeguarding the musculoskeletal health of the workforce.

Summary

The reduction of postural stress is fundamental to ergonomic workstation design. A multifaceted approach is needed to arrive at appropriate workstation designs for different workers. The requirements of tasks and the characteristics of users need to be considered in relation to the options for workstation design. It is not appropriate to talk about an 'ergonomically' designed chair in isolation since the appropriateness of any design depends on its use and the other equipment used to carry out a task. When designing or selecting office and industrial furniture the workstation/task must be the basic level of analysis.

Table 4.9 summarises general considerations for the design of workstations for standing and sitting.

Criteria for the evaluation of static work can be found in ISO 11226 and ISO 9241: Ergonomic requirements for office work with visual display terminals.

Table 4.9 General guidelines for workstation design

- Provide clearance under desks and benches so that foot and knee position are not constrained. Do not use the area around the bench or the desk as storage space.
- The feet must never be confined to a small area as this degrades balance and shifts the load of postural adaptation to vulnerable structures higher up the kinetic chain.
- For every task, find an optimal visual and manual distance that will minimise forward flexion of the trunk and flexion of the neck.
- Offer pregnant workers seats with a forward slope of 10 degrees.[a]
- Provide footrails for standing workers and footrests for seated workers.
- Enrich or enlarge the job to increase postural variety. Differentiate new tasks by introducing configurations demanding alternate postures.
- Activate the venous muscle pump. Require walking, wherever possible; 5 minutes per hour may be optimal.
- Design sit-stand workstations. Allow standing workers to sit and work for 50% of the day.
- If multiuser worksurfaces are of fixed height, provide height-adjustable platforms for standing workers. For seated workers, choose a height that will fit taller workers and provide footrests for shorter workers. The height of floors is always adjustable in the upward direction.
- Alternating between asymmetric postures is a strategy for minimising static loading. Design workstations and jobs to encourage this strategy.
- Redesign jobs if head inclination is[b]

   >85 degrees
   <0 degress without full head support

   Redesign jobs if neck flexion angle is[b]

   >25 degrees
   <0 degrees

[a] Lee *et al.* (1999).
[b] See ISO 11226 for further information.

Essays and exercises

1. Carry out a survey of the furniture in a modern office. Measure furniture dimensions and relate them to user anthropometry, commenting upon any anthropometric mismatches you observe. Find out whether the furniture is adjustable and whether users know how to adjust it to optimise their working posture.

   Examine and classify the furniture and accessories used by different categories of staff and different levels of management. Are the different types of furniture used appropriate for the different users and their task requirements? Is new furniture needed? Are adequate accessories provided?

2. (a) Use a protractor to measure the trunk–thigh, knee, trunk flexion and pelvic tilt angles in Figure 4.1. Plot the results on a graph to illustrate how the postural angles change as the trunk–thigh angle changes.

   (b) Analyse the posture of the monk in Figure 4.5 using the framework in Table 4.3. Describe the main areas of postural stress in terms of static postural and task workload. Assess the trunk posture using the criteria of ISO 1226.

3. You are the Facilities Manager of a major financial institution. You need to spend $50 million to upgrade the company's head and regional offices worldwide. Write a report to top management explaining what constitutes good office furniture and why it is essential in the automated office.

4. Obtain a copy of the display screen equipment regulations. Carry out an assessment of an office of your choice and determine whether any improvements are necessary.

   (i)   Having identified any improvements, determine the cost of implementing them by obtaining quotations from local suppliers.

   (ii) Obtain feedback from employees on any problems in the work environment that influence their productivity.

   (iii) Obtain information from management on absenteeism and any other work losses.

   (iv) Make a case for the cost-effectiveness of implementing the changes.

5. Use the material in this book to prepare a VDU ergonomics training programme for office workers. Assume that your students have no knowledge of ergonomics or of the human sciences and are completely unaware of any of the issues. (Hint: have several sections beginning with the basic concepts, followed by self-help skills and some practical demonstrations.)

5   Design of repetitive tasks

Doctors can only diagnose diseases that they know . . . This is why new diseases, after their first description, are seen everywhere – Aids, for example. If the physician is unaware that a certain constellation of symptoms and signs indicates a particular disease, there is no possibility of a diagnosis being reached.

(Skrabanek and McCormick, 1990)

Most repetitive tasks require a combination of both static and rhythmic muscle activity. In manual work, postural stabilisation of the hands and arms is essential for carrying out all but the grossest movements in a purposeful way. This stabilisation is provided by muscles farther up the kinetic chain, muscles that cross the elbow and shoulder joints and have their origins in the cervical spine and thoracic regions. If task demands are excessive, pain may be experienced in the muscles providing the stabilisation or in the muscles and joints of the effectors, or in both. Over time, a medical condition may develop.

Introduction to work-related musculoskeletal disorders

The relationship between task demands, ergonomics and musculoskeletal disorders is of a probabilistic nature and is confounded by the fact the disorders can arise as a result of many activities of daily life, both at work and elsewhere. The disorders may present as co-conditions of other diseases. Disorders such as cervical spondylosis, carpal tunnel syndrome and tennis elbow are very common, particularly in older people. The usual pattern is for symptoms to appear before objective signs of degeneration or disease. By the age of 50 years, 50% of studied populations have had neck, shoulder or arm pain (Lawrence, 1969). This number increases steadily with advancing age.

There appear to be two main camps in the debate about the work-relatedness of musculoskeletal disorders. There are those who believe that if a person experiences pain at work it must be the work that caused it and that the pain itself is evidence of an underlying medical condition, caused by the work. People in this camp draw little distinction between the words 'pain', 'disorder' and 'injury' and use terms such as RSI (repetitive strain injury) and CTD (cumulative trauma disorder) as collective nouns to describe any kind of musculoskeletal problem, irrespective of whether it falls into a medical diagnostic category or not. Those in the second camp believe that there is a lack of evidence that work causes musculoskeletal diseases, apart from

Some terms used in epidemiology

Prevalence: The number of cases of a disorder at any particular time

Incidence: The number of new cases over a time period

Odds ratio: A statistic that describes the likelihood of a disorder developing given exposure to a risk factor. In the table below, there are some imaginary data on the prevalence of shoulder problems in a group of workers. The **occupational exposure** is having to load objects on to and off high shelves. The **health outcome** is the presence of a shoulder condition.

		Shoulder problems		
		Yes	No	
Exposed to				Row total
loading high	Yes	20	20	40
shelves?	No	5	80	85
Total		25	100	125

The data can be set out, as above, in a 2×2 table. The total number of workers surveyed is 125. Of these, 25 have shoulder problems and 40 carry out work involving high shelves. At first glance, it appears that people who work with high shelves are four times as likely to have shoulder problems than those who do not (of the 25 with shoulder problems, 20 work with high shelves and 5 do not). However, this ignores the null cases – people in the sample who do not have any shoulder problems irrespective of whether they work with high shelves or not. Clearly, in any assessment of risk, we need to take these individuals into account as well.

The odds ratio calculation does this elegantly, as follows:

Let A = Those **with** shoulder problems **and exposed** to high shelves = 20

Let B = Those **without** shoulder problems **and exposed** to high shelves = 20

Let C = Those **with** shoulder problems and **not exposed** to high shelves = 5

Let D = Those **without** shoulder problems and **not exposed** to high shelves = 80

The odds ratio (OR) = $(A \times D)/(B \times C) = (20 \times 80)/(5 \times 20) = 1600/100 = 16$

The overall prevalence rate for the shoulder disorders is 25/125 or 20%, so there is a high rate of shoulder disorders in the sample. The prevalance rate for those exposed to the risk factor is 20/5 or 4. We can conclude that there is a very greatly increased risk of developing the shoulder disorders given exposure to the risk factor: people are 16 times more likely to develop the problem.

Note that the size of the odds ratio in this example is very strongly influenced by the large number of people who do not have shoulder problems and are not exposed to the risk factor (cell D). Thus we can say that the shoulder problem is very *specific* to the exposure (it is rare in those not exposed). For many musculoskeletal conditions, the odds ratios are much smaller. One of the reasons for this is the lack of specificity – many non-work-related factors can cause the condition as well (cell C is larger and cell D smaller).

minor aches and pains that are reversible. Different jobs simply disclose differences in the health of people, which is why some complain and others do not. People in this camp avoid the use of superordinate categories in favour of distinct medical entities or confine the discussion to symptoms. For them, the observation that carpenters get more arm pain than controls is is merely a fact of life, rather than evidence of a disorder. Only if there is evidence of tissue pathology or overt disability (such as pain divorced from activity) will there be any suggestion of a disorder.

The entire field is marked by a lack of scientific rigour and poor terminology. In the present discussion, the term 'work-related musculoskeletal disorders' (WMSDs) will be used (after Kuorinka and Forcier, 1995). The classification scheme of Shilling and Anderson (1986) will be used to frame the discussion of the work-relatedness:

Category 1. The work exposure is a necessary cause of the disorder (as in occupational diseases such as silicosis or lead poisoning).

Category 2. The work exposure is a contributory causal factor but not a necessary one.

Category 3. The work exposure provokes reaction by a latent weakness or aggravates an existing disease.

Category 4. The work exposes the worker to potential dangers that may increase the likelihood of a disease developing (such as alcoholism in liquor industry workers).

The main work-related musculoskeletal disorders are reviewed below. The function of the ergonomist is not to attempt diagnosis of these conditions but to take any reported symptoms, such as pain, into account when evaluating workplaces. Irrespective of whether there is an underlying condition, the goal should be to design or redesign the workplace to enable people to work more comfortably.

Work-related musculoskeletal disorders (WMSDs)

Musculoskeletal conditions that are considered to be work-related are summarised in Table 5.1, using the categorisation scheme for diagnosis of the International Classification of Diseases.

In the past, conditions such as pre-patellar bursitis were named after occupations in which they were common – 'housemaid's knee', for example (see Pheasant, 1991). Quinter (1989) reviewed the history of occupational arm pain. Diffuse pains in the hand and wrist that later spread up the arms as the condition became chronic were known in the nineteenth century as 'occupational neuroses' and were found among writers and telegraphers. Both of these groups had jobs that demanded a combination of static fixation of the limbs and fine repetitive actions requiring a high degree of motor control. There was an absence of muscle wastage and no loss of sensation in the affected limb. This led to different theories of causation, with some arguing that the cause was peripheral nerve damage and others that the fault lay in the central nervous system and was a fault of the high-level mechanisms responsible for coordinating the automatic performance of movement. Others preferred the purely psychological explanation that the problem was caused by a tendency to overreact emotionally to the normal challenges of life!

Musicians and sportsmen suffer a variety of musculoskeletal problems ranging from vague symptoms of discomfort to clear-cut conditions such as ganglia. Instrumentalists

Table 5.1 ICD[a] diagnoses for work-related musculoskeletal disorders

---

Nerve root and plexus disorder
- Brachial plexus lesions
- Unspecified nerve root and plexus disorder

Mononeuritis of the upper limb and mononeuritis multiplex
- Carpal tunnel syndrome (median nerve entrapment)
- Lesions of the ulnar nerve (cubital tunnel syndrome)
- Lesions of the radial nerve
- Mononeuritis of the upper limbs (unspecified)

Other peripheral vascular disease
- Raynaud's syndrome
- Raynaud's phenomonen (hand–arm vibration syndrome)

Arterial embolism and thrombosis
- Arteries of the upper extremities (ulnar artery thrombosis)

Other disorders of the cervical region
- Cervicobrachial syndrome (diffuse)
- Unspecified musculoskeletal disorders and symptoms referable to the neck

Peripheral enthesopathies and allied syndromes
- Disorders of the bursae and tendons in the shoulder region (rotator cuff syndrome, suprasinatus syndrome, bicipital tenosynovitis)
- Enthesopathy of the elbow region (medial and lateral epicondylitis)
(Enthesopathies are disorders of the peripheral ligamentous or muscular attachments.)

Other disorders of the synovium, tendon and bursae
- Synovitis and tenosynovitis
- Trigger finger (acquired)
- Radial styloid tenosynovitis (deQuervain's)
- Other tenosynovitis of the hand and wrist
- Specific bursitides
- Unspecified disorder of the synovium, tendon and bursa

Disorders of muscle, ligament and fascia
- Unspecified disorder of muscle, ligament and fascia

Other disorders of soft tissues
- Myalgia and myositis (fibromyositis)
- Other musculoskeletal symptoms referable to the limbs (cramping, swelling)
- Other, unspecified, disorders of soft tissue

---

[a] International Classification of Diseases.

suffer the highest prevalence of such disorders (Brandfonbrener, 1990). The main problem areas are the small joints of the fingers and the wrist and the risk factors are thought to be the number of repetitive movements, the playing posture, the requirement to support the instrument while playing it and while carrying it, and the resistance of the keys or strings. Personal factors such as hypermobility of the finger joints (joint laxity) are involved and appear to be genticially determined. Usually benign, they may even have been a factor in the individual's early success. Brandfonbrener reports the results of survey of musculoskeletal conditions in 2212 orchestral musicians. Keyboard players predominated, followed by string players. Both groups are required to perform a high number of repetitive movements in their work. Wind and brass musicians, owing to the physical effort of playing their instruments, cannot endure such long practice hours and rest more frequently, reducing the likelihood of an 'overuse' injury.

Most conditions are localised and of a peripheral nature. The exception is *focal motor dystonia*, normally pain-free but consisting of cramps that interfere with coordination or cause involuntary movement. No problems are experienced until the musician attempts to play. The problem is thougt to originate in the central nervous system, possibly the basal ganglia.

Palmer *et al.* (2001) analysed data from 12 262 people registered with 163 general practitioners in the UK. Data on pain and sensory symptoms such as numbness and tingling in different body sights were captured. Significant associations were found between symptoms in the shoulders, wrist and hands and older age, smoking, headaches and tiredness/feeling stressed. Controlling for these factors, a significant association remained between the symptoms and keyboard use of > 4 hours per average working day. Keyboard users were approximately 1.2 to 1.4 times as likely to experience these symptoms than were non-users.

LeClerc *et al.* (2001) carried out a survey of carpal tunnel syndrome, lateral epicondylitis and wrist tendinitis in 598 workers in a variety of industries. The prevalence of these conditions at the start of the study was quite high – 21.9%, 12.2% and 11.2%, respectively. The findings illustrate very well the multifactorial nature of these conditions and the influence of male/female differences. For carpal tunnel syndrome, the main associated factors for men were tightening with force, holding in position and not pressing with the hand. For women, increase in body mass index and low job satisfaction were the main factors. For lateral epicondylitis, age and the job demand of 'turn and screw' were associated as were the number of other disorders and the presence of depressive symptoms. Wrist tendinitis was associated with the presence of somatic problems at the beginning of the study (e.g. problems with sleep, headaches, personal worries), lack of social support, increased body mass and having to hit repetitively. Younger workers were at increased risk, suggesting a 'survivor effect' in which only those without symptoms stay in the job.

A multifactorial problem

Clearly, WMSDs are the result of the interplay of many different variables. Although workplace ergonomic factors play a role, wider work organisation issues, social aspects of work and the health of the workers themselves are important. It is incorrect to talk about the 'prevention' of these disorders through ergonomic redesign of workspaces at the level of the ergonomic factors identified, since the evidence that ergonomic exposures are the cause is lacking. The current state of knowledge is of a web of factors that are associated with musculoskeletal outcomes. The outcomes themselves are often defined subjectively, inferred from questionnaire responses, and little precise information is captured on the magnitude of the ergonomic exposures, making it difficult to estimate dose–response relationships.

Factors associated with adverse outcomes

The main occupational factors associated with musculoskeletal conditions at work are

- Force
- Posture
- Repetition
- Duration

Bernard reviewed the literature available in 1997 (National Institute of Safety and Health, 1997) and concluded that there was evidence that most of the conditions were associated with one or more of the above factors. In several cases, exposure to more than one factor caused a large increase in the prevalence of the disorder. Bernard concluded that there was evidence that many of the conditions were caused by the work exposures based on the following criteria:

- *Temporality*: Prospective studies that show that the exposure precedes the outcome in time (cross-sectional studies cannot demonstrate cause and effect for this reason).
- *Strength of association*: The larger the odds ratio, the stronger the association between the outcome and the exposure and the less likely it is that the findings are due to confounding factors.
- *Consistency*: Where several studies are done, the same associations keep emerging.
- *Specificity*: The outcome depends on exposure to specific factors.

Models of the development of WMSDs

Armstrong *et al.* (1993) have developed a model of musculoskeletal disorders that emphasises exposure, dose, capacity and response. These are summarised in Table 5.2. Exposure refers to work demands such as posture, force and repetition rate that have an effect (the dose) on the internal body parts. Metabolic changes in the muscles, stretching of tendons or ligaments, compression of the articular surfaces of joints are examples of what is meant by a dose. The dose may produce a response such as a change in the shape of a tissue, the death of cells or accumulation of waste products in the tissues. These primary responses can be accompanied by secondary responses such as pain or a loss of coordination. As can be seen, a response (such as pain), can be a dose that causes another response (e.g. increased muscle contraction).

Capacity refers to the individual worker's ability to cope with the various doses to which his musculoskeletal system is exposed. An individual's capacity is not fixed. According to the model, it may change over time as the person ages or the development of skill may improve the ability to generate large forces with less effort. Training can increase strength or endurance, whereas the development of scar tissue to replace injured muscle tissue may impair strength or endurance. Armstrong *et al.* point out that muscles can adapt to work demands faster than tendons and that this may lead to reduced (relative) tendon capacity. We might speculate that one of the dangers faced by bodybuilders and others who use illegal anabolic steroids to produce rapid increases in muscle bulk is injury to the tendons because tendon strength does not have time to 'catch up' with the increased muscle strength.

Some common musculoskeletal conditions are described in the following sections. Their work-relatedness is summarised following the conclusions of Bernard (1997) and subsequent literature. Ergonomic interventions are described.

Table 5.2 Key elements of Armstrong *et al.*'s (1993) model of the development of work-related upper body musculoskeletal disorders

Element	Examples
Exposure	Physical factors • workplace layout • tool design • size, shape, weight of work • objects Work organisation • cycle times • paced/unpaced work • spacing of rest periods Psychosocial factors • job dissatisfaction • quality of supervision • future uncertainty
Dose	Mechanical factors • tissue forces • tissue deformations Physiological factors • consumption of substrates • production of metabolites • ion displacements Psychological factors • anxiety
Primary responses	Physiological • change in substrate levels • change in metabolite levels • accumulation of waste products • change in pH Physical • change in muscle temperature • tissue deformation • increase in pressure
Secondary responses	Physical • change in strength • change in mobility Psychological • discomfort
Capacity	Mechanical • soft-tissue strength • bone density/strength Physiological • aerobic capacity • anaerobic capacity • homeostatic control Psychological • self-esteem • tolerance of discomfort • tolerance of stress

Injuries to the upper body at work

The most clear-cut work-related upper body injuries occur as a result of accidents at work and many of them occur when hand tools are being used. Aghazedeh and Mital (1987) carried out a questionnaire survey to determine the frequency, severity and cost of hand-tool-related injuries in US industry and to identify the main problem areas. The hand-powered tools most commonly involved in injury were knives, hammers, wrenches, shovels, and ropes and chains. The powered tools most commonly involved were saws, drills, grinders, hammers, and welding tools.

Of the main incidents that precipitated an injury, the majority involved the tool striking the user. This was the case with both powered and non-powered tools. However, a significant minority of injuries were caused by overexertion (approximately 25–30%). The upper extremities were the body area most commonly injured and the most common injuries were cuts and lacerations, followed by strains and sprains. A *strain* may be defined as overexercise or overexertion of some part of the musculature whereas a *sprain* is a joint injury in which some of the fibres of a supporting ligament are ruptured although the ligament itself remains intact.

Many powered tools can cause strains or sprains because of the reaction force they exert on the user, particularly if these forces are unexpected or occur suddenly as a result of irregularities at the interface between the tool and the workpiece. Percussive tools, such as paving breakers, exert a reaction force that has to be opposed by the user. In practice, if the tool is well-designed and used on a flat surface, the weight of the tool will dampen much of this force. Rotary powered tools such as drills, sanders and screwdrivers can exert a reaction torque on the user that may force the wrist into ulnar or radial deviation causing a strain or sprain. The design of handles for holding powered tools in place has received the attention of ergonomists, as is described below. It should not be forgotten, however, that additional handles may need to be fitted to provide the user with sufficient mechanical advantage to overcome the reaction torque of the tool or to carry it. Bone (1983) reports that in the USA General Motors specifies a maximum allowable torque for freely hand-held powered tools and fits torque-arresting arms or slip clutches to more powerful rotary tools. Such modifications reduce the risk of injury and may enhance the usability of the tool.

It appears that there are several different classes of hand-tool-related injury and that several different approaches for prevention may be needed. The most common injury would seem to be of a catastrophic nature in which the tool itself suddenly strikes the user, causing a laceration, bruise or sprain. A second, more pernicious, type of injury involves sprains or strains that appear to result from the handling of the tool itself over longer periods of time. A third type of injury occurs to the skin in the form of blisters due to pressure 'hot-spots' caused by poor handle design. Attempts to prevent the first type of injury might emphasise training workers safe tool-handling techniques and to think ahead – to recognise potentially dangerous situations and to prepare the workplace to minimise the likelihood of unforeseen events. Attempts to prevent the second type of injury might concentrate on the redesign of the tool itself and training workers to recognise the onset of fatigue and avoid stressful work postures. Handle redesign can prevent the third type of injury, as can increasing the task variety of the job.

Prevention usually requires a multilayered approach involving training, safety propaganda and workspace design. Of particular relevance to the present discussion is the ergonomics of equipment design in relation to the prevention of strains and sprains. This is discussed in a later part of the chapter.

Review of tissue pathomechanics and WMSDs

The mechanism of WMSDs is thought to be repeated microtrauma at the cellular level. Repair capacity is exceeded owing to a lack of rest during the day and repeated daily exposure (Pitner, 1990).

Muscle pain

In general, human muscle has excellent endurance capacity for loads less than 15% of the muscle's maximum voluntary contraction (Putz-Anderson, 1988). Above this threshold, rest periods are needed, if acute or chronic problems are to be avoided.

Pain due to the accumulation of waste products in the muscles is called cramp and can be accompanied by muscle weakness or spasm (the muscle may temporarily lose up to 50% of its normal strength when fatigued). Cramp in the hand or forearm is known to be more common in those whose jobs involve prolonged handwriting, typing or other repetitive movements. Cramp is more likely when extreme postures have to be adopted since these postures weaken muscle joint systems. Patkin (1989) reports that cramp can be caused while using badly designed ball-point pens that require undue pressure to write well. The use of fountain pens is compulsory for schoolchildren in some countries because these pens can be used with lower forces.

DOMS (delayed-onset muscle soreness) is a natural response to abnormal loading with pain, due to inflammation, appearing up to 12–24 hours after the exposure, and peaking 1–3 days after, before gradually decreasing. DOMS is indicative of muscle damage. It can occur after exposure to sudden high forces, particularly during eccentric contractions (as, for example, when trying to hold a falling object or resist a sudden reaction torque of a powered tool). At the tissue level there may be evidence of damage such a muscle fibre z-line rupture, ragged type 1 (red) muscle fibres, decreased intracellular adenosine triphospahate (ATP) and reduced local blood flow (Hales, 1994). Normally, the muscle will recover and even strengthen, but some researchers believe that chronic exposure to static load prevents proper recovery, leading to permanent damage.

This primary response may be accompanied by a feeling of soreness in the muscle, which diminishes as the damaged muscle fibres regenerate (Armstrong *et al.*, 1993). In properly designed weight training programmes, the goal is in fact to cause such damage, because the body responds by increasing the size of the contractile elements in the muscle, resulting in improved strength. However, a rest period of at least 48 hours is usually an intrinsic part of strength training programmes, the time needed to allow the exercised muscles to recover. The pattern of activity in many industrial jobs bears little resemblance to that in proper muscle training programmes. One of the main differences is that rest periods are far more frequent in muscle training regimes than at work, which is why work does not normally have the same beneficial effects as exercise or training. Damage to muscle tissue on a daily basis may exceed the repair capability, leading to a decrease rather than an increase in strength or endurance and to chronic pain in the muscle (myalgia). It is thought that the tissue change responsible for chronic pain, in the absence of inflammation, is due to an increase in the extracellular matrix (Cutlip and Marras, 2000). The matrix is a barrier to nutrients and contains pain receptors. There may also be an increase in collagen in the muscles, causing an increase in muscle stifness. The end result is a loss of strength, endurance and the ability to absorb shocks or vibration without further damage.

In a conscious person, skeletal muscles always have a certain degree of 'tightness'. There is a baseline level of muscle fibre recruitment even during relaxation. This is known as muscle tone and it is controlled by the central nervous system and by a feedback system involving the spinal cord and the muscle spindles. Muscle tone is essential for the maintenance of posture. People under mental stress may, without realising it, develop increased tension in their muscles that they cannot control. At work, the pre-stressed body part may be a source of pain even though the task loading is mild. A chronic, stereotyped pattern of recruitment of motor units may be the dose that leads to damage of the muscle tissues.

Exertional compartment syndromes normally occur in the lower limbs, with dull aching in a given muscular compartment and increased pressure in the muscle. The pain is triggered by activity. Muscle is known to increase in volume by up to 20% during exericse (Pitner, 1990) and it is thought that the accompanying pressure increase is sufficient to degrade blod flow through the muscle. The pain subsides after cessation of the activity.

Tendon pain

Tendons have a tensile strength of 50–100 MPa (megapascals). Tendons consist of bundles of collagen microfibrils with many cross-linkages betweem them. At rest, the fibrils have a 'crimped' appearance. Under tension, the tendons elongate non-linearly as the fibrils straighten. Under further tension, the fibrils elongate by 1–5% of their length. Beyond this, individual fibres fail, putting more stress on those remaining. Under constant load, tendons exhibit 'creep' and viscous lengthening takes place.

Pain arising from inflammation of tendons ('tendinitis') is known to be work-related when it occurs in the hands and wrists (National Institute of Safety and Health, 1997). There is evidence that force, posture and repetition are all associated with the disorder and the evidence is even stronger when workers are exposed to these stressors in combination.

When highly repetitive movements are required, the increase in blood supply to the muscles may be associated with a decrease in blood supply to the adjacent tendons and ligaments of the associated joints. As Hagberg (1987) put it, the muscles 'steal' blood from the insertions of the ligaments and tendons. 'Policeman's heel' (caused by the repetitive microtrauma of walking long distances every day) is an example. Problems of this kind are sometimes referred to as *insertion syndromes*.

Impaired blood supply to the tendons is thought to be the cause of much occupational shoulder pain because it increases the rate of cell death within the tendon. This is thought to provide sites in which calcium carbonate (chalk) is deposited. It seems that increased tension in tendons reduces their blood supply, which may explain why static work positions are associated with tendon problems. Armstrong *et al.* (1993) describe an interesting hypothesis that the accumulation of dead cells in tendons can cause an inflammatory response in the tendon by the immune system. Inflammatory responses normally occur when there is an injury such as a cut; the blood supply to the affected region is increased to attack any incoming foreign bodies and the part of

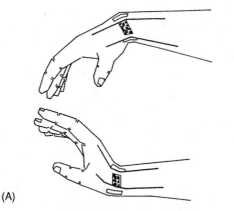
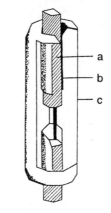

Figure 5.1 Wrist posture and tendon function. (A) Extreme postures can preload the finger tendons (B) Simplified view of a flexor tendon of the fingers. The tendon (a) is surrounded by a synovial sleeve consisting of an outer (c) and an inner layer (b). When the tendon moves, the inner layer glides over the outer layer, lubricated by synovial fluid. If the layers become inflamed or scar tissue builds up, the layers cannot glide smoothly over one another. The tendon then behaves like a rusty brake cable and smooth, pain-free movement is impossible.

the body in question normally feels hot and swollen. According to Armstrong *et al.*, if the person already has an infection such as influenza, the immune system will have been activated and the local inflammatory response described above is more likely to be triggered by the accumulation of dead cells in the tendon. This may underlie the popular conception that we are more prone to injury when suffering from colds or flu or other diseases.

Frequent mechanical loading can cause tendinitis or inflammation of the cartilage surrounding a joint. Extreme positions of the wrist can press the flexor tendons of the fingers against the bones of the wrist, increasing the friction in the tendons. Rapid, repetitive movements of hand or fingers can cause the sheaths surrounding tendons to produce excess synovial fluid (Figure 5.1). The resultant swelling causes pain and impedes movement of the tendon in the sheath. This is known as *tenosynovitis*. Repeated exposure can ultimately leave scar tissue that impedes movement of the tendon in its sheath and thus degrades function. Joint structure may be degraded by the formation of bony spurs around damaged areas. Reduced mobility, pain and weakness may result (Chaffin, 1987). Sudden, large forces may cause tendons to separate from bones.

Trigger finger (stenosing flexor tendinitis)   Stiffness and 'snapping' of fingers during volitional flexion is thought to be caused by thickening of the fibro-osseous canal through which the finger flexors pass. A higher prevalence has been found in certain meat-packing jobs that require static grasping of powered knives activited by a trigger (Gorsche *et al.*, 1998). However, non-occupational factors such as thyroid disease, diabetes and arthritis are usually the cause (Trezies *et al.*, 1998).

De Quervain's tenosynovitis   Characterised by pain on the thumb side of the wrist and impaired thumb function, this is more common in the preferred hand of middle-aged women, suggesting that it is activitiy-related (Moore, 1997). Activities that require heavy use of the thumbs seem to be associated with the disorder (fitting rubber rings on a pipe, sewing, weaving and cutting). The avoidance of ulnar-deviated wrist postures when operating tools would seem to be indicated.

Bursitis

A *bursa* is a sac containing viscous fluid situated at places in tissues where friction would otherwise occur (bursa is Latin derived from the Greek for wine skin and is related to the English word purse). There are about 150 bursae in the body and they act like cushions that protect muscles and tendons from rubbing against bones during movements of the body. Overexertion and injury can cause inflammation of bursae, or bursitis. 'Housemaid's knee' is a well-known type of occupational bursitis. Bunions are also a form of bursitis caused by wearing ill-fitting shoes: friction of the shoe on the bursa on the joint of the big toe causes it to become inflamed. Bursitis can be distinguished from tendinitis anatomically and because of the dull, aching pain that accompanies it, in contrast to the sharper pain of tendinitis.

Neuritis

Repeated or prolonged exertion can cause damage to the nerves supplying a muscle or passing through it. This can cause sensations of numbness or tingling (pins and needles) in areas of the body supplied by the nerve. The model of Armstrong *et al.* states that the response to a given dose can itself be a dose that leads to a response. In the case of nerves, overexertion can cause increased pressure in a muscle due to oedema or scar tissue formation. The increased pressure can itself be a dose that results in impaired nerve function. Impaired nerve function, destruction of fibres or damage resulting in reduced nerve conduction velocity may cause muscle weakness. All of these problems are more likely to occur if the joints are held in an awkward posture (at the extremes of the ranges of movement) since this 'pre-loads' tendons and ligaments and stretches muscles and nerves.

There are several common 'compression neuropathies' for example, 'bowler's thumb' involving the ulnar digital nerve of the thumb and 'handlebar palsy', which involves the ulnar nerve in cyclists. The relative roles of interference with blood supply and direct mechanical trauma to the nerve are not well understood but the pressure threshold limit for nerve viability is approximately 40–50 mmHg. Rydevik *et al.* (1981) found that blood flow in nerves ceased at a pressure of 60 mmHg. Furthermore, following 2 hours' compression at 40 mmHg, blood flow in the compressed segment was severely reduced up to 7 days afterwards. Any activities or conditions that increase the pressure are likely to degrade nerve function.

Peripheral neuropathies and non-localised arm pain

Several researchers have suggested that much of the diffuse and difficult to diagnose arm pain that is sometimes labelled 'RSI' may really be peripheral nerve pain (Quinter and Elvey, 1993). The pain is the result of a disorder of the pain receptors and occurs independently of the person's mental state. Unlike in neuritis, there is no inflammation of the nerve itself. Disorders of nerve function are known as neuropathies.

Chronic abnormal inputs from peripherally damaged nerves can sensitise nerve cells within the spinal cord, resulting in the hypersensitivity to painful stimuli characteristic of 'RSI'. According to the theory, 'RSI' is really a form of peripheral neuropathy in which 'damaged nerves can come to contribute actively to chronic pain by injecting abnormal discharge into the nervous system and by amplifying and distorting naturally generated signals'.

Butler (1991) points out that nerves are 'blood thirsty' structures: the nervous system constitutes 2% of body weight and consumes 20% of the oxygen in the blood. The nervous system is also the most extended and connected system in the body. This means that nerves have to accommodate postural movements. When, for example, a person flexes the elbow, the nerves on the flexion side of the joint shorten and may be pinched by other tissues, and nerves on the opposite side are stretched. It is conceivable that both these accommodations may interfere with the nerves' blood supply. According to Butler, movements at joints such as the ankle can increase the tension in the nerves in distant parts of the body. For example, the angle through which the straight leg of a person lying supine can be flexed depends on whether the person's neck is flexed. Cervical flexion pre-tenses the nervous system and hastens the onset of pain during the straight leg raising manoeuvre. In these situations, the nerve is susceptible to injury, not just at the joint that moves, but at any point along its length where adverse neural tension occurs. Vulnerable areas are where nerves pass through tunnels, where the nervous system branches and is less able to glide over surrounding tissues, where the nerve is relatively fixed (at some points in the spine, for example) and where the nerve passes close to unyielding surfaces. It is interesting to speculate whether the flexed cervical postures of many office workers predispose them to pain in the upper limbs through the mechanism of adverse neural tension.

Another puzzling feature of some activity-related upper limb pain is its task specificity. The person may complain of pain only when carrying small repetitive motions such as typing or playing the piano but be perfectly capable of carrying out similar activities that require a wider range of joint movement. Butler (1991) uses the term *activity-specific mechanosensitivity* to describe this phenomenon. It is hypothesised to occur when, during movement, a small region of scarred nerve tissue moves, in a particular direction, against a damaged or pathological surface such as bony outgrowth.

Bones and joints

Repeated, heavy loading is essential for the proper formation and maintenance of bone. Wolff's law states that bone grows in proportion to and in opposition to the forces imposed on it. Under repeated loading, particularly of the lower extremities, stress fractures can occur. Stress fracturing, like many WMSDs is a process rather than an event (Pitner, 1990). Repeated loads (below the threshold for acute fracture) damage bone, eventually leading to acute fracture. It is thought that these injuries are more likely to occur if the loading is accompanied by muscle fatigue.

Mechanical trauma seems to be a contributory factor in the development of osteoarthrosis in joints. Osteoarthrosis is a non-inflammatory disease characterised by degeneration of the articular cartilage, hypertrophy of bone and changes to the synovial membrane that cause stiffness and pain in the joints. There is plenty of evidence that joint diseases in later life are more common in some occupations than others (see Kuorinka and Forcier, 1995, for a review). Retired farmers and dockyard workers, for example have a higher prevalence of knee and hip joint problems (and indeed of surgical joint replacements) than do office workers. The risk factor seems to be working on different levels and jumping from one level to another.

When a muscle–joint system is placed in an extreme posture, the muscles on one side of the joint will be lengthened and their antagonists will be shortened, resulting in a strength imbalance in the antagonistic pair. The ability of the muscles to protect the joint against external forces is degraded and the joint itself is more easily damaged when the limb is exposed to high forces. An analysis of joint posture is particularly important when evaluating the design of hand tools, particularly in heavy work where the joints may be exposed to high forces.

Disorders of the neck

Evidence for work-relatedness

Ariens *et al.* (2000) reviewed recent literature on the work-relatedness of neck pain. Relationships were found between neck pain and neck flexion, arm force, arm posture, duration of sitting, twisting or bending of the trunk, hand–arm vibration and workplace design. However, owing to the low quality of many of the studies, the only factors for which there was firm evidence were sedentary posture and twisting and bending of the trunk. Cote *et al.* (2000) carried out a population-based survey of neck pain among 1131 randomly selected people. Unfortunately, their survey instrument did not contain items relating to occupational exposures. There was a high prevalence of neck pain – 54% had experienced neck pain in the previous 6 months. A history of having injured the neck in a motor vehicle accident was found to be strongly associated with pain in the previous 6 months (odds ratios of between 3 and 5 depending on pain severity). Other factors positively associated with pain included co-morbidities (headaches, cardiovascular and digestive disorders), suggesting that chronic health problems tend to cluster in some individuals and reinforcing the view that personal characteristics also play a role. Tola *et al.* (1988) compared the prevalence of neck and shoulder symptoms in operators of powered vehicles and compared it with that of carpenters and sedentary workers. The former group of workers were exposed to vibration, static loading due to prolonged sitting, and rotation of the back and neck when reversing. Symptoms increased with age in all groups but the vehicle operators always had more symptoms than workers of the same age in the other occupational groups. Thus, occupational factors can increase the prevalence of neck symptoms in all age groups, but older workers are more susceptible to occupationally induced symptoms because of the degenerative changes that have already occurred in their necks.

Jonsson *et al.* (1988) investigated the development of cervicobrachial disorders in female workers in the electronics industry over a 2-year period. Severe symptoms were present in 11% of subjects initially and in 24% after 1 year. Symptoms declined in the second year if workers were reallocated to different work and increased if they persisted with the same work. Previous heavy work, high productivity and previous sick leave were predictors of deterioration. Satisfaction with work tasks and an absence of shoulder elevation were predictors of remaining healthy. Reallocation to different work and spare-time physical activity were predictors of an improvement in condition. Reduction of the monotony of work and job dissatisfaction seem to be important factors in the control of neck problems at work.

It is known that dentists have a high incidence of cervical spondylosis. Milerad and Ekenvall (1990) compared cervical symptoms in a group of dentists and a group of pharmacists. Although both groups had symptoms of the disorder, the frequency was highest in the dentists and was more often accompanied by shoulder/arm pain. As noted previously, degenerative changes in the cervical spine can interfere with the spinal cord and nerve roots, causing referred pain in the arms (Smythe, 1988). *Numbness and tingling in the hands from the same cause might be mistaken for carpal tunnel syndrome.*

Possible causal pathways

The cervical spine has several functions, principally to support the weight of the head and to provide a conduit for nerves and attachment points for the muscles that control the position of the head. It consists of seven vertebrae designed to permit complex movements of the head. The first two cervical vertebrae (known as the atlas and the axis) are different from other vertebrae in the spinal column. The remaining vertebrae have the same general structure as vertebrae in other parts of the spine and are surrounded by anterior and posterior ligaments. The cervical spine consists of vertebral bodies and intervertebral discs, facet joints, bony processes for the attachment of ligaments and muscles, and the intervertebral foramen through which passes the spinal cord.

The head can be thought of as being balanced on top of the cervical spine with the fulcrum directly above the first cervical vertebra The head can be considered to be in balance when a person looks directly forward. Because the COG of the head lies in front of the cervical spine, the head has to be held erect by contraction of the posterior neck muscles. These powerful muscles are true postural muscles – they are essential to the maintenance of the erect posture and constantly work to prevent the head from falling forwards owing to gravity. The role of the posterior neck muscles in the maintenance of posture becomes clear when it is recalled how a sitting person's chin droops forwards onto the chest when the person is overcome by sleep. It can be appreciated that in ordinary standing and sitting postures, the structures of the cervical spine are 'pre-stressed' by the need to maintain the head in an erect position. They are therefore prone to overexertion by any additional stresses imposed by work tasks. In the balanced position, a cervical lordosis is present.

Movements of the head are accomplished by the muscles attached to it and to the surrounding parts of the skeleton. The arrangements of the muscles, like the movements of the head, are complex and will only be discussed in greatly simplified form. The deep, short muscles of the neck serve to stabilise the individual vertebrae, whereas the longer, more superficial muscles produce movements of the spine and head as a whole. The posterior muscles that extend the neck are stronger than the anterior muscles, which flex the neck, because the latter are assisted by gravity whereas the former have to work against gravity.

The erector muscles of the neck produce extension of the head and neck if they contract bilaterally (i.e. together). If they contract unilaterally (on one side only) lateral flexion and rotation of the head is produced. The trapezius muscle plays a very important role in many work activities. Because of its oblique orientation, it produces extension, lateral flexion and rotation of the head towards the side of contraction. It is also involved in elevating the shoulders. Static contractions of the trapezius muscle as low as 10% of maximum voluntary contraction appear sufficient to cause electromyographic changes in the muscle indicative of muscle fatigue (see National Institute of Safety and Health, 1997).

According to Kapandji (1974), prolapse of the intervertebral discs of the cervical spine is rare. However, the discs can certainly degenerate, as can the intervertebral joints, and this can cause irritation of the nerve roots in the cervical spine. Pain in the neck and shoulder may result. Degeneration of the cervical spine, sometimes known as *cervical spondylosis*, can have serious consequences. Compression of the spinal cord at the level of the cervical spine can take place, resulting in weakness and wasting of the upper limbs. This may then spread to the lower limbs.

As is the case with the lumbar spine, some of the degeneration of the cervical spine is part of the natural process of ageing. According to Barton *et al.* (1992), by the age of 65 years, 90% of the population have radiological evidence of cervical spondylosis. Cervical spine degeneration is a potential cause of neck pain due to the mechanical changes that occur as a result of age-related degenerative processes.

Static flexion of the cervical spine increases the moment arm of the head according to the sine of the angle of flexion. This increases the load on the soft tissues in the cervical region and the posterior neck muscles are placed under increased static load in order to maintain the forward-flexed head in equilibrium with gravity. According to de Wall *et al.* (1991), the increased static load on these muscles may cause pressure ischaemia and starve the muscle tissues of fuel and oxygen. Pain in the neck and shoulders may result, causing *muscle spasm* (reflex contraction of the muscles). This, in turn, may exacerbate the pain and lead to a vicious circle. The forward-flexed position may subject the cervical intervertebral discs to increased compression and the posterior ligaments to increased tension. Poor workspace design, if it requires workers to adopt flexed cervical postures, may be a cause of reversible pain or may amplify pain due to existing degenerative changes. Highly repetitive, low-load exertions may cause a gradual deterioration of tissue strength, eventually resulting in deformation of the tissues and pain on use.

Control of neck problems at work

Grandjean (1987) concluded that the head and neck should not be flexed forwards by more than 15 degrees if undue postural stress is to be avoided. There is considerable evidence that frequent or sustained flexion of the head and neck beyond this is related to chronic neck and shoulder pain. This is exacerbated if the flexion is accompanied by rotation of the head and if the shoulders and arms have to be held in an elevated position at the same time (as is common in certain occupations such as dentistry and hairdressing).

Bendix and Hagberg (1984) compared the trunk posture of subjects sitting at horizontal desks and desks sloping at 22 and 45 degrees. A more upright trunk posture was adopted when a sloping desk was used. The authors suggested that reading matter should be placed on a sloping surface and writing should be done on a horizontal surface; in otherwords, desks should have different sloped surfaces to account for differences in visual and manual requirements of reading and writing.

De Wall *et al.* (1991) investigated the effect of using a 10-degree sloping desk and found average reductions in cervical and thoracic spine load of 15% and 22%.

Elbow rests Elbow rests appear to be beneficial for both sitting and standing workers: by stabilising and supporting the arms, they reduce the load on the shoulder musculature (by supporting the weight of the arms, elbow rests would also reduce the load on the lumbar spine).

Monitor heights de Wall *et al.* (1992) found that more upright postures were obtained if the monitor was raised such that the middle of the monitor was at eye height rather than at 15–25 degrees below eye height as is usually recommended.

Carpal tunnel syndrome

Evidence for work-relatedness

Carpal tunnel syndrome (CTS) is associated with forceful and repetitive work alone or in combination with other factors (National Institute of Safety and Health, 1997). Vibration of the hand and wrist is also associated with the condition, but extreme postures on their own are not. Combined stressors such as force and extreme posture or repetition are strongly associated with CTS. As with tennis elbow, it seems that CTS is more common in 'hand-intensive jobs' in fish processing, supermarkets (checkout workers), etc. Roquelaure *et al.* (1997) found five occupational risk factors for CTS: exertion of force over 1 kg.; length of the shortest elementary operation <10 seconds; lack of a change in tasks or lack of breaks for >15% of the day; manual supply of parts and equipment to the workstations; and lack of job rotation. The only personal factor associated with CTS was having more than three children (for women). Odds ratios of 5–6 were found when three risk factors were present, rising to >90 when all six were present.

Possible causal pathways

The muscles that flex the fingers lie in the forearm and have long tendons that pass through a narrow opening in the wrist before inserting into the fingers. This opening, known as the carpal tunnel, is also traversed by the nerves and blood vessels of the hand (Figure 5.2).

An increase in the pressure in the carpal tunnel can cause carpal tunnel syndrome if it affects ('entraps') the median nerve or reduces the blood supply to the nerve by compressing the capillaries, resulting in nerve damage and reduced conduction velocity of neural signals. The result is a sensation of tingling and numbness in the palm and fingers. In severe cases, surgery may be required to relieve the pressure. Carpal tunnel syndrome has been reported in jobs requiring rapid finger movements, such as typing, and is found among professional musicians. However, Barton *et al.* (1992), in their review of the literature, concluded that the majority of cases of carpal tunnel syndrome are not caused by work. Carpal tunnel syndrome can have many non-occupational causes. It is common during pregnancy and may be a co-condition of a range of other disorders as diverse as diabetes, high blood pressure, kidney disorders, use of oral contraceptives and arthritis (Hales, 1994).

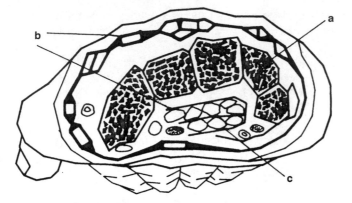

Figure 5.2 Carpal tunnel. Section through the wrist showing (a) carpal bones, (b) tendons, (c) carpal tunnel containing finger flexor tendons, blood vessels and nerves. If the wrist is held in extreme postures, the tendons rub against bone when the fingers move and friction is increased. (Supplied by Dr Douglas A. Bauk of IBM Brazil.)

Sensory conduction of the median nerve   Carpal tunnel syndrome is associated with slower median nerve conduction. Nathan and his colleagues have carried out numerous studies of the relationship between median nerve conduction velocity, carpal tunnel syndrome and the factors that are associated with slowed nerve conduction.

Nathan *et al.* (1987) were unable to find an association between slowed median nerve conduction and occupational hand activity or with length of employment in the current job. Furthermore, Nathan *et al.* (1988) found that slowing of median nerve conduction was age-related in a non-pathological way, although the frequency of CTS did also increase with age. The evidence for age-related slowing of median nerve conduction was similar in women and men despite the fact that CTS is often perceived to be more common in women. Nathan *et al.* (1992a) found that nerve conduction velocity was associated with body mass index (weight (kg)/height^2 (m^2)), being significantly slower among obese compared with slender subjects. Other factors associated with nerve conduction were the wrist depth/width ratio and exercise level. Finally, in a follow-up of their longitudinal study, Nathan *et al.* (1992b) confirmed the relationship between CTS, slower nerve conduction and age and the non-relatedness of occupational factors. This led them to conclude that the health of the median nerve is linked to the health of the rest of the body and that median neuropathy is closely related to lifestyle and only peripherally to work activities.

Classification and diagnosis of CTS   Other studies *have* found evidence of a relationship between symptoms of CTS and work. The disparity between these findings and those of Nathan and his colleagues may be due to the use of symptoms (which are subjective) rather than objective nerve conduction measures to classify sufferers and non-sufferers. This problem (i.e. of diagnosis and categorisation) currently pervades all research on the work-relatedness of musculoskeletal disorders. Cervical spondylosis and stenosis (narrowing) of cervical structures, as well as nerve entrapment in the arms, may give rise to symptoms that mimic carpal tunnel syndrome. This illustrates that the diagnosis of all arm and hand problems is best left to the experts because pain in these regions may, in reality, be due to other causes.

Loslever and Ranaivosoa (1993) investigated the prevalence of carpal tunnel syndrome in light industrial work. The syndrome was found to occur twice as often in both hands as in either the preferred or non-preferred hand, which, they suggest, is evidence that non-occupational factors are more important than occupational factors. However, the prevalence for both hands was found to correlate positively with measures of wrist flexion and high grip forces.

It seems likely that occupational factors such as repetitive work, wrist posture and force as determined by task and tool design may provoke symptoms of CTS, even if they are not necessary causal factors. In fact, there is evidence that CTS sufferers may fall into a viscious circle of disorder and dysfunction in which the condition causes a loss of tactile sensitivity. Sufferers' grip and pinch forces become excessive in relation to task demands, causing further increases in the pressure inside the carpal tunnel and excessive tension in other structures, thereby accelerating the progression of the disorder (Lowe and Frievalds, 1998).

It seems reasonable to conclude, then, that redesign of equipment to minimise exposure to vibration, highly repetitive work and extreme postures in combination with these factors may be of value in lowering the prevalence of CTS symptoms in the workplace, even if these measures cannot reduce the incidence of the disorder itself.

Tennis elbow (epicondylitis)

Evidence for work-relatedness

Overexertion of the extensor muscles of the wrist can lead to a condition known as 'tennis elbow' (lateral humeral epicondylitis) (Figure 5.3). In severe cases, the muscle and tendon may separate from the bone. The risk of injury is said to be increased by activities requiring large grasping forces. Bernard (National Institute of Safety and Health, 1997) concluded that there was insufficient evidence to support an association between repetitive work, posture and tennis elbow, although there was evidence

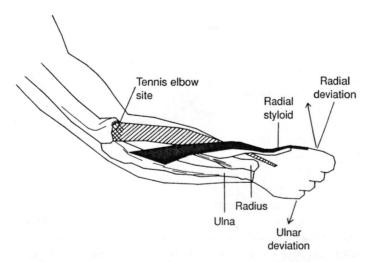

Figure 5.3 General view of the forearm, wrist and hand, showing injury sites.

for an association between tennis elbow and forceful work. There was strong evidence for an association between combined stressors (e.g. force and posture) and tennis elbow. Elbow problems are found among mechanics, butchers and construction workers.

Possible causal pathways

The elbow can be considered a weight-bearing joint. Many of the muscles that control movements of the hand are extrinsic to it and have their origins higher up. Some cross the elbow joint. The act of grasping and holding objects is only possible if the wrist is stabilised by the muscles of the forearm, many of which originate at the elbow. For example, when the finger flexors contract to enable an object to be grasped, the wrist extensors also have to contract to prevent the wrist itself from flexing. These contractile forces are transmitted across the elbow to the distal end of the humerus where the wrist extensors originate. The elbow joint is compressed and the tendons are under tension and may become swollen at the point they insert into the humerus. Clearly, any activity that requires a strong grip to be maintained for long periods will place a high load on the elbow joint and associated structures.

Dimberg (1987) found that 35% of cases of epicondylitis were work-related and 27% were due to leisure activities. Tennis accounted for another 8% of cases and the cause of the rest, was unknown. Work-related sufferers were found to have jobs higher in elbow stress (e.g. driller, carpenter, polisher) than people who suffered for other reasons or than the workforce as whole.

It seems to be the case that elbow problems are not particularly prevalent compared to other conditions and, although they can be caused by occupational exposures, they are not specific to them. In some studies, however, Bernard reports that the odds ratios for combined stressors are quite high – over 6 for hand-intensive jobs such as fish processing. It is clear that people with symptoms of the disorder will be at increased risk, irrespective of the original cause, if they are required to carry out jobs high in elbow stress, and the risk will increase greatly in the presence of combined stressors. From the evidence available it would seem that a 'high-risk' job for epicondylitis will be one that requires high grip forces combined with repetitive work in extreme postures or pronation and supination of the wrist, essentially factors that increase the stress on the tendon insertions. Low-friction handles and contaminants such as sweat, oil and lard that reduce friction (Bobjer *et al.*, 1993) would seem to increase the risk by causing people to grip harder. Perhaps this is why the disorder has been found in occupations such as fish processing where several different factors operate together.

Disorders of the shoulder

Evidence for work-relatedness

Punnet *et al.* (2000) found an increased risk of shoulder disorder when the shoulder was abducted or flexed more than 90 degrees, with the risk increasing in proportion to the percentage of the work cycle during which the arm was held in that position. The odds ratios for these exposures ranged from 1.5 to 6.5, but high risks were found to occur when the flexion or abduction was held for more than 10% of the work cycle. Bernard (1997) concluded that there is evidence for a positive association between highly repetitive work (cycle times less than 30 seconds or spending more than 50% of the time doing the same task) and shoulder problems and between repeated or static shoulder postures (more than 60 degrees of shoulder flexion or abduction with increased risk when posture is combined with other stressors such as holding a tool). There is evidence that shoulder problems increase with the duration of employment and with the length of the workday. There was insufficient evidence to link vibration and force to shoulder problems. Age and participation in sports activities are known to be confounding factors that reduce the specificity of occupational exposures.

Possible causal pathways

Most work involving hand tools imposes a combination of repetitive and static loads on the body, which usually involve the shoulder, if only indirectly. There is no single shoulder joint: the arms join the scapula at the gleno-humeral joint and the scapula joins the body at the scapulo-thoracic joint. Not surprisingly, the shoulder 'joint' is the most mobile in the body and, together with its related soft tissues, is particularly prone to injury in any activities where the arms are held above the horizontal. Working with the hands above shoulder height is stressful and may increase the risk of developing the so-called 'impingement syndrome', otherwise known as 'swimmer's shoulder', 'pitcher's arm' (Wieder, 1992) or rotator cuff syndrome. The disorder is known to be more common in sportsmen who use high overhead actions.

Figure 5.4 depicts in broad outline the anatomy of the shoulder joint and related structures. The shoulder joint is a kind of ball and socket joint, but the ball part, the head of the humerus or upper arm bone, represents only a third of the surface of a sphere when it engages the socket. The socket (the glenoid cavity of the scapula, or shoulder blade) is correspondingly shallow. The head of the humerus has to be held in place by ligaments and tonic muscle activity. This explains why the shoulder joint is so easily dislocated. This can be contrasted with the much more stable hip joint, where over 50% of the femoral head is enclosed by the acetabulum.

That the shoulder joint requires muscle activity to be held in place might alert the ergonomist to its likely susceptibility to rapid fatigue and damage when exposed to static loads or repetitive actions. One of the simplest ways to reduce occupationally induced shoulder stress in many jobs may be to provide armrests, slings or other means of supporting the weight of the arms to enable the shoulder muscles to relax. Conversely, whenever the hands or arms are used, muscle activity is necessary to keep the humerus in its socket and to hold the scapula in place on the thorax. The stabiliser muscles of the scapula are at a great mechanical disadvantage when the arms are held forward of the body (or cantilevered) and static muscle contractions are needed. These contractions increase the pressure in surrounding tissues, impairing circulation in the tendons. Together with increased load, this may cause supraspinatus tendinitis. Ischaemia in the muscle may cause increased cell death and chronic inflammation experienced as pain.

Above the humeral head lie the acromion and the coracoacromial ligament and in between lies a bursa. In this narrow space pass many tendons, nerves and blood vessels. The space can easily become taken up by the growth of bony spurs, by bleeding or by soft tissue swelling due to overexertion (the subacromial bursa can

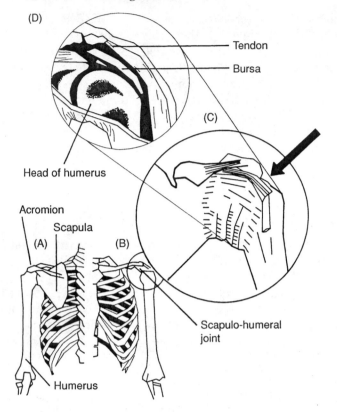

Figure 5.4 Shoulder joint: (A) rear view; (B) front view; (C) expanded view; (D) scapulohumeral joint itself.

become inflamed, for example). If this happens, the range of motion of the shoulder joint is reduced because the impingement of the subacromial structures causes pain when the joint is moved. Sufferers cannot raise their arms above shoulder height. It is likely that jobs that require the hands to be chronically elevated above elbow height can cause short-term changes that, over time, may ultimately lead to disorders of the shoulder joint. Shoulder pain can also be caused by localised muscle fatigue, particularly if the arms have to held above shoulder height for long periods of time (as in painting a ceiling or pruning a tree). Some recommendations for preventing shoulder pain are given in Table 5.3.

Lower limbs

There is an extensive literature on lower limb injuries in sports. The main occupational causes of lower limb injury (excluding falls) appear to be walking while carrying heavy loads and jobs that require excessive use of the knees. (Kirkesov and Eenberg, 1996). Reynolds *et al.* (1999) followed infantrymen on a 160 km march over 5 days. The soldiers carried heavy packs. Thirty-six per cent suffered one or more injuries and 8% were unable to complete the march. The main injuries were blisters (48%) and foot pain (18%). Other injuries included ankle and knee strain,

Table 5.3 Methods of reducing shoulder stress[a]

1. If possible work with the hands near waist level and close to the body.
2. If the hands have to be positioned above shoulder level, their elevation above the shoulders should be no more than 35 degrees. Hand loads should not exceed 0.4 kg and the posture should be held for no more than 20 seconds for each minute of work.
3. Avoid shoulder flexion/abduction >90 degrees for >10% of cycle time.[a]
4. Select taller workers for workplaces that cannot be modified.
5. Take regular rest breaks.
6. Minimize hand-held weight.
7. Provide external support for the weight of the arms (slings, ledges, etc.).
8. Confine work objects within the zone of convenient reach.
9. Provide wrist rests for keyboard workers.

[a] Punnett *et al.* (2000), see also Sommerich *et al.* (1993).

foot swelling, hip and thigh pain and metatarsal stress fracture. Risk of injury was higher among smokers and lower among older people.

Knee disorders, including osteoarthritis and bursitis, are associated with squatting and with heavy physical work (Jensen, 1996). The 12-month prevalence of knee problems in the Danish population was reported to be 19%. Unsurprisingly, kneeling work was associated with bursitis ('housemaid's knee'). Jobs that require jumping from one level to another or climbing ladders and stairs (>10 flights/day) are all associated with an increased prevalence of knee problems. High-risk occupations include, tiling, carpet laying, shipyard work, mining, bricklayering and pipe fitting. Osteoarthrosis of the knee in later life is more common in farmers (males and females) and construction workers. Among women, physically demanding work at home is associated with knee osteoarthrosis (odds ratio 2.2). Heavy work at home, such as caring for an elderly or handicapped person, is implicated as a risk factor for women (Sandmark *et al.*, 2000).

Ergonomic interventions

The design of tools and workspaces can have a profound effect on the posture of the body. Of particular interest is the posture of the shoulders, elbows and wrists and its relation to pathological musculoskeletal changes.

The control of finger movements depends on many small muscles that can easily become fatigued, particularly during prolonged work with inadequate rest periods and poorly designed tools (for example, when writing an examination). One of the most fundamental problems in hand tool design is to optimise the dimensions of the tool in relation to the hand anthropometry of the population under study.

Ducharme (1977) found that many women working in previously male craft skills in the US Air Force were dissatisfied with the design of the tools they used. Crimping tools, wire strippers and soldering irons were said to have grips that were too wide or required the use of two hands to operate. Other tools were said to be too heavy and too awkward to use. Pheasant (1986) suggests, on anthropometric grounds, that the tools were probably inadequate for male use as well but that more males were able to overcome the design deficiencies owing to their greater strength. Pheasant and Scriven (1983) report that automobile wheel braces provide inadequate mechanical advantage for both male and female users. Although a greater proportion of females than males are affected, both sexes would benefit from improved designs offering greater mechanical advantage.

Handle design

Pheasant and O'Neill (1975) investigated handle design in a gripping and turning task (such as using a screwdriver). They found that strength deteriorated when handles greater than 5 cm in diameter were used and that, to reduce abrasion of the skin, hand–handle contact should be maximised. Knurled cylinders were found to be superior to smooth cylinders because of the increase in friction at the hand–handle interface. The authors concluded that, for forceful activities, the size of a handle rather than its shape was most important. A useful rule of thumb for evaluating handle diameters is that the handle should be of such a size that it permits slight overlap of the thumb and fingers of a worker with small hands.

One of the most important criteria for handle design is provision for sufficient hand–handle contact. The larger the handle diameter, the bigger the torque that can be applied to it, in principle, but people with small hands must be able to enclose the handle with their fingers. A handle diameter of 40 mm seems to be the maximum for male users, smaller still if gloves are to be worn. Cylindrical handles are better than handles with finger grooves since these cause pressure 'hot spots' and blistering of the skin of hands they do not fit. Handle lengths should be at least 11.5 cm plus clearance for large (95th percentile) hands. An extra 2.5 cm should be added if gloves need to be allowed for (NIOSH, 1981). Some example of screwdriver handles designed for one-handed and two-handed operation are shown in Figure 5.5.

Grip strength depends very largely on the posture of the wrist. When the wrist is extended, the finger flexors are lengthened and can therefore exert more tension resulting in a stronger grip. When the wrist is flexed, the opposite occurs and grip strength is severely weakened. This underlies the method taught in self-defence classes for disarming an attacker. The hand and wrist holding the weapon are twisted (pronated) and the wrist is flexed before any attempt is made to remove the weapon

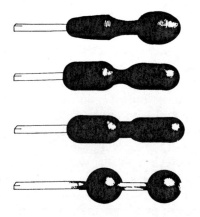

Figure 5.5 Some innovative screwdriver handle designs for one- and two-handed operation (designed by the Ergonomic Design Group, Stockholm, Sweden).

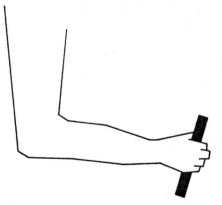

Figure 5.6 Angle of grip when the wrist is in the neutral position.

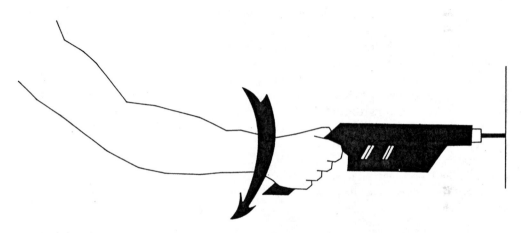

Figure 5.7 When working at a vertical surface an in-line grip reduces wrist strain. Note the large moment of ulnar deviation when the tool is not engaged with the workpiece. This can be minimised by locating handles below the centre of mass of the tool.

from the attacker's hand. The pronation and flexion of the wrist shortens and therefore weakens the finger flexors. A general requirement of handle design is that the wrist joints should be kept in a neutral position (in the middle of their ranges of movement) when tools are being used. Pheasant (1986) describes how the axis of a handle is at an angle of 100–110 degrees with respect to the forearm when the wrist is in a neutral position (Figure 5.6).

Tools such as saws and pliers can be designed with obliquely set handles to enable the wrist to be maintained in the neutral position (Tichauer, 1978). Tools such as soldering irons can be redesigned using a 'pistol grip' handle rather than the more traditional straight handle for the same reason (Figure 5.7). When using straight-handled tools there is a tendency for the wrist to be bent outwards ('ulnar deviated'). This stretches the tendons of the forearm muscles on one side, causing them to rub against a bony protrusion on the thumb side of the wrist (known as the radial styloid). Repeated exposure can cause the sheath (synovium) within which the tendon runs to become inflamed (*DeQuervain's syndrome*). Inflammation of tendons and

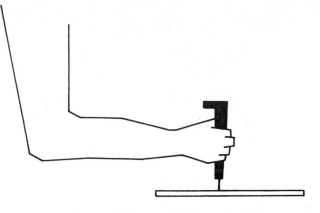

Figure 5.8   When working at a horizontal surface, a perpendicular grip may be better.

tendon sheaths can occur in other parts of the hand and other body structures. In the long term, permanent damage to the tendon and its sheath may result. The build-up of scar tissue in the tendon may ultimately reduce the range of movement of the wrist.

The idea of 'bending the handle instead of the wrist' is a valid one and has many potential applications. It has been incorporated into the design of such diverse products as hammers (Knowlton and Gilbert, 1983) and even tennis rackets and table tennis bats (Chong and McDonough, 1984). Konz (1986) sums up the use of bent handles in tool design as follows:

> . . . When a tool, gripped with a power grip has its working part extend above the hand, then a curve in the handle may be beneficial . . . A small bend (5 to 10 deg) seems best.

For powered tools, pistol grips seem most appropriate when the task is oriented vertically with respect to the operator (as when drilling a hole into a wall). When the task is a horizontal one (as in fastening a screw into a horizontal desktop with a powered screwdriver), an in-line tool may be better (Figure 5.8).

Powered tools tend to be considerably heavier than their non-powered counterparts and a potential source of wrist strain comes from the weight of the tool itself, particularly if the handle is placed at one end of the tool rather than in the middle. Having to hold a heavy drill with a pistol grip while positioning the drill bit exemplifies this problem. Wrist loading can be reduced by fitting the handle at the tool's centre of mass so that the tool is counterbalanced. Johnson (1988) investigated the design of powered screwdrivers in relation to operator effort and concluded that grip diameters should be at least 5 cm. Vinyl sleeves fitted over too-narrow handles were effective in reducing effort. A 'biomechanical' brace was designed to fit over the screwdriver handle and run along the palmar (inside) side of the forearm. The brace transmitted the reaction torque to the user's forearm and enabled the tool to be used with reduced grip force.

Finally, tools can be redesigned and fitted with longer handles, or handle extensions can be fitted to increase the worker's vertical reach, obviating the need for the hands to be raised above shoulder height. Some examples are handle extenders for

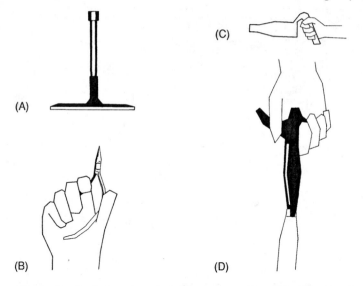

Figure 5.9 Redesigned tools with improved hand–handle interfaces. (A) Socket wrench with T-bar for greater mechanical advantage. (B) Pliers with bent handles to maintain neutral wrist posture. (C) Drill with handle at appropriate wrist angle. (D) Paint scraper with thumb-stall to relieve pressure on the palm of the hand and prevent blistering of the skin on the palm.

paint brushes when painting ceilings and long-handled secateurs for pruning the higher branches of trees. Figure 5.9 shows some redesigned tools with improved hand–handle interfaces.

Keyboard design

With old-fashioned mechanical typewriters, the typist had to stop to manually return the carriage after each line. Errors had to be manually corrected using erasers or correcting fluid and the paper had to be changed after each page had been typed. All of these secondary tasks provided changes of posture and broke up the continuity of the typing task. Brief periods of rest were intrinsic to the operation of mechanical typewriters. With word processors, the secondary tasks are carried out automatically or via special keys on the keyboard, so the work is intrinsically less varied and more likely to cause fatigue. Before the introduction of desktop computers, almost the only people who ever used keyboards were typists who had received special training in keyboard skills (e.g. how to use the fingers most efficiently and to type without looking at the keyboard). A very large proportion of today's desktop computer users have never undergone formal training and do not posses these keyboard skills. Patkin (1989) has suggested that poor motor skill leads to excess co-contraction of muscles and temporary muscle aches that may be mistaken for tenosynovitis (which is rare among typists).

Several researchers have focused on the design of keyboards as a means of reducing musculoskeletal problems in keyboard operators. Zipp *et al.* (1983) investigated the posture of the hands and wrists, noting marked ulnar variation and fatigue (Figure 5.10)). They concluded that keyboards should be designed with separate banks

Figure 5.10 Conventional keyboards cause ulnar deviation.

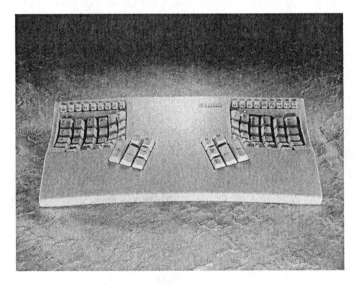

Figure 5.11 Separate banks of keys for each hand may relieve wrist strain. (Photograph kindly supplied by Kinesis Corporation, Bellevue, Washington, USA.)

of keys (one for each hand) each bank being inclined and contoured to be compatible with the functional anatomy of the hand. Keyboards based on this design are commercially available (Figure 5.11) are perceived as comfortable to use and give fair performance. Fernstrom *et al.* (1994) measured EMG activity (the electrical activity in the muscles) from the forearm and shoulder muscles of eight typists using mechanical, electromechanical and electronic typewriters, a PC keyboard and a PC keyboard with the keys angled at 20 degrees to the horizontal. The mechanical typewriter required a higher keystroke force, which placed greater strain on the muscles investigated (although the difference was small). The electronic typewriter placed more strain on the right shoulder than did the mechanical typewriter. This appeared to be due to the low keystroke force that made it impossible to rest the fingers on the keyboard without typing a character. A keystroke force of about 0.5 N is recommended.

Tittiranonda *et al.* (1999) compared several alternative geometry keyboards and a placebo over a 6-month period. Although there was no improvement in clinical findings, self-reported pain severity was lower and hand function improved with the Microsoft Natural Keyboard compared to the placebo.

Gerard *et al.* (2002) provide evidence that conventional computer keyboards can be improved by enhacing the auditory feedback when keys are depressed. Their experimental keyboard was designed to produce a tone at 1.2 dB above the ambient noise whenever a threshold value of either keying force of finger flexor EMG was exceeded. The tone increased by further increments of 1.2 dB the higher the keying force or when EMG was above threshold. The feedback produced a 10–20% reduction in 90th percentile keying force in under 3 minutes. Further exposure to the feedback did not bring about more reductions in keying force. These findings *may* mean that augmented auditory feedback will help to prevent upper extremity discomfort in keyboard users. A controlled trial would be needed to determine whether these reductions are sufficient to reduce the incidence of disorders. Feuerstein *et al.* (1997) found that subjects with more severe upper extemity symptoms typed with higher finger forces than did those with less severe symptoms. Keying rate was the same for both groups and both groups used 4–5 times the minimum force needed to activate the keys.

Work design, education and training and job design

Disorders associated with repetitive work have existed both in and out of industry for many years – biscuit packing, fruit wrapping and chicken plucking are well-known examples of high-risk tasks. The same disorders are also associated with leisure activities such as playing musical instruments or electronic arcade games, solving Rubik's cube and marathon running (Brown *et al.*, 1984). Kemp (1985) describes several approaches to their management and prevention. Task rotation and scheduling of work–rest cycles to minimise overexposure to the repetitive elements of a task are examples. Careful analysis of individual tasks is essential to identify high-risk areas.

Ergonomists usually advise correct design of tools, equipment and workspaces for the prevention of musculoskeletal pain in the upper body. Workspaces must be designed to allow flexibility in carrying out a task by means of a wider repertoire of postural behaviours to vary the load on different parts of the body. Training programmes to increase a worker's capacity have received little attention in ergonomics, although they are regarded as fundamental in sports science to avoid the very similar types of musculoskeletal problems that can occur when engaging in sport. In addition, sports scientists emphasise the importance of self-preparation immediately before any arduous activity is undertaken. This normally includes proper nutrition and fluid intake and a practised set of stretching and warm-up exercises.

Table 5.4 summarises complementary approaches to the reduction of work-related musculoskeletal disorders.

Table 5.4 Simple measures for preventing work-related musculoskeletal disorders of the upper limbs[a]

Tool design and posture
Bend the handle (5–10 degrees) not the wrist.
Avoid excessive use of 'pinch grip' (thumb and fingers).
Maintain neutral wrist posture.
Reduce required grip forces:
  Use high-friction materials for handles.
  Longer handles increase mechanical advantage.
Maximum allowable grip force (holds for less than 3 seconds):[a]
  Hand grip: 26.2 kg
  Pinch grip: 5.8 kg
Maximum allowable grip force (holds for more than 3 seconds):[a]
  Hand grip: 15.6 kg
  Pinch grip: 3.6 kg
Add handles for carrying tool and resisting reaction torque.
Damp vibration from powered tools.
Encourage use of large muscle groups.
Upper arm posture to be <15 degrees of flexion and <10 degrees of abduction (Aaras, 1994).

Workstation design
Damp jerk and impact forces.
Position work objects so as to eliminate static neck postures.
Eliminate static shoulder elevation or provide elbow rests or slings.
Provide vices, clamps or other aids to minimise the need for sustained holding of work objects.
Use arm supports in the form of slings or balancers to lower EMG activitiy in the trapezius and deltoid muscles (Feng *et al.*, 1999).

Task/job design
Limit repetitive movements to 2000 per hour or less.
Eliminate highly repetitive (<30 second cycle time) jobs.
Design out movements requiring rapid acceleration of body parts.
Spend no more than 50% of the time doing the same task.
Rotate workers between high-repetition and low-repetition tasks.
Avoid mental stress:
  Unrealistic deadlines/production targets
  Machine pacing with fast cycles
  Excessive supervision
  Piece rate systems of remuneration
Design in 'micro-breaks' (2–10 seconds rest every few minutes).
Repetition frequencies greater than 1 Hz may pose a risk to joints or soft tissues.
Frequencies less than 1/6 Hz may cause fatigue through the mechanisms of static loading.[b]
Increase task variety (e.g. integrate routine maintenance, inspection and record keeping with basic production tasks)

Management of working conditions and worker exposure
Avoid repetitive work in cold conditions.
Use ergonomic criteria when making equipment procurement decisions.
Redesign tasks before a discomfort becomes a disorder.
Eliminate excessive overtime working.

[a] US Mil Std-1472.
[b] Solomonow and D'Ambrosia (1987).

Trends in work-related musculoskeletal disorders

Brogmus *et al.* (1996) analysed trends in WMSDs from the 1980s to the 1990s using data from the Liberty Mutual Group workers' compensation claims and US Bureau of Labor Statistics. Analysis of both data sets confirmed a steady increase in cases and claims, with WMSDs rising from 1% of the total claims in 1986 to 4% in 1993. A number of potential causes were investigated to determine the reasons for the upward trend. Although there was some evidence that increased productivity and exposure to ergonomic stressors play a role, psychosocial and legal factors are of great import-ance. The following factors were investigated by Brogmus *et al.*:

- Increased reporting
- Expansion of workers' compensation to cover WMSDs
- Increased productivity
- Shift to service industries
- Growing proportion of women in the workforce
- Increasing use of VDUs

Greater awareness of WMSDs in relation to work demands was called for by NIOSH in 1985 and this, in itself, may well have caused increased reporting of problems. In particular, carpal tunnel syndrome was mentioned specifically in the case of VDUs (for many years, CTS was regarded as a disease associated with ageing and other medical conditions and the assumed link with work was not nearly so explicit as it is today). Brogmus *et al.* also cite the expansion of workers' compensation laws to include WMSDs in nearly all US states as another factor contributing increased reporting. Although increased productivity is often cited as another cause of WMSDs, the link is not so clear. Automation increases productivity but usually reduces ergonomic stressors as factory jobs become more supervisory in nature. Increases in reporting have increased steadily over the years, whereas productivity has risen sharply in some years but not in others.

Although there has been a shift to service industries since the 1950s, the mapping between the two trends is not good, i.e. increases in WMSD do not correspond with periods when the services sector grew rapidly. Furthermore, WMSDs are more com-mon in goods-producing than in service industries. Similarly, the increasing partici-pation of women in the workforce cannot be used to explain the upward trend in WMSDs because, although women are more likely to suffer from these disorders for other reasons, they are also likely to occupy jobs with risk factors for the disorders. Finally, although there has been a great deal of discussion of the risk of WMSDs posed by VDU work, both the Bureau of Labor and the Liberty Mutual statistics show that VDU-related claims are a small percentage (less than 20%) of the total and that over 80% of WMSDs are spread over a whole range of mainly goods-producing jobs.

Psychosocial and legal factors underlying the trends

Legislation and individual and mass perceptions undoubtedly work in a symbiotic way to influence the reporting of problems at work. Skrabenek and McCormick (1990) highlight the power of new verbal labels (such as 'RSI' or 'CTD') to influence perceptions of the world. Awareness campaigns (and other types of propaganda), function, according to this argument, by telling people about something and describing it in a certain way (in this case, as an occupational disease or injury). A convenient verbal label such as 'RSI' acts as a kind of 'cognitive coat hanger' with which to interpret new experiences in the light of information stored in memory.

Hadler (1996) and others have suggested that people have always had symptoms of WMSDs both at work and elsewhere. What has happened is that they have been told that these experiences are a disease or an injury and have been given a name for it. When they also learn that the disease is compensatable, they naturally make claims. A positive feedback loop occurs when successful claims cause policy makers to recognise other claims in this category as compensatable. The media amplify the feedback and more people make claims. Thus, at a national level, an 'epidemic' occurs with a sudden increase in claims as long-term sufferers suddenly file for compensation. Once the backlog of unclaimed compensation is absorbed there is a peak in the data and the situation levels off as only new claims enter the system (Brogmus *et al.*, 1996).

This explains why, for example, there was a sudden peak in 'RSI' in Australia in the 1980s. A similar phenomenom occurred in the USA in 1971 and 1981–1984 when new legislation appeared regarding occupational hearing loss. Peaks in claims occurred following the introduction of more stringent legislation. Brogmus *et al.* conclude, after examining the US trends, that WMSDs have always existed in industry, primarily in manufacturing, but tended not to be regarded as occupationally related. There are no data to support the idea that they are primarily related to VDU use. Furthermore, since reporting of disorders is a psychosocial and legal phenomenon that depends very much on people's perceptions, and because the high reporting may well subside, it would be unwise to overemphasise WMSD and WMSD control as industrial priorities, nor, indeed to introduce too-strict standards for controlling exposure to ergonomic risk factors.

WMSDs in perspective

Webster and Snook (1994) analysed Liberty Mutual Insurance Company claims data to arrive at cost estimates for WMSDs during 1989. According to US Bureau of Labor Statistics, WMSDs accounted for 61% of occupational illnesses in 1991, but occupational illnesses represented only 5.8% of illnessess and injuries combined. Clearly, the practice of reporting WMSD incidence as a percentage of occupational illnesses, rather than as a percentage of all injuries and illnesses, greatly exaggerates the seriousness of the problem.

Occupational low back pain is classified as an injury in the USA and by itself accounted for 16% of all claims and 33% of all claims costs for Liberty Mutual in 1989. In contrast, WMSDs accounted for 0.83% of all claims and 1.64% of all claims costs, an order of magnitude less. Webster and Snook concluded that the costs to society of WMSDs do not appear to be as significant as is often suggested. However, the authors caution that WMSDs may represent a significant problem in particular industries and occupations where highly repetitive or forceful movements or awkward postures are required.

Table 5.5 Per capita annual injury and illness costs in high-medium- and low-cost jobs[a]

Rank	Occupation	Average Cost/Employee ($)
1	Timber cutting and logging	5733
2	Production helpers	3679
5	Millwrights	1670
10	Grinding, buffing, polishing machine	1250
25	Sheet metal workers	839
50	Telephone installers and repairers	567
75	Automobile body repairers	435
90	Drywall installers	353
95	Supervisors, production occupations	337
99	Painters, construction and maintenance	311
100	Stock and inventory clerks	310

[a] From Leigh and Miller (1997). Ranking indicates rank from 1 (highest cost job) to 100 (lowest cost job).

Effectiveness and cost-effectiveness

There is a great deal of data on the costs of WMSDs and good case study material for the effectiveness of interventions. Leigh and Miller (1997) ranked over 100 occupations according to the cost of job-related injuries and illnesses using US Bureau of Labor Statistics. Table 5.5 presents data to illustrate high-, medium- and low-risk jobs.

Webster and Snook (1994) estimated that the mean cost per case for upper body WMSDs was $8070 and the median cost was $824 (which suggests that the distribution of costs is skewed to the right by a small number of very expensive cases). Medical costs were 32.9% of the total costs and idemnity costs 65.1%. The total compensable cost for the USA was $563 million (this does not include production losses or any indirect costs associated with the WMSDs).

Managing musculoskeletal pain in aircraft assembly

Melhorn *et al.* (2001) evaluated a risk management programme for new workers. The programme combined traditional occupational medical assessment and treatment of workers with musculoskeletal pain with medical data collection, surveillance to identify problem areas, employee education, and ergonomic, engineering and design changes. The risk assessment and medical cost per injured worker in the programme was $2468 compared to $3800 in a matched group of controls (the company-wide average cost was $2691). The programme yielded a saving of $1332 per patient.

Reduction of WMSDs at the Ford Motor Company

Chatterjee (1992) carried out a survey of WMSDs at a Ford Motor Co. facility in Basildon, UK. The factory produced mainly small electrical and mechanical components and there were many short-cycle jobs involving the use of powered, vibrating hand tools. The main disorders identified in the workforce were

- Shoulder tendinitis (swelling of tendons in the shoulder)
- Tennis elbow
- Carpal tunnel syndrome (numbness and tingling in the hand/fingers)
- De Quervain's syndrome (tendinitis in the wrist/thumb area of the hand)
- 'Trigger' thumbs and fingers (pain and stiffness on flexing the fingers)

Approximately 54% of the injured workers needed hydrocortisone injections or surgery and 51% of them were not able to continue with their existing jobs. The median time to develop these problems was 9 years on the job. The ergonomic intervention programme that was set up was similar to those described in other parts of this book and involved educational, medical and engineering interventions. A number of specifically ergonomic interventions were also carried out:

- Seating was improved and more leg space was provided to improve operator access to workstations.
- Reach distances were too large for the small women workers. Work surfaces were too wide, conveyors too far away and component trays were badly positioned. This was corrected.
- Bad postures, excessive use of the 'pinch' grip, wrist pronation, elbow flexion, shoulder rotation, neck flexion and ulnar deviation of the wrist were identified and found to be consistent with the problems identified. For example, carpal tunnel syndrome was associated with exposure of the hand to vibration.
- Many of the jobs required more than 25 000 hand motions per shift. This was reduced by implementing job rotation, giving more rest pauses, and automation.
- Vibration was reduced by substituting automated ultrasound welding for manual assembly and placing plastic sleeve guards with collars. This damped down some of the vibration and reduced grip pressure when applying a downward force.

Engineering redesign commenced in 1987 and by 1990 the incidence rate of WMSDs had fallen from 2.7% of the workforce to 0.1%. The author concluded that it seemed to be that general factors, such as age, initiate the pathological processes that cause the disease, local factors (such as task demands) determine the site of the injury, and the onset of symptoms is triggerred by sudden exposure to high forces or by excessive load. No data concerning the cost of the programme or the savings achieved were given, but it appears, in the light of the discussion above, that the reductions were substantial.

Controlling WMSDs in the telecommunications industry

McKenzie *et al.* (1985) describe an intervention programme at a Western Electric Factory employing 6600 workers. Musculoskeletal disorders were the main cause of lost time and the biggest workers' compensation cost in 1979: 2.2 cases were reported every 200 000 work hours resulting in 1001 lost workdays per year. Some sections of the facility had rates as high as 4.6. A task force was set up consisting of managers, industrial engineers, industrial hygienists and human factors engineers. The task force established engineering controls, trained supervisors and plant engineers in ergonomics, analysed medical records and managed injured workers to ensure that they were relocated to non-injurious tasks.

The engineering controls consisted, essentially, of ergonomic redesign of tools and equipment:

- Rubber sleeve guards with a hand-supporting collar were fitted to all air screwdrivers at a cost of $5.16 per year.
- Radial torque arms were fitted to larger screwdrivers. No vibration or torque is transmitted to the operators and the tools can be operated with minimum force.
- 'Wire wrap' guns were fitted with 'pneumatic triggers' to prevent 'trigger finger' injuries (the gun activates when an orifice in the handle is occluded by the finger). The trigger requires no force to activate it.

The task force discovered that 68% of all WMSDs occurred in only four departments; of 944 cases, 48% were assemblers. Once high-risk jobs and high-risk departments had been identified, a focused intervention strategy was implemented. The biggest risk factors were excessive use of vibrating tools with the hand in a flexed or extended posture, and too much repetitive grasping, squeezing or clipping when using hand tools.

The intervention programme was implemented in 1981, and by 1982 the incidence rate of WMSDs had dropped from 2.2 to 0.53 cases per 200 000 work hours. Lost workdays due to WMSDs dropped from 1001 to 129.

Training to prevent WMSDs

A study was carried out in the Electrolux Factory, Husqvarna, Sweden in 1988 to determine whether sickness absence in assembly workers could be reduced by ergonomics training. New workstations with adjustable equipment were installed as a part of a change in production methods and a decision was made to investigate the feasibility of preventing arm–neck and shoulder disorders in the workers. Thirty-three newly hired and 60 experienced workers participated in the study. In both groups, some workers received ergonomics training and the others were given no specific training. In addition, workers in the 'training' group were monitored at work using electromyography. This was used to assist workers in adopting low-stress working postures and movements (so that the postural exertions made when working were 10% or less of the person's maximum).

Table 5.6 presents absenteeism data after a follow-up period of 48 weeks. As can be seen, among new workers the ergonomic training bought about significant reductions in both musculoskeletal sick leave and sick leave for all reasons. That the

Table 5.6 Sickness absence of new assembly workers and average piece rate earnings

	Training Group (n=15)	*Comparison Group (n=18)*
Age (years)	28.1	28.2
Sick leave days (due to upper extremity problems)	8.4	18.2
Total sick leave (all reasons)	23.7	46.1
Average hourly piece rate earnings (Swedish Kroner)	42.2	42.1

ergonomic training reduced total sick leave suggests that the workers were better able to tolerate the demands of work and could cope better with the discomfort of minor illnesses. The average hourly earnings of the two groups suggest that they were working equally hard and that the reduction in sick leave among those who received ergonomics training was not due to a reduction in work performance in an effort to reduce musculoskeletal stress. Niether is there evidence that the ergonomic methods of working were less efficient than the conventional methods.

For experienced workers, however, the ergonomics training was ineffective. Total sick leave and musculoskeletal sick leave were similar in both trained and untrained workers. It is likely that experienced workers have built up and established motion patterns that are not susceptible to alteration by training.

As a result of the study, the company decided to provide all new workers with ergonomics training. No data on the benefits of the programme were given by the authors, but if (using the Oxenburgh model) 1 day's sick leave is equivalent to 3 days' pay in a Swedish company, then the programme saved the equivalent of 30 days for 15 workers. Up to 2 days could be allowed and the training would be cost-beneficial over 1 year.

Research directions

Although some progress has been made in demonstrating a link between ergonomic exposures and musculoskeletal outcomes, the data set is nowhere near refined enough to enable ergonomists to make specific recommendations that will make dangerous jobs safe. The main limitations of the current data are a preponderance of cross-sectional studies that cannot, in principle, demonstrate cause-effect relationships; a lack of validated methods to measure ergonomic exposures; and imprecise measurement of outcomes (e.g. regional pain reporting as opposed to specific diagnoses). More longitudinal studies are needed to demonstrate cause and effect and establish dose–response relationships using valid tools for exposure assessment and specific diagnoses of outcome. More longitudinal studies to validate ergonomic interventions are needed. Finally, the role of fatigue in the aetiology of WMSDs is not well understood. There is a lack of data to support the prescription of suitable rest periods, their length and frequency, etc., to prevent injury.

Summary

Despite automation and computerisation of work, WMSDs are still prevalent in society. Repetitive work is common. The characteristics of the person, including age and skill level, can interact with the requirements of tasks and the design of tools, leading to excessive demands being placed on the musculoskeletal system. Neck and shoulder strain can be reduced at work by appropriate design of the visual requirements of tasks. Any beneficial effects are likely to be greater in older workers, where the background prevalence is higher and symptoms and more likely to be amplified by task-induced stress.

Muscle and tendon problems such as cramp, tenosynovitis and tendonitis have been shown to be associated with highly repetitive activities. Musculoskeletal stress can be reduced and the efficiency of task performance increased by careful task and tool design. Researchers have recently become interested in problems of the wrist, elbow and shoulder and a number of well-documented syndromes exist that are often, but not always, associated with work activities.

Essays and exercises

1. Carry out a survey of the handle design of a variety of commercially available hand tools for use in the home. Pay particular attention to

    • Handle circumference and length
    • Handle shape and contour
    • Mechanical advantage of handle
    • Position of handle with respect to tool body

    Obtain anthropometric data on hand dimensions and decide how satisfactory is the design of these tools
2. Obtain a number of commercially available screwdrivers and a large piece of hard wood. Using friends or colleagues as subjects, time how long it takes them to screw in a standard size screw with each screwdriver. Use 5 screws per screwdriver, per person, and take the average time for each person–screwdriver combination (use a different screwdriver after each screw to balance out fatigue effects). Can any differences be attributed to handle or blade design? Is previous experience in the use of handtools important?
3. You are a consultant to the production manager of a large motor company. Draft guidelines to assist them with the procurement of hand tools.

6 Design of manual handling tasks

Low Back Pain remains the most prevalent and costly work-related injury.
(Liberty Mutual Research Center Research Report, 1998)

A large proportion of the accidents that occur in industry involve the manual handling of goods. In the USA, about 500 000 workers, suffer some type of overexertion
injury per year. Approximately 60% of the overexertion injury claims involve lifting
and 20% pushing or pulling (NIOSH, 1981). In the UK, more than 25% of accidents
involve handling goods in one way or another (Health and Safety Commission, 1991).
A 10% reduction in manual handling injuries would save the British economy some
£170 million per annum.

The relation between low-back injury and workplace ergonomics is supported by
the findings of epidemiological surveys. Hoogendoorn *et al.* (2000) found an increased risk of low back pain in workers who lifted a 25 kg load more than 15 times
per day. Magora (1972) found that low back symptoms were more common in
workers who regularly lifted weights of 3 kg or more than in those who sometimes
lifted such weights. Interestingly, low back symptoms were even more common in
those who *rarely* lifted weights.

Anatomy and biomechanics of manual handling

When carrying out manual handling tasks, the weight of the load being lifted is
transferred to the spinal column in the form of compression and shear forces. The
compression and shear are greater when the load is lifted quickly because higher
forces are needed to accelerate the mass from rest, according to Newton's laws of
motion.

Thus, larger forces are required to lift an object quickly, rather than slowly, and
these are transferred to the spine. Additional loads are placed on the spine owing to
posture – the more 'off-balance' or assymmetric the posture, the greater the muscle
forces needed to counteract the pull of gravity. Combining accelerations of the trunk
and the load with asymmetric postures introduces a requirement for high antagonistic contractions in the muscle groups around the trunk: one set of muscles acts to
accelerate the load and another to maintain the integrity of the spinal column and
control the acceleration and deceleration of the trunk itself. The result of all this
contraction and co-contraction is increased compression and shear on the vertebral
motion segments (see Granata and Marras, 2000).

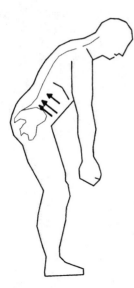

Figure 6.1 The abdominal mechanism in lifting.

The abdominal and thoracic muscles play a major role in stabilising the spine when a weight is lifted according to Morris *et al.* (1961). In relaxed standing, these muscles exhibit little activity. When a person leans forwards to lift a weight, a moment of flexion is placed on the spine. The heavier the weight, the greater the flexion strain. The back muscles contract to resist the flexion (by exerting a moment of extension about the spine) and this is accompanied by antagonistic contraction of the abdominal and thoracic muscles. These muscles pressurise the contents of the abdomen and thorax, converting them into approximate hydraulic and pneumatic splints that are thought to oppose the flexion moment (Figure 6.1). The precise role of intra-abdominal pressure is unclear because, when the abdominal muscles contract, they increase the compression on the spine. Aspden (1989) suggested that intra-abdominal pressure compresses the convex surface of the lumbar lordosis, causing it to stiffen (in the same way that placing a load on a masonry arch increases its stability). Whatever the reason, the reflex increase in intra-abdominal pressure when a person lifts a load appears to be a normal response.

Spinal compression is increased when loads are lifted and is increased even more when they are lifted quickly and when the posture is assymmetric. High load moments and high postural moments increase the forces even more. Lifting technique does influence manual handling efficiency and it is recommended that the hip extensors should play the major role in powering the lift. According to Gallagher and Hamrick (1991), the gluteal muscles can generate an extensor moment about 5–7 times greater than the lumbar erector spinae, which is why trying to 'lift with back' is both inefficient and hazardous.

Back injuries and lifting and carrying

According to Grieve and Pheasant (1982), the trunk can fail in three ways when a weight is lifted:

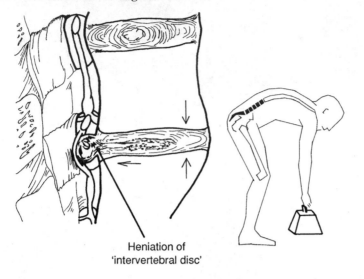

Heniation of
'intervertebral disc'

Figure 6.2 Basic mechanism of a slipped disc. The annular fibres rupture and the nuclear
material is extruded posteriorly. (Adapted from Keegan, 1953, with permission.)

1. The muscles and ligaments of the back can fail under excessive tension.
2. The intervertebral disc may herniate as the nucleus is extruded under excessive compression.
3. The abdominal contents may be extruded through the abdominal cavity owing to excessive intra-abdominal pressure.

These injuries are often referred to colloquially as 'muscle strains or tears', 'slipped discs' and 'hernias'.

The term 'slipped disc' is particularly misleading – the correct term is 'prolapse' as the nucleus pulposus is extruded postero-laterally through ruptured annular fibres. Because the posterior and lateral fibres of the annulus fibrosus are weaker than the anterior fibres (Adams and Dolan, 1995), herniation usually occurs postero-laterally. It would seem that bending to the side to lift objects would place the discs at increased risk as lateral flexion loads the weaker parts of the disc. Added forward flexion would seem to increase the risk even more. As the load is grasped, the spine is subjected to compression and when the load is picked up, the spinal extension may 'trap' some nuclear material posteriorly, stretching the posterior ligaments and pressurising nerve roots. This may serve as a basic description of the mechanism of disc herniation (Figure 6.2). The injury is one of prolapse of the annular fibres, rather than slippage of the body of the disc itself; to prevent it, from first principles, it would seem best to avoid complex spinal postures and motions in which forward flexion, lateral flexion and rotation occur simultaneously. There is some evidence that this view is correct. Gagnon *et al.* (1993) investigated a lifting task in which subjects had to bend forward and laterally and twist. They measured trunk postures and load moments when the task was carried out in this way and when a 'pivoting' technique was used. The pivoting technique involved minimising lateral flexion and rotation by making more use of the feet to (turning with the lower half of the body). The trunk posture was safer when the pivoting technique was used and twisting moments were reduced. *For the technique to be usable, free space for the feet is needed in the entire lifting zone.*

A catastrophic injury such as a disc prolapse is not simply caused by a sudden event such as lifting a heavy weight. It is usually the end product of years of degeneration of the disc and surrounding structures. Over time, the layers of cartilage in the annulus fibrosus develop micro-tears and weaknesses, the nucleus may lose fluid and the discs may become thinner, causing the vertebral bodies above and below the disc to move closer together. Parts of the vertebral bodies may then be subject to increased load as the degenerated discs take less of the load. According to Wolff's Law, bone adapts to the mechanical demands placed upon it (being laid down where needed and reabsorbed where not needed) and bony spurs (or osteophytes) may grow around the intervertebral foramen. Excessive loading of the facet joints may also cause problems. This process of degeneration may ultimately result in herniation of an intervertebral disc (Vernon-Roberts, 1989).

A prolapsed intervertebral disc usually causes severe pain for a considerable time. Owing to the visco-elastic nature of the intervertebral disc, the prolapsed material does not return to its original position as soon as the load is released. Sometimes treatments such as lumbar traction can assist in returning the prolapsed material and also relieve pain in the process. Because the intervertebral discs have no direct blood supply in the adult, recovery from injury is slow and it cannot be said that the damaged tissues ever heal completely as previously healthy tissue is replaced by scar tissue (Ring, 1981).

Manual handling injuries seem to be associated with unexpected events such as slips, trips and falls. Marras *et al.* (1987) found that spinal loads were significantly (up to 70%) greater when the spine was exposed to sudden, unexpected loading. The lack of opportunity for physical preparation may cause the back muscles to overcompensate or increase the required forces to prevent loss of balance under sudden, unexpected loading. In either case the implications are clear: workers should be allowed plenty of time to carry out manual handling tasks and ensure that they are not distracted or pressurised by conflicting task demands or exposed to slip, trip or fall hazards.

Prevention of manual handling injuries in the workplace

The perspective of ergonomics is to design tasks to make them safe, but the most common approach in most industries is to train workers to lift safely. Figure 6.3 depicts two lifting techniques. Safety propaganda directed at workers usually describes these techniques as 'unsafe' and 'safe', respectively.

The notion that it is safer to 'lift with the knees and not with the back', that people can be trained to lift safely and that injury will be prevented, is deeply ingrained despite the large number of studies that have shown no benefits of the training. Snook *et al.* (1978) compared three approaches to low back injury prevention: pre-employment/pre-placement selection (medical history taking, medical examination, low back x-ray); training in lifting techniques; and job design (whether the work could be done by more than 75% of the workforce without overexertion). The findings showed no difference in the proportion of injuries in companies that did or did not train their workers in lifting techniques, nor were there any effects due to selection based on medical screening. Significantly fewer back injuries were found in companies where the loads were acceptable to more than 75% of the workforce. Snook *et al.*

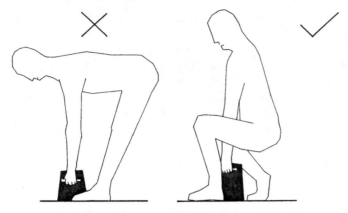

Figure 6.3 'Safe' and 'unsafe' lifting techniques according to much safety propaganda. But are the 'safe' techniques safe enough to reduce the rate of occupational back injury?

concluded that workers are 3 times more likely to hurt their backs when performing exertions acceptable to less than 75% of the workforce. Furthermore, there was scope for a 67% reduction in injuries through job redesign. Stubbs *et al.* (1983b) found little relationship between the point prevalence of back pain among nurses and the time spent training in safe lifting techniques. They concluded that if the lifting task is intrinsically unsafe (and in the case of manual handling of patients this appears to be the case), no amount of training will solve the problem. A better approach would therefore be to do away with the lifting task or redesign it to make it safe.

More recently, Daltroy *et al.* (1997) reported the results of a study of 4000 US postal workers over 5 years. Workers and supervisors were given 3 hours of 'back school' training followed by 3 to 4 reinforcement sessions over the next few years. Those subsequently suffering a back injury were randomly assigned to a control group or were given further training. Although subjects' knowledge of safe behaviour was influenced by the training, *there were no overall beneficial effects* of back school training on rates of primary injury (trained workers compared to controls) or on re-injury. The training programme had *no beneficial effects* on the rate of low back injury, the median cost per injury, the time off work following injury, the rate of other musculoskeletal injury or the rate of repeated injury after returning to work. The back injury rate was 22.1 per 1000 worker-years which is about average for the USA. The rate of injuries causing lost workdays was 10.4 per 1000 worker years and was *higher* in groups of workers who had received the training, but the difference was not statistically significant. Some groups of trained workers had lower injury rates than untrained workers (e.g. 16.4 injuries per 1000 worker years versus 19.4 injuries) but the difference was not statistically significant.

'Safe' lifting techniques. Dangerous assumptions about manual handling safety

ASSUMPTION NO. 1. THE WAY THAT UNTRAINED PEOPLE VOLUNTARILY HANDLE LOADS IS UNSAFE

To justify training people in special lifting techniques, we first need to demonstrate that the way untrained people lift weights is unsafe. There is a great deal of research linking back injury to the manual handling of loads and to the postures and motions involved in carrying out lifting tasks. However, there is no evidence that people voluntarily stoop or twist and eventually injure themselves because they neglect to use the 'correct' technique. Too often, workers are constrained by the design of the task and have little choice about how to lift. A study by Allread *et al.* (2000) provides indirect support for this assertion. They investigated the variability in trunk motions with respect to individual differences and repetition. Comparing different jobs, they found that most of the variability between them was due to the design of the job rather than who was doing it or how frequently. The best way to change the lifting technique people use would seem to be to change the design of the workspace and the task.

ASSUMPTION NO. 2. THE TECHNIQUES BEING TAUGHT ARE SAFER, IN PRACTICE

Although there is some evidence that lifting from a squatting position is safer than lifting from a stooping position, squat lifting, when performed by competitive weight-lifters, uses weights specially designed to be lifted from a squatting position. Most weights in industry are not *designed* to be lifted from a squatting position, so it is no surprise that the squat technique is rarely seen. In some situations, as in lifting an unstable load such as a large bag of wet laundry, squat lifting techniques may actually increase the load moment or they may be completely impractical. Rabinowitz *et al.* (1998) compared the squat and stoop lifts in a crate handling task and found no difference in spinal shrinkage (a measure of spinal strain) between the two techniques.

ASSUMPTION NO. 3. 'SAFE' TECHNIQUES ARE USABLE AND HAVE NO 'HIDDEN COSTS'

Squat lifting techniques require greater coordination and control than the alternatives and also place a higher load on the cardiovascular system and the knees. For one-off lifts, the additional demands may be acceptable, but for repetitive lifting (e.g. unloading crates of beer) they soon take their toll. The knees weaken rapidly beyond about 60 degrees of knee flexion and the knee ligaments are at increasing risk of rupture (see Grieve and Pheasant, 1982, for further discussion). Rabinowitz *et al.* (1998) found that stoop lifting was associated with greater back pain and squat lifting with greater knee pain. Repetitive squat lifting for 15 minutes placed an escalating cardiovascular load of an extra 26 heart beats/min compared with stoop lifting. People rated the task as 'somewhat hard' compared with stoop lifting, which was rated as 'light'. As early as 1961, van Wely reported that the stoop lifting technique was physiologically more efficient when lifting loads over 20 kg. The usability of the deep squat is suspect for older workers who may well decide, quite rationally in their case, to 'lift with the back to save the knees and heart'.

ASSUMPTION NO. 4. THE TRAINING WILL TRANSFER TO THE WORK SITUATION

The author knows of only one study that demonstrates long-term (6-month) change in lifting technique as a result of manual handling training. It is an assumption not only that such training will transfer but that it *can* transfer. Ergonomists have long understood the principle that well-learnt behaviours cannot be 'unlearnt'. Although you can teach an old dog new tricks, the old ones persist in long-term memory and return to dominate behaviour as soon as we cease to consciously monitor our performance. Anyone who has found themselves in a unfamiliar make of car will have experienced the embarrassment of turning the windscreen wipers on to indicate a turn when distracted by fellow passengers or young children.

The positive benefits of not injuring oneself whenever a load is lifted may be insufficient to reinforce what was taught. The hidden costs of these techniques may negatively reinforce the training, which is quickly unlearnt in the workplace.

ASSUMPTION NO. 5. ANY REDUCTIONS IN RISK ARE LARGE ENOUGH TO PROTECT PEOPLE FROM INJURY

For most people the spine is flexed to 50% of its maximum in a squatting position (Adams and Dolan, 1995). Training people to 'make more use of the legs' does not guarantee lower back stress. Paradoxically, 'lifting with the back' or 'stoop lifting' is accomplished mainly by bending from the hips. The lift is powered by the hip extensors, muscles with an overabundance of power for returning the trunk to an erect position.

Although squat lifting may reduce back stress by lowering the load moment, it is less clear whether the reduction is sufficient to prevent injury. Spinal tissues have a compression tolerance limit or threshold (Genaidy *et al.*, 1993). A lowering of the load will only bring about a reduction in injury rates if the compressive forces are bought below threshold: if the absolute level of risk is reduced to a safe level. The point is illustrated in a study by Marras *et al.* (1999) in which different patient handling techniques were compared. Using a low back disorder model that gives quantitative estimates of risk, it was found that, although the risk differed between the techniques, they were all very likely to be of high risk.

A reduction in risk is not the same as an improvement in safety. This is clearly recognised in the European Union manual handling guidelines, which state that manual handling should be avoided as much as possible.

ASSUMPTION NO. 6. THERE ARE NO PERVERSE OUTCOMES ASSOCIATED WITH THE USE OF 'SAFE' HANDLING TECHNIQUES

There is experimental evidence that people will lift heavier weights when they feel safe than when they feel unsafe (McCoy *et al.*, 1988). Bridger and Friedberg (1999) interviewed 50 managers in a range of light industries and found that their estimates of maximum acceptable loads for their workers were over 50% higher when the managers were told that the squat lift was to be used. The implication is that manual handling training could act as a barrier to change by creating the impression that 'something has been done'.

The content of safety training programmes

One possible explanation for the dismal record of training is that the concepts of safe lifting embodied in the programmes are too simplistic and are incomplete – protection is lost because the techniques do not map well onto the actual task requirements. McGill and Norman (1992) offered some tentative lifting guidelines based on the current knowledge of lumbar spine biomechanics (Table 6.1).

Table 6.1 Tentative guidelines for safe lifting[a]

---

1. Maintain normal lumbar lordosis and use the hip joints to flex and for extend the trunk.
2. Do not lift immediately after prolonged flexion – allow time for the disc nucleus to equilibrate and for the posterior ligaments to regain their stiffness.
3. Avoid lifting shortly after rising from bed.
4. Follow the pre-stress system. Lightly co-contract the stabilising muscles of the trunk to remove slack from the system and stiffen the spine.
5. Choose a posture that minimises the load moment on the lumbar spine (i.e. that gets you close to the load) but don't compromise point 1.
6. Avoid twisting (either to reach the load or to pick it up).
7. Exploit the acceleration profile of the load (experienced lifters only). Grasp the load and extend the trunk rapidly to impart explosive momentum. The 'ballistic' effect will keep the load moving through weak postures until it can be 'caught' in a strong, upright posture.

---

[a] McGill and Norman (1992).

Abdominal belts: health or hoax?

Research is currently under way to evaluate the claimed benefits of wearing tight-fitting elasticated or rigid belts during manual handling. The twofold aim of this research is to determine

*   Whether the occupational use of the belts confers some kind of biomechanical advantage that will protect the spine from injury in the workplace
*   Whether the occupational use of the devices does not have unwanted or unexpected side effects that impact negatively on the health of the worker

The practice of wrapping materials around the waist with the aim of improving posture and poise is found throughout history and across cultures. Shah (1993), for example, reports that in Nepal most people who lift and carry heavy weights wrap a 5-metre length of cloth (called a 'Patuka') around the waist before work. Anecdotally, it is thought to reduce the prevalence of occupational low back pain. Abdominal belts are thought to protect workers by restricting undue flexion or rotation of the spine and possibly augmenting IAP (intra-abdominal pressure), which then lowers spinal compression by exerting a hydraulic tensile force on the diaphragm and a spinal extensor moment.

Alternatively, IAP may protect the loaded lumbar spine indirectly. Contraction of the abdominal musculature sets up lateral forces that act on the spine via the pelvis, ribcage and lumbodorsal fascia, acting like guy ropes that stabilise a mast inside a now rigid cylinder.

Table 6.2 summarises the possible benefits and possible negative side effects of occupational wearing of abdominal belts in industry.

DO ABDOMINAL BELTS INCREASE IAP WHEN WORN?

McGill *et al.* (1990) measured back extensor EMG and IAP when subjects lifted weights wearing a competition weightlifter's belt. Although IAP did increase when the belt was worn (from 99 mmHg to 120 mmHg), there was no corresponding

Table 6.2 Possible benefits and potential negative side effects of abdominal belts in industry

Possible benefits
1.   Increased IAP and reduced spinal compression when lifting
2.   Stabilisation of lumbar motion segments
3.   Stiffening of lumbar spine due to increased IAP
4.   Lifter 'reminded' to avoid lumbar flexion and lift correctly
5.   'Splinting' action of belt. Dangerous motions such as excessive saggital flexion and axial rotation are restricted because the belt stiffens the trunk
6.   Increased sense of security and stability

Possible hazards
1.   Increased IAP but no reduction in spinal compression when lifting
2.   De-conditioning of trunk musculature through long-term use
3.   Increased blood pressure increases risk of blackouts, stroke or heart attack
4.   Increased risk of trunk herniation
5.   Increased sense of security and stability causes workers to take unnecessary risks
6.   Managements issue belts to 'protect' workers instead of redesigning or mechanising hazardous operations

reduction in back extensor activity, suggesting that the belts might not increase IAP enough to have a supportive effect. When subjects held their breath when lifting, increases in IAP were also observed and *were* accompanied by reductions in back extensor EMG, irrespective of whether a belt was worn.

Although there is no evidence one way or the other, increased IAP accompanying belt use may increase the risk of herniation of the trunk.

DOES WEARING AN ABDOMINAL BELT REDUCE BACK MUSCLE FATIGUE WHEN LIFTING?

Ciriello and Snook (1995) measured fatigue of the back extensors in 13 male industrial workers who lifted average loads of 28.1 kg, 4.3 times per minute, for 4 hours a day. A belt was worn on two of the days. Maximum isokinetic endurance decreased by 9–11% after 4 hours of lifting. This change was not significantly different when a belt was worn, neither were there differences in the power spectrum of the EMG signal as indicated by median frequency analysis. Subjective ratings of effort were not influenced by wearing of a back belt.

DOES WEARING AN ABDOMINAL BELT HAVE AN OVERALL PROTECTIVE EFFECT?

Reilly and Davies (1995) evaluated a weightlifter's belt by having subjects lift a 30 kg weight for eight sets of 20 repetitions. Spinal shrinkage as a result of exercise-induced loading was reduced by 49% when the belt was worn (from 4.08 mm to 2.08 mm.) Perceived exertion was also lower. Magnusson *et al.* (1996) compared lifting with and without a belt when subjects lifted 10 kg from floor to desk height twice per minute for 5 minutes. Spinal shrinkage was lower when the belt was used, as was back muscle EMG (normalised with respect to each subject's maximum voluntary contraction). Miyamoto *et al.* (1999) demonstrated that abdominal belts raise the intramsucular pressure in the erector spinae muscles and stiffen the trunk, which may be beneficial during lifting and during other work where the trunk is exposed to de-stabilising forces.

DOES WEARING AN ABDOMINAL BELT GIVE LIFTERS AN INCREASED SENSE OF STABILITY AND SECURITY?

McGill *et al.* (1990), Reddell *et al.* (1992) and Magnusson *et al.* (1996) all report that wearing either competitive weightlifters' belts or abdominal belts for industrial workers increases the sense of security. McCoy *et al.* (1988) found that subjects *self-selected* weights that were 19% heavier when they were wearing belts, which may be evidence for *risk compensation* (see Chapter 15).

DO ABDOMINAL BELTS PROTECT INDUSTRIAL WORKERS IN PRACTICE?

Walsh and Schwartz (1990) divided 90 grocery warehouse workers into three groups in a 6-month investigation. Group 1 (control) received no intervention. Group 2 received a 1-hour training session on back pain prevention. Group 3 received the training and a moulded spinal orthotic brace to be worn at work. There were no statistically significant differences in injury rates or productivity between the three groups over the study period. Lost time was significantly lower in group 3, however (2.5 days lower, on average). The groups were further divided into high- and low-risk workers. High-risk workers in group 3 had significantly fewer injuries and lost time, suggesting that previously injured workers will benefit the most from this form of intervention.

Reddell *et al.* (1992) evaluated an abdominal belt and back programme among a group of airline baggage handlers. Lost workdays and back injuries were not reduced, but back injuries increased and were more severe after belt use was discontinued. Mitchell *et al.* (1994) examined belt use, training, back injury and lost time in a US Air Force base. The predictors of low back injury were as expected: time spent lifting and previous back injury. Back training programmes were found to have a small preventive effect, as did use of back belts. However, the costs of treating injuries when they did occur were found to be higher among belt wearers, leading the authors to conclude that belt use was not indicated in this type of work.

DOES OCCUPATIONAL ABDOMINAL BELT WEARING DE-CONDITION THE TRUNK MUSCULATURE?

The evidence suggests not. McGill *et al.* (1990) noted that even though belt wearing does reduce abdominal muscle EMG during lifting, the reduction is not large. Even when lifting heavy loads (above 70 kg) without a belt, peak abdominal muscle EMG levels are a small percentage of those observed when subjects exert a maximum voluntary contraction of their abdominal muscles. Therefore, the abdominal training effect of lifting is likely to be small (compared, for example with coughing or laughing). Walsh and Schwartz (1990) measured abdominal strength before and after the 6-month study period and found no evidence for a reduction in abdominal strength among belt wearers compared with controls.

IS ABDOMINAL BELT WEARING HAZARDOUS FOR WORKERS WITH LATENT CORONARY HEART DISEASE?

It is known that both belt wearing and breath holding while lifting increase IAP and intra-thoracic pressure. Hunter *et al.* (1989) had subjects hold 40% of their maximum weight in the dead lift posture for 2 minutes. Blood pressure and heart rate were higher when the belt was worn, leading to the conclusion that cardiac-compromised individuals are probably at greater risk when exercising while wearing back supports. It is known that increased intra-truncal pressure hinders venous return to the heart and is followed by a 'rush' upon pressure release that can cause unconsciousness in extreme cases. Rafacz and McGill (1996) found that abdominal belts increased mean diastolic blood pressure by 5 mmHg in a variety of tasks, and McGill and warned that belts may increase the strain on the cardiovascular system of workers. Other studies have not demonstrated increases in systolic or diastolic blood pressure. McGill *et al.* (1990) speculate that this may explain the high incidence of heart attacks among unfit people carrying out cyclic lifting activities such as snow shovelling.

DOES ABDOMINAL BELT WEARING CAUSE GASTRIC SYMPTOMS?

Over the course of an 8-hour working day, Belt wearing did not appear to cause gastro-oesophageal reflux, in people with competent lower oesophageal sphincters, as measured by objective and subjective means (Forouzandeh *et al.*, 1998).

Summary

1. Although the mechanism is unclear, belts do seem to provide some type of protection in light tasks such as grocery selection. Previously injured workers benefit the most.
2. With heavy tasks, the evidence is less clear and there are real concerns about whether belts create a false sense of security and encourage workers to lift heavier weights.
3. There are real concerns about possible side effects of occupational belt use. Although there is no evidence for deconditioning of the trunk muscles, chronically increased IAP in the workplace may bring with it cardiovascular and other risks.
4. The possible benefits of abdominal belts for occupants of vehicles, exposed to vibration and shock are unknown.

Precautions

Potential belt users are advised to follow the precautions outlined by McGill. These belts should not be made generally available to workers and should only be issued to individuals if the following conditions are satisfied:

•   Mandatory cardiovascular screening of potential users
•   Mandatory education in lifting mechanics
•   Mandatory full assessment of the candidate's job
•   Restriction of belt wearing to short-term use only

Low back disorder models, risk assessment and task redesign

Marras and his colleagues have developed a quantitative approach to back injury prevention using a device known as a lumbar motion monitor or LMM (Figure 6.4). The LMM captures data on trunk posture, velocity and acceleration in flexion,

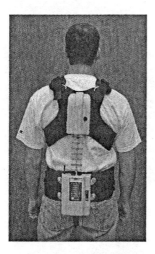

Figure 6.4 A lumbar motion monitor. (Courtesy of Professor W. S. Marras.)

lateral flexion and rotation. Combining these with data on the lifting rate and the load moment, a quantitative risk analysis can be carried out by feeding all the data in to the low back disorder model (Marras *et al.*, 1993). This model was determined empirically using data from 403 jobs differing in back injury rate. A combination of static and dynamic trunk posture variables predict risk of back injury:

- Maximum sagittal flexion
- Maximum lateral velocity
- Average twisting velocity
- Load moment
- Lift rate

Use of the model in risk assessment is illustrated in a study by Bridger *et al.* (1998). Two designs of spade were compared – a conventional spade and a spade with a second handle fitted to the neck (Figure 6.5), intended to reduce the amount of stooping needed to dig. The output of the low back disorder model, using data for the two kinds of spades, is given in Table 6.3.

The data indicate that the hazardous components of digging with the conventional spade are the digging rate used in the experiment, maximum sagittal flexion (stooping) and the load moment. Lateral flexion velocity and twisting velocity are not

Table 6.3 Biomechanical risk (probability of task being a member of a group of tasks with high risk of low back injury) associated with five components of digging

Component of digging	Conventional spade	Two-handled spade
Max. lateral velocity	0.30	0.20
Max. sagittal flexion	0.96	0.85
Average twisting velocity	0.06	0.13
Load moment	0.55	0.27
Lift rate	0.98	0.98
Overall probability	0.57	0.49

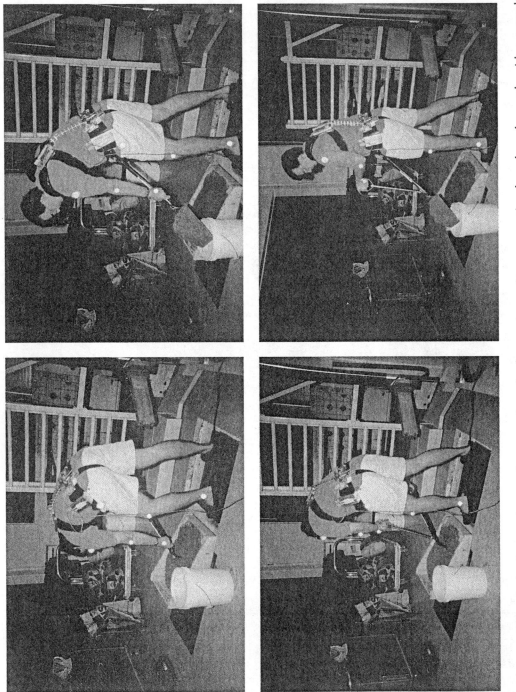

Figure 6.5 Using the lumbar motion monitor to evaluate lumbar motions using a conventional spade and a spade with a second handle, designed to reduce stooping.

hazardous. Although the redesigned spade *does* reduce the sagittal flexion, the probability of high-risk group membership is still high and the reduction is not enough to make the task safe. Interestingly, the reduction in stooping is accompanied by an increase in twisting velocity (although not to levels that markedly increase the probability of high-risk group membership), which suggests a compensatory recruitment of the trunk musculature. This example illustrates the use of the LMM and the low back disorder model to break down a lifting task into components of risk. The next step, in practice, would be to redesign the task to reduce the risk, starting with the component having the highest probability. In the digging example, this would be to lower the digging rate, followed by a reduction in sagittal flexion and load moment.

Validity of the low back disorder model

Marras *et al.* (2000) used the LMM to assess 36 jobs in 16 companies. Thirty-two jobs were redesigned. Significant associations were found between changes in the low back disorder risk values and changes in low back disorder incidence rates. The most effective interventions seemed to be the introduction of lift tables (which bring the load closer to the worker) and lifting aids (which support the load).

Design of manual handling tasks

NIOSH has produced a 'Work Practices Guide' for the design of manual handling tasks and an equation for determining safe loads (see below). In Europe, a directive for the design of these tasks has been issued and the UK Health and Safety Commission consultative document (Health and Safety Commission, 1991) provides proposals for the design of manual handling tasks. Three principles of industrial medicine (first, remove the threat; second, remove the operator; third, protect the operator) are usefully paraphrased in the context of manual handling in Table 6.4.

Table 6.4 General principles for the control of manual handling hazards in the workplace

1. Avoid hazardous manual handling as much as possible.
2. Assess any hazardous operations and redesign the task to obviate the need to move the load or automate or mechanise the process.
3. Reduce the risk by providing mechanical assistance, redesigning the load itself or redesigning the workspace.

Task requirements

Some factors that increase postural stress in lifting and carrying are given in Table 6.5.

Lifting from a seated position deserves special consideration. The flexion moment about the lumbar spine is greater in sitting than in standing (Nachemson, 1966) and thus the spine can be considered to be 'pre-stressed'. In sitting, the leg muscles cannot be used to generate a ground reaction force and more is required of the muscles of the trunk. Stooping from a seated position is particularly hazardous. It is curious that, in the light of these facts, furniture designers still design desks with the largest drawer (to contain the heaviest objects) at the bottom, close to the floor.

Table 6.5 Some task factors that exacerbate postural stress in manual handling

1.   Having to grasp or hold the load at a distance from the trunk
2.   Having to twist the trunk while supporting or lifting a load
3.   Having to lift or lower objects placed below knee or above shoulder height
4.   Having to lift or move the load through large vertical or horizontal distances
5.   Having to hold or carry the load for long periods
6.   Having to lift or carry frequently
7.   Having to lift while seated

The characteristics of the load itself should also be considered. Weight is not the only consideration. The size of the object is also important: 20 kg of lead is, in a sense, 'lighter' than 20 kg of feathers since, being smaller, it can be held closer to the body. Containers for one- or two-handed handling should be designed as small as possible so that the load is kept close to the body. The provision of handles is important: well-designed handles can facilitate secure grasping of the load and prevent sudden movements of the combined person/load COG. Conversely, slippery materials or unstable loads (lacking rigidity) can increase the risk of injury.

Personal characteristics of workers

When designing lifting tasks it is essential to consider the characteristics of the worker population. Owing to the requirements for task design described above, straightforward anthropometric considerations are not as useful in the design of lifting tasks as they are in seating, although a tall workforce will be better able to handle objects stacked higher but not low objects, and vice versa for short workers. The following personal considerations are important.

People whose occupations typically do not require them to lift objects seem to be particularly prone to injury (Magora, 1972). It is undesirable to expect office workers, drivers or professionals to lift heavy objects either occasionally or in emergencies. There is still some controversy over whether the level of fitness of workers is related to the probability of them suffering a back injury. It has been suggested that hyper- or hypo-mobility of the lumbar spine may be a risk factor for back problems, but empirical studies have failed to show a clear relationship between spinal flexibility and future risk of back trouble; flexibility would not appear to be useful for screening out workers at high risk of injury (Battie *et al.*, 1990; Burton, 1991).

Although women are generally less capable than men of handling heavy loads, sex is not always a useful consideration since there is considerable overlap in the abilities of males and females in the workforce. Age should be considered, since it is known that muscular strength declines from middle age onwards. Middle-aged people are particularly prone to low-back problems owing to the instability of their lumbar motion segments.

If the workforce contains a large proportion of women, some special considerations are relevant and, generally speaking, manual lifting of heavy loads should be avoided because it increases the stress on the trunk. According to Hayne (1981), menstrual pain may incapacitate some women and the increased intra-abdominal pressure when lifting may increase menstrual flow. It may also increase the risk of miscarriage in the first few months of pregnancy. Menopausal women often suffer

Table 6.6 Some workspace factors that exacerbate postural stress in manual handling

---

1. *Confined spaces.* The ability to exert forces decreases when space is restricted. Restricted headroom is a good example (Grieve and Pheasant, 1982). Less use of the legs is possible, which increases the load on the trunk muscles
2. *Height of object.* Only items placed between knee and elbow height should be lifted. Conveyors, shelves and palletising systems should be designed to allow for this
3. *Flooring.* Space for the feet should be provided both underneath the load and around the worker. Slippery floors should be avoided

---

from osteoporosis (demineralisation of bone leading to a loss of bone strength), particularly in the spine and pelvis. This increases the risk of trunk failure when lifting a load. A detailed discussion of the many other factors affecting women at work can be found in Chavalitsakulchai and Shahnavaz (1990).

Given that lifting is known to be a hazardous activity and that there are many good reasons why women, in particular, should not have to lift heavy weights, the special case of back pain in the nursing profession deserves some consideration. Nursing is one of the few jobs where women have to lift heavy (and awkward) weights (i.e. patients) on a regular basis. Stubbs *et al.* (1983a) reported the findings of an epidemiological study of back pain in 3912 nurses. One in six nurses attributed their back pain to a patient-handling incident; 750 000 working days were lost annually due to back pain (16% of all sick leave) and 78% of the pain was in the lower back.

Workspace design

Many aspects of workspace design can increase the risk of injury when lifting. Some of the more important considerations are given in Table 6.6.

Design of lifting tasks

Ayoub (1982) has summarised many of the guidelines for the design of lifting tasks. These are presented, in modified form, in Tables 6.7, 6.8 and 6.9.

Table 6.7 How to minimise the weight to be handled[a]

---

1. Assign the job to more than one person.
2. Use smaller containers.
3. If possible, mechanise the process.
4. Machines, rather than employees, should transfer loads between surfaces.
5. Change the job from lifting to lowering, from lowering to carrying, from carrying to pulling, and from pulling to pushing.
6. Use handles, hooks or similar features to enable workers to get a firm grip on objects to be lifted.[b]
7. Reduce the weight of containers used to transfer objects.
8. Balance and stabilise the contents of containers to avoid sudden shifts in load during a lift.
9. Design containers so that they can be held close to the body.
10. Treat work surfaces to allow for ease of movement of containers.

---

[a] Modified from Ayoub (1982) with permission of the *Journal of Occupational Medicine*.
[b] Maximum acceptable weights are about 16% lower if no handles are used (Ciriello *et al.*, 1993).

Table 6.8 How to minimise reach and lift distances[a]

1. Increase height at which lift is initiated; decrease height at which it terminates.
2. Stack objects no higher than shoulder height.
3. Store heavy components on shelves between shoulder and knuckle height.
4. Avoid deep shelves.
5. Avoid side to side lifting from seated position.
6. Provide access space around components to cut down on the need for manual repositioning.
7. Storage bins or containers should be fitted with spring-loaded bottoms.
8. Use sloped surfaces to gravity-feed items to the point of lifting.
9. Provide free space around and under the work surface to increase functional reach.

[a] Modified from Ayoub (1982) with permission of the *Journal of Occupational Medicine*.

Table 6.9 How to increase the time available for lifting[a]

1. Increase the time by relaxing the standard time for the job.
2. Reduce the frequency of lifts.
3. Introduce job rotation to parcel out lifting between workers.
4. Introduce appropriate work–rest cycles.

[a] Modified from Ayoub (1982) with permission of the *Journal of Occupational Medicine*.

The NIOSH approach to the design and evaluation of lifting tasks

NIOSH (the National Institute for Occupational Safety and Health) has developed an equation for calculating the recommended weight limit (RWL) for a specific lifting task a worker could perform for a specified period without an increased risk of low back pain. The equation has been determined empirically and specifies a weight limit as a function of the values of specified task variables. The original equation was developed in 1981. It has recently been updated to cover a wider range of tasks, including asymmetrical lifting tasks. Three criteria have been used to develop the equation (Table 6.10).

The use of three criteria is essential if a wide range of tasks is to be evaluated because different tasks impose different loads on workers. For example, infrequent handling of heavy, awkward objects may be limited by biomechanical rather than physiological factors. Repetitive lifting of light objects may be limited by metabolic stress and local muscle fatigue. Workers' perceptions of their capability are also a limiting factor. Sometimes, the different criteria will produce conflicting RWLs, in which case the lowest is used.

Table 6.10 Criteria used in the NIOSH lifting equation

Consideration	Criterion	Cut-off value
Biomechanical	Maximum disc compression	3.4 kN
Physiological	Maximum energy expenditure	9.2–19.7 kJ/min
Psychophysical	Maximum acceptable weight For 75% of females and 99% of males	

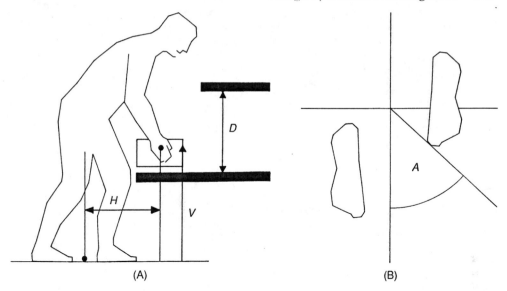

Figure 6.6 NIOSH Approach to lifting task evaluation. (A) H = distance of load from mid-
point; V = height of hands above floor; D = distance through which load is lifted:
(B) Angle of asymmetry of lift, A.

The approach taken by NIOSH has been to recommend a maximum load under
ideal lifting conditions. The RWL is taken to be a load of 23 kg, lifted in the sagittal
plane from a height of 75 cm above the floor and held 25 cm in front of the body.
The load is to be lifted no more than 25 cm vertically, there is to be a good coupling
between the load and the lifter (which is achievable using handles or occurs because
of the shape of the load itself) and the load is only to be lifted occasionally. Thus, the
conditions for lifting a maximum load are specified.

Six coefficients have been developed that reduce the RWL to account for task
factors that cause departures from the ideal situation. The RWL is multiplied by
the coefficients (which are less than 1) to arrive at a new RWL for the specified
conditions. The values of the coefficients have been determined using biomechanical
models of spinal loading and the findings of epidemiological and psychophysical
studies. The 1991 NIOSH equation is presented in Table 6.11 together with definitions of the coefficients and the terminology used in the equation (Figure 6.6).

The ratio between the load actually lifted and the RWL is known as the Lifting
Index. An index of less than 1 is believed not to increase the risk of injury. An index
of 3 or more indicates that many workers will be at increased risk and the task
should be redesigned.

NIOSH's goal in developing the equation is to specify controls on industrial lifting
that will protect healthy workers. It is not valid to apply the equation to set limits
for those with already damaged backs. Further details of the NIOSH method can
be found in the publications of the Institute (e.g. PB91-226274) as well as in Waters
et al. (1993). The above should serve to illustrate NIOSH's admirably rigorous
approach to the pervasive problem of occupational lifting. The approach has the
distinct advantage of lending itself to implementation as an interactive decision aid
for the industrial engineer or designer.

Table 6.11 1991 NIOSH lifting equation

$RWL = LC \times HM \times VM \times DM \times AM \times FM \times CM$

LC = load constant of 23 kg
HM = horizontal multiplier = $(25/H)$
VM = vertical multiplier = $1 - (0.003IV - 75I)$
DM = distance multiplier = $0.82 + (4.5/D)$
AM = asymmetric multiplier = $1 - (0.0032A)$
FM = frequency multiplier (from Table 6.12)
CM = coupling multiplier (from Table 6.13)

where

H = horizontal distance of the hands from midpoint (0) of the ankles
V = vertical distance of the hands from the floor
D = distance through which the load is lifted
A = angle of asymmetry (Figure 6.4)
F = frequency of lifting (lifts/min every 1, 2 or 8 hours)

Table 6.12 Values of frequency multiplier FM for use in the 1991 NIOSH equation for determining RWL

Frequency (lifts/min)	Work duration					
	≤ 1 hour		≤ 2 hours		≤ 8 hours	
	V < 75	V < 75	V < 75	V < 75	V < 75	V < 75
0.2	1.00	1.00	0.95	0.95	0.85	0.85
0.5	0.97	0.97	0.92	0.92	0.81	0.81
1	0.94	0.94	0.88	0.88	0.81	0.81
2	0.91	0.91	0.84	0.84	0.65	0.65
3	0.88	0.88	0.79	0.79	0.55	0.55
4	0.84	0.84	0.72	0.72	0.45	0.45
5	0.80	0.80	0.60	0.60	0.35	0.35
6	0.75	0.75	0.50	0.50	0.27	0.27
7	0.70	0.70	0.42	0.42	0.22	0.22
8	0.60	0.60	0.35	0.35	0.18	0.18
9	0.52	0.52	0.30	0.30	0.00	0.15
10	0.45	0.45	0.26	0.26	0.00	0.13
11	0.41	0.41	0.00	0.23	0.00	0.00
12	0.37	0.37	0.00	0.21	0.00	0.00
13	0.00	0.34	0.00	0.00	0.00	0.00
14	0.00	0.31	0.00	0.00	0.00	0.00
15	0.00	0.28	0.00	0.00	0.00	0.00
>15	0.00	0.00	0.00	0.00	0.00	0.00

Table 6.13 Values of coupling multiplier CM for use in the 1991 NIOSH equation for determining RWL

Coupling	V < 75 cm	V ≥ 75 cm
Good	1.00	1.00
Fair	0.95	0.95
Poor	0.90	0.90

NIOSH equation: a worked example

A worker has to unload trays of pork pies as they emerge from an oven. He picks up a tray, turns 45 degrees and places it on a conveyor. He does this 3 times per minute for 8 hours. You have been asked to investigate the task and have taken the following measurements with a tape measure.

$H = 45$ cm, $V = 60$ cm, $D = 70$ cm, $A = 45$ degrees. (estimated angle of asymmetry by visual inspection)

Step 1. Calculate multipliers
HM = 25/45 = 0.56
VM = $1 - (0.003|V - 75|) = 1 - (0.003|60 - 75|) = 1 - 0.045 = 0.955$
DM = $0.82 + (4.5/D) = 0.82 + (4.5/70) = 0.82 + 0.06 = 0.88$
AM = $1 - 0.0032A = 1 - 0.0032(45) = 1 - 0.144 = 0.856$
FM (from Table 6.12) = 0.55
CM = 1 (from visual inspection, trays are easy to hold)

Step 2. Calculate RWL
RWL = $23 \times 0.56 \times 0.955 \times 0.88 \times 0.856 \times 0.55 \times 1 = 5.08$ kg

Step 3. Calculate Lifting Index
A full tray weighs 12 kg.
Lifting Index = 12/5.08 = 2.36

Interpretation

Although less than 3, the lifting index is quite high and the task cannot be described as safe. In order to redesign the task to make it safer, we should first examine the values of the multipliers. Listing them in order from smallest to largest:

FM = 0.55
HM = 0.56
AM = 0.856
DM = 0.88
VM = 0.955
CM = 1

We find that the frequency of lifting and the horizontal distance contribute the most to the reduction in the load constant from 23 kg. Changing the lifting frequency will lower production and might be counterproductive if the weight of the trays has to be increased to compensate. However, redesign of the trays so that a full tray with the same number of pies on it can be held 25 cm from the body will change the value of HM to 1. The new RWL will then be

RWL* = $5.08 \times 1/0.56 = 9/07$

And the Lifting Index now becomes

LI = 12/9.07 = 1.32

Further reductions could be achieved by removing the asymmetry. As can be seen, the general approach to task redesign should be guided by the numerical value of the multipliers, beginning with the risk factors associated with the *smallest* multiplier values, as these will have the biggest impact.

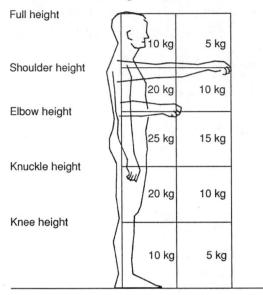

Figure 6.7 Recommended loads for lifting. (UK Manual Handling Regulations, reproduced with permission.)

Maximum loads for lifting and lowering

Figure 6.7, from UK manual handling regulations, gives recommended maximum loads for sagittal lifting. These can be used for initial screening of manual handling tasks and it is permissible to interpolate between loads if the hands are close to a boundary value, indicated by the vertical lines.

Carrying

No discussion of carrying can be complete without a discussion of postural stability ('balance') and of walking itself.

Postural stability and postural control

The erect human body is a tall structure with a narrow base of support – its COG lies at more than half its total height. In terms of physics, the body is at the mercy of environmental perturbations. In terms of physiology, however, an adult with fully developed postural reflexes is very stable given that movement is not restricted. Posture is an active process that depends on the functioning of a number of neurologically separate systems of reflexes (Martin, 1967).

The anti-gravity reflexes are centred in the hind-brain (pons and medulla) and cause automatic bracing of a limb when it is loaded by body weight. Although they are essential for the maintenance of an upright posture, they do not assist in the maintenance of equilibrium. A second set of postural reflexes (located in the basal ganglia of the mid-brain) control the posture of the various body parts in relation to each other (as in the postural fixation of limbs, for example) and of the whole body itself. The cerebellum is also involved in posture in the coordination of body movements.

The reflex control of posture can be distinguished from low-level reflex arcs such as the 'knee-jerk' reflex and from voluntary movement. Mechanisms exist to protect the body against mechanical instability using feedback from the body itself to maintain balance. Feedback is obtained from both the semi-circular canals in the inner ear and from sense organs in the muscles themselves. People with impaired somatic or labyrinthine feedback appear to exhibit little instability when standing on a stable base with visual feedback. However, when their vision is impaired, they exhibit particular disabilities (such as the inability to maintain the head erect when blindfolded; Martin, 1967). This demonstrates that although vision is of secondary importance in the maintenance of body equilibrium, it can play an important role under certain conditions.

In healthy subjects, under less drastic circumstances, the erect posture is characterised by postural sway. Small perturbations of the COG of the body are countered by an 'ankle strategy' (Duncan *et al.*, 1990). The ankle strategy underlies normal sway when someone is standing still on an unmoving surface.

A 'hip strategy' is used to correct large perturbations or when the position of the feet is constrained by, for example, lack of space. The hip strategy can be seen when someone is balancing on a narrow surface (such as a gymnast on the beam who momentarily loses balance). For posterior displacements of the COG of the body (where the line of gravity moves towards the heels), tibialis anterior and then quadriceps femoris contract in the ankle strategy. In the hip strategy, the paraspinal and hamstring muscles contract to thrust the pelvis forwards, thereby compensating for the rearward displacement of the upper body and maintaining equilibrium.

Slips, trips and falls: Catastrophic failure of the erect position

Losses of balance resulting in slips, trips and falls are a major cause of work-related injury. According to the UK Medical Commission on Accident Prevention, over 40% of industrial lost-time accidents are due to this cause alone (Porritt, 1985). Apart from trauma such as broken bones and concussion and bruising caused by the impact of body parts against unyielding surfaces, some of the worst back injuries are precipitated by sudden impacts transmitted through the musculoskeletal system when normal postural control mechanisms break down.

Because we are not normally aware of the workings of our own postural control mechanisms, we often fail to appreciate *the enormous forces which are unleashed when slipping, tripping or falling*, except in exceptional circumstances – the shock of discovering that there was one more stair to descend when walking down the stairs while reading a newspaper is a dramatic example. The impact is sufficient to cause fracture of the neck of the femur in elderly subjects, according to Citron (1985).

Walking itself is a form of controlled falling (Figure 6.8). The body teeters on the brink of catastrophe as each leg swings through to save the ever-falling mass. Trips occur when the swinging leg is prevented from reaching its destination, and slips occur when there is insufficient friction between the weight-bearing leg and the floor. Both problems are a result of a violation of a person's expectations or assumptions by some aspect of the workplace – steps, ramps or ridges may be in unexpected places, there may be differences in the depth or height of the steps in a flight of stairs,

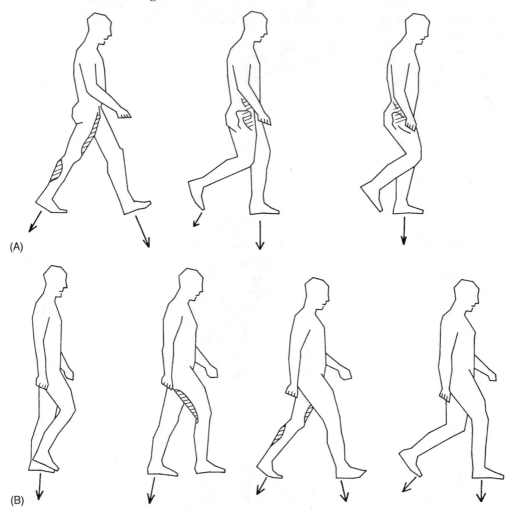

Figure 6.8 Mechanism of walking. The calf muscles plantarflex the foot and the quadriceps straighten the knee at 'toe-off' to propel the body forwards. The weight shifts to the left foot. The iliopsoas muscle pulls the trailing right leg forwards and its knee bends passively. The hamstring muscles decelerate the swinging right leg so that the heel can be planted on the ground. The body weight passes along the length of the foot from the heel at 'heel strike' to the big toe at 'toe-off', two critical stages of force transmission at the foot–floor interface. It is critical for stability that the swinging leg overtakes the body and is planted ahead of it and that the heel does not move as it strikes the ground. Throughout the gait cycle, the reaction force must be approximately in line with the supporting leg(s) or the walker will lose balance. (The drawings are adapted from the photographs of Eadweard Muybridge (1884).)

or there may be loose carpeting or unexpectedly wet or highly polished floors. All of the these factors can increase the risk of slipping, tripping and falling.

Several researchers have attempted to establish optimum dimensions for the design of stairs. Irvine *et al.* (1990) concluded that riser heights (the height of the step) below 152 mm and above 203 mm should be avoided and that run depths (the distance from the front to the back of a step) less than 254 mm and more than 330 mm should also be avoided.

The foot–floor interface

One of the most important factors influencing the incidence of slipping and tripping is the design of the foot–floor interface. Floor materials such as concrete, steel, earth, tile and rubber matting differ in their frictional properties, as do shoe materials such as synthetic rubber, natural leather, etc. The static friction between the shoe and the floor depends on the frictional properties and on the contact area. High heels are more hazardous than flat shoes for this reason. Rubber and synthetic soles have higher friction on dry floors than do leather soles but not on wet floors. Higher friction is required on slopes, which suggests that, in addition to warning signs, slopes should be designed with high-friction materials or lateral ribbing. Concrete floors can be laid with ribbing perpendicular to the intended direction of travel. NIOSH recommends shoes and worksurfaces be matched to result in a coefficient of static friction of 0.5 and that there be smooth transitions between areas with different surface frictional properties. Swenson *et al.* (1992) estimate that the threshold coefficient of friction for walking is between 0.2 and 0.4. Good housekeeping is needed to clear up spills, replace worn carpeting or missing tiles and repair loose floorboards or cracked or pock-marked concrete surfaces.

Changes in floor materials or slope can be signalled with warning signs. Yellow-on-black stripes are often used to indicate hazards and can be painted on the floor at or close to the transition zone or along the length of ramps. English (1994) makes the very important point that possibly the major determinant of slips and falls is the person's knowledge of the surface. People are perfectly capable of walking on ice *if they know about* it and of carrying out complex movements on the deliberately slippery surfaces found in bowling alleys and on dance floors.

Agnew and Suruda (1993) report that from 1980 to 1986 there were 43 505 fatal work injuries to US males, 4179 of which were from falls. Falls from ladders accounted for 20% of fatal falls in workers over 55 years, compared to 9% of all fatal falls. Older workers seem less able to survive falls, as indicated by a lower impact energy associated with the fall. English notes that the slowing of reflexes and increased skeletal fragility that occur with age make corrective postural strategies less effective during a slip and intensify the injuries associated with non-fatal falls in older people. Post-menopausal, osteoporotic women are particularly at risk.

Other factors influencing postural stability

Many factors influence postural stability and sway in healthy subjects (Ekdahl *et al.*, 1989). Age has been related to sway in some studies but not in others. Trips are common in older people because of age-related changes in gait that resulting in the feet not being lifted as high as they used to be and because of an increase in postural sway over the age of 60 years (Overstall *et al.*, 1977). Pyykko *et al.* (1990) investigated the postural stability of 23 subjects over 85 years of age. When compared with a control group, the elderly subjects had significantly higher sway velocities even during non-perturbed conditions. Visual deprivation was found to have a large effect on the elderly subjects, contributing to about 50% of postural stability. The authors concluded that postural control in elderly subjects was reduced owing to deterioration of somatic feedback from stretch reflexes in the muscle spindles. The loss of balance in older people seems to be due to increased reliance on visual feedback, which is slower than intrinsic feedback from the muscles. Another cause of falls is the muting of and delay in initiating the startle response. This increases the likelihood of collisions. The startle response is linked at a neurological level to the emergency reactions involved in recovery from losses of balance (Bisdorff *et al.*, 1999). Delayed response to loss of balance may well make recovery less likely and increase the risk of injury should a fall occur.

Approximately 6000 people die annually in the United Kingdom due to injuries caused by slips, trips and falls. Two-thirds of these are elderly. The importance of visual feedback for postural control in the elderly has clear implications for the design of the visual environment in retirement villages, hospitals and the workplace, particularly in those countries undergoing demographic aging. *Good lighting on stairwells, hallways, and so on is clearly a high priority as is a consistent approach to the placement of light switches inside and outside buildings.*

Ekdahl *et al.* also observed that females have better balance than males, possibly owing to the fact that their body COG is lower. Of the various balance tests investigated by these authors, standing on one leg while blindfolded was the most difficult (impossible for subjects over 55 years of age) and standing on one leg with a blindfold was more difficult than standing with the feet together with a blindfold. These results are not surprising in view of the reduction in the size of the base of support and increased postural load on the supporting leg. Standing on one leg appears to increase the relative importance of visual feedback for postural control – even in young subjects – and has interesting practical implications. For example, *operating foot pedals from a standing position would appear to be contraindicated, particularly when lighting is poor.*

The design of carrying tasks

It is generally believed that loads should be positioned as close to the body as possible in order to minimise energy expenditure while carrying. As soon as the load is placed away from the body (in the horizontal plane), destabilising forces will be exerted that have to be counteracted by static contraction of appropriate muscle groups on the contralateral side of the body. During sustained carrying, this muscle contraction will be a likely source of local discomfort and fatigue.

Carrying increases the load on the body in two ways. Firstly, the increase in weight increases the physiological cost of walking and the load on the muscles of the legs that propel the body forwards (the plantarflexors of the ankle joint and the quadriceps). Secondly, the method by which the load is held or attached to the body can be an additional source of postural stress. Fatigue, discomfort or injury may arise from either or both sources of loading. The second source is of particular interest in ergonomics since, through the design of more efficient carrying methods, the total postural load can, in principle, be reduced. From an energy expenditure viewpoint, an efficient method of load carriage is one that imposes little additional postural load over and above the extra energy cost per kilogram of load (Soule and Goldman, 1969).

Klausen (1965) investigated several methods of holding a 40 kg load and measured electromyographic (EMG) activity in various muscle groups. He also compared spinal and pelvic posture in free standing and when holding a load and noted any load-induced changes. Holding 20 kg in each hand was found to increase the electrical activity in the thoracic portion of the back muscles. It also places a static load on the muscles controlling flexion of the fingers used to grip the weights. Holding the weights by means of a yoke over the shoulders caused a slight forward inclination of the spine and stooping. This was accompanied by an increase in back muscle EMG to counter the forward shift in the line of gravity of the superincumbent body parts.

The effects of carrying the load high and low on the back were also investigated, using a carrying frame. When the load was carried high on the back, similar results to that using the yoke were observed. The body was inclined forwards and back muscle activity increased to exert an extensor moment on the trunk to counteract the flexor moment introduced by the load. When the load was carried low on the back, the subjects still adopted a forward inclined trunk posture, presumably to maintain the combined COG of the load plus body weight over the feet. However, decreased EMG activity was observed in the back muscles but increased activity was observed in the iliopsoas muscles. The interpretation of these findings was that the carrying the load low on the back introduced an extensor moment on the trunk (i.e. it tended to pull the trunk backwards). Although this would reduce activity of the long back muscles, the forward inclined position of the trunk would only be maintained if increased activity in the iliopsoas muscles occurred to counteract the backward pull of the load. Thus, carrying a load low on the back would seem to be a relatively efficient method that decreases activity in the back muscles. This, in turn, would be expected to reduce the stress on the lumbar spine.

Cook and Neumann (1987) investigated the effects of load placement on low back muscle EMG when carrying and compared EMG activity when walking without an external load. They concluded that carrying with a back pack required the least additional muscular effort of the methods investigated and that carrying a weight in front of the body required the most effort. Carrying a weight in one hand induced asymmetrical loading on the spine, decreasing back muscle activity on the side of the load and increasing it on the unloaded side.

It seems that backpacks provide a low-stress method of carrying loads if the load is carried low in the pack. Soule and Goldman (1969) concluded that the energy cost per kilogram of load carried efficiently on the back is identical to the energy cost per kilogram of body weight, i.e. no additional postural load is imposed. Alternatively, carrying loads in one hand or in front of the body should be avoided because it increases the load on the back muscles (Neumann and Cook (1985) cite a special case in which asymmetrical load carriage may be advantageous. Workers with osteoarthritis of a hip joint may benefit by carrying the load on the same side of the body as the 'bad' hip because carrying a load on one side of the body reduces the forces acting on the hip joint of that side, particularly when it is the weightbearing hip during the swing-through phase of gait. This is because the load assists in abducting the hip joint on the weightbearing side.)

Similar reasoning suggests that when using carrying aids such as trolleys, pushing rather than pulling would also minimise any additional back muscle load and there-fore be preferable. When pushing a trolley, the reaction tends to extend the trunk and is resisted by the abdominal muscles and the hip flexors. When pulling a trolley, the reaction tends to flex the trunk and is resisted by the back muscles. Laboratory investigations confirm the superiority of pushing compared to pulling (Lee *et al.*,

1991). The compressive force on the L5/S1 disc was two or three times larger in pulling than in pushing. Body weight itself contributes more to the compressive load on the disc when pulling a load than when pushing it.

Carrying a load in front of the body may also obscure the carrier's view of the floor in front and of potential hazards such as steps or obstructions, thereby increasing the risk of slips, trips and falls. Thus, for many applications, loads should be carried low on the back and placed symmetrically. Finally, it is worthwhile to consider the practice of carrying loads on the head, which is the way most people in the world carry heavy objects. In many industrial societies headload carrying is no longer practised but it can be readily observed in rural areas of Southern Europe and in many developing countries, where it is used mainly by women to carry heavy objects such as containers full of water, firewood and cumbersome objects such as large baskets. Datta and Ramanathan (1971) in an investigation of different methods of load carriage found that carrying 14 kg on the head was approximately half as demanding as carrying 7 kg in each hand in terms of oxygen consumption. Soule and Goldman (1969) reported that headload carrying was only slightly more demanding than carrying a load on the torso at the same speed. The extra cost was due to the increased shoulder and neck muscle activity to stabilise the head. The limiting factor in headload carrying appears to be the mechanical loading tolerated by the neck musculature rather than the total energy cost. Headload carrying appears to be one of the more efficient of all methods of load carriage.

Effectiveness and cost-effectiveness

The high costs of low back injury are confirmed by a study by Webster and Snook (1990), who analysed data from the Liberty Mutual Insurance Company in the USA. The mean cost per back pain case was $6807 and this cost appears to have remained fairly stable since the 1960s (taking inflation into account). Medical costs were 31.5% of the total and indemnity costs 67.2%. The total direct cost (i.e. compensation) to the US economy was estimated to be $11.1 billion.

Several studies, varying in scale, are presented in support of the view that ergonomic interventions can be cost-effective. Teniswood (1987) describes how a reduction in the weight of cement bags used on a construction site (from 40 to 20 kg) paid for itself by reducing cement wastage. Benson (1986) describes the redesign of a packaging operation. Loads weighing from 25 to 45 kg were lifted from a conveyor, wrapped and packed into cartons, which required the worker to lift the load and rotate 180 degrees. The cartons were then manually stacked on pallets. This task produced about six back injury claims per year. After redesigning and mechanisation of the most hazardous aspects of the operation, no back injury claims were made for 2 years (the payback period had been estimated at 3 months assuming that two claims would be prevented). In another example, back injuries were halved when a manual palletising task was redesigned using a palletising stand with a turntable top to eliminate bending, twisting, lifting and lowering motions. In another example, a clothing manufacturer replaced large boxes, which weighed about 35 kg when full of clothes, with smaller boxes. Back injury claims were eliminated in the year following the change. Benson (1987) gives further examples of effective ergonomic redesigns. In all cases, the cost-effectiveness of these efforts was evaluated by monitoring the reduction in back injury claims. The stated payback periods probably overestimate the time taken for the improvements to pay for themselves because changes that improve manual handling safety often increase output as well.

Mitchell *et al.* (1994) present cost data from their study of the cost-effectiveness of issuing back belts to workers involved in manual handling. They concluded that the belts were minimally effective in preventing injury but that injuries, when they did occur, were more expensive when belts were worn. The cost of medication, x-rays, physical therapy and specialist referral were all higher as were the total costs: the total cost of back injury from 1985 to 1991 was $395 880 for 1000 workers who had been issued with belts and $240 321 for 1000 workers who had not been issued with belts.

Garg and Owen (1992) evaluated the effects of an ergonomic intervention programme on the incidence of back injuries in nurses. Stressful tasks were identified and evaluated. Patient handling aids were identified (a mechanical hoist for lifting patients out of bed and a 'walking belt', a way of putting 'handles' on the patient). These were assessed and nurses were trained in their use. Modifications were made to toilets and showers to improve ease of handling. These interventions were found to be acceptable to employees and the back injury rate dropped from 87 to 47 per 200 000 work hours.

Prevention of falls

Garrone (2001) provides evidence for the cost-effectiveness of fall prevention programmes in retail operations where 17.1% of worker's compensations and 22.4% of costs were due to 'same level' falls in 1996–1999. Such programmes are of general interest and are of particular relevance to the prevention of falls during manual handling operations. The main risk factors for falls in retail operations are

- Floor surfaces (when dry and when wet)
- Footwear (in particular the interaction of footwear and flooring)
- Water/objects on floor (water, snow or ice from entrances, drinking fountains, or leaks in ceilings; clips, bags or packaging/refuse)
- Obstructions/poor design (transitions from tiles to carpeted areas or from one level to another; display pedestals too low and below the field of view; obstructions that obscure other hazards)
- Distractions (displays designed to attract attention)

The aim of the prevention programme was to identify and eliminate fall hazards. Employees were educated about the causes of falls and made accountable to keep their workspaces free of hazards. Feedback was provided as improvements were made. Additional safety measures included

- Extending mats at entrances on rainy days to prevent the ingress of water
- Placing mats below drinking fountains
- Enforcing a minimum 75 cm distance between display racks
- Reducing the use of electrical extension cords and running them through covers
- Providing paper towels for employees to clear spillages
- Monthly self-inspections using a checklist to assist in the identification of hazards

The programme was implemented in one retail outlet in 1997 at a time when there were 0.7 falls per $1 million worth of sales costing the company $17 million in accident losses. In the following 2 years, the fall rate dropped to 0.62 falls per $10 million and then to 0.48 falls per $10 million, yielding an estimated savings of $3 million.

Research issues

Much progress has been made in the drafting of guidelines to reduce the risk of workers developing back pain or injury. It should be noted that the 1993 NIOSH lifting equation (the current 'state of the art'; Waters *et al.*, 1993) is likely to be further improved in the future as research continues (e.g. Ciriello *et al.*, 1993). Further research is also required to demonstrate the effects of industry adopting these controls and to express them in cost-benefit terms.

The development of new devices such as the lumbar motion monitor enables researchers and designers to quantify lumbar movements of workers and relate them to task requirements and workspace design. These devices will enable more accurate estimates to be made of worker exposure to spinal stress and can be used to identify and redesign high-risk jobs.

Summary

Modern thinking about the design of manual handling tasks emphasises that a detailed analysis of all aspects of the task is necessary. It is no longer considered adequate to rely on simple specifications for maximum acceptable loads. For example, the design of the load, its size and density are as important, within limits, as its weight. The design of the workspace and the environment in which the work is to be carried out are also important considerations, as are data on the characteristics of the workers themselves.

Further information for the specification and design of manual handling tasks can be found in ISO 11228 parts 1–3, 'Ergonomics – Manual handling'.

Essays and exercises

1. A worker unloads 20 kg sacks of apples from a conveyor and loads them onto a shute, from whence they are despatched. He loads for 2 hours per day at a rate of 5 sacks per minute. The height of the conveyor is 60 cm and the height of the shute is 100 cm. There is an angle of asymmetry of 45 degrees and the load is held 30 cm from the body.

    •   Use the NIOSH equation to calculate the RWL and the lifting index (LI).
    •   Comment on the safety of the task and identify the risk factors.

2. Discuss the factors that cause the trunk to fail during heavy work
3. *Non-industrial lifting.* Using a portable weighing device such as a spring balance to survey the loads lifted and carried by people in non-industrial settings. Examples include shopping bags, suitcases, babies and young children and sports equipment. Would people be allowed to handle such loads in industry?

7 Work Capacity, Stress and Fatigue

The term (fatigue) should be absolutely banned from scientific discussion.
Is a fatigue test possible? (B. Muscio (1921) *British Journal of Psychology*, 12:31–46)

Stress and fatigue are words used in a variety of ways by ergonomists. Sometimes they are used to refer to temporary states of parts of the body. At other times, they are used to refer to chronic states of the whole person. In both senses, stress, arising out of work demands, causes fatigue when work capacity is exceeded.

Stress and fatigue

Stress

Stress is often thought of in a negative way in everyday life. In technical discussion, however, it is perhaps more appropriately viewed as 'applied load' or 'task demand'. The classical physical model of stress is based on Hooke's law: if we place a load of 30 kg on horizontal beam and the beam bends under the load, the stress is 30 kg and the strain is the deformation of the beam. Once the stress is removed, the beam will resume its former shape. Under high loads, the elastic limit of the beam will be exceeded and the beam will not resume its original shape when the stress is removed. Permanent damage results because the stress exceeds the load-bearing capacity.

The classical physiological model of stress is derived from the work of H. Selye espoused in his book *The Stress of Life* (Selye, 1956). Selye was interested in the endocrinological responses to life events and his key insight was that many, very different, noxious stimuli produce the same effects. This lead Selye to coin the term 'general adaptation syndrome' – a three-phase response to stress consisting of an alarm reaction when the threat is perceived, then resistance, followed by adaptation and finally exhaustion or death. To this day, researchers continue to use endocrine markers such as urinary catecholamine concentration and salivary cortisol levels to assess the overall level of work or life stress. It is believed that the former indicates the level of stress and resulting physiological arousal, whereas the latter indicates the degree of emotional response to the situation in which the stress is experienced (Lundberg, 1995). Salivary cortisol levels, taken early in the morning are used to assess recovery from stress – higher levels indicate delayed recovery, from which higher levels of work stress are inferred. Sluiter *et al.* (2000), for example, found raised levels of cortisol and adrenaline in workers with jobs that combined mental and physical stress compared to jobs where the work was mainly mental or mainly physical. Following the analogy with Hooke's law, endocrine markers and other measures such as blood pressure are indices of *strain* from which the existence of *stress* is inferred.

The recognition that workplace stress can take many forms has led investigators to develop more complex models. Kagan and Levi (1971) proposed that the stress response depends on one's psychobiological disposition (inherited characteristics, training and experience) and the stressor in combination. Professor T. Cox developed a transactional model for workplace stress in which cognitive appraisal of the task demands is the key function in determining the response. Whether a task is perceived as stressful depends on how the individual appraises the situation and this will depend on physical and psychological factors such as self-efficacy (one's perceived ability to cope with the demands).

Fatigue

There are at least three different meanings to the term 'fatigue'. Sometimes it is used to mean sleepiness (fatigue as a result of sleep deprivation or disruption of circadian rhythms). It is also used as a synonym for 'tiredness' (e.g. after running a marathon or lifting heavy weights). Finally, it is used when referring to the kind of habituation to a mental task that occurs after prolonged execution that manifests itself as a desire to do something else. This kind of 'mental fatigue' is task-specific – after a driving for many hours, our brains are quite capable of processing the information required to understand a book or enjoy a symphony.

Fatigue is usually inferred from its effects: most directly, decline in physical or mental task performance. Like stress, it is a term that is used in everyday life and its value as a scientific construct has long been questioned. Ream and Richardson (1996), offer the following definition, however:

> Fatigue is a subjective, unpleasant symptom which incorporates total body feelings ranging from tiredness to exhaustion, creating an unrelenting overall condition which interferes with individual's ability to function in their normal capacity.

Localised muscle fatigue occurs as a result of the depletion of nutrient stores, first in the muscle itself, then in the liver, which has a finite capacity to store glycogen. Central fatigue is thought to occur due to changes in the concentration of a neurotransmitter known as 5-HT, which is thought to be involved in the control of tiredness. The reader is referred to Newsholme *et al.* (1992) for further discussion.

In essence, fatigue manifests itself as an increasing resistance to continuing with a task. As long as this resistance can be overcome, performance continues, but with subjectively greater effort. In this chapter, we will review some fundamental aspects of human physical function and the relation of work demands, work capacity and physical fatigue. Discussion of psychosocial factors will be resumed in the final chapter.

Muscles, structure and function and capacity

Muscles, through the tension they exert when contracting, make physical work possible. There are three types of muscle in the body:

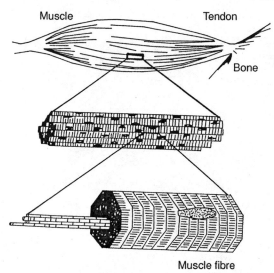

Figure 7.1 Basic structure of skeletal muscle. (Adapted from Noakes 1992, with permission.)

1. Skeletal muscle
2. Smooth muscle
3. Cardiac muscle

Smooth muscle is found in the intestines and makes possible the movements essential for the digestion of food (peristalsis). It is also found in the walls of blood vessels, where it is involved in the regulation of blood pressure and blood flow. It is not normally considered to be under conscious control. Cardiac muscle has a special structure and constitutes the bulk of the heart. The present discussion is limited to work involving skeletal muscle – muscle that is connected to the bones of the skeleton and passes over joints enabling the bones to act like levers when the muscle contracts.

The energy required for muscle contraction is obtained from phosphate compounds in the muscle tissue. These compounds are formed from the breakdown of food. A brief review of muscle structure and function is appropriate (more detailed accounts can be found in Astrand and Rhodahl, 1977; Withey, 1982; and Noakes, 1992). The contraction of most skeletal muscles is under conscious control and the rate and strength of contraction can be varied at will. A complete muscle is made up of many bundles of fibres (cells) arranged side by side and covered by connective tissue sheaths. Nerves and blood vessels are located in the connective tissue. Each muscle fibre consists of many smaller myofibrils, which consist of an alternating series of sarcomeres and Z-bands (Figure 7.1).

A myofibril is split up into a number of sarcomeres arranged in series (Figure 7.2). A sarcomere consists of many filaments layered over each other in alternating bands. There are two types of filaments. Thick filaments consist of about 300 molecules of the protein myosin. Thin filaments consist of a globular protein called actin. The filaments are the true contractile elements of a muscle. Muscles can be likened to bundles of string all joined together. Each individual string (muscle cell) is made up

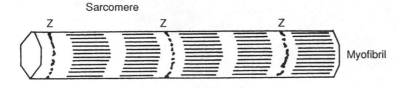

Figure 7.2 A myofibril.

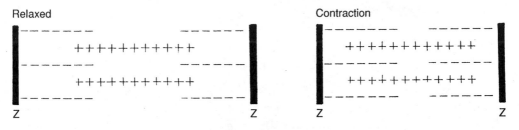

Figure 7.3 Sliding filament theory of muscle contraction.

of fibres (myofibrils), each of which is constructed from many alternating bands of actin and myosin. The whole structure is bathed in intracellular and extracellular fluid and is permeated by blood vessels and nerves.

The mechanism of muscle contraction consists of the actin filaments sliding over the myosin filaments. Since the actin and myosin filaments are arranged in overlapping, alternating bands like a multilayered sandwich, sliding of the former over the latter causes the sarcomeres to shorten. Note that the filaments themselves do not shorten (Figure 7.3). The primary stimulus for muscle contraction is the release of calcium ions stored in the sarcoplasm. The calcium ions bind to the actin, which increases the affinity of myosin for actin.

The fundamental physiology of muscle contraction is still not completely understood and much of it is beyond the scope of the present discussion. However, the main steps in the process are summarised in Table 7.1.

According to Guyton (1991), the duration of the calcium 'pulse' (from release to reabsorption of calcium ions) is about 1/20 of a second. Thus, in order for continuous muscle contraction to occur, a new impulse must reach the muscle after the calcium

Table 7.1 Steps leading to the contraction of skeletal muscle

1. A nerve impulse travels along a motor nerve to the motor nerve end plates in the muscle fibres.
2. A small amount of neurotransmitter is secreted and travels to the membrane of the muscle fibre, initiating an action potential in the muscle fibre.
3. The action potential travels along and permeates the muscle fibre until it reaches a structure known as the *sarcoplasmic reticulum*.
4. The sarcoplasmic reticulum, which acts as a store of calcium ions, releases calcium ions into the myofibrils. This increases the attraction between the actin and myosin filaments causing the sliding action of one filament over another.
5. The calcium ions are then actively pumped back into the sarcoplasmic reticulum. ATP is needed for this. Muscle contraction ceases.
6. Further nerve impulses cause the process to repeat.

'pump' has removed the calcium ions. Smooth movements of the body are, in reality, the result of many individual contractions and relaxations of contractile elements throughout the muscle.

The amount of muscle contraction that can take place depends on the total change in length of all of the sarcomeres in the muscle. As was discussed in an earlier chapter, muscle contraction is a biochemical process and need not result in shortening of the muscle itself. It is more accurate to say that muscle contraction consists of the actin filaments *tending* to slide over the myosin filaments and that this will result in shortening of the muscle under appropriate circumstances. Further details of this 'sliding filament theory' of muscle contraction and how it underpins the length–tension relationship observed in muscle contraction can be found in the classic paper by Gordon, Julian and Huxley (1966).

Energy for action

Energy for muscle contraction (and for many other bodily processes) comes from the breakdown of a substance known as ATP (adenosine triphosphate). By the breaking of one of the phosphate bonds, ATP is converted to ADP (adenosine diphosphate) and energy is made available inside the cell. Astrand and Rhodahl (1977) liken ATP to a rechargeable battery pack – a short-term store of directly available energy. In order for the cell to continue functioning, the ADP must be reconverted back to ATP so that energy can continue to be made available when required. A second phosphate compound known as creatine phosphate acts like a 'back-up' energy store to 'recharge' the ADP to ATP.

In order for muscle filaments to slide backwards and forwards over one another, energy is required. In fact, the actin and myosin filaments are always mutually attracted and energy is required to break this attraction and allow the muscle to lengthen again. During the relaxation phase of muscle contraction, ATP is needed to weaken the attraction between the actin and myosin filaments. Endurance athletes sometimes experience muscle cramps when a muscle becomes fatigued during an event (Noakes, 1992) because the ATP needed to break the actin/myosin attraction is depleted.

The creatine phosphate and ATP stores are of very limited capacity and are reduced within seconds or minutes. The creatine phosphate system can provide energy for maximal muscle activity for about 8–10 seconds and therefore has to be continuously replenished. This takes place in structures inside the cells known as *mitochondria*. If the ATP and creatine phosphate stores are depleted, the muscle will go into rigor (shortly after death, a state of *rigor mortis* sets in through the loss of all ATP).

In muscles, the mitochondria are placed close to the filaments. Carbohydrates and fatty acids (derived from fat) are broken down – ultimately into carbon dioxide and water – and energy is released to form ATP. The basic reaction for the liberation of energy involves the oxidation of glucose as shown below:

$$C_6H_{12}O_6 + 6O_2 \longrightarrow 6CO_2 + 6H_2O + energy$$

Carbon dioxide and water (so-called 'metabolic water') are the products of this chemical reaction.

The carbohydrates and fatty acids used by the mitochondria are derived from the food we eat. They are obtained either from the bloodstream or from stores in the

Table 7.2 Differences between static and dynamic work[a]

	Static work	Dynamic work
1.	Sustained muscular contraction	Repetitive muscle contraction–relaxation cycle
2.	Reduced muscle blood flow	Increased muscle blood flow
3.	No increase in muscle oxygen consumption	Increased muscle oxygen consumption
4.	Oxygen-independent energy production	Oxygen-dependent energy production
5.	Muscle glycogen → lactate	Muscle glycogen → CO_2 + H_2O; muscle up takes glucose + fatty acids from blood

[a] Noakes (1992).

muscle. Glucose is stored as glycogen in the muscles and liver. It is an important store of energy in endurance sports. Oxygen is obtained from the air ventilating the lungs and is transported to all parts of the body by the blood. The body has a very limited capacity to store oxygen, although muscles do contain a substance known as myoglobin, which is a short-term oxygen store.

To use an electrical analogy, if the bloodstream is likened to an electrical reticulation network, the mitochondria can be considered as transformers, plugged in to the mains, that keep the creatine phosphate and ATP 'battery packs' fully charged.

Oxygen-dependent and oxygen-independent systems

If insufficient oxygen is available, the mitochondria are no longer able to convert the carbohydrate and fatty acids to produce energy and a second, 'oxygen-independent', system becomes the main source of ATP in which enzymes in the intracellular fluid surrounding the myofilaments and mitochondria use glycogen and blood glucose (but not fat) to produce energy in the absence of oxygen. This back-up system operates at all times, even at rest, but it is inefficient and produces much less energy per glucose molecule than the oxygen-dependent system. Furthermore, it produces waste products that cause the acidity of the muscle cells to increase. This, in turn, reduces the affinity between actin and myosin filaments which, in turn, weakens the muscle. Oxygen is required to remove these waste products. If it is not available when the person is working, the waste products will accumulate. Under these circumstances it is said that the person has built-up an 'oxygen debt' which must be 'paid-back' when work ceases.

As was described in an earlier chapter, most tasks involve a mixture of static and dynamic work and the latter can usually be sustained longer than the former. Table 7.2 summarises some of the physiological differences between static and dynamic work.

Implications

The oxygen-independent system can form ATP molecules about 2.5 times as fast as the oxygen-dependent system but can only provide energy for about 1.3 minutes of maximum muscle activity. It is a valuable system that enables work to be carried out

Table 7.3 Limiting factors on muscle contraction

1. Demand for energy exceeds supply. This can occur when chemical fuel stored in the muscle is exhausted and the rate of replenishment of oxygen or glucose is inadequate. The supply of fuel to working muscle depends on the capabilities of the circulatory system.
2. Mechanical capabilities are exceeded. The force output of a muscle is finite and depends on the number of contractile elements (the maximum strength of a muscle is proportional to its cross-sectional area). A task may exceed the muscle's mechanical capacity. Strength training increases the cross-sectional area of a muscle and the ability to recruit muscle fibres to carry out muscular work.
3. Accumulation of waste products impairs muscle function. This can occur during static muscle activity as described in previous chapters.
4. The rate of heat production exceeds the body's thermoregulatory capacity. Excess heat can no longer be dissipated and body temperature increases. Cardiovascular capacity is impaired.

at a high rate for short periods of time interspersed with rest. Clearly, it is unrealistic to expect workers to carry out sudden, highly demanding tasks for more than a minute or so without rest and, in planning for high-workload emergency situations, back-up must be provided to enable rapid alternation of work teams. The oxygen-dependent system can function for as long as nutrients are available.

Efficiency of muscle contraction

The muscular system uses oxygen to convert chemical energy from foodstuffs (stored in the tissues or delivered by the bloodstream) into mechanical energy via the sliding filament mechanism. This process has a certain efficiency. Muscles are, at best, only 20% efficient in converting chemical reactants to mechanical output, and heat is the main by-product of the process. This heat is produced largely owing to the viscosity of body parts and overcoming friction of blood flowing through capillaries and tendons sliding over joints. Metabolic heat production has important implications for the design of work in extreme climates, as will be seen in a later chapter. The body's system for converting chemical into mechanical energy can malfunction for a number of reasons (Table 7.3).

Muscle function and fatigue

The function of skeletal muscles is to exert tension between the bony points to which they are attached. Tension is exerted when a muscle changes from its resting to its active state in response to impulses from the central nervous system. The maximum tension a muscle can exert depends on its maximum cross-sectional area and also its length (as described below).

The term 'muscle contraction' refers to the physiologically active state of the muscle, rather than its physical shortening. Muscles are able to contract eccentrically, isometrically and concentrically:

1. *Eccentric contractions.* The muscle lengthens while contracting.
2. *Isometric contractions.* The muscle length remains constant during contraction.
3. *Concentric contractions.* The muscle shortens while contracting.

These different contractions can be illustrated by considering the action of the elbow extensors (such as the triceps muscles) of a person doing push-ups. When these muscles contract concentrically, the body is raised off the floor because the shortening of the muscles causes the elbows to extend. They contract eccentrically when lowering the body to the floor. The eccentric contraction acts like a brake. It controls the rate at which the muscles increase their length and at which the elbows flex. This enables the body to be lowered smoothly to the floor in a controlled way. If the person pauses half-way, isometric contraction is required to counteract the downward pull of gravity and maintain the position of the body in space.

A close relationship exists between the length of a muscle and the tension it can exert. If a muscle is removed from the body, it assumes a resting length that depends on its own internal properties. If placed in an apparatus suitable for manipulating its length and for measuring the tension required to maintain the muscle at a given length, a length–tension curve may be plotted. It is observed that as the muscle is artificially lengthened, increased tension is required to overcome the elastic resistance of the connective tissue that surrounds the individual muscle cells and holds the muscle together (Grieve and Pheasant, 1982). Muscles can exert tension both when they are in the active state and passively by their resistance to being lengthened.

If an isolated muscle is stimulated electrically to cause it to contract, the active length–tension relationship can be investigated (Gordon *et al.*, 1966). Experiments demonstrate that the tension actively produced depends on the length of the muscle when it is stimulated to contract: muscles have an optimum length at which they are capable of exerting their maximum tension (Figure 7.4).

In living organisms, muscles form part of a larger system of joints and ligaments and the tension they exert is transmitted via their bony attachments. This results in a turning force or torque being exerted around the joint that the muscle spans. The magnitude of the torque depends not only on the length of the muscle but also on the geometry of the joint and the mechanical advantage of the muscles at a particular

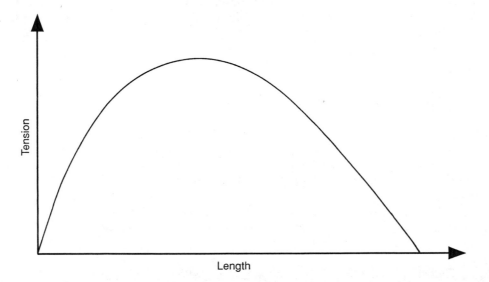

Figure 7.4 Active length–tension relationship. A longer muscle generates more tension when it contracts until its physiological length is exceeded.

joint angle. Angle–torque curves can be plotted for different muscle–joint systems to identify those joint angles at which maximum torque can be exerted. Thus, manual tasks can be simulated and 'strong' and 'weak' body postures can be identified. For example, the flexion torque around the elbow joint is greatest in the middle portion of the joint range because the elbow flexors (such as the biceps muscles) have their greatest mechanical advantage in this position.

Control of muscle function

Skeletal muscle contains afferent and efferent fibres from the central nervous system. Some of these form complex feedback loops. Receptors in the muscle body provide information about the length of the muscle and its tension (Figure 7.5). This is conducted to the spinal cord and to higher levels of the central nervous system. The 'knee-jerk' reflex is a well known example of a feedback loop, which is made possible by a neural circuit between the stretch receptors in the muscle spindles and motor neurons in the spinal cord. When the muscle is lengthened, the stretch reflex comes into operation causing the muscle to contract – that is, the reflex opposes the lengthening of the muscle. Muscle tension is detected by an organ known as the Golgi tendon organ, which is found at the point where small bundles of muscle fibres fuse with tendon fibres. The Golgi tendon organ forms a negative feedback system that prevents damaging levels of tension from building up in the muscle and possibly damaging the muscle fibres or the insertions of the tendons.

The neural circuits formed by these sensory organs in the muscles provide the central nervous system with information about the state of the muscle and the posture of joints. This is known as proprioceptive feedback and is essential for all but the

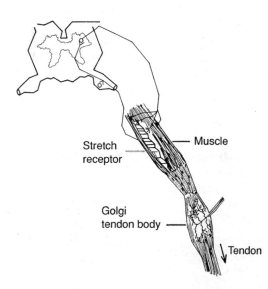

Figure 7.5 Positive and negative feedback comes from receptors in the muscles and a feedback loop at the spinal cord. A contraction/relaxation reflex exists between antagonistic muscles: when one contracts, the other relaxes to allow movement at the joint. The reflex depends on the feedback from the muscle spindles and Golgi tendon bodies, respectively.

most primitive of movements. Skeletal muscle also contains nerve endings that, when stimulated, lead to the sensation of pain. These pain receptors can be stimulated by chemical, thermal or mechanical stressors. Nerve impulses induced by any of these causes can give rise to reflex muscle contraction via neural circuits in the spinal cord. This can further exacerbate the cause of the pain and lead to a positive feedback situation in which the muscle goes into spasm, a topic that is returned to in a later chapter.

For present purposes it can be concluded that muscles are not just simple mechanical actuators like motors or springs – they are reactive tissues. Their tension, length, thermal, mechanical and chemical state are integrated and controlled by means of complex neural circuitry at different levels of the central nervous system.

Fatigue and discomfort

A major goal of research into posture is to develop principles for the design of working environments that impose low postural stress on workers. The use of these design principles, it is hoped, will reduce the incidence of fatigue and discomfort in the workplace.

Discomfort is difficult to define because it has both objective and subjective elements. Branton (1969) suggested that discomfort results in an 'urge to move' caused by a number of physical and physiological factors. Pressure on soft tissues can cause ischaemia (depletion of the local blood supply to the tissues) resulting in a shortage of oxygen and a build up of carbon dioxide and waste products such as lactic acid. This is known to lead to sensations of pain or discomfort.

Active muscles convert glucose and oxygen to carbon dioxide and water, liberating energy in the process. They require a regular blood supply to replenish fuel and remove the waste products. During rhythmic exercises involving eccentric and isometric contractions, blood flow is facilitated by the 'pumping' action of the muscle. Fatigue may ultimately occur owing to the depletion of metabolites (glucose, or glycogen which is stored in the muscle and may be readily converted to glucose). Depletion of muscle glycogen is one of the main causes of fatigue in endurance athletes such as marathon runners. When muscles contract, they occlude the blood vessels within them and thus diminish their own blood supply. During sustained isometric contractions, the muscle is starved of oxygen and waste products accumulate as oxygen-independent metabolic processes take place. Discomfort and fatigue occur rapidly during sustained isometric contractions for this reason.

Kroemer (1970) has shown that muscles fatigue rapidly under conditions of static loading even at low workload (Figure 7.6). As can be seen, the endurance time for maxiumum contractions is of the order of a second; 50% of maximum can be held for a minute or less and this is the rationale underlying ISO 1226. However, even though a static exertion such as standing with the arm fully extended at 90 degrees to the torso only requires 10% of maximum, musculoskeletal complaints will occur if the the action has to been sustained all day at work (Parenmark *et al.*, 1988).

Fatigue after prolonged exertion

According to Rhomert (1960) the value of an individual's maximum static exertion, in a particular posture, can be used to estimate their endurance time for a submaximal exertion in that posture using the equation below:

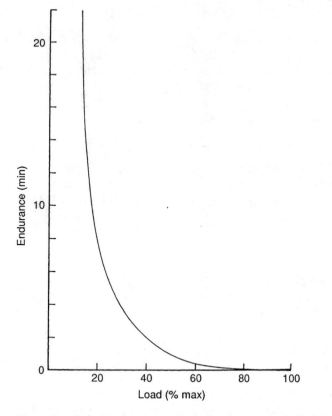

Figure 7.6 Relationship between the load on a muscle and the endurance time. Loads that require less than 20% of a muscle's maximum can be endured for minutes or hours. Loads of 80% or more can only be endured for a few seconds.

$$T = -90 + \frac{126}{P} - \frac{36}{P^2} + \frac{6}{P^3}$$

where

P = the required exertion, expressed as a decimal fraction of the maximum exertion
T = the predicted endurance time in seconds

For example, if an operator had to exert a grip force of 25 kg while carrying a milk crate and his maximum grip force was 50 kg, then the endurance time would be given by

$$T = -90 + 126/0.5 - 36/(0.5)^2 + 6/(0.5)^3$$
$$= -90 + 252 - 144 + 48 = 66 \text{ seconds}$$

where $P = 25/50 = 0.5$.

Fatigue and pain

Skeletal muscle can be regarded as the largest organ in the body. It is essential for all activities involving voluntary movement. Repeated or sustained activities, rapid movements and large forces can stimulate pain receptors in a muscle. Since skeletal muscle makes up 40% of the tissues of the body, it should come as no surprise that many of the aches and pains we experience in our daily lives are of muscular origin.

The tension exerted by muscle can serve to cause movement of a body part in carrying out work activities; it is also essential for resisting externally imposed forces that might otherwise destabilise and damage a joint. The loss of muscle function that occurs with fatigue therefore implies increased risk of musculoskeletal injury. This indicates why a consideration of work–rest cycles, work durations and forces is an essential part of the risk assessment process in job evaluation and injury prevention.

Electromyography

Discomfort is a subjective experience that can result from a combination of physiological and psychological processes including muscle fatigue. Muscle fatigue is a physiological phenomenon that can be observed directly using techniques such as electromyography (EMG). Muscle cells are arranged in functional units known as 'motor units'. Each motor unit is connected to a single nerve fibre and all of the cells in that unit fire in response to an impulse from the fibre on an 'on–off' basis. The number of muscle cells per motor unit varies considerably between muscles: calf (soleus) muscle motor units contain many hundreds of muscle cells, whereas those in the eye muscles contain as few as five. The latter muscles are capable of much finer, coordinated movement because of their superior innervation.

Muscle tension can be varied by increasing the number of motor units firing at a single time or by increasing the firing rate. Electrical activity in muscles can be detected either using surface electrodes placed on the skin overlying the muscle or by needle electrodes inserted into the muscle body. Soderberg and Cook (1984) provide a useful introduction to electromyography. EMG activity from fresh muscle resembles electrical noise – a smooth or 'buzzy' trace usually appears on the oscilloscope. It has long been known that the frequency content of the myoelectric (ME) signal shifts towards the lower frequencies as a sustained contraction progresses to exhaustion. Stuhlen and DeLuca (1982) reviewed theories of muscle fatigue and concluded that compression of the ME spectrum occurs as a result of a decrease in the conduction velocity caused by a decrease in muscle pH due to the accumulation of lactic and pyruvic acids. They used the median frequency of the ME signal as an index of muscle fatigue. The initial median frequency (of the fresh muscle) was taken as a cut-off and the ratio between the rms (root mean square) voltages below and above the median was used to monitor the progress of fatigue. This parameter increases in magnitude as the muscle fatigues. In severely fatigued muscles, the availability of fresh motor units becomes limited and visible tremor sets in (Figure 7.7).

Electromyography is used by ergonomists to detect workspace and task factors that cause unnecessary or rapid muscular fatigue. It complements subjective techniques in which workers are asked to indicate on body diagrams or by questionnaire the location and severity of musculoskeletal pain.

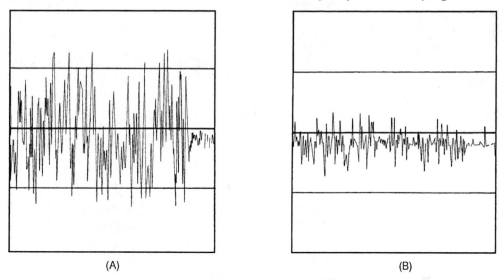

(A)   (B)

Figure 7.7 EMG activity from (A) fatigued and (B) fresh muscle. The 'buzzy' trace from fatigued muscle is due to desynchronisation of motor unit firing.

The cardiovascular system

The cardiovascular system performs a number of functions. It collects oxygen from the lungs and food from the gut and delivers them to the cells of the body. It also collects waste products and delivers them to the excretory organs. Hormones secreted by the endocrine glands in various parts of the body are transported to target cells in other organs by the blood. Finally, the cardiovascular system is involved in thermoregulation – the control of body temperature.

An average adult has about 5 litres of blood. Approximately 2.75 litres consist of plasma and 2.25 litres of blood cells. There are three main types of blood cells. Red cells (erythrocytes) transport oxygen, white cells (leukocytes) fight infection, and platelets are required for blood clotting. The red cells are formed in bone marrow and contain a red pigment (haem) that is bound to a protein molecule (globin). Haemoglobin combines with oxygen to form oxyhaemoglobin, the form in which oxygen is carried in the blood. Plasma consists of 90% water and 10% solutes (food, proteins and salts). Haemoglobin absorbs oxygen (a process known as oxygenation) in the lungs and the oxygenated blood returns from the lungs to the left atrium of the heart via the pulmonary vein. It passes to the left ventricle where it is pumped via the aorta to the tissues of the body through a series of ever-branching blood vessels. Most tissues of the body are permeated by capillaries, tiny blood vessels with internal diameters not much larger than a red blood cell.

The system pumps oxygenated blood under pressure such that it permeates the tissues of the body. Oxygen and food are delivered to the tissues and carbon dioxide and waste products are removed. After passing through the bed of capillaries that permeates the tissues, the deoxygenated blood drains into the veins (similar to small streams or tributaries flowing into a river). The veins contain one-way valves that enable the blood to flow in only one direction – back to the heart. The output of the heart increases from around 5 litres of blood per minute when at rest to 25 litres per minute during heavy work. As workload increases, the volume of blood pumped per stroke (heartbeat) can be increased as well as the heart rate itself. Furthermore, the flow of blood to the body can be increased on a selective basis by constricting and dilating different blood vessels. During physical work, more blood is directed to the muscles and to the skin (to dissipate the excess heat that is a by-product of muscular contraction). After a meal, extra blood is directed to the intestines. The selective control of blood flow to the muscles is controlled by the nervous system. Changes in acidity and the local concentration of metabolites provide feedback to the neural systems that control blood flow.

The respiratory system

The respiratory system functions as a gas exchanger. It supplies the body with oxygen and removes carbon dioxide. It consists of the nose, pharynx, larynx, trachea, bronchi and lungs. The nose filters incoming air, warms and humidifies it, and tests for substances that may be harmful to the respiratory tract. The nasal epithelium is lined with mucus. Small particles in the turbulent incoming air become trapped in the mucus, which flows under the action of cilia to the pharynx, where it is swallowed. Before reaching the trachea, incoming air is warmed almost to body temperature and humidified to almost 100%.

The pharynx (throat) is a muscular tube that serves as the entrance or 'hallway' for the respiratory and digestive tracts. It is plays an important role in the forming of vowel sounds in speech (phonation). The larynx lies at the upper end of the trachea and just below the pharynx. It is part of the open passage through which air travels on its way to the lungs and protects the lungs against the ingress of solids. It is also the organ of voice production.

The trachea (or windpipe) is a tube of cartilage and muscle about 2.5 cm in diameter extending from the larynx (in the neck) to the thoracic cavity (or chest) where it divides into two smaller tubes – the primary bronchi. The bronchi divide many times into increasingly smaller tubes that finally terminate in microscopic sacs or 'alveoli'. The alveoli are in contact with capillaries containing blood pumped from the heart via the pulmonary artery and it is here that gaseous exchange takes place.

Physical work capacity

The basal metabolic rate (BMR) is the rate of energy consumption necessary to maintain life. Individuals differ in their BMR. The relative BMR of a child is about twice that of an adult. The decrease in BMR with age is one of the reasons for adult 'middle-age spread'. The BMR can drop by about 20% in chronically malnourished individuals.

Physical work capacity refers to a worker's capacity for energy output. This will depend primarily on the energy available to the worker in the form of food and oxygen and the sum of the energy provided by oxygen-dependent and oxygen-independent processes. The rate of energy consumption during physical work is the sum of the basal energy consumption and the metabolic cost of the work in terms of energy consumption.

For continuous work at moderate intensities, oxygen-dependent processes usually make the major contribution to energy output. For each litre of oxygen consumed,

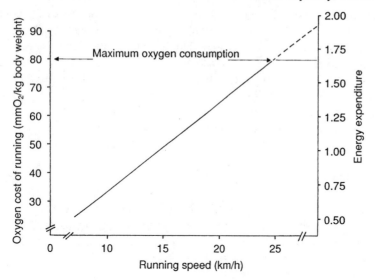

Figure 7.8 The relationship between workload and oxygen uptake.

about 20 kilojoules of energy is released. Work capacity depends on the ability to take up oxygen and deliver it to the cells for use in the oxidation of foodstuffs. The ability to work at a high rate is associated with a high oxygen uptake.

Maximum oxygen uptake

Exercise physiologists and sports scientists have used the term 'Vo_2 max' to describe an individual's capacity to utilise oxygen (aerobic capacity). Vo_2 max has tradition-ally been estimated by having subjects run on a treadmill or pedal a bicycle ergometer while their oxygen uptake is measured. The running or cycling speed is increased in an incremental manner and oxygen uptake is measured approximately every 3–5 minutes after the subject has adapted to each new work rate. As might be expected, it is observed that oxygen uptake increases as the work rate is increased. The rela-tionship is approximately linear, as is illustrated in Figure 7.8.

Clearly, oxygen consumption and heart rate cannot continue to increase indefin-itely. In any work situation, a point is reached at which a person cannot increase the work rate any more. There are limits to performance of all human activities, be they running, cycling or loading boxes onto a conveyor belt in a factory. For many years it has been believed that the factor that limits a person's work rate is the inability of the heart and lungs to supply oxygen to the muscles at a sufficiently fast rate to meet the energy requirements of the work. At this point, the person will reach his or her maximum work rate. Individual differences in maximum work rate can be traced back, according to this view, to differences in the ability to supply oxygen to the muscles.

Strictly speaking, the view that oxygen uptake limits work capacity can be criti-cised on the grounds that it is equally valid to say that the a high oxygen uptake is merely a consequence of the ability to work at a high rate. Cause and effect are more difficult to distinguish than they initially seem to be. However, the idea that limitations on oxygen supply place a limit on maximum work rate underlies the classical concept of Vo_2 max that has been used in much research. This states that a point is reached at which increases in work rate are no longer accompanied by increases in oxygen uptake – the individual is assumed to have reached his maximum level of oxygen uptake and cannot sustain a harder pace of work. Further increases in work rate can only be achieved if energy is provided by oxygen-independent processes. This can only be maintained for a short period because these processes cause waste products to accumulate in the blood and tissues. Shortly after a person reaches a work rate which exceeds their Vo_2 max, performance will decline dramatically. There is great inter-individual variability in Vo_2 max, which is thought to be due largely to genetic factors. Training can improve an individual's Vo_2 max, although only by about 10–15%.

Noakes (1988) has reviewed the evidence for the idea that the maximum work rate depends on the capacity to deliver oxygen to the muscles. He notes that in many of the individuals tested, the heart rate/Vo_2 relationship does not reach a clear plateau, even though the person exercising reaches a point of maximum performance. Noakes has suggested that the limitation on maximum performance may not be the ability to supply oxygen to the muscle tissues but a failure of muscle contractility (or muscle power). This implies that in groups of athletes Vo_2 max may not be the best method of predicting maximum performance.

Removal of waste products

In addition to the capacity to utilise oxygen, sustained performance at a high level depends on the ability to remove waste products of metabolism from the tissues. Endurance athletes do not perform at Vo_2 max intensities but at a lower level (about 80% in marathon runners according to Reilly (1991)). Successful performance de-pends on the ability to remove lactate from the body tissues as well as aerobic capacity. The accumulation of lactate in the blood of an athlete signals the upper limit to which endurance exercise can be maintained (Reilly, 1991). Unlike Vo_2 max, the blood lactate response can be improved considerably by training.

Successful performance of endurance tasks also depends on the ability to dissipate heat and the ability to utilise fat as a source of fuel.

Vo_2 max and fatigue

The equation below describes the relationship between the length of time (T in minutes) a task can be carried out continuously, until exhaustion, and the demands of the task (P) expressed as a percentage of Vo_2 max during the task (Louhevaara *et al.*, 1986).

$$T = 5700 \times 10^{-0.031 \times P}$$

Table 7.4 summarises endurance times for continuous aerobic work as a function of an individual's Vo_2 max.

Vo_2 max and industrial work

It is generally believed that individuals can work continuously over an 8-hour shift at a rate of 30–50% of their maximum capacity, depending on the frequency and length of rest periods (see below). Industrial jobs must be designed so that they can

Table 7.4 Time to exhaustion when working continuously at a percentage of Vo_2 max

Relative work demand (% Vo_2 max)	Maximum work time
10%	>12 hours
20%	>12 hours
30%	11 hours
40%	$5^1/_2$ hours
50%	$2^2/_3$ hours
60%	$1^1/_3$ hours
70%	40 minutes
80%	18 minutes
90%	9 minutes

be carried out by the less-fit members of the workforce. For example, suppose that two workers have Vo_2 max values of 6 litres and 2 litres of oxygen per minute, respectively, and that the task requires the expenditure of 2 litres of oxygen per minute. The first worker will have a greater work capacity because he will be able to work comfortably all day at only 33% of his Vo_2 max. His energy will be met by oxygen-dependent processes, which means that he will be able to work in a physiologically steady state and obtain energy from the oxidation of fat as well as from carbohydrates (recalling that oxygen is required to metabolise fat in the mitochondria). The second worker will be required to work at 100% of Vo_2 max and will tend to utilise oxygen-independent processes. He will only be able to work for short periods and will require frequent rest breaks to metabolise the accumulated metabolic waste products. He will also be dependent on carbohydrates as a source of fuel.

It will be readily appreciated that in heavy manual work that is remunerated on a piece rate basis (i.e. in which the amount each worker earns depends on how much he produces) individuals with a high Vo_2 max will be at an advantage since they will require fewer rest breaks. Astrand and Rhodahl (1977) report that this appears to be the case among lumberjacks: high earners tend to have a higher Vo_2 max.

Various attempts have been made to classify the severity of work in terms of oxygen uptake, heart rate and energy expenditure. Table 7.5 (adapted from Astrand and Rhodahl, 1977) is a representative example. Some examples of activities with different oxygen requirements are given in Table 7.6.

NIOSH (1981) has published data concerning the maximum aerobic capacity of US workers. The 50th percentile male and female capacities, in terms of energy output, are approximately 63 and 44 kJ/min, respectively. The lower 5th percentile

Table 7.5 Severity of work in terms of Vo_2, heart rate and energy expenditure

Work severity	Vo_2 (litres/min)	Heart rate (beats/min)	Energy expenditure (kJ/min)
Light work	<0.5	<90	<10.5
Moderate work	0.5–1.0	90–110	10.5–21
Heavy work	1.0–1.5	110–130	21–31.5
Very heavy work	1.5–2.0	130–150	31.5–42
Extremely heavy work	>2.0	150–170	>42

Table 7.6 Oxygen uptake and physical activity

---

1. *Rest*. Basal metabolism requires approximately 0.25 litres of oxygen per minute.
2. *Sedentary work*. Office work, for example, requires an oxygen uptake little over resting levels (0.3–0.4 litres/min).
3. *Housework*. Housework includes several moderate to heavy tasks (requiring about 1 litre/min of oxygen). It is unusual for high work rates to be maintained for any length of time, though.
4. *Light industry*. Oxygen uptakes from 0.4 to 1 litre/min are required.
5. *Manual labour*. Oxygen uptake may vary from 1 to 4 litres/min. The workload can depend greatly on the tools and methods (for example, the design of a shovel, the method of load carriage).
6. *Sports*. Endurance sports can have very high oxygen requirements. Oxygen uptakes of over 5 litres/min have been recorded in cross-country skiers, for example.

---

capacities are 52.3 and 33.5 kJ/min. For continuous work, NIOSH states that energy expenditure should not exceed a value of 33% of an individual's maximum capacity (or 21 kJ/min for men and 14.6 kJ/min for women) over an 8-hour shift. It can be seen that a 50th-percentile male worker would be able to carry out heavy work for 8 hours but not very heavy work. A 5th-percentile worker would only be able to carry out moderate work for 8 hours. By determining the energy requirements of different jobs, it is possible to identify jobs that can only be performed by the fitter members of the workforce and, where possible, redesign these jobs to lower the required energy expenditure, thus making them available to a wider range of workers.

For occasional lifting tasks (one hour or less per day) NIOSH recommends that energy expenditure rates not exceed 37.5 kJ/min for physically fit men and 27 kJ/min for physically fit women. The upper limit of energy expenditure is approximately 67 kJ/min for men and 46 kJ/min for women for 4 minutes.

Absolute Vo_2

For some tasks, the energy demands are fixed and can be expressed in absolute terms, as a requirement for oxygen consumption. However, values of Vo_2 max are usually normalised for body weight and can be misleading for the following reasons. Suppose we wish to select workers to carry a 40 kg rucksack over rugged terrain. Worker A, weighing 100 kg, has a Vo_2 max of 40 mlO_2/kg/min, and worker B, weighing 60 kg has a Vo_2 max of 50 mlO_2/kg/min. The first worker has a lower aerobic capacity in relation to his own body weight, but greater absolute capacity (4 litres of O_2 per minute as opposed to 3 litres per minute). If the carrying task (with no load) requires 50% of Vo_2 max, worker A has 2 litres spare capacity compared to worker B with 1.5 litres spare capacity. Suppose that when the loaded carrying task is performed, an additional 1 litre of oxygen per minute is needed. Worker A will then be working at 75% of his Vo_2 max (the oxygen requirement of the carrying task is 25% of his Vo_2 max) whereas worker B will be working at 83% of his Vo_2 max. The example illustrates that, with fixed loads, the larger worker will have greater endurance. This illustrates why care has to be taken when selecting workers according to their level of fitness if fitness is assessed using a measure such as Vo_2 max normalised for body weight. It also explains why body size is often a good surrogate means of estimating an individual's capacity for physical work.

Table 7.7 Factors affecting physical work capacity

Personal	Environmental
Age	Atmospheric pollution
Body weight	Indoor air quality
Gender	Ventilation
Alcohol consumption	Altitude
Tobacco smoking	Noise
Active/non-active	Extreme heat or cold
Lifestyle	
Training/sport	
Nutritional status	
Motivation	

Factors affecting work capacity

Many factors can influence a person's capacity to carry out physical work. Table 7.7. summarises some of the more common factors.

Body weight

Body weight (particularly the percentage of body tissue that is composed of fat) influences all activities in which the worker has to move his own body (e.g. walking, cycling, climbing ladders or stairs). In exercise physiology and sports science, it is usually more meaningful to express Vo_2 max in relative terms by expressing a person's oxygen consumption in terms of their body weight (litres of oxygen per minute per kilogram of body weight). This takes into account the fact that the leaner runner, for example, will be at an advantage over his plumper rival, all other things being equal. In this sense, it is possible to increase one's relative Vo_2 max by shedding excess kilograms of fat.

Age

Age has a significant effect on work capacity. Vo_2 max declines gradually after 20 years of age. A 60-year-old has an aerobic capacity of about 70% of a 25-year-old. This is due to a reduction in cardiac output. Current thinking stresses that the fundamental ageing phenomenon is due to a loss of muscle function. Since the heart is essentially a muscle, this explains the loss of aerobic capacity with age.

Sex

Women have a lower Vo_2 max than men and usually have a higher percentage of body fat. They also have less haemoglobin than men. The cardiac output per litre of oxygen uptake is higher in women than in men. For a woman, the heart must therefore pump more oxygenated blood than for a man in order to deliver one litre of oxygen to the tissues. Apart from those that are obvious, few of the differences between men and women have any ergonomic implications. Most of the difference in work capacity is really due to differences in body size. In general, women do appear to have lower upper body strength, controlling for body size. Aerobic capacity is also lower, on average, owing to a larger proportion of subcutaneous fat in women. However, for most aspects of physical performance, males and females should be considered as overlapping distributions and selection tests should be based on the ability to carry out critical job functions.

Alcohol

Alcohol may increase cardiac output in submaximal work, thereby reducing cardiac efficiency. It also affects liver function and can cause a predisposition to hypoglycaemia (low blood glucose).

Tobacco smoking

Tobacco smoke contains about 4% by volume carbon monoxide (CO). CO has an affinity for haemoglobin (combining to form carboxyhaemoglobin) 200 times as powerful as that of oxygen. Smoking therefore reduces work capacity by reducing the oxygen carrying capacity of the blood. It also causes chronic damage to the respiratory system, which impairs the ventilation of the lungs and the transfer of oxygen from the air to the blood. Tobacco smoke also contains a very large number of toxic and carcinogenic chemicals that are likely to have a generally depressing effect on the physical capacity of smokers.

Recent evidence suggests that non-smokers who work in the same room as smokers may suffer some of the same effects as smokers themselves by breathing in the smoke-laden air. Of particular importance is the so-called 'side-stream' smoke that is emitted from the burning tip of the cigarette. Side-stream smoke contains a higher proportion of toxic substances and gases than exhaled smoke because it has not been pre-filtered by the smoker's lung tissues.

Training

Work capacity can be enhanced by physical training (to increase a worker's Vo_2 max) and job training in more efficient work methods (to obtain more output per litre of oxygen consumed by the worker or to enable the worker to safely exert larger forces by using better techniques). Specific training regimes can be developed to strengthen particular parts of the musculoskeletal system with the goal of improving performance or preventing injury. Strength training requires that the body part in question be exercised at near-maximal levels. Over a period of several months, the muscle fibres increase in size owing to an increase in the number of myofibrils, and an increase in strength is observed.

Nutritional status and general health

A balanced diet is important to ensure adequate amounts of necessary foodstuffs and to minimise the accumulation of excess body fat. Excess body fat lowers a person's relative Vo_2 max as was described above.

In developed countries, many people eat a diet high in saturated fats. Such a diet causes raised plasma concentrations of cholesterol. Tiny crystals of cholesterol are deposited on the inside walls of the arteries and this eventually leads to a disease of the arteries known as atherosclerosis. The continued accumulation of cholesterol forms deposits, called plaques, that eventually reduce the cross-sectional area of the artery and thus impede blood flow. Additionally, the arteries lose flexibility (atherosclerosis is sometimes called 'hardening of the arteries' for this reason). These changes in the structure of arteries can impede the flow of blood to the muscles and to the heart itself, resulting in decreased performance and increased risk of heart attacks.

Muscles with a high proportion of slow-twitch fibres use less ATP per unit of isometric tension than those with a high proportion of fast-twitch fibres. Undernutrition may bring about a reduction in the proportion of fast-twitch fibres in voluntary muscles (Ulijaszek, 1995), lowering the energy cost of daily activities. However, this would reduce the capacity for fast, dynamic work and may be an important consideration in the design of work in developing countries where malnutrition is endemic.

Food intake and food supplements

The problem facing people in many Third World countries is that they are not able to satisfy their energy requirements and are forced to subsist at extremely low levels.

A number of researchers have investigated the effects of food supplementation on the work output of malnourished workers. Diaz *et al.* (1989) investigated the work performance of a group of Gambian labourers over a 12-week period during a time of natural food shortage (the wet season). The labourers were split into two groups, one receiving food supplements for the first 6 weeks and the other receiving supplements during the last 6 weeks. Both groups gained weight during the supplementation period and lost weight when there was no supplementation. However, food supplementation had no effect on worker productivity despite the negative energy balance of unsupplemented workers. The fact that the workers were paid on a piece-rate basis may explain the constant productivity, which was maintained even at the expense of body weight loss. In such harsh situations as these, workers may maintain their level of output at work but reduce the energy devoted to leisure activities to compensate. When this happens, one of the costs of work is reduced leisure activity.

Motivation

Motivation is an extremely important determinant of work capacity. For present purposes, it may be noted that a worker's level of motivation may be affected by intrinsic factors such as personality, personal and career goals, need for achievement at work, and so on, and extrinsic factors such as work organisation, method of remuneration and the availability of alternative forms of employment. Piece rate systems (where the worker is paid according to how much is produced) may motivate workers to work at an increased rate but have been associated with increased risk of accidents and of developing musculoskeletal ailments. Motivation is discussed in more detail in other chapters of this book.

Several environmental factors can also influence work capacity.

Air pollution

Air pollution may increase the resistance to air flow of the respiratory airways and, in the long term, cause damage to the lungs, permanently reducing the worker's capabilities. If the source of the pollution involves the combustion of organic compounds, carbon monoxide may be one of the by-products and decrements in work capacity may result. When the concentration of carbon monoxide exceeds 6.5 parts per million it begins to accumulate in the blood during submaximal exercise. To put this into perspective, carbon monoxide concentrations from 37 to 54 parts per million occur in urban traffic. The work capacity of people doing heavy manual work in urban areas may well be degraded because of this.

Climatic factors

Extreme environments can have significant effects on work capacity, as is described in a following chapter.

Noise

Noise is a stressor that can elevate the heart rate and reduce cardiac efficiency.

Altitude

The capacity for sustained work is reduced at high altitude because the partial pressure of oxygen is less (the air becomes 'thinner' with increasing altitude). Less oxygen is available per unit volume of air and thus the functional Vo_2 max is reduced. Some adaptation to work at altitude can take place after continual exposure for several weeks. Short term, high-intensity activities (such as lifting a heavy weight) are not influenced by altitude because their performance does not involve oxygen-dependent processes (Kroemer, 1991). In practice, the capacity for maximum muscular work remains unchanged up to an altitude of 1500 metres above sea level. Above 1500 metres, maximal work capacity decreases by about 10% per 1000 metres. Submaximal work capacity is not affected below 3000 metres above sea level but does require a larger percentage of the total work capacity compared to that at sea level.

Protective clothing and equipment

Protective clothing and equipment can effect work capacity in several ways, depending on the application. Bulky clothing may limit the range of motion of certain joints and hamper work movements. It may decrease stride length and therefore reduce walking and running speed and efficiency. Heavy clothing assemblies or devices that have to be carried will reduce work output for the same level of effort in activities such as climbing or carrying. Protective clothing assemblies that trap heat may increase heart rate for the same level of effort, thus reducing work capacity or hastening fatigue.

Breathing equipment almost always imposes an additional physiological cost on the user because of the resistance to air flow caused by pipes, valves, etc. The particular cost depends on the application. In the nuclear power industry, the main consideration is to protect workers from dust by providing a separate breathing system for the provision of oxygen and the removal of carbon dioxide. Firefighters need protection from noxious gases such as carbon monoxide and the provision of oxygen, which the fire tends to consume. Pilots of high-altitude aircraft need a breathing mixture enriched with oxygen because the ambient partial pressure of oxygen at high altitude is not insufficient to 'push' oxygen through the alveoli. In closed breathing systems, only oxygen is provided and wearers re-breath their own air after it has passed through a carbon dioxide absorber such as soda lime or barium hydroxide. A disadvantage of these systems is that water vapour can accumulate and in the cold it can freeze and block the system. Open systems supply the wearer with fresh air and expired air is released to atmosphere via a one-way valve. In this respect, they waste oxygen and are inefficient.

Breathing equipment can affect performance in several ways. Firstly, it must provide the user with an adequate 'lungful' of air at each breath. This varies depending on the conditions. At rest, 10–15 litres of air are needed per minute. During heavy exercise, 40–50 litres per minute may be needed. The dynamic aspects of air flow are also important. During speech, high acceleration and peak flows are observed, which the system must cater for (up to 200 litres per minute). Resistance to inspiration can be exacerbated by any factors that increase turbulence in the flow of incoming air: rough inner surfaces of pipes, sudden bends, joints or changes in bore. Resistance to expiration depends on the valve characteristics. Under heavy resistance, the breathing pattern may be characterised by 'panting' – shallow breaths repeated frequently. This brings with it the threat of *hypercapnia* (accumulation of carbon dioxide in the blood). The acidity of the blood increases and the heart rate increases.

Certain designs of breathing equipment impose a physiological cost to the user owing to the 'dead space' through which air has to be breathed. An infinitely long pipe would eventually cause suffocation as the user continually rebreathed their own air. Shorter lengths (200 mm to 2 m) can have significant effects on physiological variables both at rest and when working as part of the expired air is rebreathed. Deadspace increases heart rate, tidal volume and partial pressure of carbon dioxide and lowers work capacity.

Sulotto *et al.* (1994) investigated the energy expenditure of railway workers wearing an air-purifying respirator in which inspired air first passed through filters designed to remove organic vapours and dust. Exercise using the respirator was characterised by reduced frequency of breathing, ventilation rate, oxygen uptake and maximal oxygen uptake and carbon dioxide production. The mean drop in maximum oxygen uptake was about 9%. The recommended work load for an 8-hour day of 40% of maximum aerobic capacity needed to be reduced to compensate for the lowered maximum value when the respirator was used. Similarly, for unfit people, a level of 30% of the *maximum uptake when wearing the respirator* was recommended for an 8-hour shift. A maximum inspiratory mouth pressure of 20 mmH$_2$O was recommended for respirators designed for work of this intensity and duration.

Pacing/time pressure

Paced work is often a feature of production line manufacturing systems and the operator has little control over the work rate. It is thought to be more stressful than unpaced work and it has long been known that older workers have difficulty tolerating this kind of work. Belbin and Stammers (1972), in a study of paced work in car production, found that labour turnover among workers in their late thirties and forties was 20% – more than double the rate for this age group in other types of work. Time pressure has mental and physical effects. Birch *et al.* (2000) found that time pressure in performing a low-precision task with low mental demands caused increased EMG activity in the shoulder and forearm muscles.

Effectiveness and cost-effectiveness

In general, modern industry relies less and less on heavy manual labour, but there are exceptions, particularly in the construction and forestry industries and in certain industrial situations where people have to handle very high loads or work at a fast pace. Under these circumstances, output is limited by cardiovascular capacity more than anything else and the need to provide rest periods reduces labour productivity. One traditional way around this is to select large workers with a high level of cardio-vascular fitness, although such practices are being challenged increasingly in many countries following the introduction of legislation for equal opportunities. Self-selection occurs particularly in manual occupations paid on a piece rate basis. Those unable to work at a sufficient pace to earn high-enough rewards find alternative employment. In developing countries, self-selection often happens in unskilled agri-cultural jobs such as sugar cane cutting.

The ergonomic approach is to redesign tasks or provide job aids to reduce the load. Drury *et al.* (1983) evaluated the effectiveness of a palletising aid to improve perform-ance on a stacking task. Sixteen identical 10 kg cartons were stacked onto pallets from a supply conveyor that was 900 mm distant. They were delivered to the operators at a height of 810 mm. It was observed that operator's heart rates were considerably higher when they were stacking boxes onto the empty pallet than when they were stacking the higher layers (more stooping is needed to stack, initially). This suggested that a palletising aid that maintained a constant box loading height of about 1 metre would be effective. The palletising aid was a stand that rotated so that the worker did not have to reach over already stacked boxes. Drury and his colleagues evaluated the design to determine its effectiveness in physiological and biomechanical terms.

The recommended maximum heart rate for continuous work by industrial workers is 110–155 beats per minute (Brouha, 1960). When the boxes were stacked without the palletising aid, the mean heart rate was 129 beats per minute. When the task was done using the palletising aid, the mean heart rate dropped to 111 beats per minute (the task was now easier because there was no need to reach or stoop to stack the cartons onto the pallet. Drury estimated that subjects' oxygen consumption dropped from 48% of maximum to 37% when the palletising aid was used. These reductions can now be put into economic terms by considering the length of time an average worker could stack the boxes without having to rest to ward off exhaustion. Without the aid, a rest every 10 minutes would be needed. With the aid, a rest every 31 minutes would be needed. In both cases, the length of the rest period would be 7 minutes to allow full recovery. Without the palletising aid, subjects would only work for 59% of the day. With the aid, the actual work time would be 82% of the work day, a reduction of 0.23 workers if manning is continuous. In 1983 dollars, this was estimated to represent a saving of $3000–4000 per annum, about the same as the cost of the palletising aid.

Another benefit of the palletising aid was that its use improved working posture and this reduced the load on the back. The average reduction was from 250 kg of spinal compression to 150 kg. This would reduce back injuries, resulting in a saving of around $374 per man-year in direct injury costs.

Productivity improvements in developing countries

One of the main factors that limits the output of workers in developing countries is the shortage of food, particularly food high in carbohydrates since this provides the energy to carry out physical work. Rogan and O'Neill (1993) point out that land preparation and weeding require considerable time and energy and impose severe constraints on the productivity of resource-poor farmers in the tropics. There is evidence that better-designed tools can increase the physiological efficiency of these tasks (Yadav *et al.*, 1976; Tewari *et al.*, 1991). Clearly, in many parts of the world, the provision of a range of better designed, more efficient tools, would enable workers to be more productive or leave them more energy for other tasks or for leisure activities.

Brun *et al.* (1991) investigated the energy expenditure of women in the Sahel. The two main tasks were working in the fields and drawing water. The women's capacity for physical work was limited by the low-energy diet. Water (15–30 litres, several times a day) was drawn using a jar attached to a rope. The energy cost of this activity was about 5 times the energy cost when resting and the daily energy requirement was equivalent to the energy obtained from eating about 40 g of cereals. A trial was done in which the traditional method of drawing water was compared with that using hand-operated pumps. The pumps were found to improve the efficiency and speed with which water could be drawn, thus saving both time and energy.

In another study, in Burkina Faso, the time and energy used to prepare flour from sorghum and millet was investigated. Women spent about one hour per day making flour by hand grinding of cereals. The energy expenditure required to grind the flour was about 10% of the total daily energy expenditure used to produce cereals. The energy from about 70 g of sorghum out of a daily consumption of 470 g went on flour production for human consumption. Nineteen million people in the region subsisted on hand-ground flour, consuming 3.5 million metric tons of flour per year. Clearly, improvements in the efficiency of flour production would have enormous benefits on a national scale.

In Indonesia, the productivity of women sorting coffee beans was improved by job redesign and the provision of ergonomic furniture (Priatna, 1987). Women sorted coffee beans while sitting on the floor. The task consisted of four subtasks: separating the bean from the shell, removing the seed, removing spoiled seed, and removing the outer skin layer. The job was redesigned and ergonomic furniture designed to suit the workers' dimensions was provided (chairs seats 30 cm high and table 54 cm high). The women were reorganised into teams of four, with each person carrying out one subtask. Productivity increased from 20 kg of beans per worker per day to 25 kg. No data for the time taken to recoup the cost of the furniture are given. Fewer complaints of fatigue, back pain and headaches were made after the redesign.

Oberoi *et al.* (1983) compared the energy expenditure needed when washing clothes manually and when using a machine. Very few Indian households used a machine according to the authors and it was therefore appropriate to identify those situations in which high energy expenditure was required by the manual methods. Washing clothes in a squatting position was found to impose the highest energy expenditure and the lowest energy expenditures were recorded when subjects sat on a low stool or stood. Interestingly, the latter two work positions imposed no greater workload than that required to wash clothes by machine. Certain supposedly labour-saving domestic devices may, in fact, impose a higher workload than is readily appreciated owing to the tasks required to set-up and operate the device.

Any new design of hand tool that can be shown to enable the same amount of work to be done with a lowered rate of energy expenditure would be of particular interest for application in developing countries where there is still concern about the workload of daily tasks in relation to the aerobic fitness of the population (Varghese *et al.*, 1995) and the availability of food. Rogan and O'Neill (1993) point out that human effort provides some 70% of the energy required for crop production in developing countries.

Pedal power and its uses in developing countries

Numerous intermediate technologies are available to partially mechanise demanding/ laborious tasks in regions lacking access to power supplies or to the infrastructure needed to maintain machinery. Technology has been developed using bicycles to power a variety of machines including maize shellers, winches, cable cultivation (in which the plough or other tool is dragged across the land on a cable by a stationary, pedal-powered drive), and trikes for transporting goods and a variety of workshop tools including bandsaws and drills. Many of these have direct physiological savings; others enable humans to generate more power (and hence more output per unit time) by making use of the legs instead of the arms to provide the energy. There are also the well-known 'wind-up' radios and torches that do not require batteries (Baygen Inc. Cape Town, South Africa).

Research issues

The factors that affect cardiovascular capacity in the general population are now quite well understood. Catastrophic failure of the cardiovascular system can be avoided if the individual is not exposed to risk factors such as tobacco smoking, unbalanced diet, lack of exercise and stress. General recommendations for exercise, diet, etc. can be given, but further research will lead to more specific recommendations for maintaining physical condition.

Although there is a certain amount of data on the aerobic capacities of US workers, the population is undergoing demographic ageing and these data may well need revision. Generally, more data concerning the work capacities of US workers and workers in other countries – particularly industrially developing countries – is needed. This will enable ergonomists and industrial engineers to make better decisions about what constitutes an acceptable workrate for a given task when designing production facilities worldwide.

Summary

Physiological mechanisms set limits to the worker's capacity for physical work. Some simple, quantitative techniques exist for predicting fatigue and endurance and these can be use to specify the design of tools, tasks and work practices. Physical stress and fatigue at work can be prevented using such methods. Excessive work demands eventually lead to poor performance. Table 7.8 summarises the characteristics of poor performance associated with exposure to excessive work demands.

Table 7.8 Some characteristics of poor performance resulting from excessive work demands*

1. Slower reaction time
    Increased time to complete task

2. Increased time needed to learn new information and procedures
    Increased forgetfulness

3. Failure to take all information into account when making decisions

4. Increased need to repeat steps in a chain of reasoning
    Inability to concentrate

5. Lapses in attention
    Missing or misinterpreting signals
    Omitting steps in a procedure

6. Slowness in noticing changes in task requirements
    Slowness in initiating new tasks as required
    Increased need for instructions

7. Narrowing of attention
    Increased concentration on individual subtasks
    Neglect of other tasks

8. Erratic operation of controls
    Mistiming of actions
    Inappropriate guaging of responses

9. Making unnecessary responses
    Carrying out irrelevant tasks

10. Increasing tolerance of own errors
    Acceptance of lower performance standards

* Compiled by R. Strong

Essays and exercises

1. Discuss what is meant by 'fatigue'. How would an ergonomist go about investigating complaints of fatigue in manual work?
2. A hand-operated, high-pressure water jet for blasting rocks down mine stopes is operated by pushing together, with both hands, two handles, shoulder width apart and directly in front of the operator at chest height. The design engineer wants to know how long the operator will be able to operate the jet if the required force is 15 kg. You get data on the maximum force required by simulating the task using a bathroom scale sandwiched between two piles of books. Volunteers exert a maximum push force which is read off the scale. The maximum push forces (kg) for 10 male volunteers are 25, 17, 30, 23, 23, 24, 19, 35, 34, 25.
    Calculate the 5th and 95th percentile endurance times using Rhomert's equation.
3. Discuss whether it is possible for an organisation such as the army to introduce an entrance standard based solely on aerobic fitness? What other characteristics might be relevant.
4. The use of physical selection tests implies the failure of ergonomics. Discuss.
5. The terms fatigue and stress should be absolutely banned from scientific discussion. Discuss.

8 Industrial applications of physiology

Recognition of the role of exercise in health is changing as robots and automation now perform most of the laborious tasks that used to be done by muscle power. A cartoon in an American newspaper recently quipped, 'Time was when most men who finished a day's work needed rest. Now they need exercise!' It is encouraging to see this new wisdom grow in popular acceptance.

(R. S. Paffenbarger, quoted from Noakes, 1992)

Central to the drive for productivity improvement is the challenge of finding an acceptable work rate for a given job. Industrial engineers have developed methods for designing manual jobs in a systematic way (Barnes, 1963). These techniques enable them to specify standard times for the completion of tasks and to describe the physical load of tasks by means of performance rating, providing industry with a way of organising work in which standard levels of production can be defined and output can be monitored. These methods rely on observation of worker behaviour by the industrial engineer rather than objective measurement of physiological variables.

In this chapter, the application of physiological methods in industry will be described. In addition to the measurement of workload itself, physiological methods also offer the possibility of investigating mental stress and dealing with wider issues such as nutrition and employees' levels of fitness for a given job.

In principle, any increase in oxygen uptake over and above that required for basal metabolism can be used as an index of physiological cost to an individual. When an individual begins a work task from rest, heart rate and oxygen consumption increase to meet the new demands. Because this response is not instantaneous, the immediate requirements for energy are met by local (i.e. muscular) energy stores. When work stops, heart rate and oxygen consumption return to their initial levels slowly because extra oxygen is still required to replenish the muscle stores and to oxidise the waste products of oxygen-independent processes (Figure 8.1) – to pay back the oxygen debt.

In many industrial tasks there is a warm-up period in which physiological processes adjust to meet the new demands. This is followed by a period of steady-state work and a recovery period (Manenica and Corlett, 1977). The total physiological cost of work includes the energy expenditure during both the working and the recovery periods. Assessment of the physical demands of work using physiological measures must begin with the subject completely at rest and continue until the physiological variables return to the resting levels measured previously.

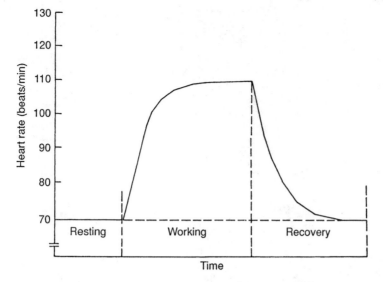

Figure 8.1 Physiological response to work. After the warm-up period, a steady state is reached where task demands match energy output. When work ceases, physiological variables return to resting levels during the recovery period. (From Barnes, 1963.)

Measurement of the physiological cost of work

The classical method of determining energy expenditure at work involves the measurement of oxygen uptake using the Douglas bag (Figure 8.2). The subject inhales air from the atmosphere and exhales it through a mask connected by tubing to a large bag. As the subject carries out the task, the initially empty bag is filled with expired air. After about 50 litres of air has been collected, the task is terminated or the expired air is diverted to a second empty bag. The volume of air in the filled bag is calculated and its gaseous composition is analysed using electronic gas analysers.

The oxygen content of the air in the bag can be compared with that of the atmosphere to determine the amount of oxygen metabolised by the subject. If the time taken for the subject to fill the bag is known, the rate of oxygen uptake can be calculated. From this, the rate of energy expenditure can be calculated. Further details can be found in standard work physiology texts such as that of Astrand and Rodahl (1977). The energy costs of some common daily activities are given in Table 8.1.

The Douglas bag method is well established but can be inconvenient and interfere with performance of the task Owing to the bulky nature of the equipment required. More compact instruments have been developed to facilitate the measurement of oxygen consumption of moving subjects. The 'Oxylog', for example, measures the volume of inspired air as it passes through a turbine flow meter in a mask placed over the subject's nose and face. Some of the expired air passes through the Oxylog itself, where its oxygen content is measured. Harrison *et al.* (1982) report that the Oxylog is sufficiently accurate to yield reliable estimates of oxygen uptake in the working environment.

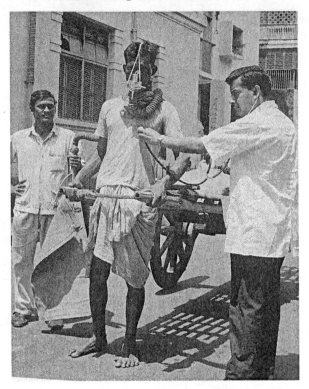

Figure 8.2 Measuring oxygen consumption of a cart puller ('thelawalla') using a Douglas
bag to collect the expired air. (Photograph courtesy of Professor S. R. Datta.)

Table 8.1 Energy costs of some common daily activities

Activity	Work rate (kJ/min)
Sitting	1.25
Kneeling	2.10
Squatting	2.10
Standing	2.50
Stooping	3.35
Walking unladen	8.80
Walking with 10 kg load	15.10
Cycling 16 km/h	21.80
Climbing stairs (30 degrees)	57.33

Indirect measures of energy expenditure

Heart rate increases as a function of workload and oxygen uptake. Because it is more
easily measured than oxygen uptake, heart rate is often used as an indirect measurement of energy expenditure.

Heart rate can be likened to a signal that integrates the total stress on the body.
Heart rate measurement can therefore be used as an index of the physiological workload. However, as was described in the previous chapter, maximal oxygen uptake
varies between individuals. Individuals can have the same heart rate but completely

different levels of oxygen consumption; on its own, heart rate cannot be used to estimate the energy requirements of a job.

To evaluate physiological workload using heart rate, it is necessary to determine, for each worker, the relationship between heart rate and oxygen uptake. Both variables have to be measured simultaneously in the laboratory at number of different submaximal workloads. This is known as calibrating the heart rate/Vo_2 relationship for a worker. Since the relationship between the two variables is linear, a worker's heart rate, when it is subsequently measured in the field, can be converted into an estimate of oxygen uptake by reference to the laboratory data. Estimates of energy expenditure during work can then be calculated from the oxygen consumption data.

Several researchers have validated this method of using of heart rate data to estimate the energy expenditure of calibrated subjects (Spurr *et al.*, 1988; Ceesay *et al.*, 1989) by comparing the heart rate estimates with estimates obtained using whole-body calorimetry (the 'gold-standard' for estimating energy expenditure). Close correlations between the two methods of estimation suggest that heart rate measurement of calibrated subjects can give valid estimates of energy expenditure.

Subjective measures of physical effort

The most common method of obtaining subjective estimates of physical effort is by use of the Borg Rating of Perceived Exertion (RPE) scale (Borg, 1982). Subjects rate their perceived level of exertion during or after performing the task on a scale from 6 to 20, corresponding to heart rates of 60 to 200 beats per minute. High positive correlations between heart rate and the RPE are usually found (Table 8.2). The Borg scale is normally used with other measures, typically heart rate and oxygen consumption. It is not normally used as a substitute for these measures but is used in a variety of research areas, very often as part of a battery of subjective measures such as rating of regional body pain using body maps.

In general, subjective measures are best used with objective measures because, on their own, they are susceptible to experimenter effects (sources of bias caused by the

Table 8.2 The Borg RPE scale

Rating	Interpretation of rating
6	No exertion at all
7	
8	Extremely light
9	Very light
10	
11	Light
12	
13	Somewhat hard
14	
15	Hard
16	
17	Very hard
18	
19	Extremely hard
20	Maximal exertion

experimenter or subject's perception of the experimental 'demands'). Levine and Simone (1991) report that there are gender differences in pain reporting and that male subjects report less pain when the researcher is female. Mital *et al.* (1993) found that experienced subjects tended to underestimate the workload in demanding experimental tasks. However, subjective measures certainly have their place, particularly in complex situations where people are exposed to a variety of different demands. The Borg scale has the advantage of being a global measure – the ratings are influenced by all the demands in the work situation, whereas physiological and subjective measures vary according to a subset of these demands.

Applied physiology in the workplace

Physiological methods have been used to evaluate physical workload in a variety of jobs both in industrialised and in developing countries.

Investigations of forestry work

Tomlinson and Mannenica (1977) compared physiological methods of evaluating the workload of forestry workers with the observational methods traditionally used in production engineering. They hypothesised that the physiological methods would be superior because

1   They take into account the physiological capacities of individual workers and are therefore more valid.
2   They are sensitive to extra stress over and above that due to the work itself (e.g. stress due to heat or humidity).
3   They are more objective and more reliable since, if correctly used, they do not depend on the judgement of the person making the measurements.

The workload of a given task cannot always be inferred directly from measures of output. In unstructured manual work (such as forestry work) it is not always valid to assume that the cost of work will bear a one-to-one relationship with the observed rate of working or the amount produced because the effort involved in carrying out a particular task may vary depending on the particular circumstances. Physiological methods reflect the effort that the worker puts into the work system rather than the output of the system itself – they are indices of the effect of work on the worker rather than the effect of the worker on the output of the work system.

The researchers calibrated the experimental subjects (a group of forestry workers) in the laboratory before taking measurements in the field. The subjects carried out a standard task in the laboratory at different, submaximal levels while oxygen consumption and heart rate were measured simultaneously. For each subject, the oxygen consumption was measured over a range of heart rates. In the field, when only heart rate was measured; oxygen consumption could be estimated by extrapolation from the laboratory data.

Maximum oxygen consumption (V_{O_2} max) was estimated using the data on the heart rate/oxygen consumption relationship. Each subject's maximum heart rate was estimated using the formula

Maximum heart rate (beats/min) = 200 − 0.65Age (years)

Thus, a subject of 45 years of age would have a maximum heart rate of 170 beats/minute. From this, a value of Vo_2 max could be obtained by extrapolation.

The researchers then measured the heart rates of the forestry workers while they were working – felling trees with a chain saw and trimming branches off of the felled trees. The heart rate data were used to estimate oxygen consumption when carrying out the work.

It is generally held that individuals can work at a level of 40% of their Vo_2 max for 8 hours without suffering undue fatigue. Each subject's oxygen consumption when carrying out forestry work was compared with a value of 40% of his Vo_2 max. It was concluded that the physiological cost to the workers was greater than that which is appropriate for an 8-hour day.

These findings indicated that

1   The tasks should not be performed continuously without rest.
2   An appropriate work–rest cycle should be established such that the average work load (both work and rest) does not exceed 40% of Vo_2 max.

The researchers then compared the physiological estimates of workload with work study officers' ratings of workload. These ratings were obtained partly on the basis of production requirements and partly on the basis of the work study officers' perceptions of worker effort. The findings indicated a lack of consistency between the ratings of the work study officers themselves and a lack of consistency between the physiological measures and the ratings; the raters tended to underestimate the workload.

Subsequent research has confirmed that forestry work imposes a high workload on operators. Kukkonen-Harjula and Rauramaa (1984) measured the oxygen consumption and heart rates of lumberjacks in a variety of tasks. Oxygen consumptions of between 1.9 and 2.2 litres/min were observed. Even though the workers were found to have high values of Vo_2 max, frequent rest pauses would be required.

Calculation of rest periods in manual work

Murrell (1971) proposed that rest periods can be calculated according to the empirical formula given below:

$$\text{Rest allowance} = \frac{w(b-s)}{b-0.03}$$

where

w = length of the working period
b = oxygen uptake
s = 'standard' uptake for continuous work

Using the forestry workers as an example, if a worker spends 0.5 hour felling a tree at an oxygen uptake of 2.64 litres/min and the standard is taken to be 1 litre/min, the rest allowance is given by

$$\text{Rest allowance} = 0.5(2.64 - 1)/(2.64 - 0.3) = 0.35 \text{ hour}$$

Workload in non-industrial settings

Developing countries rely very heavily on manual labour. Workload measurement is therefore important to identify unduly heavy tasks, to evaluate traditional work methods and to arrive at more efficient methods of work. Numerous studies have been carried out in India. Sen *et al.* (1983) investigated the workload of tea-leaf pluckers. They concluded that the leaf-plucking task itself was a light job in terms of energy expenditure and moderately light in terms of heart rate, although severe local fatigue of the finger and forearm muscles was common. A secondary carrying task was found to add to the daily energy expenditure and it was recommended that carrying of full baskets be reduced by increasing the number of weighing stations. The authors also evaluated the effectiveness of a hat to shade the worker from the sun. It was found to significantly reduce heart rate.

In another investigation, Datta *et al.* (1983), used the Douglas bag method to investigate the energy cost of pulling indigenous handcarts (see Figure 8.2). Even pulling an empty handcart was found to constitute moderately heavy to heavy work. Loaded handcarts required very heavy and unduly heavy work.

Some interesting studies of domestic work may be found in Grandjean (1973). An example of one of its many practical conclusions is that vacuum cleaners are a major labour-saving device (they save time and effort) whereas dishwashers save time but not effort (due to the added physiological cost of stooping to load and unload the dishwasher).

Assessment of job aids

Physiological methods can be used to assess the benefits of ergonomic interventions and assistive devices designed to reduce work load. Two examples are given below.

Hiking poles   Knight and Caldwell (2000) evaluated the benefits of hiking poles used by backpackers. Backpackers typically complain of pain in the lower limbs, swollen feet and a flexed trunk posture due to the weight of the pack. Hiking poles, adjustable ski-poles held in each hand, are thought to alleviate these complaints by providing an alternative pathway for the transmission of ground reaction forces. It was thought that the use of the poles would reduce some of the alterations to the walking cycle that are typically found during load carriage and possibly improve walking efficiency. Measurements were made of heart rate, oxygen consumption, muscle activity (EMG), joint kinematics (joint angles throughout the walking cycle) when walking for one hour with and without the poles and carrying a 22.4 kg pack. The results indicated that subjects did use the poles to support body weight; trunk flexion and extension velocities were lower when the poles were used; and the pattern of knee joint motions more closely resembled normal walking. Energy expenditure was the same when the poles were used but heart rate was increased (113.5 beats/min compared to 107 beats/min), probably owing to the increased involvement of the upper limbs. Ratings of perceived exertion were lower when the poles were used, suggesting that the partial normalisation of walking could be achieved without extra energy expenditure.

Spade design   Bridger *et al.* (1997) compared the physiological demands of digging with a conventional spade and the 'levered' spade described in Chapter 6. Subjects transferred 1.8 tonnes of sand in a laboratory task. Heart rate and oxygen consumption were measured throughout. It was hypothesised that the levered spade would be more energy-efficient because it reduced the need to stoop and the need to flex and extend the trunk. The findings illustrate some important considerations in work physiology, principally the need to relate the physiological demands to the work outputs. Spade design had no influence on oxygen consumption but heart rate was higher when the levered spade was used, because of the more upright posture (the heart has to pump blood against gravity to the head). However, the digging task took longer and required more scoops with the levered spade, suggesting that the improvement in posture was associated with a reduction in digging efficiency. It is likely that the upright posture was physiologically more efficient but that some of the benefits were offset by the addition of the second handle, which increased the weight of the spade.

The main conclusion was that the postural benefits of the levered spade were offset by the reduction in digging efficiency and the increased weight of the spade. Overall, there were no extra physiological costs.

Evaluation of non-physical stress

The above research involves the direct application of physiology to the analysis of work in situations where performing the work has direct physiological effects on the worker. These effects can be measured and used as indices of physical workload. In developed countries, much industrial work is light – it has no direct consequences for the worker, at least none greater than many other activities of daily living carried out outside of work. However, physiological methods can also be applied to the investigation of light work, particularly to detect the presence of mental stress. It is known that the levels of several physiological variables, including heart rate, increase when a person is under mental stress.

Khaleque (1981) investigated heart rate, perceived effort and job satisfaction in a group of female subjects in a cigar factory. The subjects were operators of cigar wrapping machines and the work involved simple, repetitive movements according to a 3-second work cycle paced by the machine. Machine-paced work such as this is generally held to be stressful, particularly in older subjects.

Job satisfaction was measured using a questionnaire (the 'Brayfield–Rothe' scale) and perceived effort by the Borg scale. Heart rate was measured using an optical pulse wave detector and appropriate circuitry. In this method, a clip is attached to the subject's ear lobe such that light from a small lamp on one side of the clip is detected by a photoelectric cell on the other side, the ear lobe being interposed between the lamp and the detector. With each pulse wave, the translucency of the ear lobe changes (as the flow of blood through the ear changes). This is detected as a change in the current produced by the photoelectric cell.

The subjects were split into two groups according to their level of job satisfaction. The satisfied group was found to have an average heart rate of 81 beats/min and the dissatisfied group an average of 91 beats/min. Perceived effort did not differ between the two groups nor was there a relationship between perceived effort and heart rate. The author concluded that emotional factors probably played a role in elevating the heart rate of the dissatisfied workers: job dissatisfaction is assumed to lead to negative emotional responses that increase heart rate.

Physiological methods have long been used in the aviation industry and in military applications as indices of mental stress. When a person is subjected to a high mental workload or to an emotionally intense experience, the level of cortical arousal (arousal of the cortex of the brain) is hypothesised to increase. There are many physiological correlates of the increased arousal, including increased heart rate, changes in the electrical resistance of the skin (the 'galvanic' skin response) and changes in the concentration of certain hormones in the blood (particularly the 'stress' hormones such as adrenaline). These physiological measures can be used to detect the presence of mental stress (some of these methods are reviewed in more detail in a later chapter).

Although many jobs require great concentration it is not necessarily the case that they impose negative emotional stress on the worker. Becker *et al.* (1983) measured heart rate and oxygen consumption of surgeons while operating. The heart rate/ oxygen consumption relationship was measured using a cycle ergometer and each surgeon's maximum oxygen uptake was predicted. Heart rate was also measured with the subjects at rest, when sleeping, and when doing ward and theatre work. Oxygen consumption during operations was found to be low (0.26–0.41 litres/min). Heart rate during surgery was commensurate with the oxygen consumption and with the isometric arm and other work required during the performance of surgical tasks. The authors were therefore able to conclude that there was little evidence of increased heart rate due to mental stress.

Heart rate and heart rate variability have long been used to asses the workload of pilots. Jorna (1993) presented data on heart rate and heart rate variability of pilots undergoing conversion training to a new aircraft. When coming into land, novice pilots' heart rate increased steadily but heart rate variability (instantaneous interbeat intervals) decreased. At the completion of training, the same pilots experienced a lower heart rate when landing but greater heart rate variability. Pilots undergoing examination had very high heart rates and almost no heart rate variability. Roscoe (1993) concluded that, of the available psychophysiological variables for assessing pilot workload, heart rate seems the most useful. Heart rate profiles of pilots approaching and landing at two airports differing in difficulty showed similar patterns of a steady increase during the approach followed by a decrease after landing. The overall heart rate was higher when landing at the 'difficult' airport. A reduction in heart rate was found when pilots landed under 'autoland' rather than manually. Thus, heart rate and heart rate variability can be used to distinguish between different mental states bought about by task demands in tasks with low physical workloads.

Under stressful conditions, breathing exceeds metabolic requirements. According to Schleifer and Ley (1994), this causes a drop in the carbon dioxide concentration in the end-tidal expired air (the last air to be expired when breathing out). Stress-induced hyperventilation decreases the concentration of carbon dioxide in the blood. This has a number of physiological effects. Vasoconstriction of blood vessels reduces the flow of blood to the heart and brain and the amount of oxygen that haemoglobin in the red blood cells can release to the tissues. This is accompanied by feelings of dizziness and heart palpitations. Theoretically, end-tidal carbon dioxide measurement should be a powerful way of measuring non-physical stress. Unlike heart and ventilation rate measurements, which increase both under stress and during exercise, end-tidal pCO_2 measurements decrease when the person is under stress but not when physical work is carried out (breathing does not exceed physical work requirements under these conditions). The authors compared these stress indices during self-relaxation, relaxation using a stress-management technique known as progressive relaxation and during data entry work at a VDU. As predicted, end-tidal p_{CO_2} more closely tracked self-reported mood states than the other variables. It seems to be a useful means of investigating the stress of VDU work and might be used as biofeedback in stress-management programmes.

Fitness for work

The motivation for the research cited above has been to ensure compatibility between the physical demands of the work and the physiological limitations of the worker. A related question is the assessment of a worker's fitness for doing the work at all.

Fitness is a subject about which even the experts disagree. Performance of a job depends on the level of skill and the motivation of the worker as well on physical fitness. Definitions of fitness can be problematic. In sports and exercise physiology, fitness is clearly event-specific. The cardiovascular robustness of the marathon runner can be contrasted with the explosive power of the weight lifter and the poise and flexibility of the gymnast. Even concepts such as V_{O_2} max have an event-specific quality to them. Reilly (1991) points out that V_{O_2} measurements on athletes are best tailored to the particular endurance event. Long-distance runners are best tested on the treadmill, cyclists on the bicycle ergometer, and athletes whose sport involves mainly upper body activity on arm ergometers.

Fitness and health

Fitness, according to Reilly, is a multivariate concept, although a distinction can be drawn between fitness and health. Medical doctors tend to pronounce a patient healthy if they can find no evidence of disease. In fact, many athletes, while having a high level of fitness for their chosen sport, may actually be very unhealthy and suffer from frequent bouts of colds and flu as well as musculoskeletal and other ailments. Conversely, few healthy people are fit to run a marathon or to lift heavy weights without risk of injury unless they have been undergone a prior programme of training.

In jobs that require prolonged manual labour, assessment of cardiovascular capacity can be used to determine a worker's level of fitness for work. V_{O_2} max can be measured directly or indirectly (from heart rate and oxygen consumption data at three or more submaximal workloads). Alternatively, one of a number of simple 'step tests' can be used. These tests have been developed to enable fitness to be assessed objectively and without the need for sophisticated equipment. It appears to be the case that physically fitter individuals have a lower heart rate when resting and when working at a predetermined level than the less fit. Furthermore, their heart rate returns to its resting level more quickly after a period of exercise.

In the gold mining industry a step test has been developed in which subjects step on and off a step 24 times per minute for 9 minutes. The heart rate is then measured and the worker is placed into one of three categories. Category A men (heart rate less than 120 beats/min) can be allocated to strenuous work especially in hot conditions, whereas category B men (121–140 beats per minute) are given less arduous tasks and are not expected to work in hot environments. In several industries and occupations (e.g. fire fighting, mining) it is not possible to redesign the working environment or equipment so as to 'fit the task to the man' and physical selection tests have a valuable role to play in ensuring compatibility between worker characteristics and task demands.

Assessment of physical work demands: Ambulatory monitoring

Ambulatory monitoring techniques are used to assess physical activities and demands in everyday work situations. Heart rate monitors and pedometers (to measure walking) have been developed that can be worn while working. Uiterwaal *et al.* (1998) describe an accelerometer-based system designed to measure walking, standing, sitting, lying and playing (in children). The system allows the investigator to estimate the amount of time a person spends in each of these activities over an ordinary work day and can be used to assess occupations such as nursing as well as managerial and technical jobs that are physically varied.

Data from ambulatory monitoring can be entered into energy expenditure models such as that developed by the Center for Ergonomics at the University of Michigan. Such models enable the overall metabolic energy requirements of a job to be estimated from data on the energy requirements of the basic work activities. Such models can be used to detect the highly demanding aspects of a job in job assessment and job redesign studies. They can also be used to assist decision making about return to work (RTW) in the case of sick or injured employees. The energy demands of the job can be expressed as a percentage of the individual's maximum aerobic power at the time the decision about RTW has to be made. If less than 33% of maximum, the individual is regarded as fit enough to return to work. Alternatively, less demanding RTW options have to be sought (Caple, 1988).

Physical fitness and everyday life

The relationship between physical fitness and customary daily activity appears to be well established. Ballal *et al.* (1982) investigated the physical condition of young adults in the Sudan. Soldiers, physical education students and urban workers were found to be in better physical condition than rural villagers and medical students. The differences could not be attributed to body size, body composition or sex and the authors suggested that the results were due to differences in the level of customary activity. Interestingly, the two most unfit groups were in other ways different: the villagers were smaller and less muscular than the other groups owing to disadvantageous features of their environments. The medical students had a greater proportion of body fat and low levels of physical activity, possibly due to their more affluent lifestyle.

Workload, physical fitness and health

The goal of ergonomics in the design of the physical component of jobs is to minimise unnecessary and possibly harmful stress. However, this is not to be interpreted to mean that all stress is bad and should be minimised. There is a growing body of evidence that people who do physically demanding jobs are fitter and healthier than their less active colleagues. Conversely, the risk of fatal heart attack is greater in those with less demanding jobs.

In an early study of the relationship between occupation and fitness, Morris *et al.* (1953) compared the incidence of heart disease in bus conductors and bus drivers

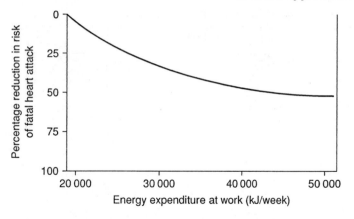

Figure 8.3 Relationship between energy expenditure at work and risk of fatal heart attack.

and in mailmen and sedentary postal clerks. In both groups, the incidence of heart disease was significantly lower in the active rather than the sedentary subgroup. The bus drivers were more overweight, had higher blood pressure and cholesterol and smoked more than the conductors. To complicate interpretation, though, it appeared that these differences had existed before the individuals entered employment, suggesting that there was an element of *self-selection*. That is, people who are more active anyway choose active occupations and vice versa.

In a second study, Morris *et al.* (1973) investigated 16 882 British Government employees. All of them had sedentary occupations and were comparable in terms of coronary risk factors. The group was then subdivided into those who did or did not participate in vigorous leisure-time activities. The heart attack rate of the active group was one-third less than that of the inactive group. Even smokers who exercised were less at risk than smokers who did not exercise.

In the USA, Paffenbarger *et al.* (1984) studied the risk of fatal heart attack in San Francisco longshoremen. Longshoremen who performed heavy manual labour were less at risk than less active colleagues. The risk of fatal heart attack was up to 50% less in the most active compared with the inactive groups. Figure 8.3 summarises these findings.

Cady *et al.* (1985) found that in a group of 40 to 49-year-old workers, those with the lowest physical work capacity had three times the injury cost for back injuries than those with the highest physical work capacity.

It seems fairly clear then that daily energy expenditure is an important factor in safeguarding the health of the population. Those in strenuous manual jobs will get some degree of protection by virtue of their work. However, the trend is for physical workload to be reduced, which suggests that Paffenbarger's view is valid. Table 8.3 gives the daily energy expenditure of a variety of occupations.

Work hardening programmes

Work hardening programmes are a type of physical training programme aimed at improving employees' physical capacity for a particular task or set of tasks with the aim of reducing the risk of physical injury. According to Guo *et al.* (1992), to be effective, work hardening programmes must

Table 8.3 Energy cost of some common occupations[a]

Occupation	Energy cost (kJ/day)	
	Men	*Women*
Bookkeeper	10 000	8 400
Secretary	11 300	
Bus driver	12 500	10 500
Letter carrier	12 500	10 500
Machine operator	13 800	11 500
Stonemason	15 100	12 500
Labourer	16 300	13 600
Carpenter	16 300	13 600
Ballet dancer	16 300	13 600
Coal miner	17 600	
Lumberjack	18 800	
Athlete	20 100	17 800

[a] From Woodson (1981).

- Be of greater intensity than that experienced on the job
- Be based on the techniques of exercise physiology described in the literature (e.g. Noakes, 1992)
- Use exercises that closely resemble the movements made on the job
- Be evaluated using tests that simulate the activities carried out during training

In many industries, work hardening takes place informally. Management frequently expect a 'break-in period' of a few weeks before a new employee can work at the required pace. This is particularly true of jobs involving heavy or repetitive work.

Evidence from a variety of sources suggests that low back sufferers have less trunk flexor and extensor strength, diminished spinal extension, overpowered flexors and tight hamstring muscles (e.g. Beiring-Sorensen, 1984; Pope *et al.*, 1985). Animal and human studies have shown that endurance training causes an increase in muscle stiffness (the ratio of tension developed in a muscle when it is stretched to the amount by which it lengthens) as a result of increased cross-linking of collagen fibres in the muscle (Herbert, 1988). Animal studies have also shown an increase in the amount of stiffness and collagen in the muscles of elderly and sedentary rats. A sedentary lifestyle appears to reduce hamstring distensibility in the general population (Milne and Mireau, 1979). A lack of exercise combined with age-related changes in muscle structure may reduce flexibility, predisposing workers to injury. Badly designed exercise programmes that increase endurance or strength but not flexibility may not have the expected beneficial effects. At the other extreme, it is known that joint hypermobility (which seems to be the result of hereditary factors, rather than training, in successful ballet dancers according to Grahame and Jenkins, 1972) is a risk factor for joint dislocation and premature osteoarthrosis.

The three main components of a work hardening programme are strength, flexibility and endurance. Depending on the requirements of the task in question, the programme must aim to achieve an *appropriate balance* between these components. This is particularly important when the training is part of a rehabilitation programme for previously injured workers.

Guo *et al.* (1992) compared the effects of flexibility and strength/flexibility training programmes on middle-aged maintenance employees. Subjects trained 5 days per week for 30 minutes per day over 4 weeks. Improved physical capacity was obtained using flexibility exercises only by progressively increasing either the time spent in a particular posture or the number of repetitions of the exercise. Increases of 28% were found for truncal rotation and low back flexibility and 39% for dynamic strength. The strength flexibility programme resulted in increases of 52% for dynamic strength, 150% for endurance time and 27% for low back flexibility. However, during a 4-week follow-up period in which the exercises were performed twice per week, most of these gains were lost. It would seem that once a degree of work hardening has been achieved, more regular training than this is needed to maintain it.

It is thought that abdominal muscle exercises are of value in improving the stabilisation of the lumbar spine and thus protecting workers from injury. Richardson *et al.* (1990) suggest that those exercises involving rotation seem best since they induce a pattern of muscle stimulation involving co-contraction of the lumbar muscles.

Work hardening programmes and rehabilitation

Robert *et al.* (1995) evaluated the effectiveness of a 6-week work hardening programme on patients diagnosed with low back dysfunction. The programme consisted of cardiovascular fitness training (subjects could choose between a variety of exercises including a bicycle ergometer, treadmill, stepping machine, etc.), weight training, stretching and work-simulation activities.

Mean Vo_2 max increased from 30.49 ml/kg/min at the beginning of the programme to 39.09 ml/kg/min at the end, a 28% improvement overall. Subjects exercised at a level of 50–60% of maximum heart rate. This appears to be the minimum threshold above which gains in cardiovascular fitness will be obtained. Strength training con-sisted of two sets of 12 repetitions of exercises to train the major muscle groups of the body (leg press, leg curl, leg extension, calf raise, chest press, incline press, fly, triceps press, triceps extension, lateral pull, rowing, shoulder press, pullover, back extension, abdominal flexion and lateral raise) as well as four sets of 10 box lifts at different heights (floor to knuckles, floor to waist, waist to overhead and floor to overhead). Static pull improved by 57%, static push by 89% and static arm lift by 76% over the 6-week training period.

Clearly, a 6-week programme is sufficient to produce large gains in cardiovascular and muscular strength. However, the authors were unable to find any difference between the fitness levels of the 50% of patients who returned to work after the programme and those who did not. This suggests that factors other than fitness also influence return to work. These factors include time off work since injury (longer times are associated with decreased likelihood of returning to work), availability of similar work, job satisfaction and the patient's understanding of their condition.

Maintaining fitness at work: VDUs

Exercise programmes for VDU users have been developed for use while at work. Their aim is to be complementary to the design of the job so as to increase physical variety and thus counteract the stresses of static VDU work. A review of physical exercise for VDU users can be found in Lee *et al.* (1992). Office exercise programmes must be usable; since many VDU workers occupy open-plan offices, they may be reluctant to perform exercises that are conspicuous or embarrassing. Lee *et al.* found that many of the recommended exercises were inappropriate because they actually reproduced the demands of VDU work, or were contraindicated for physical conditions such as osteoporosis, or were dangerous. Any exercise that requires use of office furniture as a prop is potentially dangerous since the person or the equipment may not be as stable as expected.

The general principles for VDU exercise programmes are, according to Lee *et al.*:

*   Chronically shortened and tense muscles must be stretched and relaxed.
*   The spine must be mobilised to reduce and redistribute compressive forces.
*   Chronically stretched or relaxed muscles must be contracted or strengthened to maintain posture.
*   Exercises must involve the lower limb to promote venous return.

VDU users often sit in the upright 'slumped' posture described in Chapter 4. The neck is extended and there is static tension in the extensors. These muscles require stretching and relaxation in flexion. The scapular elevators are also chronically tensed and need stretching and relaxation. The lumbar and thoracic extensors are passively stretched and need activation by actively extending the lumbar spine. The anterior thoracic muscles are relaxed and shortened and require stretching (possibly by grasping both hands behind the back). The forearm flexors are chronically tensed and shortened and require lengthening and relaxation.

Accidents and fatigue

Acute or chronic undernutrition is often a cause of low blood sugar and is accompanied by feelings of tiredness and irritability. Pheasant (1991) has reviewed the evidence for a connection between low blood sugar, low performance and increased accidents. Studies have shown poorer performance in workers who skip breakfast than in those who do not. Mid-morning and afternoon snacks seem to help workers sustain performance. Blood sugar is often found to be low in victims of freeway automobile accidents. Other studies have shown that foundry workers have fewer accidents if given a high-energy glucose drink before work. Pheasant concludes that for maximum work efficiency, calorie intake should be spread over the day to include a light breakfast, mid-morning and afternoon snacks, lunch and dinner.

Designers of production facilities in developing countries, where the nutritional status of workers may be poor, might make allowance for this in the design of on-site facilities, particularly if shiftwork is involved. High calorie meals and snacks acceptable to local tastes should be provided at no or minimum cost during plant operation.

Effectiveness and cost-effectiveness

Interest in the general fitness of the working population has increased over the last few years and it has been suggested that a fitter workforce will be more productive and have lower absenteeism. This is based on the notion that increased fitness will 'inoculate' workers workers against physical and non-physical job demands by improving the ability to cope physiologically.

Promote health

Hardman *et al.* (1989) investigated the effects of a programme of brisk walking on a group of previously sedentary middle-aged women. The women followed a progressive programme of brisk walking building-up to a pace of 1.87 m/s after 12 months. They averaged 16–17 kilometres per week, which took about 155 minutes and refresented approximately 60% of their predicted Vo_2 max.

The programme was found to enhance their tolerance for exercise and their metabolic and cardiovascular responses to it. Favourable changes in the ratio of total cholesterol to high-density lipoprotein cholesterol concentration were also observed when the walkers were compared with a control group. It would appear that for otherwise sedentary and inactive individuals even regular brisk walking may be beneficial. There may be some justification, then, for designing buildings and work environments to encourage this type of activity and even to attempt incorporating it into otherwise low-activity jobs.

Goetzel *et al.* (1994) describe an evaluation of IBM's corporate education programme, 'A Plan for Life'. Program participants and non-participants were compared over a 5-year period. At the end of the period, the proportion of programme participants no longer at risk was greater than the proportion of non-participants as measured by various risk indices such as blood pressure, total and non-high-density lipoprotein cholesterol and smoking cessation. Knight *et al.* (1994) evaluated a similar programme (the 'Live for Life' comprehensive health promotion programme) at Duke University, NC, USA. The programme consisted of health screening, smoking cessation, weight control, stress management, nutrtion education, fitness and blood pressure intervention. The programme was free of charge except for a $10 monthly membership fee for the fitness club. After 3 years of programme availability, programme participants had 4.6 fewer hours absent from work than non-participants (which was statistically significant) and there was a significant linear trend in absenteeeism reduction with years of programme participation.

Cox *et al.* (1981) evaluated the effects of an employee fitness programme in two Canadian assurance companies (an experimental and a control company). Programme participants in the experimental company exhibited increases in Vo_2 max, a loss of body fat and more positive attitudes to their employment. Labour turnover and absenteeism were lower in the experimental than in the control company. The authors concluded that a potential 1% reduction in company payroll costs was possible through the implementation of the programme, although the findings of the study could not be attributed directly to increased fitness.

These findings suggest that health promotion programmes can be effective in improving employee health with positive outcomes for the employer. However, Heaney and Inglish (1995) report that, for maximum effectiveness, programmes should attract employees at high risk for coronary heart disease whereas, in practice, health status seems to have little influence on employees' willingness to participate. Participation seems to be biased in favour of high-pay versus low-pay employees and white-collar versus blue-collar jobs. Furthermore, some interesting sex differences influencing participation in fitness programmes were observed. Males who were 20% or more overweight were more likely to participate than males who were 0–10% overweight; the reverse was true for females. Males with elevated diastolic blood pressure; were less likely to join than males with normal blood pressure; the reverse was true for women. Low baseline physical activity in females was associated with participation but not in males. These findings led the authors to conclude that males join fitness centres to maintain their health and women join to improve their health (except that being more than 20% overweight may discourage some women from exercising in public).

Target high-risk groups

Although there is an economic case for worksite health promotion, simple statistical measures may overestimate the potential benefits of health promotion. This is because injury and illness claims are not evenly distributed in the workforce. In fact, the data are highly skewed, with a small percentage of the overall workforce accounting for a large percentage of the claims. In one study (Kingery *et al.*, 1994), it was found that 27% of employees made claims costing more than $500 per year as opposed to 73% who claimed less than that amount. The authors examined volunteers from the workforce to identify those with at least one of the following risk factors: cholesterol, high blood pressure, cardiovascular fitness, body fat and smoking status. It was found that 43% of the cost of medical claims was associated with an elevated health risk. High-risk employees cost the company an average of $1341 more per person per year than did low-risk employees. The results suggest that in order to maximise the financial benefits of interventions aimed at improving the health of the workforce, programmes should begin by identifying those employees with health risks and targeting the interventions at them in the first instance.

Research issues

Escalating health care costs in developed countries have provided a stimulus for the investigation of factors influencing the health of the working population. Since work constitutes the major portion of an individual's waking life, it seems logical to evaluate the activities carried out at work and to suggest ways of incorporating health-promoting features into working life. Fitness programmes and in-house gymnasia are obvious examples.

Before the advent of the automobile and public transport systems, it was common for people to walk 10–20 miles per day. Nowadays, many people probably walk less than this in a week. Further studies of the influence of activities of daily living on health and fitness would be of interest.

As described in Chapter 7, the research issues that centre around work and health in developing countries are quite different and much work remains to be done.

Summary

Physical workload can be evaluated in terms of the physiological cost to the worker of carrying out the work. Oxygen consumption and heart rate are objective measures of workload. Since people differ in their capacity for aerobic work, it is appropriate to consider the relative workload of a task; an individual's oxygen consumption while working can be expressed as a percentage of his Vo_2 max. Workloads of 30–40% of Vo_2 max can be sustained for an 8-hour shift. The task of the ergonomist is to design the work so that it can be performed safely and without undue fatigue by the largest possible number of people in the workforce. This implies designing for individuals with lower aerobic capacity (such as older workers). If this is not possible, a physiological approach is needed to specify a minimum standard of fitness for the job.

Heart rate is influenced by factors other than workload. This is both an advantage and a disadvantage for the experimenter. If heart rate is used to estimate oxygen consumption by the extrapolation method, erroneous estimates will be obtained if extraneous variables increase the heart rate over and above the level needed to supply oxygen to the tissues. Some factors that can cause such an elevation include mental stress, drinking tea or coffee, working directly after a heavy meal and working in the heat. By the same token, heart rate may be of more use as an index of the total workload because it is affected by variables such as mental stress and heat, which do increase the load on the worker. It is in this sense that heart rate integrates the total stress of a job.

Despite the global trend to replace manual work with mechanised and automated systems, physiological aspects of work design are of great importance in view of the modern interest in eliminating avoidable ill-health. In low-technology jobs, the problems of avoiding excessive physical stress remain. In high-technology jobs, the problem is to avoid mental overstress and find ways of increasing the physical activity of the, predominantly, sedentary workers. This is particularly relevant when it is remembered that post-industrialised countries have an ageing population.

Essays and exercises

1. Try out step test described in the chapter (page 223) on yourself. Calculate your heat rate, measured immediately after the test, in beats per minute.
    The height of the step is determined as follows:

    $$\text{Step height (cm)} = \frac{542 \times 2.54}{\text{Your body mass (kg)}}$$

    Step for 9 minutes at a rate of 24 steps per minute.
    Which fitness category do you fall into?

2. Carry out the following activities for 5 minutes. Measure your pulse immediately afterwards to estimate your heart rate. Lie down for 5 minutes prior to the activity and take your resting heart rate before starting.

    *   Jogging
    *   Walking up a flight of stairs briskly
    *   Doing squats
    *   Raising and lowering a 20 kg load from knee height to chest height at a rate of 1 lifting cycle every 5 seconds
    *   Standing still
    *   Sitting
    *   Walking
    *   Walking with a 10 kg weight in each hand

- Walking with a 20 kg backpack
- Walking with a 20 kg load on your head.

Plot the heart rate data in graphical form to compare these different activities.

3. Distinguish between health and fitness. What measures can organisations take to improve the health and fitness of their employees?

9 Heat, cold and the design of the physical environment

The incidence of heat-related illness is as high as 1 in every 100 men working for 1 year ... 50 US football players are said to have died of heat illness in the 10 years to 1975 ... Heat strain occurs whenever the body generates more heat than it can lose ... even in cold conditions ... A British soldier marching in outdoor temperatures close to 12 degrees Celsius died of heat-related illness.

(G. W. Crockford, 1999)

Because the rates of most biochemical reactions are temperature dependent, it is advantageous to control the rates of these reactions by means of a thermoregulatory system.

Fundamentals of human thermoregulation

Humans have a remarkably well-adapted ability to tolerate heat compared with other primates. This applies equally to 'Eskimos' as to tropical rainforest dwellers, despite small differences. This is because humans are hairless and have a large proportion of high-capacity sweat glands – known as eccrine glands – in their skin.

Thermal balance

Thermoregulation is achieved by balancing the two main factors that determine body temperature – the metabolic heat produced and the rate of heat loss. The thermoregulatory goal is to maintain the core temperature at approximately 36–37°C. Core temperatures over 39.5°C are disabling and over 42°C they are usually fatal. The lower acceptable limit is 35.5°C and 33°C marks the onset of cardiac disturbances. Further drops in core temperature are extremely dangerous and temperatures as low as 25°C are fatal. The temperature of the peripheral body tissues, particularly the skin, can safely vary over a much wider range. From a thermal point of view, the body can be considered to have a warm core where much of its heat is produced. This is surrounded by a shell of cooler, insulating tissues, particularly subcutaneous fat.

The principal sources of heat are the liver, the brain, the heart and the working muscles. Muscular work is a source of heat because the mechanical efficiency of muscles is only about 20%. Little heat transfer from the interior to the surface of the body takes place via conduction: the body tissues are poor heat conductors. Heat is transferred to the skin from the deep body tissues by convection. Blood is an ideal

Table 9.1 Basic equation of human thermal balance

$$S = M - E \pm R \pm C - W$$

where

   S = Heat gained or lost by the body
   S = 0 when the body is in thermal balance with the environment
   M = Metabolic energy production
   E = Heat dissipated through evaporation (sweating)
   R = Radiant heat to or from the environment
   C = Convection to or from the environment
   W = Work accomplished by the worker

---

medium of heat convection because its specific heat capacity and thermal conductivity are high.

Core temperature can only be maintained within its narrow range of values if there is a state of heat balance with the environment. The fundamental thermodynamic processes involved in heat exchange are described by the equation given in Table 9.1

Metabolic heat production can be measured directly using a method known as calorimetry. A resting person is placed in a sealed chamber, the temperature of which is maintained at a constant level by passing the air in the chamber through a cooling system (pipes in a water bath). The increase in temperature of the water in the bath is directly related to the metabolic heat production of the person in the sealed chamber. Direct calorimetry is really only suitable for research purposes and another method known as indirect calorimetry is often used in its place. Because metabolic processes for energy production and basal metabolism require oxygen, the oxygen uptake of a person at rest can be measured and used to estimate M (metabolic energy production). The energy equivalent of metabolising 1 litre of oxygen is approximately 4.8 Calories (although it varies depending on whether glucose, fat or protein is being used as fuel). However, if the person is working or exercising, not all of the oxygen consumed will give rise to heat – some will be used to carry out the work. If the amount of physical work (W) can also be measured, then the heat gained due to metabolic processes can be calculated (from the total metabolic energy production less the energy used to carry out the work).

Heat may also be gained from the environment or lost to the environment by convection (C) or radiation (R). If the skin of a resting person is warmer than the surrounding air, the air at the skin surface is warmed by the conduction of heat from the skin. Since warm air rises, a flow of air around the person is established and heat is convected away from the person's body. Cool air moves closer to the skin to replace the rising warm air. This air flow is known as a convection current. In practice, a resting person has a layer of slow-moving warm air at the surface of the skin. Movement of the body or the presence of wind can disrupt this warm layer and increase the rate of heat loss. Depending on the temperature of the air, this will be perceived as a chilly draught or a cooling breeze.

All objects at temperatures greater than absolute zero emit infrared (IR) radiation. This is sometimes referred to as 'radiant heat'. The human body is no exception to this and if its temperature is greater than that of its surroundings a radiant heat loss takes place. If the surroundings are hotter than the body, it experiences a net heat gain. Inside buildings, the walls and ceiling are common sources of radiant heat, whereas the sun is the major source out of doors. In steel and glass manufacture, furnaces are a major source of radiant heat.

Sweat production and evaporation (*E*) is a mechanism by which heat is lost to the environment. Since the body tissues are composed largely of water, it is unsurprising that water is lost to the environment by diffusion from the skin, from the lungs and from the sweat glands. If the temperature of the surroundings is greater than that of the body, no heat can be lost by convection or radiation and evaporative heat loss by sweating is essential to maintain thermal balance.

When the body is in a state of thermal balance, the terms on the right-hand side of the equation in Table 9.1 cancel each other out. This can occur in several ways. For example, in a cold environment, a state of thermal balance can be achieved either by increasing the rate of metabolic heat production (by shivering or by carrying out some fairly strenuous activity) or by reducing the heat lost by convection and radiation (by wearing more clothes). Metabolic heat production can be increased by a factor of about 3 when a person is working as opposed to resting. In cold conditions, a person would therefore require three times as much clothing insulation when resting than when working. Clothing reduces heat loss to the environment by trapping air in the clothing fibres and between the garment and the skin. This disrupts the flow of convection currents and maintains a layer of warm air around the skin surface. Clothing also absorbs some of the radiant heat from the body. Both of these effects reduce the temperature gradient between the skin and its immediate surroundings and therefore the rate of heat loss. Water practically eliminates the insulating property of clothing because of its high specific heat capacity.

In manual work in hot conditions (in a foundry, for example) there is a high rate of metabolic heat production as well as heat gain from the environment (by radiation). A state of thermal balance can be obtained either by reducing the rate of metabolic heat production (by reducing the work rate or introducing rest pauses) or by protecting the worker from the radiant heat (by means of a protective suit, for example).

The design of clothing assemblies for work in extreme environments is more problematic than it might, at first, seem. A clothing assembly for cold work, if it traps air too-well may trap sweat as well. If the sweat accumulates the insulation provided by the clothing will be degraded. Suits to protect workers from radiant heat may also trap air around the worker's body. Since the rate of evaporation of sweat depends on the humidity of the air, evaporative heat loss may steadily decline as the trapped air becomes more saturated with water vapour and less sweat evaporates.

Measuring the thermal environment

Dry-bulb temperature (DBT)

DBT is the temperature of the constituent gases of the air. Although DBT indicates the thermal state of the air, other factors have an equally important effect on the heat gain or loss.

Relative humidity and wet-bulb temperature (WBT)

At all temperatures above freezing, water tends to evaporate into the air. The rate of evaporation increases with temperature. At any given temperature, a point is reached

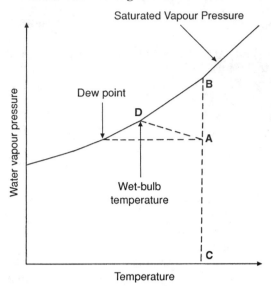

Figure 9.1 Relationship between temperature, vapour pressure, dew point and wet-bulb temperature. The dew point is the temperature at which the vapour pressure at **A** would equal the saturated vapour pressure. Relative humidity at **A** is given by **AC/BC**. The wet-bulb temperature is the temperature that would be reached if water was evaporated into the air. (From, Kerslake. 1982. © Cambridge University Press, 1982, reprinted with the permission of Cambridge University Press.)

where no more water vapour can evaporate. At this point, the air is *saturated* (with water vapour). When the water vapour pressure of the air reaches its maximum for a given temperature, evaporation from a wet object ceases. Evaporation resumes if the temperature increases. The saturated water vapour pressure of air increases with temperature. If warm air is cooled, a point will be reached at which the water vapour pressure at the original temperature is equal to the saturated water vapour pressure at the new temperature. This the dew point, when water condenses out.

Relative humidity is the water vapour pressure at a given temperature expressed as a percentage of the saturated water vapour pressure at that temperature. Figure 9.1 illustrates these terms.

Thermal assessment requires three measurements of the air temperature: dry bulb temperature (DBT), wet bulb temperature (WBT) and globe temperature (GT). WBT is measured with a thermometer. A wetted cloth 'sock' is placed over the bulb and the measurement is made after allowing the thermometer and sock to stabilise at the ambient temperature. WBT depends on the DBT and on the relative humidity of the air. Evaporation of water from the sock cools the thermometer bulb. The rate of cooling depends on the humidity of the surrounding air. WBT and DBT can be used to calculate the relative humidity using 'psychrometric charts' (e.g. Kerslake, 1982).

When the air is completely saturated (i.e. the relative humidity is 100%), no water will evaporate from the wet-bulb thermometer, no evaporative heat loss will take place at the bulb and the wet and DBTs are the same. In all other circumstances, WBT will be lower than DBT.

Globe temperature (GT)

GT accounts for the effects of radiant heat. Traditionally, the bulb of mercury in glass thermometer is placed in a metal sphere, painted matt black. Radiant heat (from the sun or from hot objects) is absorbed by the sphere and heats up the thermometer. Radiant heat can prove to be a significant source of heat load on the worker (consider the beneficial effects of shade on a hot, sunny day). GT measurement is often essential if the true nature of the thermal environment is to be evaluated.

Air movement and wind chill

When evaluating the effects of temperature on the worker, it is important to take air movement into account. Air movement moderates the effects of high temperatures and exacerbates the problems of low temperatures (causing 'wind chill'). Air movement can be measured using anemometers (in which the rate of rotation of a vane or the cooling of a hot wire is proportional to the velocity of the air flow).

Heat stress is the combination of all those factors, both climatic and non-climatic, that lead to convective or radiant heat gains by the body or prevent heat dissipation from the body (Leithead and Lind, 1964).

Modern heat stress monitors make use of thermistors (electrical transducers) instead of mercury-in-glass thermometers. DBT, WBT and GT can be measured by the same instrument and air movement can be taken into account. These devices often provide combined measurements of temperature known as 'heat stress indices'. The 'wet bulb GT' index (WBGT) is often used. Commercial heat stress monitors calculate the WBGT index as follows for measurements made outdoors and indoors respectively:

$$\text{WBGT (out)} = 0.7\,\text{WBT} + 0.2\,\text{GT} + 0.1\,\text{DBT}$$

$$\text{WBGT (in)} = 0.7\,\text{WBT} + 0.3\,\text{GT}$$

WBT is weighted most heavily in the WBGT index, demonstrating why measurements of DBT alone cannot provide an accurate measure of the stress involved in working in a hot environment. Figure 9.2 shows a modern heat stress monitor for measuring WBGT.

Thermoregulatory mechanisms

A number of physiological mechanisms exist for maintaining heat balance.

Peripheral vasomotor tone

Heat production and heat loss can be balanced within a fairly narrow range of skin temperatures by adjustment of peripheral vasomotor tone. In hot environments, peripheral vasodilation occurs – the arterioles dilate and capillaries at the skin surface open, blood flow increases and heat is conducted to the skin over a small distance from where it is dissipated to the environment. In the cold, vasoconstriction occurs, reducing blood flow in the cutaneous circulation. This increases the conductive distance and the insulation of deeper body tissues. Less heat is lost at the skin surface.

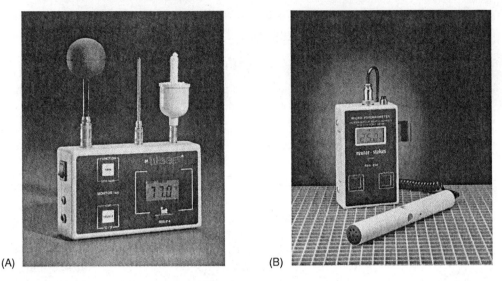

(A)   (B)

Figure 9.2 Modern instruments for thermal environment evaluation.
(A) Heat Stress Monitor (note the three separate sensors for measuring DBT, WBT and GT).
(B) Micro-psychrometer that gives direct readings of WBT and DBT temperatures, relative humidity and dew point.

Thus the skin can function both as an insulator and as a radiator of heat depending on the peripheral vasomotor tone.

An indication of the effectiveness of this system can be obtained by comparing its insulating capacity with that of clothes. The CLO is a common unit of clothing insulation. If a nude person has a thermal insulation of zero CLO, a person wearing a business suit (plus shirt, underwear, etc.) has 1 CLO (Parsons, 1993). The insulation provided by the vasomotor system is effective over the range of approximately 0.2–0.8 CLO.

Countercurrent heat exchange

Countercurrent heat exchange is a method of conserving heat, involving the exchange of heat between arteries and veins supplying the deep tissues. Arterial blood is pre-cooled before it reaches the extremities and venous blood is warmed before it returns to the vital organs. The efficiency of this system depends on the anatomical distribution of veins and arteries in the various bodily structures. Some animals have very impressive heat exchange systems; penguins, for example, can spend long hours standing on ice at many degrees below freezing and maintain a very high temperature gradient between their deep body tissues and the soles of their feet.

Sweating

There are about 1.6 million high-capacity sweat glands on their skin. These glands are only found on the hairless regions of other mammals that sweat principally through low-capacity (apocrine) glands, associated with hair follicles. Humans can lose about 500 grams of sweat per square metre of skin per hour, whereas horses and camels can only lose 100 and 240 grams per hour, respectively (Hanna and Brown 1979). Heat tolerance is well-developed in all human populations.

Sweat is a dilute solution of various electrolytes, principally sodium, potassium and chloride. Passive diffusion of water through the skin occurs most of the time, as does 'non-thermal' or 'non-sensible' sweating, so called because people are not aware that they are sweating.

Sweating becomes sensible when the rate of sweat production exceeds about 100 grams per hour. Sweat rates of 1–2 kilograms per hour are possible. Sweat cools the body when it evaporates (owing to the latent heat of vaporisation of water), but in humid environments evaporation diminishes and cooling efficiency is lost even though sweat continues to be produced. Another problem in humid environments is a kind of *reverse sweating* in which atmospheric water vapour condenses onto the skin, releasing its latent heat of condensation and warming the skin.

Profuse sweating has two important disadvantages:

- Dehydration may occur if more water is lost than is replaced.
- Salt may be lost.

The adrenocortical hormone *aldosterone* is involved in the conservation of sodium, causing it to be actively reabsorbed in the sweat glands and in the kidneys. As the rate of sweat production increases, less sodium ions can be reabsorbed. Lower sodium concentration in the blood may result in increased heart rate, lower blood pressure and a reduction in peripheral circulation.

People do not voluntarily replace all of the lost water when working in the heat, even if an ample supply of fluid is available. *In hot industrial work, plenty of fluid must be made readily available at all times and workers must be encouraged to regularly replace lost fluid to avoid both dehydration and overhydration.*

Shivering

Involuntary shivering is a thermoregulatory mechanism involving active heat production. Groups of motor units act out of phase with one another and muscles act against their antagonists. Almost no movement occurs and the result is a high level of heat production.

Voluntary movement also increases heat production but breaks up the insulating layers of air around the body, which increases the rate of heat loss. Both shivering and voluntary movement increase oxygen uptake and cardiac output and can lower work capacity.

Work in hot climates

Heavy physical work in the heat imposes *conflicting demands* on the cardiovascular system. Peripheral vasodilation requires an increase in blood flow to the skin (up to 10 litres/min). However, working muscle also demands increased blood supply (up to 25 litres/min). Up to a point, blood flow to the skin and muscles can be increased by diverting blood from the viscera, but this reserve capacity is limited. Since cardiac output cannot exceed venous return and the average maximum cardiac output is about 25 litres/min, cardiac capacity is a limiting factor for muscular work in the heat (Kroemer, 1991). The cardiovascular system may be placed under considerable strain when a person is working in the heat as output rises to meet the demands of both physical work and bodily cooling. A dangerous condition can arise if the heart is no longer able to meet both demands.

Heat illnesses

If a worker becomes dehydrated, sweat production decreases and core temperature increases. Rapid elevations in body temperature increase the metabolic rate. Since heat is a by-product of metabolism, the heat load is exacerbated and a positive feedback situation may develop. In extreme situations, thermoregulation breaks down. If core temperature rises above 42°C, blood pressure drops and insufficient blood is pumped to the vital organs, including the heart, kidney and brain. Some of the main conditions that can arise when the body is unable to cope with the thermoregulatory challenge are discussed below.

Heat hyperpyrexia (heat stroke)   Thermoregulation fails, sometimes suddenly, and core temperature exceeds 41°C. The condition may be potentially fatal if untreated. The individual may collapse and will be disorientated. The skin is hot, red and dry. Active cooling on rescue from the heat is needed.

Heat exhaustion   When thermoregulatory strain combines with cardiovascular strain, heat exhaustion can occur. The person feels weak and may be uncoordinated, breathing may be shallow with a rapid, weak pulse. Dehydration contributes to heat exhaustion. Heat stroke can follow if the dehydration is not corrected. Removal from the heat, removal of clothing and fanning may restore normal functioning.

Heat syncope   Fainting due to inadequate venous return. It is more common in unacclimatised individuals and can be fatal if the sufferer cannot lie down (when working in a confined space, for example).

Heat hyperventilation   This can occur when working in the heat while wearing protective clothing. Hyperventilation results in an excessive loss of carbon dioxide. It is treated by making the sufferer breathe into a small bag for a few minutes.

Prickly heat   This manifests as a fine, superficial skin rash associated with excessive sweating. It often occurs on areas of skin that are covered by clothing or protective equipment.

Relative humidity

Whether workers can tolerate a hot environment depends on several classes of variables (Table 9.2). For example, DBTs of more than 38°C can be tolerated if the relative humidity is less than 20% because, at such low humidities, the cooling efficiency of sweating is high. However, relative humidity of 90% at 32°C is only tolerable with air movement and if a low level of work activity is required. Work in hot environments can be made more tolerable by introducing job aids or rest pauses to reduce metabolic heat production or by periodically removing the worker from the environment. Finally, workers differ in their ability to tolerate heat stress. At one extreme, hyper-heat-tolerant individuals can work in hot environments without the need for prior acclimatisation. Heat-intolerant individuals are never able to work safely in the heat. Identifying such individuals is an important part of worker selection in industries such as deep underground mining.

Heat acclimatisation

Heat acclimatisation is a physiological process of adaptation rather than a psychological adjustment to life in a hot environment. It involves an increase in the capacity to produce sweat and a decrease in the core temperature threshold value for the initiation of sweating. The maximum rate of sweat production can double from 1 litre/hour in an unacclimatised person to 2 litres/hour in an acclimatised person. A state of acclimatisation is best achieved by exercising in the heat and drinking plenty of fluid.

Increased sweat production enhances evaporative cooling of the skin and thus improves heat transfer from the deep body tissues to the periphery. The risk of dehydration and salt depletion is reduced in an acclimatised person owing to expanded blood volume and a reduction in the salt concentration of the sweat. Furthermore, acclimatisation reduces the skin's blood flow requirements, which reduces the cardiovascular load during work in the heat.

Heat acclimatisation occurs naturally but it may also be induced artificially. In deep mines, manual work has to be carried out in WBTs up to 33°C. In some mines, workers are artificially acclimatised to heat as a requirement of their employment. Surface acclimatisation chambers are used and workers exercise (by performing step-ups) in temperatures of 31.5°C WBT and 33.5°C DBT. Rectal temperatures are monitored and rest periods are given in response to temperature increases over 38.3°C. Other methods of acclimatisation include exercising while wearing a vapour barrier suit (which prevents sweat evaporation) and resting in baths at 41°C.

It has been suggested that an individual's ability to produce sweat depends on the climate they experienced in the early years of life (Diamond, 1991). It appears that the sweat reflex develops in the first few years of life (Hanna and Brown, 1979). Although we are all born with a similar number of sweat glands in the skin, the number that develop or are 'switched-on' depends on the conditions experienced early in life. People who grew up in a hot environment develop a larger number of functional sweat glands than those who grew up in a cold environment, and can tolerate hot conditions more easily.

Heat tolerance

Apart from acclimatisation, several personal factors can influence a worker's ability to tolerate hot conditions (Table 9.2).

Age

Young children have less sweating capacity than adults. Older men are less able to tolerate high heat stress partly owing to their higher skin temperature threshold for the onset of sweating. The increased mortality of the elderly during heat waves is thought to be due to increased cardiovascular load rather than heat stress.

Table 9.2 Some factors influencing the ability to work in the heat

1. The characteristics of the worker	Physiological heat tolerance
	Age
	Aerobic capacity
	Degree of acclimatisation
2. The thermal environment	Relative humidity
	GT/radiant heat/shade
	Wind speed
3. The requirements of the task	Work rate
	Provision of rest pauses
	Provision of protective clothing

Sex

There are no qualitative differences between men and women in their response or acclimatisation to heat. There is some evidence that women begin sweating at a higher skin temperature and sweat less than men. Women have a higher proportion of body fat than men, which may also play a role.

Physical fitness

Physical fitness improves heat tolerance because both characteristics rely on cardio-vascular function and sweat production. Physically fit workers are less stressed by hot conditions even if they are accustomed to a temperate climate.

Body fat

Excess body fat degrades heat tolerance by increasing the mass-to-surface area ratio of the body and reducing cardiovascular fitness. Adipose tissue contains less water than other tissue and has a lower specific heat capacity. Thus, the same heat load will cause a greater temperature increase in obese compared to lean individuals.

Heat stress management

Some approaches to the elimination of heat stress are summarised in Table 9.3. Table 9.4 gives threshold WBGT values at which these precautions should be implemented.

Figure 9.3 describes the relationship between the ability to carry out work at different intensities for 20–30% of the day and WBGT (Kjellstrom and Dirks, 2001). As can be seen, as temperatures exceed the 26–30°C range, work ability starts to diminish rapidly for all but the least physically demanding tasks.

Work in cold climates

Core temperature can be maintained in the cold if the person is working and suitable protective clothing is provided. When manual work is not being performed, less metabolic heat is generated and adequate insulation becomes of increasing importance. Metabolic heat production increases by a factor of 3 when a person is working

Table 9.3 Some basic steps in heat stress management

1. Reduce high relative humidity using dehumidifiers.
2. Increase air movement using fans or air conditioners.
3. Remove heavy clothing; issue loose-fitting overalls.
4. Reduce the work rate.
5. Include frequent rest pauses.
6. Introduce job rotation.
7. Carry out outdoor work at cooler times of the day (e.g. early morning).
8. Allow 2 weeks for acclimatisation.
9. Enforce rest breaks and provide drinking water or other fluids.
10. Provide shade to reduce radiant heat load (plant trees, build awnings, issue wide-brimmed hats).
11. In factories, build cool spots and refuges to lower worker exposure.

Table 9.4 Recommended threshold WBGT for heat stress management[a]

	Air velocity	
Work rate	<1.5 m/s	≥1.5 m/s
Light	30.0	32.0
Moderate	27.8	30.5
Heavy	26.0	28.9

[a] From OSHA.

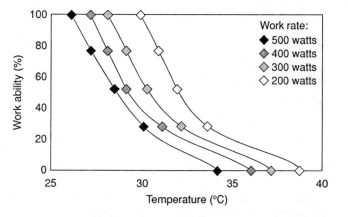

Figure 9.3 Limits for physical work in the heat at different work rates (redrawn from Kjellstrom and Dirks, 2001.)

compared to resting. When resting, therefore, three times the amount of clothing insulation is required. Adequate hydration is essential when working in a cold environment. Sweating may still take place and moisture is lost as the lungs are ventilated by the cold, dry air. In extreme circumstances, a form of dehydration known as cold dehydration may occur. Cooling of the nasal epithelium, particularly during heavy breathing, may cause nosebleeds. Cold diuresis may also occur, increasing the loss of fluid.

Core temperature

If the core temperature of a person in a cold environment drops below 33°C, central nervous system temperature control becomes disrupted. At 29°C, hypothalamic core temperature control breaks down completely. Cooling of the core tissues lowers the metabolic rate and therefore the amount of metabolic heat production. If metabolic heat production cannot be elevated while heat is being lost, a vicious circle of further heat loss and still lower metabolic heat production develops.

Peripheral temperatures and repetitive work

Cooling of the peripheral tissues, particularly in the hands and feet, causes reductions in strength and in neuromuscular control, resulting in a loss of dexterity. Oksa *et al.* (2001) compared forearm EMG activity and maximum wrist flexion force in subjects doing light repetitive work in thermoneutral conditions and at 5°C. Cold increased muscle strain by 20% and forearm flexor muscle fatigue was doubled. Rintamaki *et al.* (2001), surveyed 1490 workers in the meat processing and dairy industries. Fifty-one per cent of the workers worked at 0–5°C and 34% at 6–10°C. Complaints of cold in the hands and fingers were most prevalent (89%), followed by the wrist (58%), toe (59%) and shoulder (52%). Compared to warm conditions, the prevalence of musculoskeletal pain was greater in the hands, arms and neck, especially. Workers *with* musculoskeletal symptoms had lower skin temperatures in the neck/shoulder and forearm regions. Muscular strain, indexed by EMG, was negatively correlated with skin temperature. Low skin temperatures in the cold seem to be associated with increased muscular strain and pain in the tissues beneath the cooled skin. These findings indicate that the desirability of preventing regional cooling of extremity skin in those carrying out repetitive work in the cold.

Heus *et al.* (1995) present threshold temperatures for different tissues to ensure normal functioning (Table 9.5). These can be estimated from skin temperature.

Acclimatisation to cold?

Local acclimatisation to cold may occur in the extremities as a reduction in the peripheral vasoconstrictor response. Increased blood flow through the hands can occur after repeated exposure to cold conditions. Manual dexterity is preserved but at the cost of increased heat loss from the body as a whole. It has been suggested that

Table 9.5 Critical temperatures for peripheral tissues

Tissue	Temperature (°C)
Nerves	20
Receptors	10
Joints	24
Muscles	
Dynamic force	38
Dynamic endurance	28
Static force	28–38
Static endurance	28
Skin, local	15

people accustomed to cold conditions have an increased basal metabolic rate and the ability to sleep while shivering. It may also be the case that those exposed to a cold climate increase their daily food intake, which would increase their metabolic rate and the thickness of the subcutaneous fat layer resulting in a higher rate of heat production and better insulation of the core tissues.

The chemical processes involved in the digestion of food are also a source of heat. It is well known that BMR rises in people of all ages following a meal. Clothing provides a subtropical microclimate of warm, moist air between the skin and clothes. At rest, a skin temperature of about 33°C is perceived as comfortable; most people put on more clothes in cold conditions in order to maintain this microclimate and, for this reason, are rarely exposed to any stimulus for true acclimatisation to the cold. Behavioural adaptation to the cold, through experience, is of great importance; wearing correct clothing and keeping 'on the move' are examples. Peripheral vasoconstriction takes place in most areas of the body except the head, where up to 25% of all heat loss can take place when it is cold (Kroemer, 1991), so the head should be considered as an extremity and protected in a similar way to the hands and feet.

Immersion in cold water

Sudden immersion in cold water produces a violent reaction characterised by initial hyperventilation, tachycardia and peripheral vasoconstriction. This temporary reaction, in the first few minutes of immersion, may incapacitate the victim and cause drowning, even in competent swimmers. The hyperventilation makes it almost impossible to hold one's breath and submersion of the head for even short periods is extremely dangerous. This explains why the provision of life jackets is crucial. Even conscious victims who are competent warm-water swimmers may not survive cold water immersion. Should the victim survive the first few minutes, the next threat is from 'swimming failure', thought to be caused by direct cooling of the peripheral nerves, resulting in a loss of coordination and a failure to synchronise swimming strokes and breathing. In water at 10°C, stroke rate increases and stroke length decreases, indicative of a reduction in swimming efficiency (Tipton *et al.*, 1999). The main threat to life following cold water immersion appears to be drowning. Hypothermia only becomes a serious threat if the victim survives the first 20 minutes, and the chance of survival diminishes rapidly unless the person is rescued. Acclimatisation to cold water immersion has been shown to occur experimentally after eight repeated exposures of 40 minutes in water at 15°C and is characterised by a reduction in hyperventilation and tachycardia (see Oakley, 2000, for a fuller discussion).

Perception of cold

The perception of cold seems to depend on experience. People accustomed to a cold climate may feel comfortable when their deep body tissues are suitably insulated by clothing, despite local cooling of the skin of the cheeks, ears, nose fingers and feet. Those unaccustomed to wide disparities in skin temperature may confuse 'being cold' (i.e. having a low core temperature) with 'feeling cold' (i.e. experiencing low skin temperatures on the extremities). In fact, their core temperature remains perfectly normal and in a physiological sense they are not cold.

Cold perception also depends on the set-point of the hypothalamic temperature-regulating centre. Bacterial infections elevate the set-point for core temperature thermoregulation. The person therefore feels cold even though the core temperature is normal. Peripheral vasoconstriction takes place and the person shivers. After several hours, the core temperature reaches the new set-point and the person again feels neither hot nor cold. When the factor that initially caused the increase in set-point is removed, the person feels hot. Peripheral vasodilation and profuse sweating occur as the core temperature drops to the original set point.

Cold injury

Cold exposure causes injuries to the fingers, toes, nose, cheeks and ears. Clinicians distinguish between freezing and non-freezing cold injury.

Freezing cold injury   Frostbite occurs when the tissues freeze; the damage is caused by high concentrations of electrolytes left in tissue fluids when most of the water turns to ice (Oakley, 2000). The skin is white and hard when frozen and turns red on warming as blood flow returns. Frostnip is a mild form of frostbite; with frostnip, only the superficial layer is frozen and the tissues recover completely within 30 minutes of re-warming, with the return of sensation to the affected area. Rewarming is best achieved by placing the affected part against the warm skin of another person or under the armpit. Superficial frostbite is more serious and there is gross discoloration and sometimes blisters on rewarming. Gangrene may develop and re-warming itself may be extremely painful. Re-warming should not be attempted until it is certain that further freezing will not occur. Deep frostbite is even worse; deep frostbite is rarely seen in occupational settings and has serious medical consequences, including possible amputation of the affected limb.

Non-freezing cold injury   These injuries occur as a result of longer-term exposure to less severe temperatures and are found in military populations (e.g. 'trench foot') and in survivors of shipwrecks. The cause of the injury is not known but the result is cold sensitisation. Exposure to innocuous cold stimuli for a short period causes peripheral vasoconstriction that persists for several hours afterwards. Although the condition may resolve in the years following the exposure, it may worsen as the cold sensitisation response makes it more, rather than less, likely that the affected part will be cooled in future. In the initial stages, gradual, rather than rapid re-warming is recommended (Oakley, 2000).

That exposure to the cold can result in such serious injuries leading to chronic disability highlights the importance of ergonomic interventions to prevent exposure to cold in the workplace. The term chilblain is sometimes used to describe symptoms such as itching, redness and burning of the skin of the upper part of the fingers and toes, following exposure to damp cold.

Skin temperature

The temperature of the skin can vary over a much wider range than that of the deep tissues of the body. Burns occur at temperatures over 45°C. At lower temperatures, sweating may impair functional hand strength by reducing friction at the hand–

handle interface. Parsons (1990) provides an interesting discussion of the surface temperature of objects in relation to the risk of sustaining burns. Conduction of heat to and from the skin by a solid object depends on the thermal properties of both skin and object – a metal surface at 100°C would feel much hotter than a cork surface. The threshold at which burning would take place can be estimated for different materials assuming a momentary contact time of a quarter of a second (the reaction time to remove the part of the body touching the object). The threshold temperature for wood, for example, is estimated as 187°C compared with 136°C for brick and 90°C for metal. Similar thinking has led to the drafting of specifications for the maximum surface temperatures of domestic appliances to minimise the risk of burns (Parsons, 1990).

Wind chill, the decline in effective temperature due to air movement can have consequences for the extremities (hands, nose, cheeks and ears) particularly if they are directly exposed to the wind. Skin at 20°C is one-sixth as sensitive as skin at 35°C. At 5°C, pressure and touch receptors do not respond to stimulation. This increases the risk of frostbite as all sensation is lost at lower temperatures. In extremely cold environments, skin contact with bare metal is particularly hazardous – the skin may freeze immediately and stick to the metal causing extensive tissue damage.

The reduction of skin sensitivity and loss of manual dexterity that take place at low skin temperatures are important from an industrial viewpoint. Allowance must be made for the fact that tasks may take longer to perform than they would otherwise. Controls and equipment should be designed in wood or plastic rather than metal, and be usable with gross movements of the body.

Protection against extreme climates

When the work environment cannot be improved nor the task redesigned, methods of protecting the worker are needed.

Specify safe work–rest cycles

Welch *et al.* (1971) measured pulse rate, skin temperature, sweat loss and rectal temperature of subjects working in hot, humid conditions. Rectal temperature, which is used as an index of core temperature, proved to be the only reliable indicator of the onset of heat exhaustion. Bell *et al.* (1971) investigated the time taken for subjects to reach a state of imminent heat collapse and specified safe exposure times for men at work in hot environments. Figure 9.4 shows the limits of permissible exposure to hot conditions of workers working at different energy expenditures for a range of work–rest cycles (NIOSH, 1981).

Design 'cool-spots'

Thermal comfort can sometimes be improved by designing a 'thermal refuge' for operators. Window coverings and screens can be installed to provide radiant heat 'shade'. In very hot conditions, a screen made of vertically hanging chains can be effective at blocking radiant heat, while allowing access for those with portable protection. Sims *et al.* (1977) designed a 'cool spot' for inspectors in a glass making

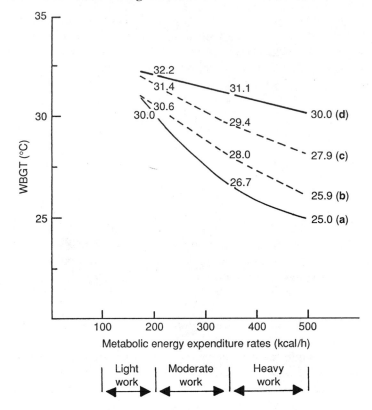

Figure 9.4 Lines of permissible exposure to work in hot conditions (NIOSH, 1981).
(a) 8 hours' continuous work; (b) 75% work – 25% rest each hour; (c) 50% work – 50% rest each hour; (d) 25% work – 75% rest each hour.

factory. Discomfort caused by radiant heat and high air temperatures was reduced using glass and aluminium radiant heat shields and low-velocity cooling air.

Ohnaka *et al.* (1993) investigated heat stress among operators in the asbestos-removal industry. These workers wear protective clothing to prevent the inhalation of asbestos and the contamination of personal clothing. The protective ensemble consists of impermeable garments and air masks, preventing heat loss and exacerbating the problems of working in hot weather. Several solutions to the problem were possible including air conditioning the work area, using air-cooled or liquid cooled suits or ice vests or constructing a cool room adjacent to the work area for rest breaks. The first option was too expensive and the second was impractical due to the restrictive nature of the clothing. The third option was trialled and evaluated using measures of heart rate, sweat rate and rectal temperature of subjects working in hot conditions, cool conditions and the 'hot-cool condition using the air-conditioned rest room. The hot-cool condition reduced physiological stress and discomfort significantly.

Issue protective clothing

Protective clothing can provide an acceptable microclimate. Liquid-cooled and air-cooled suits have been designed for pilots. Air or a special cooling liquid is circulated around the suit to maintain thermal balance. In gold mining, special cooling jackets have been designed to enable selected workers to acclimatise in the work environment itself (Schutte *et al.*, 1982). The workers are selected on the basis of a heat tolerance test carried out on the surface and are given a jacket containing dry ice to wear underground for the first few days. This reduces the risk of increased core temperature and enables acclimatisation to take place naturally in response to the demands of the task and environment.

If the external temperature is greater than the core temperature (about 37°C) additional clothing is required to protect the skin from a net heat gain from the environment. In some hot countries, people traditionally wear long, flowing robes covering the whole body, presumably to maintain a layer of cool air between the skin and the clothing and protect the skin from solar radiation.

In more severe cases, protection from radiant heat can be achieved using clothing assemblies designed to insulate the worker from the heat. Helmets with a large air space between the shell and the operator can be used in conjunction with thick woollen suits and wooden clogs. Fireproof reflective materials are used in fire-fighting.

The sun is a major source of radiant heat. Wide-brimmed hats have been used by rural people for millennia and constitute portable shade – an effective first line of defence against the potentially debilitating effects of intense solar radiation. People who live in hot deserts cover their bodies with long, black, flowing robes. Black materials absorb more radiant heat than white materials but give better protection against sunburn. In hot, humid environments, minimal clothing is worn to maximise evaporative heat loss. If there is also intense sunlight, this conflicts with the need for protection from solar radiation and the only solution may be to restrict outdoor work to the early morning and late afternoon.

Cool the extremities

A simple method of reducing heat stress in hyperthermic individuals is to immerse the hands and forearms in cold water. House *et al.* (1997) compared hand immersion in water at 10, 20 and 30°C in subjects who had exercised in the heat (40°C) for 45 minutes. All subjects had increased core temperatures (38.5°C). They then rested in the same hot environment but with the hands immersed in water. Hand immersion produced reductions in core temperature in all three conditions within 20 minutes (from 38.5 to 36.9°C in water at 10°C, to 37.3 in water at 20°C and to 37.8°C in water at 30°C). In a control condition, where subjects only rested in the heat, no reductions in core temperature were observed. Water at 10°C produced the fastest reductions in core temperature in the first 10 minutes. This simple method appears to be effective in treating mildly hyperthermic individuals in hot environments. However, the method is less likely to be effective in preventing hyperthermia in the first place. In individuals who are not suffering heat stress, the natural response to immersion of the extremities in cold water is peripheral vasoconstriction (to prevent heat loss). When core temperature rises, peripheral vasodilation occurs under central nervous system control. Thus, when the extremities of a hyperthermic individual are placed in cold water, the peripheral blood vessels remain dilated and convective heat loss takes place.

Cold climate protection

The main challenge is to provide a sufficient amount of insulation while allowing moisture, from sweating, to evaporate. Garments made of vapour-permeable materials such as Goretex® can provide insulation and allow moisture to escape. This is very important where physical work is carried out in the cold. Sweat produced when working, if it collects in the interstices of the garment, will reduce the thermal insulation properties of the garment and may even freeze within the garment when the work stops and metabolic heat production drops. The wearer will rapidly become cold. At temperatures below 0°C, gloves provide inadequate insulation and mittens are needed, resulting in a major reduction in functional manual dexterity.

Comfort and the indoor climate

The climatic conditions inside a building depend primarily on the conditions prevailing outside and the people and processes housed within. However, many additional factors can mediate the effects of the exterior conditions on those indoors. Of particular importance is the design of the building itself.

Building design and the indoor climate

Building design determines the amount of solar penetration into the building and the radiant heat gain. The amount of insulation, particularly of the roof, has a large influence on the heat exchange between the building and the environment. The construction materials influence a building's thermal performance via a mechanism known as the 'flywheel effect' (see below).

Solar heat gain through uninsulated factory roofs and walls is an important cause of overheating. This is particularly important in developing countries where cost-effective but thermally efficient building concepts are needed. In hot countries, roof insulation is an important first step in combating overheating. Parts of the building that are exposed to direct sunlight can be painted in bright colours to reflect solar energy. It has been estimated that reductions in external roof and wall temperatures of almost 20°C are obtained by whitewashing building exteriors. Awnings or covered walkways can be built around buildings to provide shade both for people and for the walls. If plentiful, water can be sprayed onto the roof and walls and adjoining courtyards and sidewalks. In developing countries, people sometimes pour water onto a nearby floor before sleeping to lower the air temperature in a room (Hanna and Brown, 1979). Screens placed over doorways or windows may be wetted to cool the incoming air.

Steel roofs in factories can be shaded with tiles, preferably with an air space between the roof and the tiles. Rectangular-shaped buildings can be constructed with their long walls facing north/south. In the summer, the sun shines on the shorter walls in the morning and evening and on the roof during the day. This minimises the degree of heating of the building by the sun. Conversely, in the winter, when the sun is low in the sky, the long walls and roof are exposed to the sun during the day, which helps to warm the building.

Sunlight entering through windows is a another major source of heat. In existing buildings, physically constrained workers should not have to work in direct sunlight for any length of time (this also applies in cold countries since sunlight is a source of glare, which degrades vision). The orientation of windows with respect to the sun is an important consideration in the design stage of a building. Special glazing, louvres or shades can be retrofitted to reduce solar penetration. There is a modern trend to build buildings with large windows to save on construction materials. This, however, can lead to faster heat loss and gain.

The heat transfer properties of building materials vary considerably depending on their mass. Lightweight materials store very little heat themselves. Heat transfer through them depends on their thermal conductivities and the temperature differential across them. Heavyweight materials have greater ability to store heat – heat gained during the day warms the material itself before being transferred to the air inside the building. Similarly, at night, heat is lost from the building materials themselves and the interior of the building remains warmer for a longer period. The building material itself can be thought of as a buffer or heat store that lies between the internal and external thermal environments. Thus, heavy materials act like a 'flywheel' to smooth out the effects on the indoor climate of daily oscillations in external temperature. To use an extreme example, the temperature inside a cave remains more or less constant throughout the year, whereas the temperature inside a tent changes rapidly throughout the day in response to changes in the external temperature. The magnitude of the flywheel effect does not depend so much on the heat-conducting properties of the materials used in construction so much as the mass of material enclosing the interior space. Thus, there is often a trade-off between the cost of constructing a new building and the later cost of controlling its internal temperature within acceptable limits. Money saved on construction materials and construction time in building a more lightweight structure may have to be spent later on additional air-conditioning (except in permanently hot, humid climates where all buildings require artificial cooling to maintain a comfortable indoor temperature).

Thermal comfort in buildings

The thermal comfort of a factory or office worker depends on there being an average skin temperature of approximately 33°C (Astrand and Rhodahl, 1977), although colder temperatures at the extremities may be tolerated. Large disparities in skin temperature may lead to complaints of discomfort even if the average skin temperature is close to 33°C. Draughts, sunlight falling on an arm or the face and sitting next to a cold wall are all causes of thermal discomfort due to uneven skin temperature distribution.

Modern approaches to the indoor climate have attempted to specify an acceptable range of conditions for the worker. ISO 9241 recommends winter temperatures of 20–24°C and summer temperatures of 23–26°C. Maximum values of relative humidity are 60–80% at 20°C, 50–70% at 22°C, 45–65% at 24°C and 40–60% at 26°C. At temperatures of 24°C and above, workers may begin to feel lethargic. At temperatures of 18°C or lower, shivering may commence in sedentary or inactive workers unless extra clothing is worn.

Thermal comfort, air quality and sick buildings

Investigations have shown that some people find temperatures as low as 18°C comfortable whereas others prefer temperatures higher than 23°C. Sundstrom (1986) suggests that people be given a certain amount of control of the temperature of their workplaces (being able to control local ventilation or to use a small fan or heater).

Comfort and well-being do not depend only on the temperature in a room. They are also influenced by the total indoor environment and this includes many additional factors over and above the climatic ones discussed above.

Unlike old offices, which were relatively small and furnished with natural products such as wood, leather and cotton or wool fabrics, large, modern office buildings contain hundreds of tonnes of human beings, several tonnes of glue and many kilograms of potentially poisonous or carcinogenic compounds such as ammonia, carbon monoxide, asbestos and glass fibre. If the concentrations of these substances increase beyond threshold levels, either locally or globally, the health of occupants will be at risk.

A great deal of attention has been paid recently to the air quality inside buildings. The term 'Building Sickness Syndrome' has been coined to describe unduly high absenteeism among the occupants of buildings with ostensibly poor air quality. Particular attention has been paid to sealed, mechanically ventilated buildings that may be either permanently or only temporarily 'sick' (Sterling *et al.*, 1983; Scansetti, 1984; Sykes, 1988).

Modern office buildings use sophisticated air conditioning systems to provide workers with a comfortable indoor climate to work in. In order to conserve energy on heating and cooling outside air, some of the air already in the building is recirculated. Under certain circumstances, this can cause the concentrations of everyday substances to reach unacceptable levels that may, in turn, cause health complaints.

The occupants of a building and the machines they use are themselves sources of heat. The body produces approximately 100 watts (about the same as a visual display unit). When introducing new equipment or processes into existing facilities (automating an office, for example), the additional heat load should be evaluated and increased ventilation or air cooling provided if required, particularly in hot weather.

All buildings enclose a finite volume of air. If they are not properly ventilated, the air quality is likely to deteriorate. Air quality can be degraded by several classes of contaminants. Carbon dioxide, carbon monoxide and ozone are examples of inorganic contaminants. Organic contaminants include formaldehyde and other hydrocarbons. Living organisms such as bacteria, fungal spores and mites can also contaminate the air.

There are several sources of indoor air pollution. The most obvious is the air entering the building from outside. In cities, pollution from factories, car exhaust emissions (carbon dioxide, carbon monoxide and lead from leaded petrol) may enter the building. The building itself may be a source of pollution emitted by the construction materials, furniture and fittings. Business machines and cleaning chemicals are another source. A final source is the occupants themselves. People give off carbon dioxide, water vapour, microorganisms, dead skin cells, unpleasant odours and, sometimes, tobacco smoke. Most people are exposed to some or all of these substances from time to time and experience no problems. Building sickness occurs when unusually high levels occur for some reason. Sterling (1990) reports that 49% of sick buildings are the result of ventilation and air conditioning problems and 28% the result of indoor air pollutants. Acute or chronic accumulation of pollutants can occur in any type of building, including residential buildings. Some of the main suggested indoor pollutants are reviewed below.

It has been suggested that photocopiers may degrade air quality because they emit ozone, an unstable molecule made up of three oxygen atoms. Ozone has a half-life of a matter of minutes and is destroyed by contact with most surfaces including the tissues inside the nose and by cigarette smoke. An ozone hazard would be unlikely to occur except locally, in a badly ventilated and heavily used photocopying room. Photocopiers might be placed by extractor fans or open windows to prevent the accumulation of ozone in the air. Ozone can cause eye irritation.

At one time, there was concern that the internal components of visual display units might emit carcinogenic gases such as polychlorinated biphenyls (PCBs). This now seems unlikely.

Humidifier fever is an acute influenza-like disorder that occurs mainly on Mondays at the resumption of work, decreasing in severity across the working week. It is apparently caused by the inhalation of spores and cysts in the air. Some humidifier systems can provide a breeding ground for bacteria and other microorganisms, as can an accumulation of water in cooling systems and air ducts, particularly if the water contains organic substances from other sources. Legionnaires's disease is a dramatic example of this type of problem. Regular system inspection, disinfection and maintenance can reduce problems of this nature.

Low relative humidity (less than 40%) may also increase the risk of respiratory infection (Sykes, 1988). Several mechanisms have been suggested. For example, dry air is said to cause microfissures in the respiratory tract that act as landing sites for infection by airborne bacteria. Formaldehyde is a common contaminant of indoor air. It is found in many manufactured wood products including particle boards and in ceiling tiles, carpets and urea-formaldehyde foam. It irritates the skin and the mucous membranes. High levels of formaldehyde 'off-gassing' can occur in new buildings, particularly in hot weather, causing eye, nose and throat irritation. Much of the dust in occupied buildings consists of dead skin cells. Clothing acts like a microscopic 'cheese-grater' that removes dead cells from the surface of the skin. Other sources of dust include dry cleaning and insulation materials. Microscopic particles of glass fibre have reportedly caused conjunctivitis in office workers. Reports of facial rashes among visual display unit users have sometimes been attributed to dust collecting on the surface of the skin. The cathode ray tube is thought to be able to induce a static electrical charge on the face of the operator, which then attracts oppositely charged dust particles.

Finally, it has been suggested that the air in modern buildings may be deficient in negatively charged small air ions. Air ions are electrically charged forms of the molecules of the various gases found in air. In offices, air ions are destroyed by contact with metal air ducting, dust, smoke and static electrical charges on visual display unit screens. Although there are marked differences in the concentrations of air ions found in city and country air and inside and outside buildings, it is not known whether artificially increasing the concentration of ions using ion generators does improve air quality. Hawkins (1984) found some evidence that the introduction of negative-ion generators into an office reduced the incidence of dizziness, nausea and headaches. However, in a subsequent study he could not replicate these findings. Many factors can degrade the air quality inside buildings. Despite sometimes sensationalistic reporting, the sick building syndrome would appear to have some physical basis.

Increased ventilation can reduce the incidence of symptoms, whereas high, uniform temperatures result in more symptoms. Wanner (1984) cites a minimum ventilation rate of 12–15 cubic metres (m^3) of air per person per hour. If manual work is carried out or if cigarette smoking is allowed, 30–40 m^3 per person per hour is needed. The personal characteristics of employees may also influence health outcomes. Some people are allergic to chemicals, bacteria or fungal spores that are found in enclosed spaces. Even minute, barely detectable quantities of these substances can cause adverse reactions characterised by headache, fatigue, muscle aches, mucous membrane irritation, chest tightness, coughing and wheezing. These reactions are known as 'hypersensitivity pneumonitis' or 'extrinsic allergic alveolitis' (Broadbent, 1989). That the symptoms are milder or disappear at weekends strongly implicates the work environment.

Kleeman *et al.* (1991) suggest a minimum of one and preferably two air changes per hour are needed to maintain air quality and that intakes should be on the roof rather than the street level of buildings to minimise the intake of already polluted air. Dull, uniform lighting combined with a lack of daylight has also been cited as a contributory factor to sick building syndrome. Although the syndrome is not psychogenic, low moral and job dissatisfaction may amplify employees' negative responses to poor indoor air quality.

When confronted with complaints about poor indoor air quality, the ergonomist might begin with an analysis of temperature, humidity and air flow. If these prove satisfactory, it is possible to test for more exotic forms of contamination. Occupational hygienists have standard equipment to test for the presence of contaminants such as formaldehyde and other airborne hydrocarbons. Samples of the air and of dust in the ventilation ducts can be taken and cultured to determine what types of bacteria are present. One of the most useful tests of indoor air quality is of the concentration of carbon dioxide (CO_2) in a room. Atmospheric air is approximately 78% nitrogen, 21% oxygen, 0.9% argon and 0.033% carbon dioxide. Carbon dioxide build-up occurs whenever ventilation is poor, and even small increases are known to make the air seem stuffy. Occupational hygienists can measure carbon dioxide concentration using commercially available meters. Elevated CO_2 is nearly always indicative of poor ventilation. In addition to improving ventilation in problem buildings, electrostatic filters (which actively remove particles in the air) and passive filters (e.g. charcoal) can be installed but require regular maintenance. Good housekeeping, the avoidance of spillage of cleaning materials, regular cleaning and careful selection of new products to be used in the building are also practical measures to ensure acceptable air quality. Kildeso *et al.* (1999) have developed a visual-analogue scale (Table 9.6) that can be used to monitor occupants' perceptions of indoor air quality over time. They suggest using it once per week while interventions to improve air quality are carried out. Up to 4 weeks should be allowed for the interventions to have an affect and for symptoms to change.

Ventilation

Ventilation is a determinant of thermal comfort and, more generally, of satisfaction with the indoor environment. The main purpose of ventilation is to provide fresh air and to remove accumulated noxious gases and contaminants. Ventilation helps to remove heat generated in a working area by convection and cools the body. It is not always possible to lay down acceptable limits from the point of view of thermal comfort. However, air speeds less than approximately 0.1 m/s second will usually

Table 9.6 Scale for detecting changes in indoor air quality[a]

How do you feel right now?	
Too cold	Too hot
Too humid	Too dry
Draught	Too little ventilation
Bad air quality	Good air quality
Too dark	Too light
Too quiet	Too much noise
Nose blocked	Clear nose
Dry nose	Runny nose
Dry throat	Normal throat
Dry mouth	Normal mouth
Dry lips	Normal lips
Dry skin	Normal skin
Dry, brittle hair	Normal hair
Brittle nails	Normal nails
Dry eyes	Eyes not dry
Stinging, itching eyes	Normal eyes
Eyes ache	Eyes do not ache
Severe headache	No headache
Pressure in the head	Clear head
Dizzy	Not dizzy
Feeling bad	Feeling good
Tired, exhausted	In shape, fresh
Hard to concentrate	Easy to concentrate
Depressed	In a good mood

Today I have been able to work

0%	100%

[a] From Kildeso *et al.* (1999).
Lines are 100 mm long in actual scale. Respondents place a cross on the line to indicate how they feel for each item.

cause a sensation of staleness and stuffiness, even at relatively low temperatures. Air speeds greater than 0.2 m/s may be perceived as draughty. In hotter conditions (with a corrected effective temperature of more than 24°C) air speeds of 0.2–0.5 m/s will aid body cooling, particularly when the relative humidity is high. It is apparent that whether or not air movement is perceived as an irritating draught or a cool breeze depends on the ambient temperature and the relative humidity. In hot countries, an important use of ventilation is to cool working areas and their contents by ventilating buildings at night, when the air outside has cooled.

Apart from its effects on the cooling efficiency of sweating, the humidity of the air influences the comfort of people working indoors. Low relative humidity (less than 30% at office temperatures) causes bodily secretions to dry up. Under these conditions, the occupants may complain of dry, blocked noses and eye irritation. Contact lens wearers may experience eye discomfort since proper adhesion of the lens to the eye depends on a continuous supply of lachrymal fluid to maintain a thin, moist film over the cornea. Static electrical charges are also more likely to build up in dry buildings, causing irritating shocks when an earthed conductor is touched. Somewhat counterintuitively, low relative humidity inside houses and offices can be a problem in cold, wet countries if artificially heated buildings are not ventilated adequately (which the occupants may be reluctant to do) and if humidifiers are not provided.

ISO standards

For further information on environmental ergonomics, the reader is referred to the appropriate ISO standards. An overview of the available standards can be found in ISO 11399 (1995) and in Parsons (1995). Exposure assessment for hot environments is dealt with in ISO 7243 and WBGT values are provided that can be used to make quick indoor and outdoor assesments as long as there is no excessive solar load. More detailed assessments can be made using ISO 7933, which deals with thermal stress in conditions likely to lead to excessive sweat loss or core temperature. It also enables required modifications to be identified such that the threat of thermal strain is eliminated. Allowable exposure times can also be specified using the standard. ISO 7730 provides a method for predicting thermal sensations in different environments and for specifying thermal conditions that will be perceived as comfortable. Required insulation for cold environments is dealt with in ISO 11079 and in ISO 9920, methods for determing the insulation value of clothing are dealt with. ISO 9886 deals with the evaluation of thermal strain using physiological measurements (core temperature, skin temperature, heart rate and body mass loss).

Effectiveness and cost-effectiveness

There is no reason why extreme thermal environments should have any direct effects on performance unless they alter the state of the individual. If thermal stress causes thermal strain, capacity may be reduced and decrements in performance may follow. The goal of cost-effective interventions to improve the thermal environment should be to eliminate thermal strain.

Physical tasks

There is a great deal of evidence that adverse temperatures have a major impact on performance. Figure 9.5 shows monthly variations in factory production levels as a function of the outdoor air temperature. In even moderately hot environments, people lower their work rate to minimise thermal strain.

Snook and Ciriello (1974) investigated the performance of manual handling tasks carried out at WBGT temperatures of 17.2 and 27°C by unacclimatised subjects.

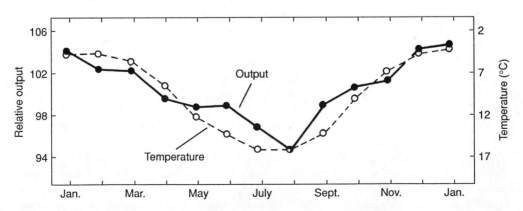

Figure 9.5 Yearly production figures from a factory circa 1918. (From H. M. Vernon, 1924.)

The hot environment caused thermal stress, as indicated by increased heart rate, rectal temperature and reduced work rate. Lifting work rate dropped by 20%, pushing dropped by 16% and carrying by 11%. However, the lower work rates were not sufficient to eliminate symptoms of thermal stress entirely – heart rate was 10 beats/min higher in the heat and rectal temperature was 0.2–0.3°C higher.

Wyon (1974) reanalysed data from the 1923 New York State Commission report on ventilation and was able to show that people consistently produced more at 20°C than at 24°C and 50% relative humidity. The task was a typing task and the research was carried out over a series of days with subjects spending whole days (up to two weeks) working in the laboratory. Furthermore, subjects were allowed to work at their own pace, in a relaxed manner, so that many aspects of the experimental design were similar to ordinary office work. Over a three-day period, subjects working at 24°C produced 43%, 18% and 49% less than subjects working at 20°C. When exposed to temperatures of 24°C, subjects who had worked at 20°C in the previous three days produced significantly less. Interestingly, subjects did not report feeling any different in the different temperatures, suggesting that they had unconsciously lowered their work rate to maintain the same thermal state. These findings suggest that even light work, such as typing, may be sensitive to moderate thermal stress, with people lowering their work rates to reduce metabolic heat production and thereby compensate for the increased temperature. A temperature of 24°C is not particularly high and may be commonplace at certain times of the year in offices that are not air-conditioned. That it can result in reductions in typing performance of 40–50% suggests that major improvements in productivity can be achieved through better building design and temperature control.

Mental tasks

There are two main theories of the effects of climate on the performance of mental tasks. Arousal theory states that for any task there is an optimum level of arousal (readiness to act) at which maximum performance occurs. Climatic extremes can increase arousal and overly comfortable conditions can lower it. If arousal deviates from the optimum, performance will suffer. An alternative theory states that climatic extremes, particularly cold, have a distracting effect – performance deteriorates because attention is distracted away from the task to the environment or to unpleasant thermal sensations.

High internal body temperatures appear to increase the speed of performance because they accelerate the body's 'internal clock'. Fox et al. (1967) showed that increases in body temperature accelerate the perception of the passage of time. Allnutt and Allen (1973) found that when their subjects' body temperatures were raised to 38.5°C the speed of performance on a reasoning test increased, even though skin temperatures were maintained at a comfortable level (suggestive of a central rather than a peripheral effect of heat). Colquhoun and Goldman (1972) investigated the ability of subjects to detect target stimuli against a noisy background at a temperature of 38°C DBT and 33°C WBT. Effects on performance were observed only when core temperature increased. Subjects were more confident that they had detected a target and made more 'false positive' errors (saying that they had detected a target when there was none). Elevated body temperatures may lead to more risky decision

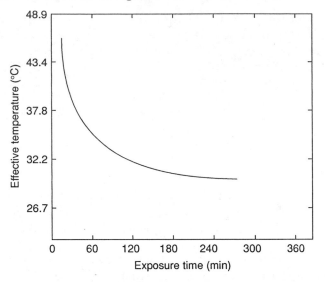

Figure 9.6 Upper limit for unimpaired mental performance as a function of exposure time to different effective temperatures. (Adapted from McCormick and Sanders, 1982.)

making and this could have implications for quality control and general safety in 'hot' industries such as steelmaking.

Azer *et al.* (1972) present more evidence that hot conditions only have effects on task performance when they cause thermal strain. Performance decrements occurred only when subjects worked at 35°C *and* 75% relative humidity. At 50% relative humidity, no performance decrements were observed. Allen and Fischer (1978) found that if relative humidity was held constant at 40%, performance on a simple learning task did not vary over the range 11–28°C. If relative humidity was not controlled, best performance was obtained at 18°C, 35% relative humidity.

Figure 9.6 gives upper limits for mental task performance as a function of exposure time.

There is an extensive literature on the effects of heat on mental task performance. The reader is advised to approach this literature with caution because many studies do not include measures of thermal strain and do not control all of the factors that may effect performance. Ramsey (1995), in a review of this literature, concluded that the literature as a whole presents contradictory and unclear findings. However, for perceptual motor tasks there appears to be a threshold level of 30–33°C, above which decrements in task performance do occur.

Cost-effective interventions

There is a great deal of evidence that ergonomic approaches to the solution of environmental problems are extremely cost-effective. Reducing workers' exposure to heat enables them to work longer or the work to be done by fewer people. In foundries, a curtain of heavy chains can be installed to block radiant heat but allow the passage of workers or materials. Fans can be installed behind the curtain to further cool workers (Teniswood, 1987). Heat stress can be eliminated using microclimate cooling garments in tasks such as underground mining. In one example, Teniswood describes the successful use of such garments by underground mining vehicle operators. Rock wall temperatures of 67°C produced a DBT of 48°C and a WBT of 35°C. An air-cooled vest in direct skin contact supplying 4 m³/min to the worker from a vehicle-mounted air supply was able to minimise heat stress. Crockford (1962) described how air-fed permeable clothing enabled workers to work for as long as 2 hours whereas unprotected workers could only work for a few minutes.

Protection against heat

Feinstein and Crawley (1968) redesigned a shearer's cab in a steelworks. The job involved detecting defects at the ends of steel slabs, prior to rolling. Defective ends were cut off. Slab surface temperatures resulted in a high radiant heat flux, so the cab was positioned 10 metres from the slab. The excessive distance, combined with an awkward field of view, meant that operators tended either to make several attempts at processing a slab (resulting in production bottlenecks) or cut off too much steel to ensure quality but resulting in unnecessary recycling of good steel. The cab was redesigned and fitted with gold-laminate glass to block the radiant heat. It was then repositioned nearer to the shearer blades and the angle of view was improved. The cost of these alterations was £10 000 and the saving due to reduced bottlenecks and a lower tonnage of recycled steel was £120 000 per annum (probably in excess of $1 million today). The redesign paid for itself in the first month.

Sterling (1990) presents data on the costs associated with 'sick building syndrome' and its solution using environmental technologies to optimise air quality as well as energy efficiency. In one Vancouver company, absenteeism rose from around 4% to around 7% when the company moved into an energy-efficient building with poor air quality. In another study, an improved environment in the Fireman's Fund Insurance Co. reduced complaints by 40%. Sterling estimated employer's total office costs (including staff costs) as approximately $1000 to $2000 per square metre of office space. Absenteeism of around 5% costs approximately $50 to $100 per square metre. Poor air quality, which may increase absenteeism by 2.5%, can cost an additional $25 to $50 per square metre. For every 10 000 square metres of office space, salaries and wages cost $10–20 million – reductions in absenteeism of 2.5% can save $0.75 to $1.5 million. Since operating energy costs are only $10–$30 dollars per square metre, increasing these costs to improve the environment can bring substantial returns if reductions in absenteeism and improved productivity result.

The cost–benefit trade-offs in constructing energy efficient buildings at low cost are worthy of consideration. Broadbent (1989) describes how a building constructed at less than $100 million developed such a bad reputation for poor air quality that employees refused to work in it and new tenants were deterred from leasing office space. The entire contents of the building had to be removed, leaving only concrete and steel, and the building completely refurbished with better services. The cost of retrofitting was $30 million.

In a second case study, Broadbent describes the problems that occurred in an office building in Washington when it was recarpeted. Work began in 1988, but employees experienced unpleasant reactions to the chemical used by the carpet manufacturer. Of the 6000 occupants 60 became ill and six were hospitalised, resulting in direct costs to the employer. Indirectly, costs increased owing to work time being used for meetings to discuss the problem, the administration of questionnaires, and other activities that arose as the work environment came to be highlighted as an industrial relations issue. Interestingly, all parties to the ensuing dispute agreed that, in future, no more carpets should be installed in the building, resulting in direct losses to the carpet manufacturer and installer.

Research directions

ISO 13732–3 is a draft standard that provides guidance on the touching of cold surfaces. Safe surface temperatures for different materials touched for different time periods are provided. Future work might look at novel ways of providing users with feedback about the state of an appliance. One example of this is the 'thermocolour' kettle made by Russell Hobbs, which is constructed of a plastic that changes colour as the water boils.

Further research is needed into the causes and treatment of thermal injuries and there is plenty of scope for developing and evaluating new methods of protecting people from extreme thermal environments.

Summary

Measurement of DBT, WBT, GT and air speed is necessary in order to evaluate the thermal environment. The heat or cold stress on the worker depends on these factors and on the workload and insulation value of clothing. Each of these factors can be manipulated to ensure a state of thermal balance between workers and their surroundings.

People feel comfortable when surrounded by a subtropical microclimate of air and an average skin temperature of 33°C. The ability to carry out manual work in extreme environments is limited. Most people are able to acclimatise to hot working conditions within 2 weeks but the ability to adapt to the cold is much more limited. For these reasons, if work has to be carried out in extreme environments the following steps should be considered:

1. Remove the operator by mechanising the task.
2. Change the task or the environment.
3. Protect the operator.

Thermal comfort depends on several factors and there are individual preferences. Temperatures ranging between 19 and 23°C 40–70% relative humidity will accommodate most workers. Low relative humidity, still air and draughts are some of the main causes of thermal discomfort in offices and factories. A lack of thermal balance between workers and their surroundings will lead to decrements in the ability to perform many tasks. However, a level of discomfort sufficient to cause complaints will occur long before any serious decrements in task performance or threats to health. If nothing can be done to eliminate thermal stress, then changes in work organisation and more rest periods are needed. Approximate thresholds at which this should be done are 28–33°C for sedentary tasks, 28–28°C for moderately hard work and 20–26°C for heavy work (Oakley, 2000).

Essays and exercises

1. Take DBT, WBT and GT measurements in the following places (measure the external ambient temperature at the same time.

   - A modern, air-conditioned office
   - A workshop of factory floor
   - A busy commercial kitchen
   - A sauna room
   - A cold room or cold store
   - A park or beach at 8:00 am, noon and 4:00 pm on a sunny, summer day
   - A park or beach at midday in the shade

   Calculate the WBGT temperature and comment on your findings
2. Measure the WBGT in a modern, air-conditioned office containing 20 employees or more. Take measurements at various parts of the office and at different times of the day. Develop a small questionnaire to assess the occupant's satisfaction with the indoor climate. Suggest possible improvements and comment on any differences in the level of satisfaction of different people.
3. Write an essay on thermal comfort in the indoor environment.

10 Vision, light, and lighting

The eyes lead the body.
(Dr J. Sheedy, School of Optometry, University of California at Berekely)

Light is electromagnetic radiation that is visible. The electromagnetic spectrum is extremely wide but the visible part is extremely narrow (Figure 10.1).

Vision and the eye

The eye is a fluid-filled membranous sphere that converts electromagnetic radiation into nerve impulses that it transmits to the brain along the optic nerve (Figure 10.2). Light enters through a transparent outer covering called the cornea. The cornea plays a major role in refracting the light. Further refraction occurs as the light passes through the lens. The pupil works like the aperture of a camera to vary the amount of light entering the eye. In bright light, the iris contracts, the pupillary diameter decreases and only the central part of the lens forms an image on the retina. In poor light, the iris expands and a larger area of the lens is used. Because the peripheral regions of the lens focus the light slightly in front of the image formed by the central part (a characteristic of all simple lenses termed 'spherical aberration') slight blurring

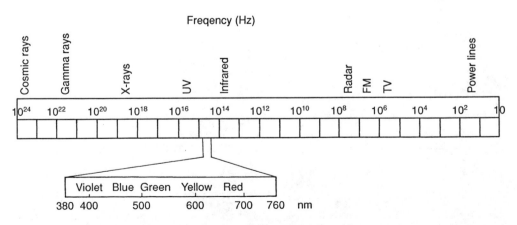

Figure 10.1 The electromagnetic spectrum. Visible light falls in the range 380–760 nanometers (nm).

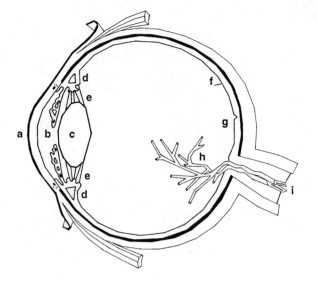

Figure 10.2 Basic structure of the eye. **a** = cornea; **b** = pupil; **c** = lens; **d** = ciliary muscle; **e** = suspensory ligaments; **f** = retina; **g** = fovea; **h** = blind spot; **i** = optic nerve.

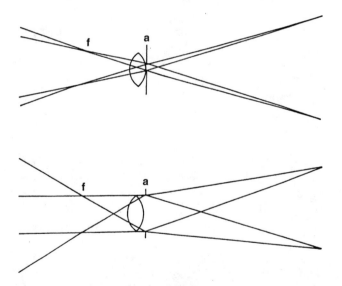

Figure 10.3 The depth-of-field phenomenon, one of the reasons we can see better in good rather than poor light. With a large aperture (a), only the image *exactly* at f is in sharp focus.

occurs when objects are viewed in poor light. This explains why the ability to discern detail (visual acuity) is reduced in poor light.

Using photographic terminology, it can be said that *good lighting increases the depth of field* of the eye. When the pupil is very small, the eye acts like a pinhole camera – all objects are focused on the retina irrespective of viewing distance. Increased depth of field reduces the need for optical adjustment by the lens. It is in this sense that it can be said that good lighting reduces the load on the visual system. Figure 10.3 illustrates the optics of the depth of field phenomenon.

The refractive apparatus of the eye

The lens is held in place by a non-rigid membranous sling. It divides the eye into two compartments. The smaller, anterior, compartment contains a watery fluid called the aqueous humour, which is secreted by the ciliary body. The posterior compartment contains the jelly-like vitreous humour. The humours help maintain the structural integrity of the eyeball and lens and supply the lens with nutrients. The surface of the eye is covered by a transparent membrane, 'the conjunctiva', which supplies the cornea with nutrients. Dilation of blood vessels in the conjunctiva as a result of injury or infection causes the characteristic bloodshot or 'pink-eye' appearance known as conjunctivitis.

The cornea, the humours and the lens are the refractive apparatus of the eye. The refractive power of the eye is measured in *diopters*[1]. A lens that can focus parallel light rays to a point 1 metre from its axis has a refractive power of 1 diopter. A lens that can focus parallel light rays to a point 50 cm from its axis has a refractive power of 2 diopters. Similarly, a lens that can focus parallel light rays to a point 10 cm from its axis has a power of 10 diopters. The eye is considered to have a single lens 17 mm in front of the retina. When focusing on a distant object, it has a refractive power of 59 diopters. About 48 diopters of the eye's total refractive power is due to the cornea, rather than the lens. This is because of the large difference between the refractive indices of the cornea and air as opposed to the smaller difference between the refractive indices of the lens and the humours of the eye. If the lens is removed from the eye, its refractive power is about 150 diopters, but inside the eye it only contributes about 15 diopters when distant objects are viewed. This explains why removal of the lens of the eye (due to the presence of opacities or 'cataracts' in the lens) does not lead to blindness: corrective lenses can be supplied to replace the lost refractive power and compensate for any optical abnormalities.

Blinking

Blinking is a reflex action that occurs every 2–10 seconds. It is also a voluntary, forced closure of the eye. The function of blinking is to stimulate tear production and flush out foreign objects (such as dust particles) from the surface of the eye. Tears, a dilute saline solution, lubricate eye movements and are mildly bactericidal. The eyelids can be likened to windscreen wipers. At the inner corner of the lower eyelid is the naso-lachrymal duct through which the tears are drained. Tasks requiring concentration can reduce the blink rate. This can cause particles of dust to accumulate on and lead to drying and irritation of the surface of the eye (particularly if the relative humidity of the air is low), causing people to complain of 'hot' or 'rough' eyes. It has been suggested that this abnormal drop in blink rate is common in work involving VDUs and is a contributory factor in the incidence of VDU-related visual problems. Tsubota and Nakamori (1995) measured the effects of ocular surface area on tear dynamics. Subjects blinked every 2 seconds when looking down, looking straight ahead and looking up. The area of exposed eye was 2.2 cm^2 when subjects were relaxed, 1.2 cm^2 when subjects read a book (with the angle of gaze cast downwards) and 2.3 cm^2 when subjects looked straight ahead to operate a VDU. The rate of evaporation of tears increased with increasing occular surface, leading the researchers to recommend that VDUs should be set at a lower height to prevent the eyes from

---

1 The reciprocal of the focal length (in meters).

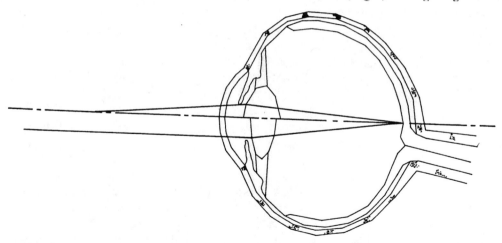

Figure 10.4 Accommodation. The lens assumes a spherical shape to focus divergent rays from near objects onto the retina. Parallel rays from distant objects can be brought into focus with a flatter lens.

drying out during VDU work. The reduced occular surface when reading a book negated the drop in blink rate, which accounts for part of the increased prevalence of visual complaints when operating VDUs compared to reading. The use of spectacles with side panels was also recommended. These recommendations are likely to be of most use in environments where the relative humidity is low.

Accommodation

Unlike the cornea, the lens has variable refractive power, enabling light from both distant and near objects to be focused sharply onto the retina, a process known as *accommodation*.

Cameras usually have lenses of fixed focal length. 'Accommodation' or focusing of a camera depends on adjusting the distance of the lens from the focal plane (the photographic film). The refractive power of the eye is varied by changing the shape of the lens. When distant objects are viewed, the lens assumes a flat, disc shape. When near objects are viewed, the lens is fatter and rounder with greater refractive power (Figure 10.4).

Most visual problems that are due to optical causes can be successfully corrected using appropriate lenses. When viewing a distant object, the incident light rays are approximately parallel. They are refracted by the cornea and lens and produce an image on the retina. In practice, any object more than about 6 metres away can be considered to be at infinity. When viewing close objects, the incident light rays are divergent and greater refractive power is required to produce a sharp image on the retina. In young people, the refractive power of the lens can increase from 15 to about 29 diopters to bring close objects into focus: the lens has about 14 diopters of accommodation in these individuals (Guyton, 1981). Although this is quite a small proportion of the eye's total refractive power, it is necessary for seeing clearly both near and middle-distance objects (such as a VDU on a desk and a notice board on a wall across a room).

The mechanism of accommodation is as follows. The lens has a naturally convex shape. It is held in place by a capsule and a muscle known as the ciliary muscle. The ciliary muscle is situated around the equator of the lens, to which it is attached by ligaments. When a near object is fixated, the ciliary muscle contracts and moves closer to the lens in a sphincter-like action. This reduces the tension in the ligaments and permits the lens to adopt its natural convex shape, increasing its refractive power. When fixating a distant object, the ciliary muscles relax and move further away from the lens. Tension in the capsule is increased by the pull of the ligaments and the lens is pulled into a flatter shape with less refractive power.

Thus, the ciliary muscles have to contract to accommodate near objects and it is in this sense that the 'visual workload' can be considered to be greater when viewing near rather than distant objects. Visual workload in close tasks can be reduced by permitting micro-breaks every few minutes in which the eyes are refocused on a distant object for a few seconds. This is known as *visual relief.*

The near point of vision is the closest distance at which an object can be bought into sharp focus. A 16-year-old can focus on an object less than 10 cm in front of the eye. However, the lens loses elasticity with age and in practice this results in a reduction in refractive power. By the age of 60 years, the near point may have receded to 100 cm. This is why older people often need reading glasses or, when reading a newspaper, for example, have to hold the paper at arm's length. By the age of about 50 years, the lens has only about 2 diopters of accommodation left. After this, it can be regarded as completely non-accommodating, a condition known as presbyopia. The result of presbyopia is that the eye becomes focused at a fixed distance, which varies between different people depending on the characteristics and condition of their eyes. Frequently, the fixed viewing distance in the presbyopic eye is intermediate between the previous near and far points and the person has to wear bifocal lenses; the upper part is set for distant vision and the lower part for near vision (mainly for reading). Bifocal or trifocal lenses can restore a kind of 'stepwise' accommodation to the presbyopic eye. In practice, if workplaces are adequately lit, the depth of field of the eye is increased and the net effect is to lower the requirements for accommodation. *This explains why good lighting is important in all facilities that are used by older people.*

A more detailed discussion of accommodation may be found in Miller (1990).

Visual defects

Asthenopia is diminished visual acuity associated with eyestrain, pain in the eyes and headache. It is common in people who carry out near visual work for long periods and naturally reverses shortly after the cessation of close visual work.

In a normal or *emmetropic* eye, there is a correct relationship between the axial (antero-posterior) dimensions of the eye and the power of its refractive system. Parallel light rays are focused sharply on the retina (Figure 10.5).

In myopia, light rays entering parallel to the optic axis are bought into focus at a point some distance in front of the retina. This can be caused by the eye being too long antero-posteriorly or due to excessive power of the refractive system. Myopia is sometimes referred to as 'nearsightedness' because the near point is closer to the eye in myopic people (for an equal amount of accommodation) than it is to a healthy eye. Myopic individuals cannot bring distant objects into focus. Temporary myopia often

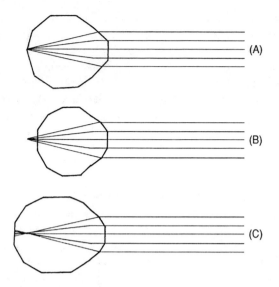

Figure 10.5 In the emmetropic eye (A), light rays are focused onto the retina. In hypermetropia (B), the eye lacks refractive power. In myopia (C), there is excessive refractive power.

occurs after a near object has been viewed for a period of time; accommodation is not instantaneous because the lens requires time to change to a flatter shape when the ciliary muscle relaxes. Myopic individuals can often carry out close tasks such as VDU work or sewing with ease but experience difficulties with tasks such as driving when target objects are more than 5–10 metres away.

Myopia has a genetic component. In the developed world, the adult population prevalence is about 25%. In late childhood, it is about 12% if both parents are myopic and only 2.7% if neither parent is affected. Myopia is rare in hunter-gatherer societies and seems to be triggered by childhood exposure to close visual work. It was originally rare in Eskimo societies but reached 25% in the years following the intro-duction of school attendance by Eskimo children. There is experimental evidence that blur is a stimulus to eye growth and can cause the system to 'overshoot', resulting in myopia (Nesse, 1995).

In hypermetropia, light rays entering parallel to the optic axis are bought into focus behind the retina. This can be caused by the eye being too short antero-posteriorly or by insufficient curvature of the refractive surfaces of the eye. Hypermetropia is sometimes referred to as 'farsightedness' because the near point is farther away from the eye (for an equal amount of accommodation) than it is in a healthy eye. Hypermetropic individuals can be said to lack refractive power and may tire quickly when carrying out work in which the viewing distance is short (such as using a VDU).

In astigmatism, there is an unequal curvature of the refractive surfaces of the eye such that the refractive power is not the same in one plane as in another. When an object of complex shape is viewed, the retinal image may be out of focus in one plane but not in others. Astigmatic individuals often perform quite well when given simple eye tests because the defect is corrected by the depth of focus of the eye. However, they may experience severe difficulties at night or when there is excessive glare.

City of Westminster College
Learning Services / Queens Park Centre

These defects can be corrected using appropriate lenses. Myopic individuals can be given diverging lenses to reduce the total refractive power of their optical systems. Hypermetropic individuals can be given converging lenses to increase their refractive power. Several problems may occur if defective vision is not corrected. Firstly, suboptimal vision may degrade performance. Secondly, excessive load on the muscles of the refractive system may cause visual fatigue. Thirdly, the worker may adopt stressful body postures to orientate the head in an attempt to see better. Neck and shoulder problems may result.

Chromatic Aberration

'White' light is a mixture of different wavelengths. A defect of all simple lenses is that light of a given wavelength is focused slightly nearer or farther away than other wavelengths. This is known as chromatic aberration. Chromatic aberration can cause the outlines of objects to have purple or red fringes (the shortest and longest wavelengths). The refractive power of the eye is about 3 diopters greater for saturated blue than for saturated red, so blue objects are focused in front of red ones and both cannot be in focus at the same time. Purple letters or characters on VDU screens may appear to have fuzzy edges for this reason. With a VDU using less saturated colours, the difference may be only 1 diopter. However, this may still be enough to cause startling visual effects such as chromostereopsis (the illusion of depth).

Since blur is a stimulus for accommodation, unresolvable blur may destabilise the accommodation mechanism making it 'hunt' in vain to resolve the blur caused by the different focal points of the different colours. The onset of visual fatigue may be hastened. Reds, oranges and greens can be viewed without refocusing but cyan or blue cannot be viewed with red (because they are at opposite ends of the visible spectrum). Chromatic aberration is exacerbated when the pupillary diameter is large (as in poor lighting). Collins *et al.* (1994) measured steady-state accommodation of subjects viewing VDU screens with different coloured backgrounds and text. They found a small, but statistically significant increase in accommodation when subjects viewed high-contrast blue-on-red and red-on-blue screens compared to monochrome screens of either polarity. This suggests that it is more visually demanding to view such screens and that red/blue colour combinations should be avoided. For clarity and the avoidance of strange visual effects, the conjoint use of saturated colours or colours from opposite ends of the spectrum should be avoided.

Convergence

The eyes are a small distance apart. When one is looking at a distant object such as a mountain several kilometres away, each eye receives a similar image because the lines of sight of the two eyes are parallel and the distance apart of the two eyes is negligible compared to the distance of the viewed object.

When closer objects are viewed, the eyes converge on the object; that is to say, the lines of sight of each eye meet at the object. Convergence decreases with distance. When viewing close objects, the two eyes view the object from slightly different angles so that the images cast onto the two retinas differ slightly. The position of the eyes in their sockets is controlled by the vergence system. Lightly pressing one eye with the finger causes the perception of double images, which is known as *diplopia*.

Long hours of close visual work may cause imbalances in the muscles controlling eye movement, a condition known as *phoria*, and increase the perceived effort required to carry out the task.

The resting posture of the eye

If people are asked to look ahead in the dark, the eyes assume their 'resting' posture. Dark vergence and dark accommodation are the levels of binocular vergence and monocular accommodation that the eyes assume under such conditions. When looking straight ahead, the resting position of vergence is about 1 metre and that of accommodation is of intermediate distance. Prolonged viewing of near objects causes temporary, proximal shifts in vergence and accommodation and these shifts are greater in people whose resting positions are farther away. The onset of these shifts is related to the onset of visual fatigue. Further exposure to high visual demands can create 'vergence disparities', which express themselves as 'double vision' caused by overconvergence to far objects and underconvergence to near objects (Heuer and Owens, 1989).

Jampel and Shi (1992) describe the primary position of the eyes as being 4 degrees, on average, below the Frankfort plane (the Frankfort plane is described by lines passing through the upper margin of each external auditory canal and the lower margin of the orbit of the ipsilateral eye). In normal standing, the Frankfort plane is approximately horizontal and the direction of gaze of the eyes in the resting position is slightly downwards. According to Jampel and Shi, the primary position of the eyes is the position to which the eyes reset automatically after we cease to view something that is elsewhere in the visual field. This resetting occurs automatically and is remarkably constant. Subjects can reset their eyes to the resting postion with an accuracy of a few degrees in lighted rooms with or without fixating objects and in dark rooms. Furthermore, the resting position of the eyes in relation to the head (brain) is unaffected by body position (head supine, prone, tilted to one side or when subjects are standing on their heads). The primary position is not affected by whether the eyes are open or closed. Subjectively, it is the natural position of the eyes and no conscious effort is required to maintain it. According to Jampel and Shi, the primary position of the eyes is determined by neurological mechanisms in the reticular activating system of the brain stem rather than by the passive response of the eyes to mechanical forces and gravity.

Although the primary position of the eyes is fixed, the resting position of vergence is not. Heuer and Owens (1989) measured changes in the resting position of vergence of subjects viewing displays from 45 degrees below the horizontal to 30 degrees above. Subjects viewed the displays either by moving the eyes or by moving the head and keeping the position of the eyes constant, with respect to the head. In both cases, the resting position of vergence increased with upward gaze (from about 80 cm at 45 degrees below to 140 cm at 30 degrees above). Furthemore, accommodative power increases with declination of the line of sight, by 20% when looking down at 40 degrees and by 33% when looking down and in at 40 degrees.

General recommendations for restful viewing

The ability to view objects up to about 1.4 metres depends on the interaction between angle of gaze and the amount of vergence and accommodation, both of which increase when looking downwards. For comfortable viewing, then:

- The most frequently viewed displays should be accessible with the head erect and the eyes in their primary position.
- Vergence effort should be minimised – the closer the display to the operator, the lower it should be (up to 30 degrees below the horizontal) and vice versa.
- To facilitate accommodation, displays that require a high degree of visual acuity should be placed up to 40 degrees below the line of sight and be directly in front of the viewer.
- To account for individual differences in the resting posture of the eyes, displays such as VDU screens should be adjustable in both the horizontal and vertical planes from eye height to over 1 metre away from the eyes.
- No displays should be nearer to the operator than the resting point of vergence – 110 cm for displays level with the eyes.
- Avoid 'accommodation traps' where objects placed close to the near point of accommodation (such as dirt on a car windscreen) cause involuntary focus and blurring of visual targets further away.

Empirical evidence for these guidelines can be found in Lehman and Stier (1961); cited in Hill and Kroemer (1986). Subjects carrying out an assembly task requiring precise vision preferred an average viewing angle of 38 degrees, half of which was obtained by tilting the head downwards. Hill and Kroemer further investigated preferred sight lines by having subjects view displays when sitting erect (90 degrees) and reclining at 105, 130 and 180 degrees, respectively. Subjects sitting erect preferred a viewing angle of 29 degrees below the Frankfort plane, increasing to 40 degrees when reclining fully. This is almost certainly to correct the recession of the near point of vergence cause by the change in head position. Target distance also caused a change in preferred line of sight and this was independent of any changes due to head position. The preferred line of sight was greater for a task distance of 50 cm than for a distance of 100 cm: 38 degrees as opposed to 30 degrees below the Frankfort plane. The increased declination in the line of sight when viewing the 50 cm display was almost certainly to compensate for the increased accommodation required to view the closer display. Hill and Kroemer concluded that when working with VDUs, 'one should place the monitor closely behind the keyboard, not on top of the disc drives'. Since display technology has improved greatly since 1986, we might add the rider, 'but only if the display demands visual acuity'. Larger monitors, requiring less acuity can be placed on the disc drives if the viewing distance is around 1 metre. In both cases, the top of the display should not be higher than a horizontal line from the viewers' eyes.

Jaschinski-Kruza (1991) investigated eyestrain in a VDT task that was carried out at viewing distances of 100 cm and 50 cm. Visual strain was greater at 50 cm than at 100 cm and was greater still the more distant the subject's dark convergence. Subjects' preferred viewing distances varied from 51 to 99 cm for screen characters 5 mm high.

The retina

The retina is the most complex part of the eye, consisting of a layer of light-sensitive cells connected to nerve fibres, and is sometimes considered to be an extension of the brain ('a little piece of brain lying within the eyeball'). Unlike photographic film, the retina actively processes incoming information before passing it on to the brain via the optic nerve.

Incident light causes chemical changes in the light-sensitive cells, which give rise to nerve impulses. The nerve fibres pass over the cells and converge to make up the optic nerve. The point at which the optic nerve leaves the retina is known as the blind spot. The retina can be likened to an array of electronic light detectors linked in complex ways that act in an 'on/off' fashion when activated by the incident photons.

The retina contains two types of light-sensitive cell, known as *rods* and *cones*. There are over 100 million rods and about 6 million cones. Rods are more sensitive to light than cones and are essential for scotopic (night) vision. Bright light bleaches the rods, which renders them ineffective and the cone or photopic system then comes into operation.

Photopic and scotopic vision differ in several other ways. The retina has an uneven distribution of light-sensitive cells. Cones predominate at the central part of the retina (the fovea). Towards the periphery, the concentration of rods increases. Furthermore, each foveal cone cell has its own nerve fibre whereas several rods may share one fibre. The structure of the retina is exceedingly complex but it can be concluded that photopic vision is less sensitive to light but more acute than scotopic vision.

Subjectively, we are aware of an area of focal attention in bright light that corresponds to the fovea or central part of the retina where there is a high concentration of cones. In dark conditions, the ability to resolve detail is lost because the cones no longer function. The peripheral areas of the retina contain more rods and are sensitive to changes of light and to movement *so warnings can be placed in peripheral areas of a panel as long as flashing light or movement is used.*

Retinal adaptation

Retinal adaptation is the ability of the retina to change its sensitivity according to the ambient lighting. When walking from a darkened cinema into a sunny street, a temporary feeling of being dazzled is experienced. The diameter of the pupil decreases to reduce the amount of light entering the eye, the rods are bleached and the photopic system quickly comes into operation. On going from a bright area to a dark one, the pupil increases in size and chemical changes take place in the retina as the rods slowly come into operation. Full adaptation to dark conditions can take up to 20 minutes, so it is important not to expose the dark-adapted eye to sudden, bright lights since even brief exposures will degrade scotopic vision for some time afterwards. This can happen to motorists driving along unlit roads at night when oncoming cars have badly adjusted or undimmed headlights. Extra light is usually provided at the entrances and exits to tunnels for similar reasons, to smooth the transition between light and dark and provide more time for retinal and behavioural adaptation.

Figure 10.6 depicts the sensitivity of the photopic and scotopic systems to light of different wavelengths. The systems are differentially sensitive to light of different wavelengths. Both are maximally sensitive to light in the middle of the spectrum (which is perceived as blue-green, green and yellow). Violet and red are less readily sensed, which implies that in order for a red object to appear as subjectively bright as a green one, more illumination is required. It is unfortunate that red is normally used to signal danger since the retina is less sensitive to red than to other colours at the same illumination level.

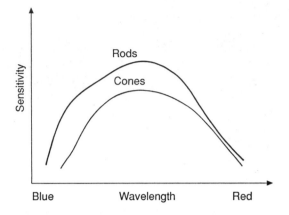

Figure 10.6 Sensitivity of rods and cones.

The scotopic system is several orders of magnitude more sensitive than the photopic system to shorter wavelengths. However, the two systems have a similar sensitivity to the longer wavelengths. When the eyes are dark-adapted, long-wavelength (red) light should be used to illuminate objects temporarily. The cones will be sensitive to these wavelengths at an intensity sufficiently low so as not to bleach the rods and degrade their state of dark adaptation. Night watchmen, sentries, etc. can be issued with torches containing long wavelengths for use, for example, when inspecting documents at night.

In practice, a stage intermediate to purely photopic or scotopic vision can occur. This is known as 'mesopic' vision and has characteristic features of both of the others.

Colour vision

The output of the photopic system is experienced as colour vision. Three types of cone cells with different spectral sensitivities have been identified. Because the absorption spectra of their light-sensitive pigments overlap, a given wavelength of visible light will cause the three types of cones to respond in varying degree. The result is a *pattern of cone outputs* which is the signature of the particular wavelength. Different wavelengths give rise to different cone output patterns, which are interpreted as different colours by the brain. That the cone output patterns are transmitted to the brain where they give rise to the colour perception seems straightforward and compatible with our everyday experience. That white light consists of light of many different 'colours' (i.e. wavelengths) supports the component process explanation of colour vision. For example, red and green light sources can be combined to give a yellow colour, which matches monochromatic yellow even though the wavelength which that rise to the perception of yellow is not present. These *metameric matches* are evidence for a component process view of colour vision. The cone output pattern caused by the mixed-wavelength stimulus is identical to that caused by the monochromatic source, so the same colour is perceived. According to this view, colour vision is purely the result of the cone output pattern caused by the incident light: different wavelengths have characteristic cone output patterns and if a combination of different wavelengths impinges on the retina a cone output pattern intermediate between the two wavelengths and in proportion to their respective intensities will result. This type of reasoning is used to explain how people can distinguish between colours such as 'bluish-green' 'greeny-blue', etc.

Certain visual phenomena are more difficult to explain using simple colour mixing theories. If a bright light is fixated for several seconds, a retinal afterimage remains when the viewer looks away. Blue lights produce yellow afterimages and vice versa, and red lights produce green after images. It has been suggested that at the retinal level, a component process exists with functionally different cones responding maximally to different wavelengths. The outputs of this system are thought to fire into a so-called opponent process system. There are said to be three opponent process systems – a blue/yellow system, a red/green system and a light/dark system. When long wavelengths are incident, the red/green system increases its firing rate above its spontaneous level and red is perceived. As the incident wavelength is shortened, the system is inhibited and its firing rate decreases until it drops below the spontaneous level and green is perceived. Thus, after viewing a bright red light for several seconds, a green afterimage of the object is seen as a result of a sudden drop in the output of the opponent process system. Certain cells in the brain appear to act like opponent process cells, increasing and decreasing their firing rate in this way.

Colour vision is a complex phenomenon that can only be dealt with here at a superficial level. However, these theories of colour vision can have implications for ergonomics. The notion that red and green and blue and yellow are related neurologically has interesting design implications. Generally, these opponent colour combinations should be avoided in display design because of the afterimage problem, particularly in active displays such as VDU screens and if the colours are highly saturated. A saturated colour is 'pure' in that it contains only those wavelengths which give rise to the perception of the particular colour: 'pillar box' red is an example. Unsaturated colours are those that have been 'diluted' with white. Bright blue letters on a yellow background may make an attention-getting sign but would be unsuitable for use in a visual display screen. Sharp discontinuities between bright colours may produce startling visual effects – red letters on a blue background may appear to 'vibrate' for example. Colour should not be used for borders or to display messages on the periphery of a display since cone density is very low at the periphery of the retina. The use of flashing lights or a moving cursor would be more effective. Some general guidelines for the use of colour in visual displays are summarised in Table 10.1. Further details can be found in Christ (1975).

The dark-adapted eye is more sensitive to shorter wavelength (blue) than longer (red) light and most sensitive to green. These colours should be used to design displays or notices that have to be read in dark conditions. The light-adapted eye is most sensitive to middle wavelength (yellow/green) light. In designing passive (reflective) displays for light conditions, maximum colour contrast can be obtained by using yellow on black.

Not all colours can be duplicated by mixing three primary hues in the laboratory (brown, silver and gold are examples). In colour television sets, three-colour mixing is used yet viewers do perceive these otherwise unattainable colours. In laboratory conditions, it is possible to illuminate a dark object such as a piece of coal so that its luminance is greater than a piece of white paper. However, subjects will say that the paper looks brighter as long as they can see enough to recognise the objects as paper

Table 10.1 Guidelines for colour selection[a]

In general
1. Choose compatible colour combinations. Avoid red/green, blue/yellow, green/blue, red/blue pairs.
2. Use high colour contrast for character/background pairs.
3. Limit number of colours to 4 for novice users and 7 for experts.
4. Use light blue for background areas only.
5. Use white to code peripheral information.
6. Use redundant coding (shape or typeface as well as colour); 6–10% of males have defective colour vision.

For visual display units
1. Luminosity diminishes in the order white, yellow, cyan, green, magenta, red, blue.
2. Use white, yellow, cyan or green against a dark background.
3. For reverse video use nothing (i.e. black), red, blue or magenta.
4. Avoid highly saturated colours.

---

[a] Adapted from Durrett and Trezona (1982) and van Nes (1986).

and coal. This demonstrates that higher learning processes are involved in making judgements about colour. Some colours are associated with familiar objects (as is indicated by sayings such as 'as black as coal') and judgements are influenced by other factors such as intensity and the reflectance of the object.

Colour and visual acuity

When stationary objects are being viewed, acuity is least if the object is blue. Long and Garvey (1988) investigated colour and dynamic visual acuity. When moving targets are being viewed under photopic conditions, their colour does not seem to influence visual acuity. However, dynamic visual acuity under scotopic/mesopic conditions is affected by target colour, and blue targets are much more easily resolved than other colours, probably owing to the differential sensitivity of the rod system. Long and Garvey suggest that we should be careful not to overgeneralise the usual ergonomic recommendations about colour. Under some lighting conditions where there is a moving display or a moving observer, colours such as blue (which are commonly avoided for stationary targets) might be advantageous. There are possible implications here for the design of roadway warning signs which will be used primarily at night.

Measurement of light

Measurement of light is essential in the design and evaluation of workplaces. Because the eye adapts to light levels, automatically compensating for any changes in illumination, subjective estimates of the amount of light in a work area are likely to be misleading. Data concerning the visual response of the eye have been used to define lighting measures. The radiant flux arriving at a surface is weighted according to the eye's sensitivity to each of a number of wavelength intervals. The total incident radiant flux after weighting is known as the luminous flux.

The measurement of light is known as photometry. The main photometric units are luminous intensity, luminous flux, luminance and illuminance. Their definitions

Table 10.2 Main photometric terms

Luminous intensity	The power of a source or illuminated surface to emit light	candela (cd)
Luminous flux	The 'rate of flow' of luminous energy	lumen (lm)
Luminance	The light emitted by a surface	cd/m^2
Illuminance	The amount of light falling on a surface	lux (lx)[a]
Reflectance	The ratio of the luminance and illuminance at a surface	

1 lumen = 1 cd/m^2 in SI base units

[a] 1 lux = 1 cd/steradian/m^2 in SI base units

are given in Table 10.2. The SI unit of luminous intensity is the candela (cd). An imaginary point source of luminous intensity 1 cd will emit light in all directions. A source of greater intensity will emit more light. In both cases, we can imagine a sphere of light spreading out from the source. Clearly, the intensity of the source itself does not depend on the distance from which it is viewed. However, the 'strength' of the light at the edges of the imaginary sphere will depend on viewing distance.

The surface area of a sphere is 4π steradians of solid angle. A steradian is defined as the solid angle that encloses a surface on a sphere equivalent to the square of the radius. By definition, a point source of luminous intensity 1 cd will emit a total luminous flux of 4π lumens or 1 lumen/steradian.

Illuminance refers to the light falling on a surface. If we imagine a point source of one candela inside a sphere of 1 metre radius, the illumination on the inside of the sphere is defined as 1 lux, or 1 lumen/square metre. If we increase the radius of the sphere to 2 metres but keep the intensity of the source the same, the illuminance on 1 square metre of the inner surface will now be $\frac{1}{4}$ lux. This is because when the radius of the sphere increases linearly, the surface area increases according to the square of the radius. Illuminance therefore follows an inverse square law with intensity, decreasing with the square of the distance from the source.

The luminance of an object depends on the light it emits or reflects towards the eye. It corresponds roughly to brightness, although brightness perception depends on other factors such as contrast. In the example above, if the inside of the sphere is a perfect reflector of light then the luminance will be the same as the illuminance. The percentage of the incident light that is reflected by a surface depends on the reflectance of the material. Reflectance is defined as the ratio of luminance to illuminance. White paper has a reflectance of about 95%, white cloth about 65%, newspaper about 55%, plain wood about 45%. Matte black paper has a reflectance of about 5%. More formally, reflectance is given by

$$\text{Reflectance} = \frac{\text{Luminance}}{\text{Illuminance}}$$

If we know the illuminance of the surfaces in a room (for example, by measuring it with a light meter) we can select materials of appropriate reflectances for each of the surfaces in order to achieve a balance of surface luminances in the room and ensure that the ratio of luminances of adjoining surfaces is not excessive (as described below).

More detailed discussion of photometric terms, the relationships between them and recommendations for the design of lighting systems can be found in publications of the Illumination Engineering Society (IES) and in Boyce (1982).

Table 10.3 Examples of recommended and naturally occurring illuminances

Area/activity	Illuminance on a horizontal surface (lux)
Clear sky in summer	150 000
Overcast sky in summer	16 000
Performance of extremely low contrast tasks.	
Certain surgical operations	10 000
Textile inspection	1 500
Office work (pencil handwriting, poorly reproduced documents)	1 000
Precise assembly work	1 000
Office work (without VDUs)	500
Office work (data entry)	500
Office work (VDU, conversational tasks)	300
Heavy engineering	300
Rough assembly work	200
Minimum illuminance for manual handling tasks (NIOSH)	150
Rarely visited places where little perception of detail is required	
(e.g. railway platforms)	50
Good street lighting	10
Emergency lighting[a]	2
Moonlight	0.5

[a] Data are representative compilations from many sources. For emergency lighting recommendations, see Jaschinski (1982).

Lighting standards

Much effort has been applied over many years to the drafting of standards for the illumination of workplaces. Standards differ from country to country. Table 10.3 presents recommended illuminances for different work situations. Some other illuminance values that are found in more extreme situations have been included in the table to provide context. USA readers should refer to the *IES Lighting Handbook* for up-to-date and detailed recommendations, whereas readers outside the USA may find the *CIBSE Code for Interior Lighting* useful. It can readily be seen that the eye can operate under an extremely wide range of illuminance levels, owing largely to the differential sensitivity of the photopic and scotopic systems.

Recommended illuminance levels vary both between countries and over time. Before the invention of electric lights, indoor workers depended on daylight and building dimensions were limited because daylight does not penetrate into a room beyond about 10 metres. Sundstrom (1986) has noted that lighting standards have risen over the years and have been higher in the USA than elsewhere. In the early part of the last century, investigations by various companies in the USA and by the Industrial Fatigue Research Board in Britain demonstrated that the performance of visually demanding tasks could be improved by increasing the level of illumination. Artificial lighting became accepted in factories and standards rose under the assumption that 'more is better'. With the arrival of fluorescent lights, recommended levels could increase further because of the better luminous efficiency (lumens per watt of electricity), longer life and reduced heat production of fluorescents.

Recommended illuminance levels differ between the USA and other countries for several reasons, including differences in design philosophy. US standards have been heavily influenced by the laboratory research work of Blackwell. In other countries, a more pragmatic approach has been taken. In Britain, the view has been that daylight should be the dominant source of light with artificial light taking on a supplementary role.

A more recent trend has been to reduce the levels of illumination in workplaces, particularly offices. This has occurred partly because of the desire to conserve energy and also as a response to the introduction of VDUs into the workplace. Ergonomists have recommended that illumination levels be lower in VDU offices to avoid glare and reflection problems (see below). In practice, the choice of an appropriate level of illumination depends not only on the task but also on the distribution of objects in the visual field and their luminances.

Contrast and glare

The function of the eye is not so much to detect light but to detect luminance discontinuities between objects in the visual field. It is the difference in luminosity between an object and its surroundings that make it visible, rather than the light the object reflects. A person wearing opaque spectacles can detect the presence of light but not the presence of objects.

The retina functions more like an edge detector than a light meter. Special cells (called horizontal cells) are part of a network of cells interposed between the rods and cones and the fibres of the optic nerve. If one part of the retina is stimulated by a bright light, the horizontal cells inhibit the adjacent photoreceptors. This has the effect of increasing the difference in the firing rates of the stimulated and unstimulated parts of the retina and serves to enhance the perceived contrast between the stimulus and its surroundings, which has the effect of sharpening the contours of objects, facilitating their detection (the importance of contrast and contour detection in design is discussed below and in Chapter 13).

The direction of gaze is involuntarily drawn to bright objects in the visual field. This is known as phototropism. Jewellery shops usually display their wares on black velvet cloth under bright lights to obtain maximum contrast, the intention being to stimulate a phototropic response in passers-by.

One of the most important considerations in the design and evaluation of lighting systems is to ensure appropriate contrast between objects in the visual field. The luminance contrast between two surfaces is given by the difference between the luminances of the brighter and dimmer surface expressed as a percentage of the brighter (note that contrast does not depend on the absolute brightness of the surfaces).

$$\text{Contrast} = \frac{L_{\text{bright}} - L_{\text{dark}}}{L_{\text{bright}}}$$

Contrast percentages can also be calculated using data on the reflectances of adjacent surfaces, assuming equal illuminance of the surfaces in question or if the actual illuminances are known.

The luminance ratio is the ratio of the luminance of a work area to that of its surroundings. Recommended maximum luminance ratios have been proposed – 3:1 between a task and its immediate surroundings to 10:1 between the task and the walls, floors, etc. It is usually suggested that the task should be the brightest area in the visual field to take advantage of the phototropic response. This precludes the use of materials such as white Formica for desktop materials since the luminance ratio between white paper and the desktop will be too low. Wood or pastel-coloured finishes are preferable from this point of view.

Interior decoration should be chosen to achieve a balance of surface luminances gradually diminishing from the task to the surroundings. Brightly coloured walls, carpets, furniture and fittings are contraindicated for this reason.

Glare occurs when there is an imbalance of surface or object luminances in the visual field – the brighter sources exceeding the level to which the eye is adapted. Sources of glare include the sun, bright or naked lamps, or reflections off shiny objects. Although the retina is able to adapt to different levels of luminance so as to operate over a wide range of conditions it is not able to adapt selectively to large, simultaneous discontinuities in luminance in the visual field. For example, if the ambient luminance is high compared to the task luminance, the retina will adapt to the former rather than the latter and the task will appear dim and will be more visually demanding. This can happen when VDU screens are placed against a window such that users face the window. If the illuminance in a room is very low, the retina will dark adapt and will be more vulnerable to the effects of glare from task elements or extraneous sources.

Glare is categorised in several ways. Disability glare increases task demands, whereas discomfort glare does not. Discomfort glare may occur in offices, for example, when one or more bright objects are seen peripherally. Disability glare occurs when objects brighter than the task interfere with the detection and transmission of visual task data: extraneous light sources may increase the adaptation level of the retina, making the task appear dimmer than it really is. Intense light may be interreflected by structures within the eye itself and reduce the contrast between the retinal image and the background. Older workers may be particularly prone to this type of glare as a result of age-related changes in the refractive media of the eye. Finally, bright lights may cause retinal afterimages, which have a 'veiling' effect on the main task.

Glare may be direct or indirect; that is, it may be emitted by a source or reflected off an object. If reflected, the glare may be either *diffuse* or *specular* (mirror-like). White walls on the inside of a room may reflect sunlight into an operator's eyes and would be classed as a diffuse source. Chrome-plated controls or components such as bevels on dials and gauges may cause specular reflections. More detailed discussions of glare can be found in Cushman and Crist (1987) and Howarth (1990).

Lighting design considerations

For visual comfort and to meet visual demands the following should be considered (Grandjean, 1980):

1   A suitable level of illumination
2   A balance of surface luminances
3   Avoidance of glare
4   Temporal uniformity of lighting

The colour rendering properties of light sources might also be taken into account.

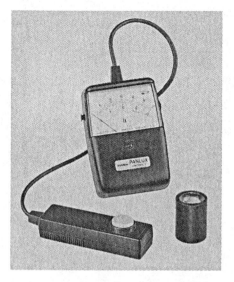

Figure 10.7 A light meter suitable for measuring illuminances in ergonomics. (Photograph kindly supplied by Messrs Gossen-Metrawatt, Numberg, Germany.)

Illumination levels

Although early research indicated that improvements in productivity were possible when illumination levels were increased, more does not necessarily equal better. In fact, high levels of illumination may increase glare and may wash out important visual details. If illumination levels are inadequate, increasing them may improve performance, but a region of diminishing returns will be reached as non-visual limits to performance (such as motivation, fatigue or manual dexterity) become increasingly important. Older workers generally require higher levels of illumination than younger workers owing to the loss of refractive power and changes in the light-transmitting media of the internal structures of their eyes.

Lighting surveys

Ergonomists are not illumination engineers but they should be able to use a light meter (Figure 10.7). Light meters can be used to measure illuminance levels on the work surfaces in offices, shops, factories, etc. and the readings can be compared with the levels recommended in published standards or in Table 10.3. Care should be taken not to stand between the light source and the meter's sensor when taking measurements.

Balance of surface luminances

In practice, a balance of surface luminances is achieved by specifying appropriate illuminances and corresponding reflectances of the surfaces in a room. Detailed specifications can be found in the publications of the IES and ANSI and examples of this information are included here. Some additional considerations are described below.

Direct lighting is often used to illuminate sculpture because the shadows produced have a 'modelling effect'. The areas of light and dark enhance the 3-D appearance of the works by emphasising differences in depth. Angled direct lighting is useful for

Table 10.4 ANSI recommended reflectances for offices

Surface	Reflectance (%)	Suitable materials/finishes
Ceiling	80–90	White paint (matte)
Furniture	25–50	Wood (matte, unpolished)
Upper walls		
(Wall/ceiling border)	80–90	White paint (matte)
Walls	40–60	Neutral, unsaturated hues (e.g. pastel shades)
Business machines	25–50	Matte grey finish
Curtains/blinds	40–60	Cloth
Floor	20–40	Carpet (choose slightly darker colours, beige, brown, grey, etc.)

enhancing the surface texture of materials such as cloth or wood and can be used in the production of these materials to aid inspection and quality control. However, direct light can reflect off desktops or surfaces and cause indirect glare and sharp luminance discontinuities between lit and unlit surfaces.

When indirect lighting is used, most of the light is directed onto the ceiling and walls, which reflect it back. Thus, objects are illuminated from many different directions at once, a smoother distribution of luminances is obtained and shadows are reduced. Direct and indirect lighting can be combined to achieve a balance of surface luminances and minimum glare. In a well-illuminated office for example, all large objects and major surfaces should have a similar luminance and surfaces in the middle of the visual field a contrast ratio of no more than 3:1; contrasts at the sides of the visual field should be avoided. A balance of surface luminances can best be achieved in practice by using materials with different reflectances in a room. The ANSI recommendations for the reflectances of surfaces in offices are given in Table 10.4.

A balance of surface luminances is also an important design requirement for lighting systems in corridors, stairwells and outdoor facilities such as railway platforms that are used at night. Since shadow is a cue to depth, patterns of shadow on the floor can be misinterpreted as changes in ground level and increase the risk of people slipping, tripping or falling.

Avoidance of glare

Glare can be reduced by choosing a suitable combination of direct and indirect lighting. With direct lighting, most of the light is directed towards the target in the form of a cone (Figure 10.8). This produces hard shadows and sharp contrasts between illuminated and non-illuminated areas. Indirect lighting is reflected off other surfaces in a room and produces a smoother transition between surface luminances and reduces shadows.

No light sources should appear within the visual field during work activities. Glare control thus depends on both the design of the lighting system and the general workspace and task design. All lamps should have glare shields or shades and the line of sight from the eye to the lamp should have an angle greater than 30 degrees to the horizontal. Overhead lamps should not reflect off desktops or work surfaces into the operator's eyes. Conversely, reflective desktops and work surfaces should not be used. Wood or matte finishes are preferable. Fluorescent lamps arranged in rows

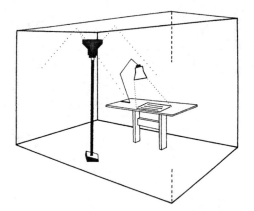

Figure 10.8 Direct and indirect lighting.

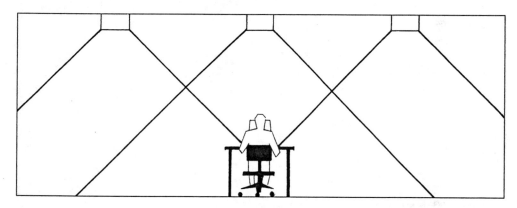

Figure 10.9 Place rows of luminaires parallel to line of sight and to the operator's side. Overlapping cones of light ensure uniform distribution of surface illuminances.

should be arranged parallel to the line of sight to present the minimum luminous surface area possible. Generally, use of a large number of low-powered lamps rather than a small number of high-powered lamps will result in less glare (Figure 10.9).

Glare and VDUs

Good lighting is particularly important in offices where visual display units (VDUs) are being used. Traditional offices designed for pencil and paper work usually have rows of fluorescent lamps designed to provide an appropriate level of illuminance on the horizontal desk tops below. Desks can usually be placed anywhere in the room since the spacing of lamps is such that the 'cones' of light from each lamp overlap to provide a uniform distribution of illuminance. Because the direction of gaze is downwards, towards the horizontal desktop, lamps on the ceiling are more than 30 degrees above the line of sight of a seated worker and do not cause glare.

Because a VDU user looks directly ahead at the screen, light from overhead lamps may cause direct glare or reflect off of the screen, having a 'veiling effect' – reducing the contrast between the characters and background (Figure 10.10). Correct positioning of workstations with respect to windows, lamps and bright surfaces is therefore very

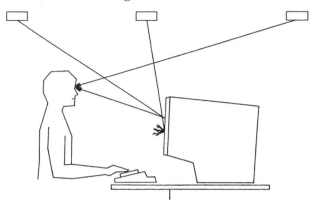

Figure 10.10 Avoid direct and reflected glare at VDU workstations.

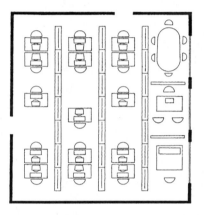

Figure 10.11 Layout for VDU and non-VDU work in an open office.

important (Figure 10.11). Rows of fluorescent lights should be parallel to and on either side of the operator–screen axis to reduce direct and indirect glare, VDU workstations should be at right angles to windows and more use should be made of indirect lighting.

At the workstation itself, sharp contrasts due to the dark screen (i.e. light characters against a dark background, as still found on specialist monitors and display used in scientific instruments) and bright source documents can be problematic. Since the VDU screen emits its own light, general illumination can be lower in VDU than in traditional offices. Low-reflectance surfaces can also be used (particularly behind the screen) and task lighting (e.g. desk lamps) can be provided at each workstation to support the visually demanding aspects of the work.

Temporal uniformity of lighting

Fluctuating luminances can be more disturbing than static contrasts. The temporal uniformity of lighting can be influenced by both the characteristics of the light source and the requirements of the task. Incandescent bulbs radiate light fairly uniformly over time, whereas fluorescent lamps are known to flicker. Fluorescent lamps work by passing an electric current through a gas in a glass tube. The gas emits visible (almost monochromatic) light and ultraviolet radiation periodically in accordance with the mains frequency. The ultraviolet radiation excites a phosphor on the inside of the glass tube, which emits visible light; different phosphors are used to produce the various shades of commercially available lamp. The flicker frequency of correctly functioning fluorescent lights exceeds the threshold for the perception of flicker. However, malfunctioning or old lamps can produce visible flicker that may cause visual discomfort. In factories, flicker is a hazard since it can have a stroboscopic effect – rotating or oscillating machine parts may appear stationary or to move more slowly if their frequency is similar to that of the flickering source that illuminates them. Phase-shifting of fluorescent tubes (using banks of two tubes and alternating the phase of each) and regular inspection/maintenance can prevent flicker problems.

Colour rendering and artificial light

The apparent colour of an object depends on the *spectral composition* of the incident light and the wavelengths the object reflects. For example, in broad-spectrum (white) light an object that reflects long wavelengths and absorbs the rest would appear red. The same object, when viewed under artificial light, deficient in long wavelengths, would appear reddish grey or brown. An object that appeared bright red under white light would appear black when illuminated by monochromatic blue light.

The light emitted by a source can be described in terms of the relative amounts of different wavelengths of which it is composed. Monochromatic sources emit a narrow band of wavelengths. Objects that reflect this wavelength will appear very bright when illuminated by the monochromatic source. Objects that do not reflect the wavelength at all will appear black. For example, sodium street lights emit a narrow band of wavelengths and appear yellow. Most objects illuminated by them appear various shades of yellow/grey and can only be distinguished by the contrast or brightness differences between them. A monochromatic light source can be said to have poor colour rendering ability because it does not reveal the way objects differentially reflect incident light according to its wavelength composition.

In order to have good colour rendering properties, a light source must contain sufficient amounts of light across the visible spectrum. The term spectral energy distribution is used to describe the relative amounts of different wavelengths that a light source emits. Figure 10.12 depicts spectra for a variety of light sources.

Daylight, incandescent light and some fluorescent tubes have broad emission spectra and therefore good colour rendering properties. The colour rendering properties of a light source cannot be determined subjectively by the appearance of the light itself. Although green and red lights can be mixed to give a yellow light of similar appearance to a monochromatic source, the colour rendering properties of the two sources are completely different. The red and green mix will render the colours of ripening apples very well, whereas under the monochromatic yellow they will appear a dull, greyish yellow.

Selection of light sources for good colour rendering should be based on objective data about the emission spectra of the source. In practice, an index – the colour rendering index (CRI) – has been devised to describe the colour rendering properties of light sources. Daylight is used as a standard and is given a value of 100. The colour rendering properties of other sources are expressed with respect to the standard (Boyce, 1982; Howarth, 1990). A high value of the CRI usually indicates that a

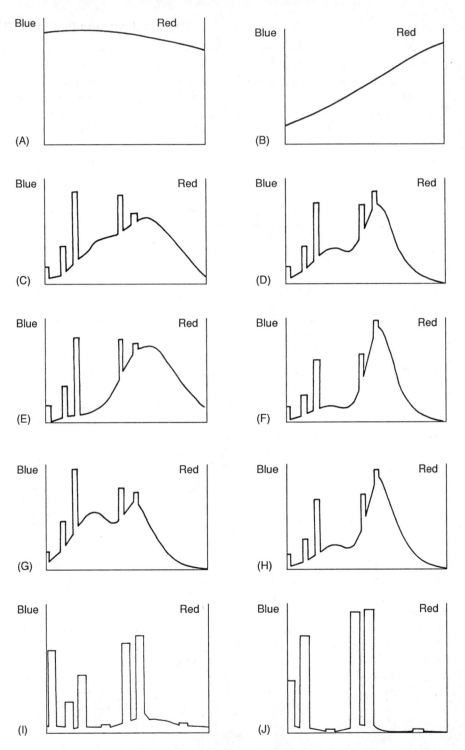

Figure 10.12 Spectra of a variety of light sources.
(A) Northern skylight; (B) incandescent lamp; (C) deluxe cool white; (D) cool white (E) deluxe warm white; (F) warm white; (G) daylight fluorescent; (H) white (I) mercury lamp; (J) mercury lamp. (Mercury lamps look white but have poor colour rendering and are deficient in long (red) wavelengths.) (Courtesy Osram SA, Pty Ltd.)

light source has good colour rendering ability. Unfortunately, there is an approximate negative relationship between colour rendering and luminous efficiency, in other words, good colour rendering costs more.

The apparent colour of an object depends very strongly on the properties of the incident light. Woodson (1981) summarises some of the effects of using 'coloured' light. Under red light, for example, white appears pink, red colours are greatly strengthened and blue appears purple. Incandescent lamps have good colour rendering properties and strengthen the appearance of red, pink, orange and yellow objects, but bring out blue and green less well. Standard cool white fluorescent lamps bring out the latter colours but not red, pink and orange. Lamp colour rendering property is an important consideration where colour identification is a task requirement as in assembly and inspection of wiring looms, fruit inspection and textile and garment manufacture. Dental laboratories, florists and paint shops are examples of places where good colour rendering light is needed. In the retail industry, lighting is used to create displays that enhance the appearance of products such as clothing or food (meat is usually illuminated using lights that strengthen red).

The colour rendering properties of lights also influence the 'atmosphere' in a room. Although there is still debate about whether environmental colour has psychological effects, the ambient illumination can have social implications because of the way different light sources render the complexion of a person's face. Incandescent lights enhance skin tones and can facilitate social interaction through the creation of a 'warmer' atmosphere. Because of their lower luminous efficacy and higher heat production, incandescent lamps are not normally used for general lighting in offices and factories. However, they may be used to advantage in interview and meeting rooms where social interaction occurs and in areas such as lifts and toilets where mirrors are often fitted onto the walls. Table 10.5 summarises some colour rendering properties and applications for different commercially available lamps.

Table 10.5 Colour rendering and luminous efficacy of lamps

Lamp	CRI	Luminous efficacy	Applications/comments
Incandescent	90–99	Low	Home, hotels, rooms dedicated to social interaction
Tungsten–halogen	85–95	Low	Have many applications where good colour rendering is needed
Mercury fluorescent	50–60	Medium	General illumination. Long life. Weakens red
High-pressure sodium	25–40	Medium–high	Background/outdoor illumination
Low-pressure sodium	Very low	High	Street lighting. Almost no colour rendering
Fluorescent:			
Colour matching	85–100	Medium	Florists, paint shops, etc.
White	50–70	Medium	General purpose. Factories, shops, etc. Reds appear orange; violet/blue is dulled
Deluxe warm white[a]	70–85	Low–medium	Violet/blue weakened. Creates relaxed, informal atmosphere similar to incandescent

[a] Fluorescent lamps last 5 times or more as long as incandescent lamps.

Visual fatigue, eyestrain and near work

Near work is thought to cause visual fatigue and occular symptoms, including temporary shifts in the near point of accommodation and the resting position of vergence. Recession of the near point suggests a loss of refractive power and is sometimes referred to as *accommodation strain*. It has been reported in microscope operators (e.g. Soderberg *et al.*, 1983) and in VDU workers (as has the time to refocus between near and far points) and is exacerbated by organisational factors such as the rigidity of work routines and the duration of work periods (Gunnarson and Soderberg, 1983).

Owens and Wolf-Kelly (1987) compared two groups of students who read from hard copy or from a VDU for one hour. Measurements of eye function were made before and after the task and subjects rated any feelings of visual fatigue at the end of the task. Significant changes in accommodation and vergence were detected for both hard copy and VDU display modes. Dark focus shifted by an average of 0.6 diopters and dark vergence shifted in the convergent direction by an average of 11.4 cm. The change in dark focus was significantly associated with decreased distance acuity ('blurred vision') at the end of the procedure, whereas changes in dark vergence were associated with higher ratings of strain. The greatest changes were observed in subjects whose initial eye postures were most distant. Eyestrain seems to be a problem of temporarily altered tonic vergence, whereas blurred vision is due to a temporary shift in dark focus. The findings suggest that it is not the VDU itself that causes acute visual problems but the requirement for near work to be sustained.

Surveys of visual function and VDUs

Yeow and Taylor (1991) report that up to 30% of the US working population is thought to have an uncorrected visual defect and that many will be at risk of visual fatigue when challenged by VDU work for extended periods. They compared 282 new users of VDUs with 96 controls who carried out similar, non-VDU, tasks over a 2-year period. Eye tests were carried out every 3 months. For the VDU users, there was no evidence of VDU-induced myopia, nor was there any increased tendency to start wearing spectacles or to replace spectacles in use at the start of the experiment. Age-related reductions in the amplitude of accommodation are to be expected and reductions of 0.3 D (diopters)/year for people between 20 and 30 years, 0.25 D/year for 30- to 40-year-olds and 0.2 D/year for people between 40 and 60 years are the norm. For the control group, the reductions in accommodation were as expected: 0.32 and 0.2 D/year for people below and above 40 years of age. However, in the VDU group, the corresponding reductions were 0.55 D/year and 0.34 D/year. These findings suggest that younger VDU users lose more accommodation than older users, probably because they had more to lose at the beginning of the trial. The effect of the loss of accommodation manifested itself as a recession of the near point of accommodation. A reduction in the near point of convergence was reported, but this was related to age and not to VDU use. Collins *et al.* (1991) investigated the effects of task variables on the occurrence of visual, occular and systemic symptoms. One hundred and fifty office workers kept diaries for 5 days, recording their VDU and non-VDU activities four times a day. Data on work activity, symptoms, work pressure and interest were recorded. Although VDU work was associated with a higher prevalence of symptoms, the relationship was weak: task variables accounted for only 11% of the variance in ocular symptoms, 7% of the variance in visual symptoms and 21% of the variance in systemic symptoms. Work pressure tended to increase the prevalence of symptoms regardless of the task and job interest tended to decrease the prevalence. It seems that, for people whose daily work involves both VDU and non-VDU tasks, the association between symptoms and VDU use is weak and that psychosocial and job design factors have a large influence on outcomes.

It must be remembered that image quality has improved over the last 10–15 years and so the findings of older studies may no longer be valid. Grignolo *et al.* (1998) performed opthalmological examinations over approximately 3-year intervals on nearly 6000 VDU users. Changes in the refractive state of the eye were mainly age-related, although there was a slightly increased tendancy to long-sightedness in those who spent more than 80% of the week working at a VDU.

Prevention of visual and upper body fatigue

Some general guidelines are summarised in Table 10.6. It seems that visual fatigue and upper body discomfort may be related: excessive loading of the accommodation and vergence systems during close visual work for long periods increases the tension in the muscles of the upper body (Lie and Watten, 1987, 1994). This connection between the ocular muscles and those that control the position of the head is not surprising when it is remembered how the eyes and head move together to locate a stimulus in space. In fact, the two systems share common neural pathways.

Experiments have shown that the wearing of lenses to distort vision also increases the upper body EMG activity. Alternatively, providing subjects with appropriate spectacles has been shown to reduce upper body EMG. These findings all serve to demonstrate the close connection between visual workload, visual fatigue and pain in the upper body, including headaches, and support the practice of providing VDU users and others who engage in visually demanding work with regular eye tests and

Table 10.6 Guidelines for the visual design of VDU tasks

---

1. Workstation and work environment: Provide 'visual relief'
   Never position the VDU against walls or screens.
   Design workstations so that users can periodically change focal length and vergence by glancing out of a window or along the length of an open plan office.
   Provide pot plants or other visually complex objects as indoor visual relief. This facilitates automatic recovery from visual fatigue when operators are not looking at the screen.
   Align VDU screens in correct relation to light sources.
2. Terminal and screen
   Use 'reverse-video' to minimise screen glare.[a]
   Reduce remaining screen glare using filters.
3. Job design
   Design-in natural rest breaks.
   Increase task variety.
4. Operator selection and assessment
   Eye-test new operators for existing visual defects.
   Regular eye tests for frequent VDU users.[b]

---

[a] I.e. Make screen brighter than glare.
[b] See Stone (1986).

well-designed workstations. Users must be able to select their own preferred viewing distance when using a VDU. Jaschinski-Kruza (1991) has shown that healthy subjects, when free to set their own viewing distance, choose distances between 51 and 99 cm for characters 5 mm high. Since reading glasses are normally prescribed for a distance of 30 cm, users may need another set of glasses for VDU work.

It is usually recommended that dedicated VDU workers take periodic rest periods throughout the day (5 minutes per hour). Jobs can be designed to provide natural rest periods if some non-VDU work is included. Even short breaks (2–60 seconds) every 10 minutes may help prevent fatigue (Pheasant, 1991). Galinsky *et al.* (2000) evaluated the effects of supplementary rest breaks on eyestrain and upper body discomfort in VDU workers. Under the conventional schedule, workers received one 15-minute break in the morning and in the afternoon, plus lunch. This was supplemented with an extra 20 minutes of breaks and workers alternated between the schedules over 4-week periods. Eyestrain and upper limb discomfort increased over the day in both schedules, but the increase was lower when supplementry breaks were given.

Psychological aspects of indoor lighting

Sundstrom (1986) has reviewed the research on lighting and satisfaction. The findings of both laboratory and field studies indicate that people are satisfied with illumination levels of about 400 lux. Further increases in illuminance bring only modest increases in satisfaction. Glare is associated with dissatisfaction.

There is also evidence to suggest that people dislike uniform lighting. Light from windows varies throughout the day. Not only does the intensity change, but also the spectral composition – sunlight differs from blue skylight, which differs from light from an overcast sky. Reflections from objects and buildings also add variety. It is often suggested that uniform lighting leads to boredom whereas changes in light have a stimulating effect.

The contribution of daylight from windows to the illuminance on indoor surfaces is usually much less than it appears to be (although direct sunshine can cause severe glare on surfaces such as VDU screens, chromed surfaces, etc.). However, anecdotal evidence suggests that windowless workspaces are associated with dissatisfaction. Many indoor workers appear to believe that daylight is superior to artificial light or that working under artificial light is unhealthy. Windows are a valuable source of visual relief and seem to have psychological value. By conveying information about changes in the weather and day/night, they provide continuity and 'contact' with the outside world.

The dominant source of artificial light in offices is provided by fluorescent lamps. Conventional lamps flicker by virtue of their mode of operation and luminaires usually consist of two phase-shifted lamps to minimise the perception of flicker. However, there are large individual differences in sensitivity to light and to flicker and complaints about the quality of fluorescent lamps are not uncommon. Lindner and Kropf (1993) compared 3030 normal subjects with 200 subjects who had complained about the fluorescent lighting (FL) in their workplaces. Those with FL complaints tended to be female and had poorer vision, higher light sensitivity and higher flicker sensitivity than those who did not complain. Wilkins and Nimmo-Smith (1989) investigated the incidence of headaches among office workers in offices lit by fluorescent lights. With high-frequency lighting (using an electronic ballast to drive the lamps at 32 kHz) the incidences of headaches and eyestrain were halved. Headache incidence also decreased with the height of the building above the ground (higher offices benefit from increased daylight). Kuller and Lake (1998) compared subjective well-being, task performance and physiological arousal when subjects worked under fluorescent lighting systems powered conventionally or by high-frequency electronic ballasts. Under the conventional system, individuals with high flicker sensitivity exhibited attentuation of the alpha wave of the EEG, indicative of high arousal. These individuals were observed to perform faster and to make more errors than those with low flicker sensitivity. The findings were interpreted as support for the use of high-frequency ballasts to drive fluorescent lighting systems. The findings imply that even barely perceptible flicker is stressful.

Depression

Light is known to have biochemical and behavioural effects on humans in addition to 'driving' vision. Vitamin D is produced in the skin by the action of ultraviolet (UV) light. Vitamin D is required for normal calcium metabolism in humans. Research has shown that people living in temperate latitudes exhibit seasonal variation in intestinal calcium absorption, with the lowest levels occurring in winter. This was one of the reasons for the fortification of dairy products with vitamin D, which was introduced to reduce the incidence of vitamin D deficiency bone disease ('rickets') in children (Hughes and Meer, 1981).

People living in temperate latitudes are subject to wide variation in daily exposure to natural light over the course of the year. In animals, the level of physical activity and behaviours such as reproduction and migration are thought to be triggered by the daily duration of sunlight. Changes in human mood and behaviour may also exhibit seasonal variation. Depression is more common in winter than in summer.

Jacobsen *et al.* (1987) describe the characteristic features of SAD (seasonal affective disorder). Patients suffering from SAD typically exhibit decreased activity, anxiety, increased appetite and sleep, work and interpersonal difficulties. Autumn/winter usually marks the onset of signs and symptoms of SAD which persist until spring. In the USA, the incidence of SAD is greater in the northern than in the southern states.

SAD patients have recently been successfully treated using phototherapy, which consists of daily exposure to bright light (2500 lux) in the morning and evening to extend the 'day' by up to 6 hours. It should be remembered that even in winter in the temperate latitudes, illuminance levels in a building can be an order of magnitude lower than those outdoors. Owing to the short period of winter daylight, many indoor workers may spend all day indoors exposed to very low levels of illumination, only emerging from their workplaces at dusk. The daily exposure to light of indoor workers may therefore be extremely low. Jacobsen *et al.* suggest that more research on how ambient light levels affect mood and productivity would be worthwhile.

Effectiveness and cost-effectiveness

It amounts to common sense to say that light is essential for the completion of all tasks that depend on vision for their succesful performance. The main economic considerations concern the cost of the lighting in relation to the benefits in terms of increased output and better quality work. There is a 'U'-shaped cost–benefit function that describes the relation between lighting and the performance of tasks that demand vision. With no light, performance is impossible. Performance increases rapidly as more light is provided but reaches a point of diminishing returns where additional light brings no further benefit. Further increases in lighting then cause glare and performance deteriorates. The recommended illuminance levels in this chapter have been designed to give optimum levels – maximum performance with no waste.

In developed countries, the cost of lighting is low in relation to the value added by the illuminated tasks and so economic benefits can usually be expected when optimum lighting levels are provided. This is one of the reasons why electric lighting came to be used in offices and factories in the early twentieth century (Sundstrom, 1986). One of the main economic benefits of electric lighting was that it freed architects from the constraints of having to build workplaces that could be illuminated only by daylight. With electric lighting, a building can be any shape; this has led to rapid advances in the design of industrial buildings, the concentration of production and the potential for high productivity, even during the gloomy winter months.

Safety

Eye injuries can be very expensive, both in terms of the disablement of workers and loss of skills and in terms of workers' compensation payments. Patel and Morgan (1991) analysed all eye injury cases presenting at a Manchester, UK, hospital. Of 258 penetrating eye injuries, 26.7% were work-related. All the patients were men, mainly machinists, engineers, builders and mechanics, and 75.4% were under the age of 40 years. Hammering and chiselling were the commonest activities at the time the injury took place. It was estimated that 88% of the injuries were preventable if eye protecton had been worn (only one patient had been wearing safety glasses when injured). The reasons for not wearing eye protection are often that the devices are uncomfortable or reduce peripheral vision, or there may be problems with misting of the lenses. In an early study of an eye injury prevention programme, Suleck (1965; quoted from Beevis and Slade, 1970) reported that the safety programme in a US factory *cost less than the cost of one compensatable eye injury*. Following mandatory eye examinations, 35–40% of workers were found to need prescription safety glasses and the rest were issued with standard safety glasses.

Illumination levels

Studies were done by the Industrial Health (Fatigue) Research Board in the UK in the early twentieth century. Weston (1921; quoted from Sundstrom, 1986) studied the hourly output of linen weavers in relation to the intensity of daylight (the workplace had no artificial lighting). Not surprisingly, production decreased markedly with decreasing daylight. Similar results were found for typesetters. Note that both of these jobs are visually demanding. The principle of diminishing returns was first demonstrated by H. C. Weston in 1945 in a laboratory task that involved the use of Landolt rings (subjects had to inspect a series of cards with circles drawn on them; their task was to detect circles with small gaps in the circumference at the '12 o'clock' position). Performance improved as the lighting levels were increased then levelled off. Thus, for any task, an optimum level of illuminance can be specified that gives maximum performance.

It is well known that good lighting is more important for older than for younger workers because it increases the depth of field of the eye and corrects optical deficiencies such as astigmatism. In an early study (Hughes, 1976, quoted from Sundstrom, 1986), two groups of office workers, aged 19–27 years and 46–57 years, worked on clerical tasks for 18 sessions with illuminance increasing from 15 to 100 to 150 footcandles (1 footcandle = 10.76 lux). Performance on a visual target detection task increased with increased lighting and the largest increase was for older workers. Interestingly, older workers rated the lower lighting levels as much less satisfactory than the highest level, whereas the younger workers rated both levels almost equally. This suggests that older workers are far more likely to express dissatisfaction with poor lighting and/or are more aware that they need good lighting to work well.

An interesting study, carried out in a cigarette factory in Indonesia, is reported by Rasyid and Siswanto (1986). Thirty female wokers engged in cigarette rolling operations worked for 3-day periods under different levels of illuminance, ranging from 100 to 500 lux and increasing in 100 lux increments. All were experienced in the task and were able to achieve an output of 2500 pieces per day at the start of the experiment. Measures of output were taken hourly and the mean hourly output was found to increase linearly with increasing illuminance from 304 pieces/hour at 100 lux to 333 pieces/hour at 500 lux, a 9.5% increase in hourly output. Interestingly, the authors concluded that the improved lighting could not be recommended on economic grounds owing to the high cost of electricity in Indonesia. Another reason is likely to be the low level of added value intrinsic to the task and the profitability of this industry in Indonesia.

Robinson (1963; quoted from Beevis and Slade, 1970) reports a study of 12 leather workers who worked for 2 years at an illumination level of 376 lux followed by 2 years under 1076 lux. Mean production increased by 7.6% under the new illumination levels. The value of the increased production exceeded the cost spent on upgrading the illumination by a factor of 13.

Lines of sight and visual access

One of the key tasks during the ergonomic design of vehicles is to evaluate the driver's line of sight – both straight ahead, to the sides and behind, including the use of mirrors. There are computer-aided design tools (such as SAMMIE CAD) that enable this to be done before the first prototypes are built. Simpson (1988) describes a study of development machines carried out in the British coal mining industry. These machines are used to bore underground roadways by cutting rock and gathering it onto conveyors. The machines studied had two main ergonomic deficiencies in common – poor control layouts and a restricted field of view for the driver. The restricted field of view meant that the driver required an assistant, who took up a position that enabled him to see what the driver could not see and enabled him to direct the driver. Redesign of the driver's cab to improve the field of view was estimated to save £8 million as a result of reduced cycle time and by dispensing with the driver's assistant.

Teniswood (1987) describes a project that produced high returns when a bulldozer driver's field of view was improved. The problem was the bad posture and excess time spent by the driver of a bulldozer when reversing. When a new bulldozer was built, the cab was angled at an angle of 45 degrees to the body of the vehicle. Little

Table 10.7 Comparison of analgesic doses in patients in tree-view and wall-view groups[a]

| Analgesic strength | Number of doses of analgesic | | | | | |
| | Days 0–1 | | Days 2–5 | | Days 6–7 | |
	Wall	Tree	Wall	Tree	Wall	Tree
Strong	2.56	2.40	2.48	0.96	0.22	0.17
Moderate	4.00	5.00	3.65	1.74	0.35	0.17
Weak	0.23	0.30	2.57	5.39	0.96	1.09

[a] From Ulrich (1984).

additional cost was involved because the machine was built this way to begin with. The machine spent less unproductive time reversing because it could be reversed at much higher speed and more time was spent bulldozing. There may have been other benefits because it is known, from research and from fundamental grounds, that reversing is a high-risk posture for the back, particularly when driving over uneven ground.

Pyschosocial factors

Ulrich (1984) analysed hospital records concerning recovery from gall bladder surgery in a suburban hospital in Pennsylvania between 1972 and 1981. Ulrich identified the room in which patients had stayed postoperatively and was able to compare recovery of patients staying in a room with a view of a brown brick wall with patients staying in a room with a view of a small stand of deciduous trees. Patients were divided into matched pairs on the basis of age and other variables that might affect recovery (sex, tobacco use, obesity, etc.). Only patients treated between May and October were treated because the trees only have foliage during those months. Forty-six patients were divided into 23 pairs and comparisons were made. Statistically significant differences were found in length of hospitalisation (7.96 days in the tree group compared to 8.70 days in the wall group) and in the number of negative comments (e.g. 'needs more encouragement') made by the nurses about each patient (1.13 comments per patient in the tree group as opposed to 3.96 in the wall group). Table 10.7 compares the analgesic doses given per patient in both groups (weak analgesics include aspirins, strong analgesics include barbiturates).

In days 0–1 after surgery, no significant differences in analgesic dosage were found. This is not surprising since the patients would still be recovering from the surgical procedure and anaesthetic and would be unlikley to pay much attention to the view. On days 2–5, the tree group were administered more weak pain killers such as aspirin and the wall group more strong and moderate pain killers. There were no significant differences in the use of anti-anxiety medication. It would be unwise to overgeneralise these findings because they may be due to extraneous variables that Ulrich was unable to control. However, they do suggest that 'soft' features of the visual environment, including contact with the outside world and exposure to daylight variation have positive effects and that these effects can result in shorter hospital stays and easier patient management during recovery.

Research issues

One of the most controversial research issues at present is whether operating a VDU has detrimental effects on vision. When this issue first appeared in the late 1970s, it was dismissed on the basis of the then current knowledge. Not enough people had been using VDUs for long enough to enable epidemiological surveys to be carried out. The design of CRT and LCD displays has advanced since, then resulting in great improvements in image quality. The current view is that no permanent degradation in vision occurs as a result of occupational VDU viewing.

Somewhat surprisingly, little work seems to have been done on the optimum size and shape for the screen of a standard VDU. CRT technology clearly limits the range of sizes and shapes that can be cost-effectively manufactured, but other display technologies may not be so limiting. The introduction of flat screens will remove one of the biggest constraints on VDU workstation design – the footprint and viewing requirements of the CRT. Flat screens can be mounted on a wall or dividing panel or placed on a document holder. They will make possible new workstation designs and increase the flexibility available to users and facilities managers.

Summary

In order to design appropriate workspace lighting, the following variables need to be considered. The amount and type of light emitted by the light sources themselves determines the illuminance delivered to work and other surfaces. The furniture and materials, by virtue of their reflectances, determine the balance of luminances seen by workers and the amount of indirect lighting and glare. The visual demands of tasks should be analysed when evaluating lighting – particularly with respect to visual acuity and the demands on accommodation and adaptation of the eye and the avoidance of 'visual fatigue'.

An ergonomic approach to lighting must, however, go beyond the purely functional specification of lighting systems and consider the psychological and biological effects of light on people.

Essays and exercises

1. Measure illuminances on a horizontal plane in the following places:

    *   Outdoors, under a shady tree, on an overcast day
    *   Outdoors on a clear night (no moon)
    *   Outdoors on a clear night (full moon)
    *   Under urban street lights at night
    *   In a fast-food outlet at night and during the day
    *   In a modern office at noon with the blinds open and the lights (a) on, (b) off
    *   On workbenches in a workshop or factory

    How do the measured values compare with your subjective impressions of light/dark?

2. Carry out a lighting survey in an office. Obtain or draw a plan map of the office showing the location of desks, luminaires and windows. Measure desktop illuminances (at least two measurements per desk) and write them onto the map.

Compare your measurements with published guidelines. At each workstation, take note of any glare sources. Ask employees whether they have any comments about the present lighting arrangements.

Combine this information into a report and make suggestions for improvement.

3. Describe the structure and function of the eye. How can this information be used to analyse practical visual problems in the workplace?
4. 'Lighting should be right, not bright.' Discuss.

11 Hearing, sound, noise and vibration

About 450 million people – 65% of the European population – are exposed to noise intensities above 55 dB, a level high enough to cause annoyance, aggressive behaviour and sleep disturbance.

(European Environment Agency Report, 1995)

Acoustic waves can be defined as pressure fluctuations in an elastic medium. Sound is the auditory sensation produced by these oscillations. In air, sound consists of oscillations about the ambient atmospheric pressure (Figure 11.1). Vibrating surfaces or turbulent fluid flow can act as the source, propagating alternately high- and low-pressure areas. The amplitude of the acoustic wave is expressed in Newtons per square metre or in pascals (1Pa = 1N/m²). The threshold of hearing (lowest amplitude of pressure oscillations in air detectable by the ear) is 0.00002 N/m² at a frequency of 1000 Hz.

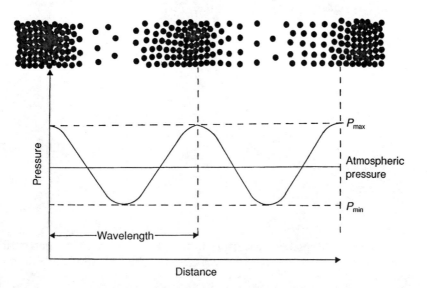

Figure 11.1 Sound is propagated in air as pressure oscillations above and below the ambient pressure.

Terminology

Frequency and *amplitude* define a pure tone. Their subjective counterparts are *pitch* and *loudness*. The perception of pitch also depends on the waveform and that of loudness depends on frequency. A healthy young person can hear sounds in the range 16–20 000 Hz. Noise is usually defined as a sound or sounds at such amplitude as to cause annoyance or interfere with communication. Sound can be measured objectively; noise is a subjective phenomenon.

The amplitude of sound is evaluated by measuring the sound pressure level (SPL). The range of SPLs to which the human ear is sensitive is so wide (0.00002 to 20 N/m²) that linear scaling would present a problem. For this reason a logarithmic scale – the decibel scale – is used. The decibel is a dimensionless unit related to the logarithm of the ratio of the measured sound pressure level to a reference level (usually taken to be the threshold of hearing). The smallest noticeable difference in intensity between two sounds is about 1 decibel. Commercial sound level meters measure and display an rms (root mean square) SPL, L_p, where:

$$L_p = 20 \log_{10}\left(\frac{p}{p_r}\right) \quad \text{dB}$$

where

L_p = sound pressure level in decibels
p = sound pressure in newtons per square metre
p_r = reference sound pressure level (0.00002 N/m²)

For example, if the rms SPL is 2 N/m², then

$$\log_{10}\left(\frac{p}{p_r}\right) = \log_{10}(2/0.00002)$$

$$= \log_{10} 100\ 000$$

$$= 5$$

and

$$L_p = 100 \text{ dB}$$

Table 11.1 depicts the range of human hearing in decibels.

The ear

The ear converts sound waves in air into nerve impulses, which travel along the auditory nerve to the brain. Further processing in the brain results in the perception of sound and the recognition of auditory patterns. A simplified diagram of the ear is presented in Figure 11.2.

Hearing requires three main steps. In each, energy is converted from one form to another:

Table 11.1 Range of human hearing

Source	SPL (dB)	
Jet engine at 30 metres	140	Ear damage
Pneumatic chipper	130	Onset of pain
Pneumatic rock drill	120	
Punch press	115	
Hydraulic rock drill	115	
Textile loom	105	
Powered lawn mower	100	
Heavy traffic		
Newspaper press	95	
Milling machine	90	
Diesel lorry	85	
Very noisy street corner	80	
Crowded room	75	
Vacuum cleaner		
Passenger car (at 15 m)	70	
Conversation	60	
Air conditioning system Sound inside car at 50 km/h		
Private office	50	
Quiet room	40	
Library	35	
Whisper 1 m from ear	20	

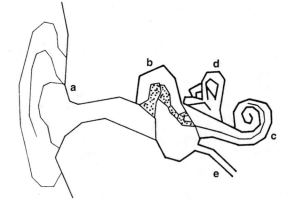

Figure 11.2 The ear consists of the outer ear (a), the middle ear (b) and the inner ear consisting of the cochlea (c) and the semicircular canals (d). The Eustachian tube (e) is also shown.

1. Pressure variations due to longitudinal waves in the environment are guided into the external auditory meatus where they cause a mechanical vibration of the tympanic membrane (eardrum). The tympanic membrane is connected to three small bones (ossicles) in the middle ear which, in turn, are caused to vibrate
2. Mechanical vibration of the auditory ossicles is converted to wave motion in the cochlear fluid at the oval window.
3. Wave motion in the cochlear fluid is converted to nerve impulses in the auditory nerve by hair cells in the cochlea.

Anatomically, the ear is divided into three parts at which these energy transformations take place.

The outer ear

The outer ear consists of the ear flap (pinna), the external auditory canal and the ear drum. The pinna is like an ear trumpet that funnels sound into the canal. The pinna is relatively small and unimportant; when one is attempting to listen to a faint sound, it is best supplemented with a cupped hand. Sound waves travel to the eardrum. The sound causes the membrane to vibrate and these vibrations are transmitted to the middle ear.

The middle ear

The middle ear is an air-filled, bony chamber connected to the outside atmosphere by the Eustachian tube, which terminates at the back of the throat. Discrepancies in air pressure on either side of the eardrum are potentially harmful. Swallowing momentarily connects the middle ear to the atmosphere enabling pressure differences to be equalised. The 'popping' sensation of the ears, experienced when changing altitude is caused by this sudden equalisation of pressure.

Vibrations of the eardrum are transmitted to the inner ear via three connected bones, which comprise the ossicular system. The malleus is connected to the centre of the tympanic membrane at one end and to the incus at the other end. Movements of the eardrum cause movements of the malleus that are directly transferred to the incus. The opposite end of the incus is connected to the third auditory ossicle, the stapes. The far end of the stapes is known as the 'faceplate'. It lies against the opening of the inner ear, the oval window.

The auditory ossicles are supported by ligaments. The tensor tympani muscle keeps the tympanic membrane under tension, like the diaphragm of a microphone. Movements of the tympanic membrane are transmitted from the auditory ossicles to the oval window of the inner ear. The ossicles play an important role in increasing the efficiency with which sound is transmitted to the sensitive hair cells of the inner ear. Since the inner ear is filled with fluid, sound energy has to be transmitted from air to fluid. The transmission loss from air to water is about 99% (which is why fishermen can listen to the radio while fishing, without frightening the fish away).

The function of the auditory ossicles is to transmit sound energy from air to water. They act as a lever system in which displacements of the malleus at the tympanic membrane cause corresponding, but reduced, displacements of the faceplate of the stapes. This reduction in amplitude is accompanied by a corresponding increase in force of approximately 1.3. The surface area at the faceplate is approximately 17 times smaller than that at the tympanic membrane. The pressure of the air against the tympanic membrane is increased approximately 22-fold at the stapes against the oval window. Without the auditory ossicles, the threshold of hearing would rise by about 30 dB (Guyton, 1991).

Volume control is achieved by the tensor tympani and stapedius muscles. Intense sound (particularly at frequencies less than 1000 Hz) causes these muscles to contract to pull the malleus and stapes away from the interfaces of the outer and inner ear. This attenuates loud noise and protects the inner ear from damage. Differences in the

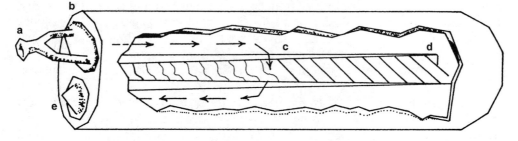

Figure 11.3 Simplified view of the cochlea 'unwound' to reveal its tubular structure. **a** = the stapes; **b** = the oval window; **c** = the basilar membrane; **d** = the helicotrema; **e** = the round window. (From Guyton, 1991, WB Saunders Company.)

effectiveness of this *attenuation reflex* may help to explain why some individuals are more susceptible to noise-induced hearing loss than others.

The inner ear

The inner ear is known as the cochlea. It can be modelled as a fluid-filled tube separated into an upper and lower half by a membrane (Figure 11.3). At the end of the tube are two windows, the oval and round windows on either side of the membrane. The structure of the cochlea is apparent if we imagine a tube folded back on itself and coiled like a shell. The central membrane is known as the basilar membrane and it ascends along the length of the coiled tube from its base to its apex where it is perforated by a hole (the helicotrema). The tube is tapered, being widest at the helicotrema end. Since the cochlea is sealed, inward movement of the stapes pushes the oval window inwards and causes a column of fluid to move inwards into the cochlea. At the other end of the cochlea, the round window moves outwards. Fibres in the basilar membrane run along its length, projecting outwards from the centre of the cochlea. At the base of the cochlea, close to the oval and round windows, these fibres are short and thick. Towards the apex, they become longer and thinner. Sound vibrations at a particular frequency are transmitted by the stapes and cause motion in the fluid column in the cochlea. This sets up a travelling wave in the cochlea. At a certain point along the cochlea the basilar membrane is just the right thickness to resonate at the particular frequency of the cochlear wave. At this point along the length of the cochlea, maximum vibratory amplitude will occur.

The basilar membrane is connected to an organ on its surface (the organ of Corti). This organ contains hair cells connected to afferent fibres of the auditory nerve. Movement of the hair cells causes depolarisation of nerve endings and the transmission of a nerve impulse to the brain.

When the stapes transmits sound vibrations to the cochlea, they pass along its length until a point is reached where the frequency of the vibrations matches the resonant frequency of the basilar membrane. At this point the movement of the basilar fibres 'short circuits' the incoming sound wave, which is converted to motion of the fibres. This causes the hair cells at that point to depolarise. Since the length and thickness of the basilar fibres varies along the length of the cochlea, different frequencies are detected at different parts of the cochlea. High frequencies cause resonance at the base of the cochlea and lower frequencies are detected higher up. This theory, known as the place-pitch theory, explains how the cochlea separates a complex sound into its spectral components. In this sense, the basilar membrane and fibres are not unlike a harp lying along the length of the cochlea. Detection of sound energy at a particular frequency depends on there being a 'string' available that will resonate at that frequency.

Discrimination between sounds of different frequencies takes place in the inner ear rather than in the brain. Supporting evidence for this comes from the observation that the cochlea is the only part of the human body that is the same size in an infant as in an adult. If the infant cochlea were a scaled-down version of the adult cochlea, infants would be deaf to low frequencies and sensitive to very high frequencies above the adult range and would probably experience difficulties in learning to speak (because they would not be able to hear their parents' voices properly). Although children are sensitive to higher frequencies than adults, the frequency range to which their ears are most sensitive is the same as in the adult because the cochlea is the same size.

Sensitivity of the ear

Auditory sensitivity is greatest between 1000 and 4000 Hz, the frequency band in which speech is transmitted (n.b. the notes on a piano range from 26 to 4096 Hz). Figure 11.4 shows the sensitivity of the ear to different audible frequencies.

The loudness of a noise depends on its frequency as well as its sound pressure level. Contours of equal loudness have been worked out using large numbers of subjects. A reference tone of 1000 Hz and, say, 60 dB can be presented to subjects. This is followed by a second tone at a higher or lower frequency. The subject then adjusts the intensity of the second tone until its loudness matches that of the reference. In this way, contours of equal loudness can be plotted for all frequencies with respect to a series

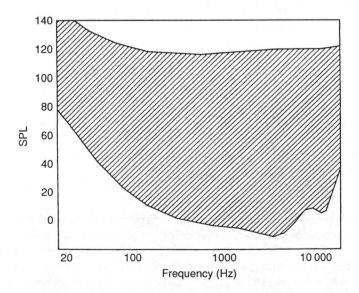

Figure 11.4 Sensitivity of the human ear. Sounds below the hatched area are inaudible; sounds above cause damage to the ear.

Table 11.2 Median threshold shifts for men at 4 kHz, due to age[a]

Age (years)	Threshold shift (dB)
35	5
45	12
50	22
65	35

[a] From Lutman (2000).

of 1000 Hz reference tones from 10 to 120 dB. For example, a tone of 100 Hz of intensity approximately 70 dB would sound as loud as a 60 dB, 1000 Hz reference tone.

The unit of loudness is the *phon* and is derived from the equal-loudness contour data. In the above example, the 70 dB, 100 Hz tone would have a loudness of 60 phons. A second loudness measure is the sone. One sone is defined as the loudness of a 1000 Hz tone at 40 dB. Because loudness perception is a logarithmic rather than a linear process, it can be said that a 10 dB increase in intensity represents a doubling of loudness. Thus a 50 dB, 1000 Hz tone would have a loudness of 2 sones.

Old age is marked by compression of the audible frequency range from 16–20 000 Hz to 50–8000 Hz, a condition known as presbycusis. The higher frequencies are usually lost first and most aged people cannot hear sounds above 10 000 Hz, which is why it is better to lower the voice when attempting to coomunicate with older people. Raising the voice increases the pitch and is self-defeating.

The mechanism of hearing loss is at the neural/cochlear rather than the ossicular level and is irreversible. Table 11.2 gives median age-related threshold shifts for men.

Noise-induced pathology of the ear

Sudden noises of sufficient intensity may produce trauma resulting in a reduction in the efficiency of sound transmission to the cochlea or in cochlear response itself. The tympanic membrane may be ruptured or the ossicular chain mechanically damaged. Surgical repair may be necessary.

Sensory neural hearing loss can also occur, particularly after chronic exposure to loud noises. The largest threshold shifts occur around 4 kHz (Lutman 2000), where the vibrational characteristics of the basilar mebrane are greatest. The loss is irreversible. In audiometry, such loss is described as a 'permanent threshold shift'. ISO 1999 gives formulae that enable one to predict hearing loss in men and women for different intensities and durations of exposure to noise.

Table 11.3 gives the median threshold shifts associated with different noise exposures. As can be seen, the loss at 80 dB is negligible but the risk rises rapidly around 85–90 dB. Exposure to levels as high as 95 dB pose a serious threat to hearing.

Audiometric testing determines the minimum intensity (the threshold) at which a person can detect sound at a particular frequency. As sensitivity to particular frequencies is lost owing to age or damage, the threshold increases. It is in this sense that hearing loss can be described as a threshold shift. Temporary threshold shifts can occur after exposure to loud noise (e.g. in clubs). Repeated exposure leads to permanent threshold shifts (noise induced deafness). Figure 11.5 shows example audiograms of noise-induced threshold shifts.

Table 11.3 Median threshold shifts at 4 kHz as a function of noise level

Noise level	Threshold shift (dB)
80	2
85	7
90	15
95	27

[a] From Lutman (2000).

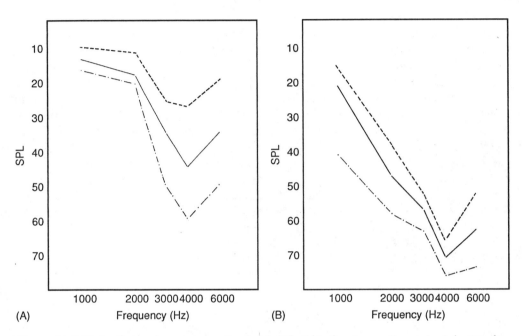

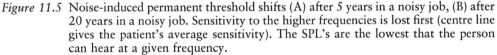

Figure 11.5 Noise-induced permanent threshold shifts (A) after 5 years in a noisy job, (B) after 20 years in a noisy job. Sensitivity to the higher frequencies is lost first (centre line gives the patient's average sensitivity). The SPL's are the lowest that the person can hear at a given frequency.

Although permanent threshold shifts occur with age, chronic exposure to noise, particularly over 90 dB(A) (the dB(A) scale is described in the next section), hastens their onset and severity (some airborne chemicals such as toluene have a toxic effect on the auditory system and can cause hearing loss, but this is beyond the present scope).

Tinnitis

Tinnitis ('ringing in the ears') is a common auditory effect of exposure to loud noise, both in the short and the long term. Phoon *et al.* (1993) found that the prevalence of tinnitus was 23% in a sample of 647 noise-exposed workers of average age 39 years (working in industries such as shipbuilding, woodworking, construction, etc.). After controlling for confounding factors such as tympanic membrane abnormalities and ear infections, such as otitis media, no significant associations were found between tinnitus and sex, age, race and exposure duration. However, tinnitus sufferers had significantly higher hearing thresholds at both high and low frequencies and 30% of them claimed that it interfered with activities such as telephone conversation and sleep. That tinnitus is common in noise-exposed workers and that it interferes with activities of daily living suggests that information about tinnitus, and its avoidance, should be included in hearing conservation programmes.

Psychosocial aspects of noice-induced hearing loss

Noise-induced hearing loss can have serious implications for quality of working life and employability. It causes social isolation, it can debar individuals from jobs where teamwork is required and it can damage promotion prospects by lessening the ability to manage others. Jobs such as sonar operation have precise hearing standards. Harris *et al.* (1979) found that high-frequency hearing was essential for passive sonar listening. They proposed a hearing threshold level of 35 dB in the frequency range 3–8 kHz. Many people with mild hearing loss would be excluded from this occupation.

Measurement of sound

Sound level meters provide several different measures of sound intensity. The dB (linear) scale is used to give the Sound Pressure Level (SPL). This is the root mean square sound pressure at the microphone. When sound is measured using the dB(A) scale, a weighting network selectively amplifies or attenuates different frequencies, similar to adjusting the tone control on a modern hi-fi amplifier. The sum of the pressure levels of the various frequencies is called the Sound Level (LS).

Many devices (e.g. electric fans or motors) produce intense sound energy at a particular frequency (centred around the 60 Hz mains frequency or harmonics of this). Although the sound pressure levels might be high, the weighted sound level is 20 or more decibels lower since the ear is not very sensitive to sound below 500 Hz. Apart from the 'A' scale, some sound level meters have 'C' and 'D' scales. These specialised scales are used to evaluate noise containing intense pure tone components (e.g. noise from jet engines).

Frequency analysis

Frequency analysis is sometimes carried out to obtain further information about the constituent frequencies of a sound source. Octave analysis involves dividing the frequency spectrum into bandwidths such that the lower boundary (f_1) is one-half of the upper (f_2),

$$f_2 = 2f_1$$

The centre frequencies of the octave bandwidths are 31.5, 63, 125, 250 . . . Hz

Further resolution can be obtained by using **half-** or **third-octave bandwidths**, where

$$f_2 = \sqrt{2}f_1 \quad \text{(half octave)}$$
$$f_2 = 3\sqrt{2}f_1 \quad \text{(third octave)}$$

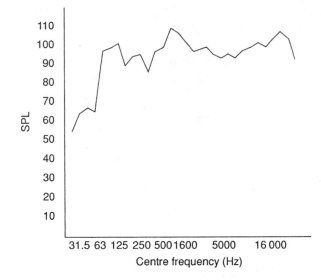

Figure 11.6 Third-octave analysis of a pneumatic hammer. Peaks at around 50 and 100 Hz correspond to mechanical impact of the tool. Most of the higher-frequency noise is due to the exhaustion of compressed air to the atmosphere.

The centre frequency of a bandwidth is given by the geometric mean of the upper and lower frequencies.

Octave analysis and its variations are logarithmic scales and are used to take account of the ear's approximate logarithmic frequency selectivity. They have disadvantages when the goal is to analyse the sound source because resolution drops as frequency increases (i.e. the bandwidths get wider and all sounds within a bandwidth are added together). Narrow-band constant-bandwidth analyses may then be used to determine intense sound at a particular frequency. Fortunately, the moving parts of many industrial machines such as pistons, fans or gears are often the primary sources of noise. Since many of these rotate or cycle at known frequencies below 1000 Hz, their contribution to the total noise can be determined fairly easily. Figure 11.6 shows a third-octave analysis of a pneumatic hammer.

Several sound sources

In practice, there are usually several sources of sound and it is the combined effects of these on the worker that is of interest. The combined SPL from several sound sources can be calculated from first principles, although it is usually estimated by calculating the difference between the two loudest SPLs and adding the amount shown in Table 11.4 to the louder of the two.

If there are more than two sources of noise, the exercise can be continued by taking the combined SPL of the two loudest sounds and comparing this with the third loudest sound. The difference between these is then added onto the estimate of the two loudest sounds to obtain a new estimate of the overall noise. This process can be repeated for other noise sources until the difference between the combined estimate and any remaining sources is 10 dB, at which point the addition of more noise sources will, in practice, have negligible impact on the overall noise.

Table 11.4 Summation of SPLs L_1 and L_2

Difference between L_1 and L_2 (dB)	Amount to increase louder of L_1 and L_2 (dB)
0.0	3.0
0.5	2.8
1.0	2.5
1.5	2.3
2.0	2.1
2.5	1.9
3.0	1.8
3.5	1.6
4.0	1.5
4.5	1.3
5.0	1.2
5.5	1.1
6.0	1.0
6.5	0.9
7.0	0.8
7.5	0.7
8.0	0.6
8.5	0.5
9.0	0.4
9.5	0.4
10.0	0.3

That the addition of two equal sources increases the total sound level by 3 decibels reflects the logarithmic nature of the decibel scale. It also applies to the subtraction of noise levels. If three adjacent machines have noise levels of 90, 95 and 101 dB, the combined noise will be 102 dB. Removal of the 90 dB machine will still leave a noise level of 102 dB whereas removal of the 101 dB source leaves a noise level of 96.5 dB. This has many practical implications for noise control, as will be seen below. In particular, it emphasises that the best approach to noise control in a room with several noise sources is to begin with the noisiest source, even if it is not the most amenable to acoustic treatment. Another practical implication is that surveys of the noise produced by machines can be carried out in the presence of background noise if the latter is 10 dB or more below the machine noise.

Measuring noise exposure

A number of measures have been developed to enable employees' daily noise exposure to be measured. The equivalent A-weighted noise level or 'LA_{eq}' has been developed for those situations where noise levels fluctuate over the course of a day. It is an integrated value, the average level of sound energy over the measuring period.

The LA_{eq} is defined as the steady-state sound level that would have the equivalent sound energy as the actual noise over the same period of time.

The LA_{eq} can be thought of as 'smoothing out' the peaks and the troughs of the daily noise exposure to give an average. If a worker's exposure times to different noise levels over the course of an 8-hour shift are known, the LAeq can be calculated as follows:

$$LA_{eq} = 10\log_{10}\left[\frac{T_1}{8}\,\text{antilog}\left(\frac{L_1}{10}\right) + \frac{T_2}{8}\,\text{antilog}\left(\frac{L_2}{10}\right) + \ldots + \frac{T_n}{8}\,\text{antilog}\left(\frac{L_n}{10}\right)\right]$$

where $T_1 \ldots$ to T_n are the exposure times (in hours) at sound levels $L_1 \ldots$ to L_n. Note that the time, T, spent at a given sound level is expressed as a proportion of the total 8-hour shift.

In practice, employees wear a small recorder or sound level meter with the LA_{eq} facility during a shift. It continuously measures the noise to which they are exposed. Because the LA_{eq} takes into account infrequent, but very loud, noises it can sometimes give a better indication of an employee's total noise exposure than a noise survey consisting of one or more short visits to measure the noise in an area during which loud, but infrequent, noises may not occur.

Safe exposure levels

In the USA, OSHA has specified 90 dB(A) as the maximum permissible exposure to continuous noise for an 8-hour shift. Many other countries, including those in the EU, also regard 90 dB(A) as the maximum permissible level. However, there is a modern trend worldwide to reduce daily exposures to below 90 dB(A). An exposure level of 85 dB(A) is regarded as the first action level at which workers must be informed and offered ear protection.

Noise dosimeters

A noise dosimeter integrates the noise measured at the microphone over a period of time and expresses it as a percentage of the daily allowable noise dose (e.g. as a percentage of 90 dB(A) for 8 hours). Dosimeters can be used to decide whether workers are being exposed to excessive noise and whether they require ear protection and also to specify for how long per day they may be exposed to a particular noisy work situation (Figure 11.7).

LA_{eq} and noise dose measurements are particularly useful in hazard surveillance of occupations where the worker is mobile because they integrate the different intensities of noise, which can vary considerably over the course of the working day. For example, garbage collectors, factory supervisors and agricultural workers are all exposed to varying noise levels during the course of their work, and an integrated value of their total noise exposure is often more meaningful than isolated measurements made during particular work activities.

Standards for noise exposure are a compromise between the desire to provide maximum protection of the hearing of workers and the cost-benefits of noise reduction. Many industrial processes are intrinsically noisy and, although practical steps can usually be taken to reduce the noise, a point of diminishing returns is reached (at around 90 dB(A)) where further reductions in noise become increasingly modest but cost much more. Also, individuals differ in their susceptibility to noise-induced hearing loss – it is often impossible to lower noise levels to ensure complete protection of all workers. A level of 90 dB(A) for 8 hours is OSHA's maximum permissible exposure level. An equal-energy rule is used to specify maximum exposure times for louder noises. For example, using a nominal halving rate of 5 dB, a noise dose of

Figure 11.7 A noise dosimeter. (Photograph kindly supplied by Messrs Bruel and Kjaer, Naerum, Denmark.)

90 dB(A) for 8 hours is held to be equivalent to 95 dB(A) for 4 hours or 100 dB(A) for 2 hours (a halving rate of 3 dB is used in some countries). Standards usually specify a maximum steady noise level (overriding limit) to which operators must not be exposed at all. The OSHA limit is 115 dB(A). It can be seen that even very brief daily exposures to very high noise levels can render an otherwise innocuous work area a noise hazard. Once again, it must be emphasised that very loud noises (over 115 dB(A)) can make an inordinately large contribution to a worker's daily noise exposure even if they are of short duration. Eliminating exposure to noise at this level is usually one of the simplest and most straightforward ways of reducing daily noise exposure.

Integrating sound level meters

To cope with the widely varying types of noise encountered in industry, commercial meters (Figure 11.8) usually have several response modes – normally 'slow' and 'fast' and sometimes 'impulse' response modes.

The slow mode is used to obtain an average reading when rapid fluctuations in intensity, measured on the fast mode, are greater than 3 or 4 dB. In fairly steady noise with fluctuations of only a few dB, the fast response may be used and a simple average can be taken by the person carrying out the measurements. For example, fluctuations from 85 to 91 dB could be interpreted as 88 ± 3 dB. If the fluctuations are greater than 6 dB, the average sound level can be estimated as 3 dB below the maximum value. Impulse noise cannot be measured accurately with basic meters because their response time is too slow. More sophisticated equipment is now available. As a rule of thumb, the impact noise level measured with a standard meter should be regarded as an underestimate of at least 5 dB.

Before the measurement of a sound source, the sound level meter should be calibrated. Meters are commonly sold with a portable calibrator (which itself should be

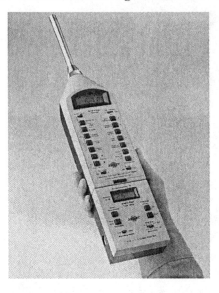

Figure 11.8 A modern sound level meter with octave filter set attached.
(Photograph supplied by Messrs Bruel and Kjaer, Naeram, Denmark.)

periodically checked). When one is measuring out of doors, it is advisable to attach a
spherical windscreen (made of foam and of about 5 cm radius) over the microphone
to prevent wind from distorting the measurements. When operator exposure to noise
is being measured, the microphone should be situated close to the operator's ear.
Octave, half-octave and third-octave sets are available that can be used in conjunction with standard meters.

   When one is carrying out investigations of the noise produced by machines in
offices and factories, supplementary information should be obtained to complement
the sound level measurements (Table 11.5).

Table 11.5 Supplementary information for noise measurement

---

 1.   A description of the space in which the measurements were made, its dimensions,
      background noise and the presence of other noise sources
 2.   A description of the source itself. For example, data from the nameplate such as the
      model number, operating speed and power, the percentage of maximum load when the
      measurement was made and, if different from this, the normal load
 3.   Calibration, weighting network and response mode of the sound level meter
 4.   Background noise level
 5.   Number and location of personnel in the area
 6.   Position of microphone with respect to the source
 7.   Extent of fluctuation of noise levels
 8.   'A' scale measurements at operator's ear level and at positions of other personnel
 9.   Time spent at machine by operator each day
10.   Results of any previous audiometric testing of workers
11.   Previous attempts at noise control
12.   Whether ear protection is available

Table 11.6 Some basic steps in the management of industrial noise exposure

1. Short-term measures	Issue ear plugs/ear muffs.
2. Medium-term measures	Reposition noisy machines.
	Soundproof noisy machines.
	Demarcate noisy areas with warning signs.
	Rotate workers between 'noisy' and 'quiet' jobs.
3. Long-term measures	Comprehensive noise reduction programme:
	Soundproof machines.
	Replace with less noisy machines.
	Change the process.
	Build 'acoustic refuges'.
	Conduct audiometric testing.
	Implement rules and procedures for the wearing of ear protection.
	Audiometric testing programme.

Noise surveys

Noise surveys are a useful method of evaluating the distribution of noise in a working area and the exposure of workers. Occupational hygienists use two methods – *area sampling* and *personal sampling*. In area sampling, the workplace is evaluated; this works well if the noise is fairly constant and workplaces are fixed. If people move around at work or if noise levels are closely tied to a worker's activity, personal sampling is necessary. In some industries, such as construction and mining, a combination of both approaches is best.

In area sampling, a plan map of the workplace is needed. Sound levels at various locations are recorded directly onto the map. Areas of high and low noise can be identified so that appropriate remedial action can be taken. Noise contours – regions of equal sound level – can be superimposed onto the floor plan. In conjunction with data about the positioning and movement of personnel, a number of steps can be taken to minimise excessive noise exposure (Table 11.6).

An example of a noise survey that was carried out in a colliery can be found in Bowler (1982). All areas where noise levels exceeded 90 dB(A) were identified and the duration of exposure of persons working therein was determined. Special attention was paid to areas with noise levels greater than 105 dB(A). Within the previously determined 90 dB(A) contour, attenuation of noise was measured by taking repeated measurements up to 1 metre from the source (intensity halved approximately as distance from the source doubled). Maximum and minimum noise levels were also measured at the operator's ear level at all fixed workstations. Together with data on exposure times, representative LA_{eq} values were calculated. Operators suspected of being exposed to noise over 105 dB(A) were issued with dosimeters so that more information on their exposure could be obtained. Noise exposure when workers were travelling to and from working areas was also measured. In this way, both high-risk areas and high-risk jobs were identified.

Ear protection

Ear muffs or ear plugs, singly or in combination, can protect the ear from excessive noise by providing an airtight (acoustic) seal between the atmosphere and the

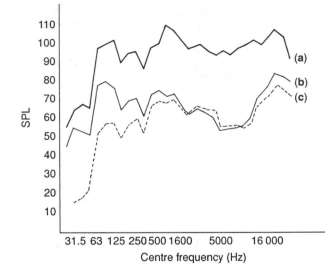

Figure 11.9 The effects of ear protection on the attenuation of different frequencies of sound from a pneumatic hammer. (a) third-octave analysis; (b) attenuation due to ear protectors; (c) A-weighting of attenuated noise.

Table 11.7 Attentuation characteristics of commercially availbale earmuffs

	Frequency (Hz)						
	250	500	1000	2000	4000	6000	8000
Attenuation (dB)	16	21	26	30	34	34	33

auditory canal. Reductions in SPL at the eardrum of up to 40 dB are possible. For very high noise levels (above 140 dB) simple ear protection is inadequate since excessive vibration can be conducted to the cochlea by bone or tissue. Figure 11.9 illustrates the effects of ear protection on the attenuation of different frequencies of sound from a pneumatic hammer. Table 11.7 gives sample data on the attenuation likely to be achieved using well-fitting, commercially available earmuffs (Melamed and Bruohis, 1996).

Even if ear protection is worn, sound may still find its way into the auditory system. The protector may vibrate or energy may pass through leaks – usually a problem with 'home made' ear plugs. Some characteristics of effective ear protectors are given in Table 11.8.

Table 11.8 Characteristics of effective ear protectors

1. Impervious to air
2. Adaptable to the shape of the user's head or external auditory canal to:
   • Form an airtight seal
   • Avoid pressure 'hot-spots' to ensure comfort
3. Remain firmly in place without causing pressure ischaemia

Sized ear plugs are reusable and are usually made of soft, flexible materials such as moulded plastic. They come in different sizes and should be cleaned with soap and water after use. Formable ear plugs are usually made of materials such as cotton or wax. The user moulds the plug into a cone shape to conform to the dimensions of the ear canal. Protection depends on the material used: cotton provides less protection than wax and certain polymers. Ear plugs have the advantages of being small and easy to carry, convenient to wear in confined spaces and cheaper than ear muffs. They are also compatible with long hair or protective headgear. The main disadvantages are that they take longer to fit than ear muffs and are not always suitable for regular use because of hygiene problems. Unclean hands may transfer dirt to the ear canal via the plug. Ear plugs cannot be worn by people with ear infections.

Ear muffs should provide a good seal between the inside of the cup and the skin surrounding the pinna. They should be adjustable for a range of head sizes. They usually provide better and more consistent protection than ear plugs and are perceived as more comfortable. Normally, a single size, adjustable muff can fit most head sizes. Ear muffs can also be incorporated into other protective head gear such as hard hats and are less likely to be forgotten. However, they can cause skin rashes in hot environments because sweat accumulates at the muff–skin interface. In confined spaces such as mine stopes, they can be dislodged too easily.

Manufacturers of ear protection sold in the USA are required to provide data on the noise attenuation provided by their products (known as the NRR, or noise reduction rating). These data are obtained under ideal testing conditions and less protection may be provided in practical settings. Casali and Park (1990) investigated ear protection attenuation performance under different fitting conditions and physical activities. When users fitted the ear protectors using only the manufacturers' instructions, protection was 4–14 dB less at 1000 Hz than when they were trained in correct fitting techniques. Muffs were less susceptible to fitting effects than were plugs. When subjects carried out tasks involving much upper torso movement and acceleration of the head, losses in protection up to 6 dB occurred. Pre-moulded plugs, muffs and muff/plug combinations were most susceptible to these losses and compliant foam earplugs were least susceptible. It seems that workers should be trained in correct fitting techniques if proper protection is to be obtained. This applies to all kinds of ear protection, particularly compliant foam plugs. For tasks requiring a great deal of movement, foam plugs may offer the best protection over the workday if workers are trained to fit them properly. If wearers cannot be trained or for occasional users, muffs may be the best choice.

Acceptance of ear protection is often a major problem in industry. Workers exposed to excessive noise frequently refuse to wear any ear protection, complaining that it causes discomfort, interferes with speech communication and, in jobs where the machine is the source of noise, degrades task feedback. While these complaints are sometimes justified, ear protectors do not necessarily interfere with speech communication because they lower the intensity of noise as well as of speech (the signal-to-noise ratio remains the same). The lowered sound levels may reduce mechanical distortion of structures in the ear and actually improve speech recognition.

Design of the acoustic environment

The alerting quality of noise makes it an ideal warning signal. By the same token, noise can distract workers and is a major source of dissatisfaction with the environment.

Table 11.9 Effects of noise below 85 dB(A)

Noise level dB(A)

---

80   Conversation difficult.
75   Telephone conversation difficult. Raised voice needed for face-to-face conversation.
70   Upper level for normal conversation. Telephone conversation difficult. Unsuitable for office work.
65   Upper acceptance level when people expect a noisy environment.
60   Acceptable level for daytime living conditions.
55   Upper acceptable level when people expect quiet.
50   Acceptable by people who expect quiet. About a quarter of people will experience difficulty in falling asleep or be woken.
40   Very acceptable for concentration. Few people will have sleep problems.
<30   Low-level intermittent sounds become disturbing.

Individuals differ widely in their attitudes to and tolerance of noise. Clearly, many factors need to be considered if an optimal auditory environment is to be designed. Woodson (1981) has compiled the findings of many studies of the effects of noise on human performance. Although noise levels below 90 dB(A) are not a serious threat to hearing, they can degrade task performance and cause annoyance. Table 11.9 summarises some of these effects.

Reverberation

The noise level in a room depends on the intensity of any noise sources and also on the *reverberation* characteristics of the room. The sound emanating from a source reaches the listener both directly and after being reflected off the walls and other objects (Figure 11.10).

Reverberation is the interreflection of sound waves inside a room. The more reflection, the longer the reverberation time. If there is continuous sound, background noise caused by interreflection builds up to a steady level, called the reverberant field. Reverberation is a very important acoustic characteristic that influences an occupant's perception of a room (Table 11.10).

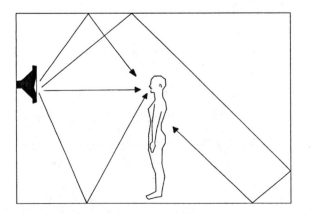

Figure 11.10 Reverberation. Sound in a room reaches the ear by direct and indirect routes.

Table 11.10 Some characteristics of reverberation in rooms

1. Reverberation time in a room does not depend on the position of the source or the position of the listener.
2. Excessive reverberation is a major cause of the blurring of acoustic signals such as speech.
3. Room shape has little effect on reverberation time.
4. The intensity of the reverberent field in a room depends on
   - The sound level of the source
   - The volume of the enclosed space
   - The amount of sound-absorbing material in a room

The reverberation time of a room is the time taken for the reverberant level to fall by 60 dB. It is a function of room volume and total sound absorption. Excessive reverberation causes problems because, as the distance of the listener from the target sound source increases, the intensity of direct sound decreases (by about 6 dB per doubling of distance). A point is reached where the intensity of the direct field is the same as the intensity of the reverberent field. This is known as the critical distance. The total SPL at this point will therefore be 3 dB greater than the direct sound and the reverberant sound alone.

Reverberation can be minimised by increasing the amount of sound-absorbing material in a room. The occupants themselves (or rather their clothes) are sound absorbers, so the reverberation characteristics of meeting halls and conference rooms will differ when they are empty compared to when they are full unless the seats are upholstered.

Reverberation can mask speech. Reverberation times of less than 1 second are recommended for rooms where speaking is a main activity. Longer reverberations times are needed for listening to music because reverberation gives 'depth'. Rooms suitable for listening to music may not be suitable for conferences and vice versa.

The auditory startle response

The startle reaction generated by unexpected auditory stimuli is common to all mammals (Valle-Sole *et al.*, 1995) and originates from the brainstem (Brown *et al.*, 1991). It is thought to be an adaptative response that results in the rapid adoption of a defensive stance with maximum postural stability. The eyes close, the neck flexes and the shoulders abduct with pronation of the forearms and clenching of the fists. The torso flexes forwards and the knees flex. Auditory stimuli are powerful producers of the startle response. The ergonomic relevance is that startle interferes with task performance and slows reaction time for other tasks. Startle is elicited by loud noises of about 100 dB with a rise time of 5 milliseconds (ms) (Matsumoto and Hallet, 1994). In the short term, operators will habituate to startling auditory stimuli after 4 or 5 exposures or over the course of a few minutes. Auditory stimuli with fast onset times have the advantage of being very attention-getting, but Sorkin (1987) cautions against onset rates of more than 10 dB/ms. *For auditory stimuli to be alerting, but not startling, onset rates of 1 dB/ms are recommended.* Sudden loud noises can have serious health consequences for sensitive individuals such as 'startle epileptics', in whom they may trigger a seizure. Posttraumatic stress sufferers may exhibit an exagerrated startle response that is easily triggered.

Industrial noise control

In the UK, the Noise at Work Regulations require that noisy workplaces be assessed and that steps be taken to reduce the risk if necessary. Similar regulations apply in the EU. If the noise level in the workplace is 85 dB(A), the 'first action level', the noise must be reduced as far as possible, those exposed must be informed and provided with ear protectors and noise hazard zones must be demarcated. If the noise exceeds 90 dB(A), the 'second action level', steps to protect the hearing of those exposed are mandatory and include provision of ear protection and reduction of exposure times. Employees who wish to undergo audiometric testing must be provided with it.

Several approaches to noise control can be identified:

1   Eliminate the threat to hearing by redesigning the machine or using a less noisy machine.
2   Remove personnel from the noisy environment.
3   Protect personnel by issuing ear plugs or muffs or build an acoustic refuge.

Many machines have both primary noise sources (such as the power unit) and secondary noise sources (parts of the machine that resonate). The relative contribution of the primary and secondary sources to the total noise depends on the particular machine. For example, pneumatic machines are inherently noisy owing to the exhaust gases. The energy loss, due to noise, of machines such as pneumatic rock drills is only about 0.08% (Holdo, 1958) because 87.5% of the noise is due to exhaust gases rather than the impact of the drill steel against the rock or the pistons and gears of the machine (Weber, 1970). Hydraulic pumps can be very quiet yet the hydraulic system itself can be a source of considerable noise due to resonance if the pump is not securely mounted and mechanically isolated from fluid lines and other structures.

Noise from the primary sources is often periodic and below 1000 Hz, whereas secondary noise is usually of a higher frequency. Velocity gradients in compressed air create random vortices, which move and are dissipated in a chaotic manner, giving rise to wide-band random noise (Beiers, 1966). Frequency analysis with the machine operating under representative loading is an important first step in machine redesign. The most intense frequencies can be identified and redesign can proceed in a systematic manner. Bartholomae and Kovac (1979) redesigned a pneumatic rotary-percussive rock drill. These machines emit noise of 115–120 dB(A) at the operator's ear level. By muffling the exhaust, redesigning the drilling mechanism using nylon instead of metal components, and acoustically treating the drill steel, noise was reduced to 95 dB(A).

Some common problem areas and control methods are described below.

Fans   Fans or blowers are much noisier when running at high rather than low speed. The sound level varies as the 5th power of the speed. Use a larger fan running at a lower speed.

Muffling   Pneumatic tools such as paving breakers, screwdrivers, and dentists' drills produce noise due to the exhaustion of compressed air to the atmosphere. This can be reduced by piping the air away from the operator to an unoccupied area or by

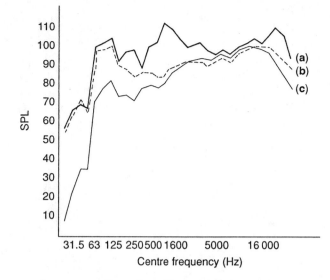

Figure 11.11 Effect of muffling a pneumatic machine. (a) third-octave analysis of un-muffled machine; (b) attenuation when exhaust is muffled; (c) A-weighting of attenuated noise.

using mufflers that reduce the escape velocity of the compressed air. Reductions of up to 6 dB are possible, but efficiency can drop by up to 15%. A third-octave analysis of a muffled and unmuffled pneumatic machine is presented in Figure 11.11.

Part ejection Pneumatic ejectors are sometimes used to remove parts from presses. Mechanical ejectors are usually quieter.

Pneumatic tools Hydraulic or electric equivalents are usually quieter.

Impact tools Large electric solenoids for tripping clutches can be sources of impact noise. Resilient bumpers fitted to the point of impact can be beneficial.

Hydraulic reticulation Cavitation (the release of dissolved air in a fluid due to sudden pressure drops) and turbulence can increase noise from pumps considerably. Cavitation can be reduced by decreasing fluid flow velocity, by 'smoothing' the system by removing sudden bends or junctions, and by avoiding sudden drops in pressure.

Vibration Machine vibration can be exacerbated by imbalance or eccentricity in rotating members, by inadequate mountings and by wear. Resilient couplings can be used to isolate a source from surrounding objects. Regular maintenance and replacement of worn parts can prevent vibration.

A change in process can also reduce noise. Riveting can be replaced with welding and dot matrix printers can be replaced with laser or ink jet printers. 'Dead' versus 'live' materials are preferable in the construction of flooring, pipes, etc. Rubber, wood and some types of plastic have high internal damping and do not transmit sound as readily as metals.

Since sound attenuates with distance, attention should be paid to the location of machines or components such as fans or pumps. Locating these at floor level rather than ear or waist height can significantly lower worker exposure to noise.

Noise insulation

If machine redesign is not possible or produces inadequate reductions in noise, an acoustic enclosure can be built around the source or an acoustic refuge built for the operator. If the source is enclosed, noise is interreflected by the walls of the enclosure, losing energy as it does so. The inside of the enclosure may be lined with sound-absorbing materials to increase attenuation as the noise interreflects off the enclosure walls. The interreflection increases the noise level in the enclosure by up to 10 dB without absorbent lining.

Sound-absorbing materials are usually porous and lightweight. As the sound waves travel back and forth within the tiny interstices of the material, their energy is converted to heat by friction. The sound-absorbing characteristics of a material are described by its absorption coefficient – the ratio of the energy absorbed to the energy striking the material. A perfect absorber has a coefficient of 1, whereas a perfect reflector has a coefficient of 0. The absorption coefficient of a material varies with frequency, so frequency analysis of the problem noise is needed if an optimal choice of material is to be made.

Sound energy does not pass efficiently across the interface of two materials differing greatly in elasticity – a principle used in the construction of acoustic refuges. Noise striking the walls of the refuge is reflected away. The best barrier materials have a high density and are non-porous. Rubber seals must be used on windows and door frames to prevent leakage of sound into the refuge.

Concrete, which has absorption coefficients of 0.01 to 0.02 at frequencies from 125 to 4000 Hz is a good reflector and therefore a good noise insulator. Fibreglass has a coefficient of 0.48 at 125 Hz rising to 0.99 at 1000 Hz and is a good noise absorber.

Partial enclosures can be effective provided that the operator's ear remains in the 'shadow' provided by the enclosure. Partial enclosures are suitable for protection against higher-frequency noise; the dimensions of the enclosure have to be several times the wavelength of the sound. Noise from daisy wheel and dot matrix printers can be reduced significantly by installing correctly designed hoods over the platen.

Screens, carpets, curtains and tiles

In offices, upholstered screens, carpets and accoustic ceiling tiles can be installed to absorb noise and block its transmission from one place to another. These materials are rated according to their noise reduction coefficient, or NRC. An NRC of zero indicates no ambient sound is absorbed by the material and an NRC of 1 means that all the sound is absorbed. Kleeman (1991) recommend a minimum NCR of 0.85 for upholsetered screens used in offices and that the absorption characterstics should be highest at the higher frequencies of noise found in offices (2000–4000 Hz) since these are the most disturbing.

Thick pile carpets with heavy underlay can have an NRC of up to 0.7 over the frequency range 125–4000 Hz. Carpets are also very effective at absorbing vibration.

Without underlay, the NRC can be halved. Pleated curtains ('drapes') can be placed over bare walls or over windows to reduce noise transmission and can be very effective because the pleating increases the surface area of material in contact with the air and hence the scope for absorption of sound and the interreflection of sound from curtain to curtain rather than curtain to air. NRCs of 0.55 to 0.60 have been reported for pleated curtains. It may also be less expensive to cover bare walls with curtains rather than with accoustic tiles.

Active noise control

Active control is only really effective at lower frequencies but has many applications because much machine noise is of a periodic/cyclic nature. Noise cancellation works well for non-random noise sources. An anti-phase version of the source, of equal amplitude, is produced. Fast processing of noise from one cycle can be used to predict the noise from the next cycle, which is propagated from a loudspeaker back towards the source. If the propagation of 'anti-noise' is correctly synchronised, noise cancellation takes place. Noise cancellation systems can be combined with ear-defenders to improve attenuation at low frequencies (Steeneken, 1998) and can also be incorporated into intercom systems. Active vibration cancellation is used, for example, in some helicopters to elimiate the transmission of vibration from the rotors to the cockpit.

Noise and communication

Most work depends on verbal communication and concentration. Noise can interfere with communication in a manner analogous to the way glare interferes with vision. Registration of target sounds will be disrupted by background noise and the disruption will be greater if the background noise shares similar frequencies to the target signal. Such disruption is known as *masking* and there is evidence that masking is greater when the signal and the noise frequencies are similar (Davies and Jones, 1982).

The auditory environment outdoors

Several factors influence the transmission of sound out of doors. Climatic factors such as temperature gradients, wind and relative humidity can attenuate and refract sound. Buildings can have a major effect by reflecting sound in a particular direction or by causing echoes (when the reflected sound reaches a listener more than 50 milliseconds after the direct sound). Such factors should be considered when, for example, positioning loudspeaker systems out of doors.

Fisher (1973) found that, owing to the increase in traffic, urban noise inside and outside buildings had increased in intensity to a point that is beyond the physiological and psychological level of adaptation of urban dwellers. Two common approaches to the reduction of traffic noise are the use of barriers or embankments to shield residential areas from busy roadways and the installation of windows with double-glazing in residential and other buildings. Measurement of the annoyance level of community noise is a complex area beyond the present scope. As well as objective measurements of ambient and peak noise, survey and questionnaire data are needed to evaluate people's attitudes and reactions to noise.

Effects of noise on task performance

It would seem obvious that noise can affect the performance of auditory tasks by masking important sounds, making the task more difficult to perform. How much more difficult the task is would depend on factors such as the intensity and nature of the noise and the importance of the information conveyed acoustically. In addition, the magnitude of decrements in task performance depends on non-auditory factors such as the difficulty of the task without noise and the motivation of the person doing the task. It has been hypothesised (Poulton, 1976a) that noise may improve task performance under some circumstances by increasing 'arousal'. Arousal is the central state of activation of the nervous system that underpins preparedness for action and is controlled by the reticular activating system (RAS) of the brain stem. Sleep is accompanied by decreased activity in the RAS. The RAS receives nerve impulses from the senses, which can cause an arousal reaction – nerve impulses from the RAS are sent to the higher centres of the brain including the cortex. Noise is known to be a particularly arousing stimulus. From a neuropsychological point of view, terms such as 'mental fatigue' and boredom really refer to a state of deactivation of the RAS, which leads to decrements in the performance of tasks or in the readiness to perform tasks. Thus, it seems possible that noise can prevent boredom or mental fatigue.

Poulton (1976b, 1977) has suggested that continuous noise can interfere with work by masking auditory feedback and inner speech. Human short-term memory typically utilises a verbal or articulatory method of coding information so it is easy to imagine how noise, particularly conversation or singing, might interfere with task performance by masking inner speech, with workers not being able to 'hear themselves think'.

Sundstrom (1986) has summarised the complex literature on noise and performance. Predictable noise, such as the background noise in a room, can reduce accuracy in clerical tasks, highly demanding motor tasks, and vigilance tasks if the noise is very loud and when two tasks are being carried out simultaneously. Unpredictable noise can also degrade performance of these tasks as well as tasks involving mental calculation or short-term recall.

There does not appear to be one simple effect of noise on performance. Haslegrave (1990) concluded her review by stating that effects are difficult to determine and are task-specific. Sundstrom concluded that laboratory findings on noise and performance have limited applicability to offices and factories.

Industrial music

In developing countries it is still common for people to sing while carrying out work. In developed countries, music is often played to workers in factories in the belief that it will improve task performance. There has been very little recent research on the effects of industrial music on performance. It does seem clear that factory-floor workers enjoy listening to music while they work, but there is no theoretical reason to think that this is related to improved output – it might equally be related to lower output. It seems likely that 'music while you work' would be appropriate for those engaged in non-verbal tasks at risk of falling asleep through boredom. In this case, a better approach would be to redesign the task or the operator's role to make it more meaningful. Improvements in the performance of tasks high in verbal mental activity

Table 11.11 Non-auditory effects of noise on health[a]

Nonspecific stress effects	On reticular activating system: activation of sympathetic nervous system, adrenal medulla, cerebral cortex.
Cardiovascular effects	Increased blood pressure.
Problems caused when communicating in noisy environments	Laryngopathies, laryngitis, vocal chord polyps. Annoyance, social isolation, impaired teamwork.
Factors increasing noise annoyance	Noise is perceived as uneccesary. Those causing the noise are perceived to be unconcerned about the welfare of those exposed to it. The noise is perceived to be harmful. The hearer has no control over the intensity of the noise. The noise is believed to be harmful to health. The noise is associated with fear.
Effects of noise on sleep	Noise may prolong sleep onset time. Noise may cause awakening once asleep. Noise may interfere with returning to sleep. Noise may cause sleep to be shallower.
Effects on mental health	Acute symptoms may be worsened. Noise annoyance appears to best predict adverse effects.

[a] After Smith (1991).

or requiring a rapid rate of communication would probably not be expected from the introduction of music. Non-verbal tasks requiring a high level of arousal, such as industrial inspection, would seem suitable for industrial music, though.

Industrial music may be contraindicated in already noisy factories if, in order to hear the music above the background noise, sound levels exceeding 85 dB(A) are necessary. There is a danger that industrial music, rather than noise from the industrial processes themselves, may present the greatest threat to the hearing of exposed employees.

Non-auditory effects of noise on health

There is some evidence that exposure to loud noise affects health in other ways, apart from damaging the ear. The quality of much of the research is poor – many of the studies lack detailed information about how the noise exposures were measured, how the health outcomes were diagnosed (if they were diagnosed and not just self-reported) and how confounding factors were controlled for. Table 11.11 summarises the main non-auditory health effects for which there is at least some evidence.

Noise and blood pressure

Talbott *et al.* (1999) investigated noise exposure and blood pressure in two groups of workers in factories with different noise levels. Cumulative noise exposure was significantly related to elevated diastolic blood pressure in workers in the noisier of the two plants (noise levels ≥ 89 dB(A)) but not in the quieter plant (noise levels ≤ 83 dB(A)).

Noise and stress

Evans and Johnsson (2000), carried out a laboratory study in which 40 clerical workers performed typing and clerical work in quiet (40 dB(A)) or noisy (55 dB(A), peaking to 65 dB(A)) offices. Noise had no effect on typing speed or error rates, but performance indicators suggested lowered motivation. Urinary adrenaline levels were elevated under noise but there was no effect on perceived stress. Workers made fewer ergonomic adjustments to their chairs, footrests, whiteboards and document holders in noisy conditions, suggesting that noisy offices may provoke postural fixity.

These findings suggest that negative health effects may occur below the threshold of awareness, even under the relatively low exposure levels found in open plan offices. An absence of complaints about noise, then, should not be taken to mean that all is well and *noise measurements should be regarded as mandatory in ergonomic surveys of offices, even when it is clear that there is no threat to hearing.*

In noisier environments, both subjective and objective changes occur, indicative of stress. Melamed and Bruohis (1996), measured urinary cortisol, subjective fatigue and post-work irritability in 35 textile mill workers under chronic exposure to noise over 85 dB(A). The workers did not normally wear ear protection but wore ear muffs for 7 days over the study period. Significant reductions in urinary cortisol at the end of the shift, post-work irritability and subjective fatigue were found when the workers wore the ear muffs. Increased cortisol levels can occur when workers are exposed to stress from many different sources, including high workload, so the findings of this study suggest that *noise is a general stressor* and the subjective reactions to it may be indicative of heightened physiological arousal. Although 51% of the workers in the study reported moderate to great discomfort when wearing the earmuffs, the inconvience was not enough to offset the physiological and psychological benefits of a reduction in noise exposure.

Noise and satisfaction

Whether sound is perceived as noise depends as much on the listener and the context as it does on the sound itself. Sound levels of 100 dB(A) are commonly found in factories and are rightly termed noise. Fisher (1973) measured sound levels of over 100 dB(A) in a busy hotel bar – whether this was perceived as 'noise' and by whom, is much more debatable.

Noise is often associated with dissatisfaction with the environment. Nemecek and Grandjean (1973) studied noise levels in 15 landscaped (open-plan) offices. Mean noise levels were low, from 48 to 53 dB(A) with peaks 8–9 dB(A) higher. However, 35% of employees complained of being severely disturbed by noise and 61% said their concentration was worse in the landscaped than in a conventional cellular office. Conversation seemed to be the most common cause of annoyance. Its content, rather than its decibel value appeared to be the biggest problem. Other sources of complaint in open-plan offices are sudden noises such as ringing telephones.

Office work usually involves conscious mental activities that fall prey to intrusion by noise. Several factors determine whether noise in offices will be a source of annoyance. Firstly, if the noise level itself is too high, telephone or speech communication will be rendered more difficult and satisfaction with the environment will be reduced. If the noise level is too low ('quiet enough to hear a pin drop') intermittent noises such as conversation, ringing telephones, loud traffic noise, etc. will be more likely to distract people. Sundstrom notes that the intensity of sudden noise above the background (the signal-to-noise ratio) is more important than the noise level itself in causing annoyance. White noise (noise in which amplitude is equal at all frequencies) is sometimes deliberately introduced into open-plan offices to reduce the intensity of noise with respect to the background (to mask conversation, telephones, etc.).

By a process known as habituation, people can block out regular noise (the experience of becoming aware of the presence of a clock only when it stops ticking indicates habituation). Inhibitory fibres exist in the neural pathway from ear to brain that inhibit specific parts of the organ of Corti (Guyton, 1981). Unexpected noise is more distracting than regular noise because it cannot be blocked by this mechanism. Attitudes towards a noise source play a major role in determining its disturbing effect. Approval of the source minimises disturbance (e.g. noise from one's own or subordinate's machine), whereas disapproval increases disturbance. The listener's control over the noise and the social distance between listener and noisemaker have a similar mediating effect on disturbance.

Vibration

A detailed treatment of vibration is beyond the present scope, although a few of the most important aspects in ergonomics can be briefly introduced. Vibration is defined as the oscillation of a body about a reference position and can be described, like noise, in terms of amplitude, frequency and phase (Cole, 1982). Of most interest in ergonomics are the effects of vibration on health, task performance and communication.

The unit of vibration is the root mean square (rms) or peak acceleration of the oscillation. Vibration dose values are time-weighted averages of vibration magnitude. Vibration is measured using accelerometers placed in the workplace or on seats where exposure measurement is required. Accelerometers are placed such that vibration can be measured in three translational axes (backwards and forwards, up and down, and side to side) and three rotational axes (pitch, yaw and roll). Vibration in the vertical plane is usually weighted most highly when the outputs of the different accelerometers are combined (see ISO 2631-1, for example).

Human responses to vibration

Table 11.12 summarises some of the main effects. See Griffin (1990) for further information.

Vibration and health

According to Griffin (1990), there is no disorder that is caused purely by exposure to vibration. Most people exposed to vibration are also exposed to other factors that may affect their health (e.g. bad posture or cold). The main disorders in which vibration is a causal factor are summarised in Table 11.13 (see Wikstrom *et al.*, 1994, for a detailed review).

Vibrations in the frequency range 4–8 Hz (the natural frequency of the trunk) are particularly hazardous. Design of seats to minimise the transmission of vibration at these frequencies can reduce the risk of back injury. The ISO (ISO 2631-1, 1985)

Table 11.12 Human response to vibration

Axis	Frequency (Hz)	Effect
Vertical	0.5	Motion sickness, nausea, sweating
	2	Whole body moves as one
		Difficulty positioning hands
	4	Vibration transmitted to head
		Lumbar vertebrae resonate
		Problems writing or drinking
	4–6	Resonance of gastrointestinal system
	5	Maximum discomfort
	10–20	Voice warbles
	15–60	Vision blurred (resonance of the eyeballs)
Horizontal	<1	Increased postural sway
	1–3	Upper body destabilised
	>10	Backrest is a prime cause of vibration transmission to body

Table 11.13 Health effects of vibration exposure

Disorder	Effect of Vibration
Low back problems	Microfracture of vertebral endplate
	Schnorl's nodes (following sudden intense shock)
	Increased disc pressure caused by higher back muscle activity
	Decreased disc height
	Increased radial disc bulge (after chronic exposure)
	Bad posture amplifies effects of vibration
Gastrointestinal problems	Increased secretion of gastric juices causing acute stomach ache
	Possible link with gastric ulcers
Urogenital problems	Slight evidence for a link with whole-body vibration in women
Cardiovascular problems	No evidence for an association
Hearing problems	Some evidence that whole-body vibration combined with noise amplifies hearing loss by 6 dB

specifies an acceleration limit of 0.32 m/s² beyond which exposure to vibration can have acute effects. Burdorf *et al.* (1993) investigated low back pain in crane drivers and concluded that elevated prevalences of low back pain were unlikely to be due to vibration of less than 0.20 m/s².

Raynaud's disease affects the blood vessels and nerves of the hands or feet. When associated with the use of vibrating tools it is called vibration white finger (VWF). The modern term for the disorder is hand–arm vibration syndrome. It is characterised by local ischaemia, pain or numbness and is exacerbated by working in cold conditions. It is caused by operating hand tools with low-amplitude vibration below 500 Hz. Paving breakers, powered sanders, grinders, chipping hammers and powered wrenches are examples of vibrating hand tools that can cause VWF. Nelson and Griffin (1989) surveyed 1200 dockyard workers exposed to vibration. The prevalence of WVF was 81% in caulkers and riveters, 33% in men working in combined trades, 16% in painters and 21% in boilermakers. The tools most likely to be associated with WVF were chipping tools, riveting hammers, grinders and pneumatic scaling tools.

Prevention of vibration white finger

If detected early, VWF is reversible. If not, it can cause permanent disability in the use of the hands. Control of employee exposure to vibration follows the same principles as for noise control. Regular surveillance of exposed employees is important for early detection of the onset of problems; exposure times can be lowered by job design; vibration from tools can be measured to identify problem tools that cause high vibration doses and these can be replaced. Changes in work design can be made to automate high-risk processes and protection can be provided in the form of rubber/foam rubber handles to damp vibration transmission at the hand–handle interface. Use of vibrating tools in the cold should be avoided. Vibration exposure can be reduced by issuing gloves, which also reduce cold exposure. Hansson *et al.* (1987) report that the level of vibration from different makes of hand tool can vary considerably and that rubber dampers placed between the tool and handle can reduce vibration by about 65%. They suggest that products should be tested and that this information should be made available to industry to assist them in making equipment purchasing decisions.

Vibration and public transport

Mention must also be made of how people respond subjectively to vibration of different frequencies and intensities. This is particularly important in the design of public transport systems if passenger comfort is to be optimised in relation to engineering constraints. Oborne has carried out several studies on the subjective response to vibration. Oborne and Boarer (1982), for example, investigated the effects of posture on subjective response to whole-body vibration. They found that vibration affected standing subjects significantly less than seated subjects but that sitting posture itself did not have a significant effect (no one sitting posture was better than the others).

A final consideration in the design of transport systems is motion sickness and its causes and cures. Individuals differ considerably in their susceptibility to motion sickness but it seems to be caused by low frequency (0.1–0.8 Hz) vibration in the vertical plane; not surprisingly, vibrations of this nature are commonly encountered at sea. Some general guidelines for the amelioration of ship motion effects are given in Table 11.14 (Bittner and Guignard, 1985).

Effectiveness and cost-effectiveness

Several studies demonstrate the cost-effectiveness of implementing sound environmental ergonomics practices with respect to noise.

Avoidance of retrofitting

Teel (1971) investigated the effectiveness of portable vapour detectors fitted with auditory alarms for use in US missile launch facilities. These alarms were needed to alert workers in the facilities of the leakage of toxic propellants from ballistic missiles. Concern had been expressed that the alarms might not be audible in the launch facilities, where ambient noise levels ranged from 73 to 89 dB. Teel carried out an

Table 11.14 Principles for minimising motion sickness

---

1. Locate critical workstations near effective centre of rotation
   - Minimise pitch and roll motions (<15 degrees).
2. Minimise head movements
   - Place tools within close reach (avoiding the need to stoop).
   - Provide head rests on seats.
   - Place controls and displays in area of focal attention (moving eye field).
3. Align operators with principal axis of vessel hull
   - Reduces complexity of vestibular inputs.
   - Minimises passive head movements.
4. Avoid combining provocative sources
   - Nauseogenic stimuli tend to summate.
   - Avoid bad odours, medication, disorientating displays.
5. Provide an external visual frame of reference
   - Provide a view of the outside world.

---

investigation in which the portable vapour detector was placed in its prescribed location in the missile silo and subjects stood in eight other locations. He recorded whether or not they had heard the alarm. The alarm was sounded during some, but not all, of three 2-minute test trials. All subjects heard the alarm every time it was sounded in all of the different silo locations. The findings were deemed to confirm that the portable vapour detectors with auditory alarms were adequate to protect the safety of personnel working in the missile silos. It was concluded that no additional steps were needed to upgrade the warning system. A total of 500 missile silos would have needed to be retrofitted had the safety of the portable detectors not been confirmed. Cost estimates for a re-engineered warning system ranged from $250 000 to over $1 million (around $1–5 million today) can be contrasted with the study cost of $1000 (around $5000 today).

Hearing conservation programmes do work

Feldstein (1993) carried out a longitudinal assessment of hearing loss in 11 435 automobile workers exposed to noise. Noise exposure in different areas of the plant was estimated from annual noise surveys. All employees were expected to wear hearing protection after commencement of the study. Audiometric tests were carried out at the start of the study and again after 1, 3 and 5 years. The hearing loss after 5 years was then calculated for noise-exposed employees in different areas of the plant and for a group of control subjects who did not work in noisy areas at all. After adjusting for each subject's hearing loss at the beginning of the study, for their age and for presbycusis, it was found that the average hearing loss over the period was less than 2 dB in most locations. This compared favourably with the adjusted hearing loss levels in the control group. In one location, significant hearing loss was discovered, a foundry where maximal time-weighted noise exposure levels of 114 dB(A) were observed. It seems, then, that hearing conservation programmes involving audiometric testing and the use of ear protection work in many noisy areas. In extremely noisy areas (over 110 dB(A)) audiometric testing and ear protection may not be enough.

Costs and benefits of hearing conservation programmes

Both the costs and benefits of such programmes vary between countries and depend on the way in which health care and worker's compensation are funded. However, there is an active debate about the cost-effectiveness of these programmes and several cost–benefit models have been put forward. Gibson and Norton (1981) in Australia put forward a simple model as follows:

$$\frac{\text{Benefits}}{\text{Costs}} = \frac{\text{Reduction in compensation} + \text{Recurrent compensation}}{\text{Investment in noise control} + \text{Hearing programmes}}$$

The benefits of implementing a noise control programme according to the model are a reduction in the number of claims and a reduction in the size of payments to workers suffering permanent threshold shifts as a result of exposure to noise at work. This includes a reduction in provision of hearing aids, for example. The costs are firstly the money spent to implement the noise control programme, which consists of acoustic engineering of workplaces and machines or the replacement of noisy machines with quieter ones. Secondly, there are costs associated with audiometric testing of employees and the provision of personal protective equipment.

According to Gibson and Norton's model, every 1.9 Australian dollars spent on prevention would save only 1 dollar in compensation costs. According to this model, almost half of the money spent on noise reduction to a maximum of 90 dB(A) is a welfare payment rather than an investment.

Hocking and Savage (1989) reviewed the limitations of the model, pointing out that it it does not take into account the fact that investments in noise reduction will take time to produce benefits since noise-induced hearing loss is an insidious process that can take place over many years. They propose a more comprehensive model that discounts costs over several years and takes into account the tax effects of paying for both the hearing programme and compensation payments.

The costs associated with hearing conservation programmes are due to

- Noise assesssments. This consists of the equipment and labour needed to carry out a survey either by in-company personnel or by external consultants.
- Personal protective equipment and its use. This includes the cost of ear muffs (and their replacement every 2 years) and ear plugs as well as a hearing conservation education programme (including instruction on the correct use of hearing protection).
- Job rotation. There may be costs associated with changing the work organisation so as to rotate workers from noisy to quiet areas (these may be offset by unexpected benefits such as reduced fatigue and fewer injuries).
- Retrofitting of noisy machines.
- Audiometric testing. The costs include the audiometrist's fees and lost production (since workers are tested during working hours.
- Administration of the noise control programme. This includes the salary of the person managing the programme plus sundry costs.

The benefits include

- Reduction in compensation claims (including the cost of hearing aids)
- Reduction in lost time
- Reduction in the costs of administering the compensation scheme
- Possible improvements in productivity (e.g. due to better communication, less distraction, ability to concentrate for longer periods of time).

Cost-effectiveness of different noise control strategies

Noise control strategies differ as do their costs and benefits. Scannel (1998) points out that, although ear plugs and ear muffs are cheap, they can also be very ineffective at reducing noise exposure. For example, if ear muffs with a noise reduction rating of 25 dB are removed for only $\frac{1}{2}$ hour a day, they provide only about 12 dB protection over a the whole day. In practice, even good ear protectors may be of little use if the noise levels are over 100 dB. If employees only wear protectors in the presence of supervisors, the protection will be negligible. Rotating employees between noisy and quiet areas only lowers daily noise exposure if the the noisier area is close to 90 dB(A) or if the time spent in it is extremely short: for example, assuming a doubling rate of 3 dB, 1 hour spent in 100 dB(A) is equivalent to 8 hours in 91 dB(A), which is still too much even if the other 7 hours are spent in total silence. Soundproof enclosures around machine *can* be effective, but even small gaps or holes lead to large reductions in sound insulation. A hole of 10% of the surface area of the enclosure reduces the maximum noise insulation from about 30 dB to 10 dB. Even a 1% hole reduces it to 20 dB (three times as much sound energy escapes from an enclosure when it has a hole in it which is 1% of its surface area!).

Noisy machines are often inefficient machines

Noise control can reduce noise levels and extend machine life, since the same processes that give rise to noise also give rise to wear. Narrow-band frequency analysis can be used to locate the part of the machine giving rise to the noise if the operating frequencies of the components are known. Scannel gives some examples of cost-effective interventions:

- The flywheel of a 10-tonne power press was found to generate bell-like tones, due to resonance. Dynamic vibration absorbers were attached to the flywheel to damp the resonant frequencies. The costs were $50 for materials and one day's pay for a fitter. The life of the machine was extended by 8 years and the noise levels dropped from about 96 dB to 86 dB.
- The aluminium castings being worked on a capstan lathe emitted 94 dB(A) at the operator's ear. The cutting tool was causing the product to vibrate, which not only caused noise but also affected the smoothness of the cut. Vibration damping straps were made at $5 each and fitted onto the castings before they were machined. The noise dropped to 78 dB(A), the tool life was increased and the machining cycle time was reduced.

Reduced noise improves productivity

A study by Broadbent and Little (1960) compared performance in accoustically treated and untreated rooms in a film processing plant. Performance was measured over four

6-week periods both before and after reducing the noise level in one of the rooms from 99 to 89 dB. Improvements in performance, possibly linked to improved morale, were observed as was a reduction in operator error (errors were reduced by a factor of 15).

Hartig (1962) carried out a study of the effects of noise on productivity in offices. In one office, the walls and ceiling were treated with sound-absorbing materials while the other was left untreated. The performance of the treated office was soon seen to be superior to that of the untreated office. Output increased by 9%, there were 29% fewer typing errors and 52% fewer errors involving calculating machines and there was 37% less sick leave and there were 47% fewer staff changes. Morris and Jones (1991) compared the performance of subjects transcribing visually presented text either in quiet conditions or when exposed to irrelevant speech, relayed through headphones. The number of omission errors was found to be greater when there was irrelevent speech and the effect was greatest when the text to be transcribed did not conform well to grammatical English. The authors suggested that the effect of irrelevant speech may be to interfere with the processing of the text that has to be stored momentarily in an auditory form in the subjects' memory while being described. This is in accordance with the long-held view that verbal tasks are particularly susceptible to disturbance by background noise, especially other people's conversations. For this reason, low-intensity white noise is usually broadcast in modern open-plan offices to enable the occupants to work without disturbing each other by masking speech.

Research issues

The technology for evaluating the auditory environment and for dealing with troublesome noise sources is well established. There is an international trend to lower industrial noise exposures beyond their present levels in order to further safeguard the hearing of the workforce and improve working conditions. Applied research to investigate cost-effective ways of achieving these reductions is appropriate. Although vibration exposure is linked to many health effects, there is a lack of evidence for dose–response relationships. This hinders the drafting of exposure limits for the prevention of specific effects.

Summary

Noise can be evaluated by measuring sound levels, noise doses and the LA_{eq}, and by carrying out frequency analyses. Unacceptably noisy machines should be quietened, replaced or isolated from occupied areas. If this is not possible, operator exposure should be reduced or ear protection provided. Further information on the physical aspects of noise and noise control can be found in the publications of the International Organization for Standardization and the American Industrial Hygiene Association and in the *Handbook of Noise and Vibration Control*.

An ergonomic approach to the auditory environment should go beyond measurement of sound levels and consider the satisfactoriness of the acoustic environment as a whole. Creative, intellectual work may suffer from noise more than routine work in which operators have a high level of skill. A knowledge of people's attitudes to noise is essential in the evaluation of physical measurements and in designing hearing conservation programmes.

Essays and exercises

1. Carry out a noise survey in a small factory. Obtain or draw a plan map of the factory floor indicating the locations of machines and other noise sources. Indicate all fixed work positions.
    Write the measured sound levels onto the map. Identify demarcated areas beyond which hearing protection should be worn. What other recommendations can you make?

2. The noise level in a bottling plant is 93 dB(A). Assuming an 8-hour exposure level of 85 dB(A), what is the maximum time an employee can work in this environment without ear protection, assuming a doubling rate of 5 dB(A), if the doubling rate was 3 dB(A), how would it affect the exposure time?

4. One group of workers in a factory are exposed to a noise level of 100 dB(A) for 8 hours. In the adjoining work area, their colleagues are exposed to 70 dB(A) for 8 hours. Management proposes introducing job rotation (everyone works for 4 hours in the noisy area and 4 hours in the quieter area) to bring the 8-hour exposure down to 85 dB(A), on average, for everyone. Will management's proposal work? If not, suggest what the new noise dose will be.

5. A set of ear muffs have a NRR of 30 dB(A) and a set of plugs have an NRR of 20 dB(A). Workers in a car factory are exposed to 110 dB(A) for 8 hours. Before advising management on whether to purchase muffs or plugs, you do a small trial on the factory floor. Workers wearing the muffs take them off for about 1 hour a day while working, because of discomfort, whereas plug wearers soon forget they are wearing them. Which device will give the best ear protection in practice?

12 Human information processing, skill and performance

If the organism carries a 'small scale model' of external reality and of its possible actions within its head it is able to try out various alternatives, conclude which is the best of them, react to future situations before they arise, utilise the knowledge of past events in dealing with present and the future, and in every way to react in a much fuller, safer and more competent manner to the emergencies which face it.

(Craik, K. 1943)

Human–machine interaction depends on a two-way exchange of information between the operator and the system. Designers usually have detailed, explicit models of *machines* and *machine behaviour* that can be used to improve human–machine interaction. According to Preece (1993), a general model of the user's *cognitive processes and cognitive behaviour* is also needed to

- Provide knowledge about what can and cannot be expected of users
- Identify and explain the nature and causes of problems
- Supply modelling tools to help build more compatible interfaces

A general information processing model of the user

According to Reason (1990), in broad terms, the human information processing system (HIP) can be thought of as

- A general-purpose pattern recogniser
- With limited information processing capacity
- Using heuristics (rules of thumb) to simplify the information processing load
- And acting as a 'satisficer' rather than an optimiser

Satisficing refers to the tendency to seek pragmatic, rather than optimal, solutions to problems by trading-off the costs and benefits of alternative actions.

In the information processing approach, flow diagrams are used to represent how the brain processes information. It seems to be the case that much of human information occurs without conscious effort. A great deal of pre-conscious (or 'pre-attentive') processing takes place. Velmans (1991) gives the following example:

'The forest ranger did not permit us to enter the reserve without a permit'.

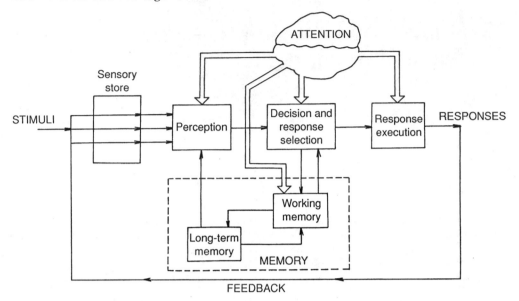

Figure 12.1 Wickens' general model of human information processing. (From Wickens, 1992. © Reprinted by permission of Pearson Education, Inc., Upper Saddle River, NJ.)

Reading the sentence silently, the word 'permit' is heard with the stress on the second syllable (per*mit*) in the first instance and on the first syllable (*per*mit) in the second instance. It happens automatically, suggesting that syntactic and semantic analysis of the sentence occur prior to consciousness.

Conscious ('focal attentive') processing occurs in novel situations or when decisions have to be made. The limit on processing capacity seems to be at the conscious level.

Wickens (1992) presented a composite, general model of human information processing (Figure 12.1). Information flows through the system. Different processes operate on the information at different stages. Information enters through the senses as a result of a physical stimulus impinging on the sense organs. A representation of the stimulus is created by the central nervous system (CNS). This representation persists for a short period but decays rapidly once stimulation has ceased (in several hundred milliseconds). In visual perception, this persistence is attributed to an 'iconic' store (Sperling, 1960). Iconic storage is automatic, in that all incoming information is stored in this way, and analogue, in that the CNS representation has the same structure as the stimulus that gave rise to it. The function of the sensory store appears to be to prolong the availability of stimuli that have been detected by the sense organs.

From sensation to perception

Sensation occurs when environmental stimuli impinging on an organism give rise to neural events. These events may or may not be processed further. Perception is the construction of a model of the event that gave rise to the initial sensation. The percept is not a replica of the external world but a representation of it that depends both on the neural 'hardware' and on previous experience. Table 12.1 depicts the

Table 12.1 Hierarchy of sensory and perceptual thresholds

Conceptual categorisation
  Naming

-------------------------------------RECOGNITION THRESHOLD-----------------------------------

Features of stimulus situation
  Shape
  Colour
  Contour

-----------------------------------IDENTIFICATION THRESHOLD-----------------------------------

'Something is there'
  Boundary information
  Brightness differences
  Movement

-----------------------------------DETECTION THRESHOLD-----------------------------------

Sensory nerve impulses

-----------------------------------PHYSIOLOGICAL THRESHOLD-----------------------------------

Environmental stimuli
  Impingement on sense organs
  Transduction and transmission of energy
  Stimulation of receptor cells
  Depolarisation of neurons

-------------------------------------ENVIRONMENTAL ENERGY-------------------------------------

---

hierarchy of stages from sensation to perception and the operation of high-level mental processes.

Sensation is characterised by a direct relationship between the intensity of the stimulus and the evoked neural discharge or inner experience. Perception, however, is mediated by individual factors such as personality and mood and by the context in which sensation takes place. For example, a pilot landing an aircraft in difficult conditions might see a series of lights below and ahead of the aircraft. Whether these are perceived as lights on the runway or streetlights will depend on internal factors such as the pilot's expectations, the pilot's level of training and its retention as well as external factors such as visibility and the visual cues that distinguish between the two types of lights.

For perceptual processing to be initiated, impinging stimulus energy must exceed physiological thresholds. Impulse conduction leads to the detection of 'something' in the environment. This is accompanied by orientation of the senses towards the source of stimulation. The identification threshold is passed when the 'something' that is detected is refined into a specific object by detection of contours and main features. Recognition occurs as the emerging percept is matched against stored percepts in memory. Further processing of the incoming information may take place at a number of levels, for example:

1. Physical level: This is the letter 'A' written in italics.
2. Phonetic level: This word sounds like 'cheese'.
3. Semantic level: This flashing red light means danger.
4. Other levels: This is a nice photograph of my grandmother.

The distinction between sensation and perception is by no means an academic one – the lower the level of processing required for a task, the quicker the operator's reaction time. The more intense the stimulus, the more likely the operator is to register it at a sensory level. In some US states and various other countries, legislation has been introduced compelling motorcyclists to use their headlights during the day as well as at night. Olsen (1989) has pointed out that although there is evidence that the use of daytime headlights on motorcycles reduces accidents, this also applies to cars and there may be other factors that contribute to the motorcycle accident figures. Two perceptual factors suggested by Olsen are that because motorcycles viewed from the front are much smaller than cars, drivers may overestimate the distance of an oncoming motorcycle and be more likely to pull out in front of it. They may also have more problems judging the speed of oncoming motorcycles compared to cars. 'Looked but failed to see' accidents are quite common. Drivers look at stationary or other vehicles, but fail to 'see' them until it is too late. Langham *et al.* (2002) found that drivers took longer to identify a parked police car as a hazard if it was parked in the direction of the flow of traffic than if it was parked at an angle. The effect was more pronounced when drivers performed a secondary reasoning task. Clearly, during tasks such as driving, sensations do not always become perceptions. The factors that affected this in Langham's experiment were whether a 'false hypothesis' was possible ('That car is pointing in the direction of traffic flow, therefore it is moving'). Langham suggested that emergency vehicles be parked at an angle to the flow of traffic and with their emergency lights switched off to convey more information that the vehicle is stationary.

'Banner blindness'

Banners are distinctive perceptual objects located on web pages to help users orientate themselves during search (Figure 12.2). According to Benway (1998), these banners are often missed by web users, particularly when they are placed on top of the

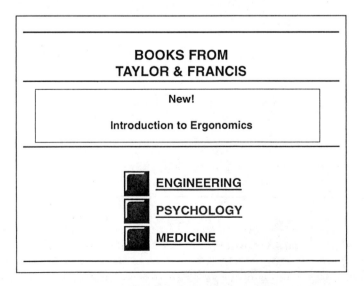

Figure 12.2 Banner blindness occurs when people look at a web page but don't see the banner – they only see the links. (Based on Benway, 1998.)

webpage, away from the links at the bottom. It seems that people don't just look at web pages, they look for things to click on and if banners are distant from links, or cannot be clicked on, their presence is not registered. Banner blindness illustrates how it is possible to look at something, but not see it (that is, processing never gets beyond the lower stages of the hierarchy in Table 12.1).

Coding

Only a fraction of the information impinging on the senses reaches conscious awareness. As soon as environmental energy impinges on the senses, it is transduced, distorted and coded by the very processes that transmit it to the higher centres of the brain. The concept of coding is central to the information processing approach. To use an analogy with computing, data are often coded in some way before being entered into the machine (names might be abbreviated, for example) according to the requirements of the software. Having been entered and stored in some way, the coded data can then be operated on by programs. Thus, what is operated on is a coded version, or representation, of the external (real world) data.

Coding and cognition

Since cognitive processes act on percepts, not on stimuli, the way the system codes a stimulus determines what can or will follow. The stimulus configuration and the coding mechanisms of the perceptual system determine what will be perceived. Memory also plays an important role. This has major implications in ergonomics – particularly in display design. The information displayed to an operator must be designed to be perceived in an appropriate way for the task. This requires that the task requirements be analysed early in the design stage and that the skills and knowledge of the operator be taken into account.

Two kinds of memory?

Psychologists normally distinguish between short-term and long-term memory. Short-term memory (STM) can be likened to a temporary store (or buffer) in which small amounts of information are briefly retained while a particular mental or physical operation is carried out. Remembering a phone number while writing it down is an example. STM contains symbols related to current processing but has limited storage capacity. STM storage limitations can cause errors – for example, forgetting important data before it can be consolidated in long-term memory (LTM) or used in decision making or acted on during a sequence of operations. Forgetting of intermediate results when doing mental arithmetic or omitting one of the ingredients of a recipe are examples of STM errors. STM is sometimes referred to as working memory. The relationship between perception, working memory, LTM and the operation of mental 'programs' is shown schematically in Figure 12.3.

LTM contains symbolic structures built up through learning in which new data can be embedded. General knowledge of the world and of life events is stored in LTM. We may sometimes be unable to retrieve this information but can be reminded of it by others, evidence that LTM storage is more or less permanent. LTM is associative in nature: new data can be represented in the context of past behaviour, but this takes time.

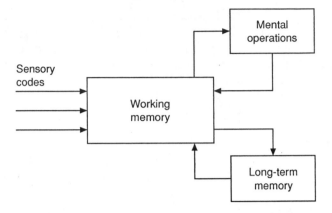

Figure 12.3 New information is kept in memory by the action of mental operations such as rehearsal and stored in LTM. Information in LTM can be written to working memory to be operated on.

To use a computer analogy, STM can be likened to the assignation of a particular value to a variable in a computer program for the duration of an operation. After the operation is completed, the value is no longer needed and is lost. The variable itself, however, remains in the program (in LTM) and new values can be assigned to it the next time the operation is invoked. Unlike in a computer, however, storing information in STM requires considerable mental resources, particularly conscious attention. Storage is fragile and the information is easily lost if the operator is distracted by unexpected events or overloaded by new information. For example, when driving home in busy traffic, a driver may forget details of traffic conditions just heard on the car radio when distracted by a call on his car phone. The STM trace of these items is displaced by new information, but the route information that is stored in LTM is not forgotten.

There is a tendency to focus on the fragility of STM. Although it may be a handicap in some situations, it is probably advantageous in daily life. We often notice things going on around us and act upon them with little thought. That STM is 'fragile' and of 'limited capacity' means that redundant or extraneous items that find themselves in STM are soon gone when we stop thinking about them, leaving us free to concentrate on the business at hand. In the sense sense, we might say that, instead of being 'fragile', STM has an automatic delete facility.

Two memory stores?

Much research has been carried out to describe information handling within LTM and STM. A number of classic experiments are described here.

Peterson and Peterson (1959) presented subjects with sets of three consonants (e.g. CMJ, PQF) and a number. After hearing the three consonants, subjects were required to count backwards from the number in threes until a light was switched on, at which time they were asked to recall the trigrams heard initially. The researchers repeated this procedure many times, varying the length of time between presentation and recall (the 'retention interval').

The probability of correct recall was found to diminish as the retention interval increased – from about *80% after a delay of 3 seconds to 10% after a delay of*

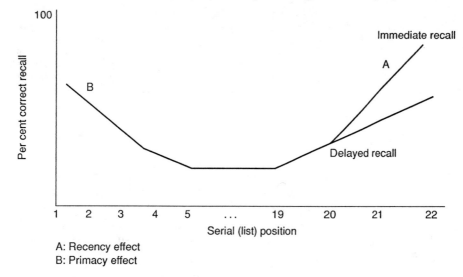

Figure 12.4 The serial position effect in short-term storage.

18 seconds. By asking people to count backwards in threes, the investigators hoped to prevent them using mental operations to keep the stored trigrams in memory (Figure 12.3). One of the simplest such operations is to code the consonants verbally and to rehearse them over and over again using inner speech. Mental operations reactivate items in STS that otherwise decay rapidly.

The drop in recall probability over time was assumed to be due to the 'decay' of the memory trace of the consonants in the short-term store (STS). This experiment demonstrates some of the storage dynamics of STS: storage time is of the order of seconds.

Murdock (1962) drew up lists of unrelated words from 10 to 40 words in length. The lists were presented to subjects who then had to recall the words in any order. Murdock found that the items at the end of the list were recalled first and with the highest recall probability. He called this the recency effect. The items at the beginning of the list were next best recalled. This was called the primacy effect. The items in the middle of the list were recalled worst. The STS concept can be used to explain these results (Figure 12.4). The recency effect occurs because the last items have only just arrived in STS and are recalled while the trace is still 'fresh' in the subjects' STS. The primacy effect occurs because the first items to enter STS get rehearsed the most and are thus more likely to enter permanent storage. As more items enter STS it 'fills-up' and less time is available to rehearse them (this happens after about the first 5 items). Thus, these items are less likely to enter long-term storage. Increasing the presentation time of each item was found to improve recall of even the middle items. Increasing list length lowered the recall probability of all words, as might be expected.

In addition to a rapidly decaying trace, the STS seems to have limited capacity. Longer-term storage of information depends on items in STS being subjected to mental operations. Miller (1956) concluded that the 'magical number 7, plus or minus 2' was the limit on the capacity of STS.

Table 12.2 summarises the differences between the long-term and short-term memory stores.

Table 12.2 Differences between long-term and short-term memory

	STS	LTS
Storage capacity	7 items ±2	Extremely large
Retention interval	5–30 seconds	Many years
Mechanism of information loss	Trace decay	Inability to retrieve item
	Displacement by new items	
Way information is coded	Phonetic/articulatory	Semantic

STM limitations have important ergonomic implications. People who have to use short-term memory in their jobs are easily distracted by interruptions of even a few seconds – particularly by conversation – or by having to carry out verbal secondary tasks (such as answering the phone while doing mental arithmetic). Tasks that require STM storage from multiple sources can be said to be of high *mental workload.*

Attention

Attention is the activity directed to facilitate processing of an expected stimulus (Valle-Sole *et al.*, 1995). In Wickens' model, it is presented as a limited resource that is channelled to 'drive' processes such as working memory, response execution, etc. It can be thought of as the spotlight that illuminates the information world or, to use a computer analogy, as the CPU (central processing unit) responsible for scheduling tasks and allocating them to subroutines.

Attention can be *selective*, as in listening to a particular instrument in an orchestra, or *focused*, as in concentrating on particular aspects of a task, or *divided*, as when reading a newspaper while watching the television. In divided attention, two or more tasks are not necessarily carried out simultaneously. A time-swapping mode can be used as attention switches rapidly between each task so that it appears that both are being carried out simultaneously.

How many tasks can we do simultaneously?

Some of the important issues concern the number of inputs an operator can attend to at once and how many operations can be carried out simultaneously. This is of importance if information overload is to be avoided, particularly in emergencies. New methods of interacting with machines, particularly the use of voice, make this an important issue for ergonomics.

A multichannel theory of attention was proposed by Allport *et al.* (1972). Humans are multimodal in the sense of being able to process and represent information in many modalities (e.g. visual, auditory, semantic) and many purposeful activities, such as walking and standing can be carried out without being consciously attended to. Attention is seen more as a problem of allocating processing resources to tasks. The multichannel model of attention is presented in Figure 12.5.

A limited-capacity processor handles incoming data but specialised subroutines requiring no processing capacity can be used to free the processor for additional tasks. The main features of the multichannel view are summarised in Table 12.3.

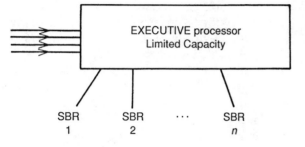

Figure 12.5 A multichannel model of attention.

Table 12.3 Characteristics of the multichannel view of attention

1. The processor and subroutines are fed by limited capacity, modality-specific channels.
2. The processor handles data in a way that roughly equates with consciousness.
3. The subroutines are built up through practice.
4. The system can break down under a number of conditions:
   - No subroutines available. Multiple inputs then compete for executive processes, which may be insufficient.
   - Competition for subroutines. More than one task is being carried out simultaneously and the tasks are similar in nature.
   - The existence of subroutines implies skill. If the processor tries to take over the operation, the subroutine is disrupted.

The model is intuitively attractive in that it can account for how we can walk and talk at the same time but might have difficulty in walking and juggling or in talking and reading simultaneously. Activities that occupy different modalities are fed by different channels to dedicated subroutines and can be carried out simultaneously. Activities that occupy a single modality can only be carried out simultaneously if channel capacity is not exceeded.

Experimental support for the multichannel view of attention comes from Allport *et al.* (1972) and Schaffer (1975). Allport *et al.* gave subjects a speech shadowing task to do and a memory task. Listening to a passage of prose and detecting embedded target words is an example of a speech shadowing task. If the memory task was presented in verbal form, performance was almost completely disrupted, suggesting that the shadowing was taking up all the processing capacity. However, if the memory task was presented in pictorial form, no disruption occurred. In terms of a single-channel view, it is difficult to explain where the extra processing capacity came from. A multichannel view would say that the shadowing task and the memory for pictures task utilise different limited-capacity channels that can operate independently of each other.

In another experiment, Allport *et al.* found that pianists could sight read and carry out a verbal shadowing task equally well simultaneously or individually. Schaffer (1975) found that typists could shadow an auditory message while copy typing visually presented text but could not shadow a visual message and audio type simultaneously. Copy typing from visually presented text does not involve the processing of verbal information (skilled typists do not read text when typing, they just type in the letters), so the copy typists had spare verbal capacity to process the auditory message. However, audio typing does require verbal processing capacity, as does shadowing a visual message, and insufficient capacity was available to do both tasks together.

Wickens (1980) has described how human attentional resources are subdivided along spatial/verbal and other lines. Vidulich (1988) has argued that the traditional all-manual human–machine interface leaves a pool of human resources untapped. New ways of enabling humans to interact with machines promise to make these resources available and lessen the conflicting demands on operators in situations where more than one task has to be carried out at a time. It is now possible to interact with a variety of machines using voice. Voice control is of benefit in tasks requiring a large amount of visual/spatial processing (such as computer-aided drafting and graphic design) but can interfere where the primary task is verbal (such as text creation).

Should people be allowed to use cellphones while driving?

Car driving often requires attention to be divided between different tasks such as lane tracking (e.g. keeping the car on course in a lane while negotiating an intersection) and visual analysis (scanning the traffic ahead and detecting and interpreting road signs, etc.). Some of these tasks are automated and require no conscious attention. Shinar *et al.* (1998) found that experienced drivers fared equally well on a signal detection task when driving a car with manual transmission compared to driving a car with automatic transmission. In terms of the multichannel theory, the experienced drivers had a subroutine for handling gear shifting, leaving attentional resources free for other tasks. Novice drivers, on the other hand, performed significantly less well on the signal detection task when driving the car with manual transmission. According-ing to the researchers, *up to 2 years' driving experience* is needed to *fully automate the skill of gear-shifting*. This has clear safety implications for drivers who change from an automatic to a manual car.

Using a cellphone might interfere with driving ability at the physical level and/or at the information processing level. The two tasks would appear to occupy different modalities (perceptual motor for driving and verbal/STM for using the cellphone). Further, experienced drivers would be expected to have most of their manual skills fully automated. Under conditions of low load, driving would be handled by subrou-tines and the cellphone conversation by the executive processor. Dual task perform-ance would seem possible. Under difficult driving conditions, when time-sharing is not possible, conscious effort is required and the two tasks might interfere with each other. The evidence suggests that interference does occur. Parkes (1991) found that the ability to remember verbal or numerical data and to interpret background infor-mation was degraded when driving. More seriously, Lamble *et al.* (1999) found that the time taken for drivers to detect that the car ahead was decelerating was reduced by 0.5 second when they carried out a memory and addition task. The time to collision was reduced by 1 second. The performance impairment of the non-visual task was similar to the impairment caused by a keying task that required them to take their eyes off the road and enter numbers into a keypad (a divided visual attention task). *Thus, it seems that even hands-free mobile phone use would degrade driver performance.* The view of the ROSPA (the Royal Society for the Prevention of Accid-ents) in the UK is that hand-held and hands-free cellphone use while driving should be banned, because drivers need to allocate 100% of their attention to driving 100% of the time. ROSPA has long issued guidelines on the use of mobile communication equip-ment in vehicles (Table 12.4), originally prompted by the use of hand-held microphones by drivers of tour buses.

Table 12.4 ROSPA guidelines for the safe use of mobile communications equipment in vehicles

Use hands-free systems only

Take incoming calls
- Only in light traffic
- Only at low speeds
- Only when there are no conflicting pedestrian or vehicle movements
- Only when there are no other distractions

Outgoing calls can be made
- Only when the vehicle is stationary in a safe place (off the road, in a lay-by or car park)
- Only when the vehicle is stationary as a result of an emergency situation (breakdown or crash)
- Only when the vehicle is stationary in heavy traffic (with no chance of moving until after the call is over)

Response selection and execution

Response selection and execution may occur automatically, as the output of a specific subroutine, or they may be initiated by the executive processor. This corresponds to the automatic or 'reflex' response of the skilled performer or the rapid avoidance behaviour that overrides all other outputs to avoid impending danger. When initiated by the executive processor, response selection and execution are more akin to a measured or calculated decision to act.

Skilled versus unskilled operators

The multichannel view of attention highlights the importance of task modality and skill. In Schaffer (1975), the copy typing (visual/manual) task did not interfere with the verbal shadowing task. Since the typists were skilled, it can be assumed that the typing was done automatically by means of a subroutine. However, the audiotyping (verbal/manual) task did interfere with the visual shadowing (visual/verbal) task. Even though the typists had subroutines for typing and for reading the visual message, the two tasks shared a verbal component that overloaded the available processing resources. It is in this sense that the requirements of tasks can be analysed to determine whether the performance of one of them will interfere with the performance of the other. Tasks that are unlikely to interfere can be said to be compatible and tasks with compatible components can be said to be of low mental workload if the capacity of either the subroutine of the processor is not overloaded.

A skilled operator has subroutines to handle the low-level aspects of task performance that spare the executive processes. The subroutines that handle motor actions (sometimes referred to as motor programmes) are of particular importance in integrating different muscular actions into a unified whole. The skilled golfer or tennis player conceives of and executes a swing or a serve as one smooth action, whereas the novice has to assemble the action out of a set of more primitive movements. For the skilled operator, detailed actions are derived in a 'top-down' fashion from higher-level representations. For the novice, purposeful actions are assembled from smaller components in a 'bottom-up' fashion.

The unskilled operator is said to be tied to the present through lack of automatic processes – the mechanics of the task require all his attention, leaving none for other activities. The skilled operator has subroutines built up to deal with the mechanics of the task. He does not have to monitor his own control actions and can predict and continually look ahead. Thus, the difference between skilled and unskilled operators is more than quantitative. The skilled operator exhibits more parallel processing of inputs and an increased ability to separate out doing the task from thinking about doing it. The distinction between skilled and unskilled users is made explicit in the multichannel view of attention and can be seen to be profound. Empirical studies have demonstrated qualitative differences between the way skilled and unskilled operators carry out work tasks. Prumper *et al.* (1992), for example, measured the number of computer errors and the time taken to rectify them of 260 clerical workers in 12 German companies. Before collecting the data, they classified the workers as skilled or unskilled. Skilled computer users made fewer errors due to a lack of knowledge than unskilled users but made more low-level errors. Experts are usually held to have knowledge that is represented in a high-level, abstract way and may thus be more error prone when the situation calls for precise, concrete actions (e.g. switching from one program to another with a different interaction style or command language notation). Unskilled users, because they lack high-level or abstract knowledge, may be more prone to error when confronted with conceptual problems in the operation of a system. Surprisingly, in Prumper *et al.*'s study, the skilled users made more errors overall than novices, which suggests that simple error counts are not sufficient to separate skilled from unskilled performance. However, the big difference between experts and novices was found in the time taken to deal with the error – experts were consistently faster. One of the interesting and important implications of this finding is that perhaps less attention should be paid to minimising the number of errors in the operation of a system and more should be paid to teaching users what types of error can occur and effective error-handling skills. In other words, design effort should not be spent only in minimising errors, since even experts are error-prone, but on error recovery and on minimising the consequences of an error. This seems to be one of the key requirements in the operation of interactive systems. *When evaluating existing systems, data are needed not only on the number of errors, accidents or unexpected events but also on their effects on system operation and the resources required to restore the system to its previous level of functioning.*

A first step in the design of any human–machine system, then, is to characterise the expected skill level of the user population since the interface design requirements may be completely different for different groups. Of particular importance are the constraints on display complexity and layout and the requirements for on-line help, training manuals and operating instructions, warning signs and labels and feedback.

Sustained attention – the vigilance paradigm

The ability to sustain attention over long periods is known as vigilance. Research on vigilance dates from the Second World War, when there was great interest in the ability of operators to monitor radar and sonar displays.

Mackworth (1948) carried out some of the first laboratory experiments on vigilance using the now-famous clock test. Subjects viewed a pointer moving around a blank-faced clock. The pointer moved once per second but occasionally made a

double jump. This was the signal that subjects were required to detect. The error rate
– at best – was about 15% (15 misses per hundred targets). In laboratory tasks, it
was found that performance declined by about 20% after about half an hour on
the task. This decrement was found to be highly reproducible and is known as the
vigilance decrement.

There has been a great deal of research on human vigilance and on the vigilance
decrement. Two main conclusions are that

- For a variety of reasons (e.g. boredom, distraction, sleepiness) operators are not
  very good at performing these tasks, particularly when they are isolated.
- The decrement appears to be an artefact of psychological research techniques
  and does not occur in real settings (e.g. Nachreiner, 1977).

Some of the ergonomic and work design factors that influence vigilance task performance are summarised in Table 12.5.

Aiding vigilance task performance

In aviation, in response to crashes caused by 'human error', attempts have been made
to use automatic systems such as the 'ground proximity warning system'. Aircraft
have crashed when, under heavy workload, the pilot stops looking out of the cockpit
and, when the autopilot disengages, crashes in a perfectly controlled manner into the
ground. The problem with automated warnings is that they may malfunction. The
result is to exchange one type of vigilance task for another. Type 1 errors (warning
you when there is no danger) and type 2 errors (not warning you when there is
danger) can occur and there is a trade-off between the two. E. L. Wiener has argued
that alerting devices are nothing more than substitute vigilance tasks.

Vigilance: cautionary notes

Much of the research on vigilance was based on a pre-ergonomic paradigm and was
human-centred rather than system-centred. The emphasis was on how to maintain
human performance over long periods. To the modern ergonomist, in a world of
zero-defect programmes and 100% quality control, it is perhaps of more interest to
consider how to change the interface or redesign the task to get an improvement in
system performance, rather than how to prevent already bad human performance
from becoming even worse. Another limitation of much traditional vigilance research
is that it was carried out on individuals, deliberately isolated in the laboratory.
In practice, operators often work together in teams. They can help each other and
flexibly share tasks, depending on the workload.

Moray *et al.* (1999), in a computer simulation of a process control task, found that
vigilance performance was less than 100% (subjects missed some of the targets) in all
conditions investigated. However, the vigilance decrement only occurred during simple
tasks requiring attention. For more complex cognitive tasks involving more than one
display, there was no vigilance decrement. So, for modern process control operations,
the focus should shift towards optimising the design of the interface, the workteams
and the work organisation to achieve better system performance.

Table 12.5 Factors influencing vigilance[a]

Signal conspicuity
Auditory signals: increased signal amplitude (signal-to-noise ratio) improves detection. Number of missed signals drops approximately by a factor of 10 when the signal is 5.1 dB greater than noise as compared to 2.1 dB (double the intensity of the signal and get a 10 fold improvement). Vigilance decrement goes away.

Signal and noise intensity
Boosting the intensity of both signal and noise in an auditory task can abolish the vigilance decrement (probably owing to the alerting aspect of noise).

Signal duration
Increasing signal duration to 2–3 seconds leads to almost 100% detection in visual tasks.

Number of events
Quality of sustained attention is inversely related to the total number of events. Mackworth suggests that when there are lots of events and only some of them are signals, we habituate to them more rapidly, which explains the poorer performance.

Knowledge of results
Feedback can abolish the vigilance decrement, and enhance the speed and frequency of detections. On simple vigilance tasks the beneficial effects of KOR are probably motivational, people know how well they are doing and do not get bored. In more complex tasks, the feedback may reinforce detection of signal cues.

Ways of sustaining attention
Synthetic targets with feedback, secondary tasks.
Provide amphetamines, caffeine, methylphenidate; but there are problems with side effects.

Other factors
Effects of elevated body temperature are equivocal. Effects of cold, noise and vibration difficult to predict in practical situations because the observer is an active participant in the environment and can take steps to modify exposure or enhance performance.

Techniques for improving performance
Reduce uncertainty (about what constitutes a target and where it will occur. If possible provide specifications – what to look for (reference standards, if possible).
Provide redundant information – e.g. visual and auditory cues.
Signal enhancement techniques: Signal processing techniques exist for modifying displayed data. For example, instead of displaying raw EEG data, you can display change in voltage over time in different sites, electrical power rather than voltage, or smooth out noise using running means or other statistical techniques that operate in real time.
Reduce spatial uncertainty: redesign display to fit into a smaller area (e.g. eye field as opposed to head and eye field).
Introduce job enlargment: For example, inspection of coins in batches can be interspersed periods of fetching and loading/unloading a conveyor. Combine radar monitoring with navigation (useful in teams where tasks can be rotated dynamically).
Provide regular short breaks of around 5 min.
Motivation: Subjects perform better when told the task is a selection test for a well-paid job than when they believe it is a psychological experiment (Nachreiner, 1977).

---

[a] See Warm (1984).

Feedback

Much research has been carried out on the factors affecting the ability of an operator to detect faults in a system. The industrial inspection task in which an operator scans newly made items for defects is a classic example. Frequently, the operator's task is to classify items as 'good' or 'faulty' according to pre-established criteria. Drury and Addison (1973) investigated the effects of feedback on the performance of inspectors at a glass-making factory. They found that increasing the speed of feedback about fault-finding performance improved the inspectors' ability to detect faulty items.

The lag between executing an action and receiving feedback is one of the most important variables affecting our ability to carry out continuous tasks. The classic experiment in which subjects are given delayed feedback about their own voice illustrates the importance of feedback in the production of speech. Even slight delays severely disrupt the ability to speak. In the field of human–computer interaction, the lag between user action and feedback is often due to the computer system's response time (the time taken to process the entry and display the result). Excessive response time can be a cause of user dissatisfaction (Goodman and Spence, 1981).

Modern approaches to user-interface design stress the importance of appropriate feedback (of the system talking to the user). For example, when executing a command, some immediate feedback must be given that the command has been received and is now being executed.

Bank statements, bills and receipts are examples of the complex array of feedback individuals receive in their everyday lives. An interesting and potentially important area, in the context of the increasingly integrated modern communication and data-processing systems, is the design of feedback about everyday behaviour. Feedback, in this context, is only useful if users recognise it as such, if they are motivated to use it and if it is in a form that they can easily map onto their own behaviour. McClelland and Cook (1979), in an early study, investigated the effects on domestic energy consumption of providing continuous feedback (in cents per hour) in 101 all-electric homes. They found an average energy saving of 12%, with the greatest improvement in the more temperate months, suggesting that the savings were due to changes in energy use other than that of energy for heating and cooling.

The design of feedback requires careful consideration. If the 'grain size' of feedback is too small, major trends may be obscured by minor variations. Alternatively, the patterning of feedback over time may be just too complex for the user to extract meaning from it. If the grain size is too large, people may not be able to map the information in the feedback onto specific behaviours amenable to self-modification. Thus, there is a trade-off between two extremes. At one extreme, detailed feedback that can be related to particular behaviours must not cause information overload. At the other extreme, general feedback may be easily grasped but, if too general, cannot be related to specific behaviours.

Returning to the domestic energy consumption example, most people do not regularly monitor how much electricity they are using. Often monthly or even quarterly bills are the only feedback they receive. Alternative energy purchasing and payment schemes have been developed, including the 'pay as you use' system in which users buy electricity 'credits' that are entered into their domestic energy meters using a plastic card similar to a credit card. The meter displays the amount of electricity remaining before more credits have to be purchased. It seems likely that users of this type of system would be much more able to modify their energy usage because of the much closer mapping between consumption behaviour and feedback.

Long-term memory

LTM contains symbolic structures, built up through learning, in which new data can be embedded. Cognitive processes – those that have to do with knowing – are of great importance for the design of usable products. For example:

1. Learning what facilities a new word processor offers and how to use them to create documents
2. Deciding which of a number of packages to buy based on an analysis of the usefulness of their attributes and cost
3. Solving comprehension problems that arise when learning to use the system

There is much anecdotal and experimental evidence to suggest that LTM is organised on the basis of meaning or semantics. If people are given lists of random words to remember, they tend to recall them in semantically similar clusters, which suggests that they have actively organised the information in this way when storing it. Bartlett (1932) was one of the first researchers to systematically investigate LTM. He showed people pictures or stories, which they repeatedly recalled up to 6 years later. He found systematic differences between the original stimuli and the recalled material and noted that people abstracted out the details, integrated them with other information and omitted 'irrelevant' information, producing a condensed form of the original material. People also made additions to the original material in an 'effort after meaning' in which 'we fill-in the lowlands of our memory with the highlands of our imagination'. This has important implications for the assessment of eye-witness testimony in court cases and the need to avoid leading questions when interrogating witnesses and designing questionnaires (a leading question is one that implies its own answer).

A fundamental distinction can be made between episodic and semantic memory (Tulving, 1972). Episodic memory is memory about personal experiences and events, whereas semantic memory contains world (or general) knowledge. The former is more easily forgotten than the latter.

Encoding factors influencing the retrieval of information from memory

The way an item is encoded in memory plays an important role in determining how easily it can be recalled. In fact, some theorists regard memory as a by-product of processing rather than a store in which items are placed.

Paradoxically, in order to retrieve something from long-term memory, we first have to know what it is we are trying to remember. A 'retrieval cue' acts like a description of the to-be-remembered item. If the cue is sufficiently detailed, recall probability is increased. In addition, the desired item has to be discriminated from other similar items in memory in order to be recalled, and this depends on how it was encoded when it was stored – similar items that were encoded in sensory or contextually different ways can be discriminated more easily than those that were encoded in a similar way.

Elaborative rehearsal (Craik and Lockhart, 1972) is an encoding method that increases the number of retrieval cues for an item. Items can be encoded multi-dimensionally (e.g. visually, verbally and semantically) to increase information redundancy and meaningfulness. According to Bower (1972), multidimensional coding works on the principle that if more features of an item are encoded it is more likely that at least one will be retrieved later on. Most *mnemonic systems* rely on some method of elaborative rehearsal to code new information.

Mnemonics, verbal elaborative processing and visual imagery

The use of mnemonics has a long history. Ancient Greek orators used a system known as the 'method of loci' to help them remember the details of a long speech. In rehearsing the speech, a familiar street would be imagined. At the start of the speech, the orator would imagine himself at the beginning of the street. Each important section of the speech would be associated with a particular part of the street such as a shop. On 'walking down' the imaginary street during the speech, each section of the speech would be recalled in the order it was memorised. The method works by linking new information (the parts of the speech) to an intrinsically sequenced set of previously memorised objects that act as retrieval cues.

Verbal elaborative encoding can be used to assist in the initial stages of learning new material. Bailey (1982) describes how the material may either be reduced or expanded. Reduction usually involves taking the first letters of each word to be remembered and using them to create a new word. The original words can be derived from their first letters, which are stored together as one item. For example, in the days before pocket calculators, many students of trigonometry used the acronym 'SOHCAHTOA' to remember the relationship between the trigonometric functions and the sides of a right-angled triangle (Sine = Opposite/Hypotenuse, Cosine = Adjacent/Hypotenuse, Tangent = Opposite/Adjacent).

Expanded mnemonics usually involve the addition of meaning to the to-be-remembered material, as when learning the lines of the treble clef in music. E,G,B,D,F can be expanded to 'Every Good Boy Deserves Food'. At first it may seem paradoxical that one of the best ways to improve the ability to remember something is to increase the amount of information to be remembered. The paradox disappears when it is recalled from memory theory that the process of storing an item in memory involves linking the new item into a previously established associative network of meaningful information. Expansion works by increasing the number of links between the new item and the network.

Visual imagery

Bower (1972) described the principles for the use of visual imagery as an aid to learning:

1. Both the cue and the item must be visualised.
2. The images must interact.
3. The cues must be self-generated.
4. The semantic similarity between the cue and the item must be minimised.
5. Bizarre images are no better than obvious relationships.

Interestingly, imagery only seems to be necessary during the initial learning of new material. With use, retrieval becomes automatic and the person may eventually forget the image itself, even though the word can now be remembered.

Information in long-term memory is stored at a semantic level but is linked to words, rules and other structures. Many concepts can be represented in terms of other symbols or sensory modalities at will. In this way, it is possible to talk about the imagery value of words. Words such as Justice and Philosophy are linked to abstract concepts that are difficult to link with information received from the senses. In contrast, words such as Balloons, Birthday Cake and Candy Floss are linked to more concrete concepts; we obtain knowledge about them directly via our senses and they evoke strong visual and other images when we hear them.

There is evidence that abstract words are more difficult to learn than concrete words (Baddeley, 1982), which may explain some of the difficulties experienced by novice computer users in learning to interact with system in which objects are represented in an abstract way.

Chunking

'Chunking' increases the amount of information that can be held in short-term memory through assembling separate items of information into larger units. A word is a way of chunking letters, for example. Codes such as telephone numbers can be printed so as to explicitly encourage people to store them in a particular way. For example, an 11-digit number such as 27214712503 appears less daunting when it is revealed to be the telephone number 27-21-471-2503 and that 27 is the country code, 21, the city code, and 471 the area code. By making the structure explicit, we reduce an 11-digit string to seven items (the three chunked codes and the unique 4-digit user number).

Chunking can also take place at a semantic level. Chase and Simon (1973) showed novice and experienced chess players a board containing about 25 chess pieces. Grand masters were able to reconstruct the board after viewing it for a few seconds, using information stored in long-term memory. Novices, however, could only replace about six pieces. The difference in performance can be explained by the notion of semantic chunking. The grand master is able to encode the positions of the 25 chess pieces in terms of a smaller number of higher-level representations of the state of play. Novices, lacking the appropriate concepts, cannot encode the board in this way and their performance is determined by the limitations of short-term memory. As might be expected, if the pieces were arranged in a random manner, rather than as a stage in a real game of chess, the performance difference between the novices and the grand masters disappeared – a knowledge of chess does not contain information that can be used to semantically code the randomly positioned pieces and the grand masters, like the novices, have then to rely on short-term memory.

When semantic chunking begins to take place vertically as well as between categories of items, we can see the development of a hierarchical system of semantic memory organisation. For example, after a walk in the country, a skilled biologist may be able to recall seeing many more types of plants than a novice owing the biologist's more organised system for categorisation, making distinctions between the various plants seen and organising the data using the existing hierarchical structure. The ability to categorise often depends on or is facilitated by the acquisition of verbal labels for the different categories (Liublinskaya, 1957), a point that will be returned to in a later chapter.

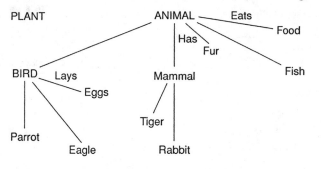

Figure 12.6 Sample representation of LTM as a semantic network.

Network theories of memory

Modern theories of semantic memory emphasise its associative nature – all items stored in memory are ultimately connected to all other items *on the basis of their meaning*. LTM is modelled as a *semantic network* (Figure 12.6). The problem facing these theories of memory is twofold: to describe the organisation of information stored in memory and to account for the ways it can be accessed (Simon, 1979).

Since all items in semantic memory are ultimately linked (memory is associative), items can be retrieved by searching along the links that connect them. Thus, in trying to remember something we may search for a cue – some other item known to us that is closely connected to the required item. A good retrieval cue can amount to a full description of the desired item of information that distinguishes it from all other items. Discriminability is a major determinant of the ability to remember something.

Memory can be modelled as a series of nodes (concepts) interconnected by relations. The nodes themselves are not words but are connected to words in a mental diction-ary. Several words may be linked to the same concept (e.g. car, automobile) or the same word may be linked to different concepts (e.g. plane).

Figure 12.7 shows an example of a network theory that was developed by Quillian (1969). In order for us to understand a sentence, each word is decoded and the appropriate concept is flagged. A parallel search then takes place in all directions. Thus, understanding the sentence 'Canaries can fly' would involve searching the network for the existence of interconnections between the different nodes. If the nodes are found to be connected, then the sentence is understood and can be verified. Despite some evidence supporting the theory (Collins and Quillian, 1972), it could not account for certain features of sentence comprehension.

More elaborate theories of LTM have been developed over the years to overcome the limitations of earlier models (e.g. Anderson, 1983). Anderson distinguishes be-tween declarative knowledge ('knowing that . . .') and 'procedural' knowledge ('know-ing how . . .'). The former is stored as a network of linked nodes, a small part of which can be active at any one time (which corresponds roughly to working memory). Activation can spread out from a node to other nodes – thinking about a particular person may remind us of other related people, or when trying to recall the details of a report we may think of a similar report whose details we do remember. The more nodes through which activation spreads, the more slowly it spreads. Since declarat-ive knowledge is said to be stored in the form of abstract propositions, the theory explains why people remember only the 'gist' of things.

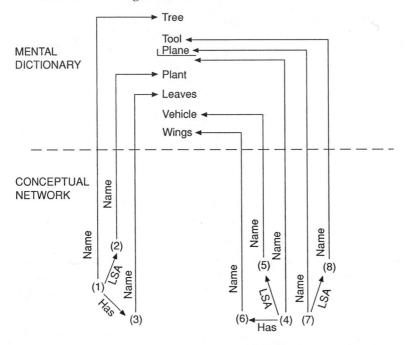

Figure 12.7 Concepts are stored in an associative manner in the semantic network. They are linked to words that are stored separately in a mental dictionary.

Procedural knowledge is stored as a set of production rules specifying what should be done if a given set of conditions becomes active (e.g. if the light is green, then cross the street).

Learning to use new systems

The general idea of knowledge being represented as a semantic network is useful in characterising and analysing many human–machine interaction issues, such as how to represent complex systems to users and how users learn to operate new systems. Existing concepts can be used to assist in the learning of new ones by drawing analogies between the new system and systems the users already understand.

Theories such as Anderson's predict that how well we remember something will depend on the processes we used to learn it, because memory is a by-product of processing information. Storage of items in long-term memory is slow because new items have to be 'linked' in with the existing network. Linking a new item such as a computer command word could be achieved by rehearsing it over and over again or by associating its sound with another known word (finding a rhyme for it). Alternatively, the user could concentrate on the meaning of the word in relation to the functions it invokes and to other command words and their functions. Semantic processing such as this would be predicted to lead to better retention. There is evidence that this is the case, that deeper or more elaborative rehearsal of new material does lead to better retention (Craik and Lockhart, 1972). This implies that, when trying to learn something new, such as the command words of an interactive system, the act of trying to understand them in relation to the system will itself lead to them

Table 12.6 Conclusions about long-term memory[a]

- Information is stored in LTM on the basis of meaning.
- Concepts are organised in terms of a network of propositions (e.g. 'A dog is an animal').
- Working memory is that part of the network that is active.
- The spread of activation from a concept depends on the strength of the links.
- Successful learning depends on
    Activating appropriate parts of the network
    Linking new concepts to existing ones
    Elaboration and explanation to build more links
    Practice to strengthen new links
- Successful recall depends on retrieval cues to direct search to appropriate parts of the network.

[a] Adapted form Anderson (1983).

being committed to memory. Retrieval of information will also be improved if we try to understand something from 'all angles', relating it to what we already know in several different ways.

There is evidence that well-learnt information really is stored in this kind of multi-dimensional manner. For example, when searching for a particular word, we may experience the 'tip of the tongue' phenomenon. Unable to retrieve the word at a lexical level, we access the code that describes how it articulated or is spoken. When unable to remember how to spell a word, we may write down differently spelt versions of it to see which one 'looks right'.

The network approach can be said to have heuristic value because it provides a general framework and vocabulary for discussing people's knowledge of systems. Some conclusions about learning and remembering are presented in Table 12.4 using network theory terminology.

Preece (1993) emphasises that learning to use a new system requires active involvement of users. Many modern computer systems are designed to enable users to carry out simple tasks with no or minimal instruction. The idea is that they will see the outcome of their actions immediately and be encouraged to do more. Learning by doing is accompanied by learning by thinking. Users will develop their own ideas about what the system is and how it works, and if the system is well designed they will be able to predict the outcome of novel actions and have the confidence to try more ambitious tasks involving goal setting and planning. In a well-designed system, tasks should be broken down into subtasks in a consistent and logical way to support this type of behaviour. When users make errors, meaningful feedback can help them learn more about the system and improve their understanding of the activity. One of the major usability problems of older computer systems was that many of the error messages (e.g. 'illegal entry' or 'syntax error') were too general and contained no information to assist the user either to understand the nature of the error or to plan a course of remedial action.

Transfer of learning from one system to another

Many mechanical or electromechanical products are being replaced by electronic products. Physical similarity between them encourages users to apply familiar ideas about the old product to the operation of the new one. Where this happens successfully and is appropriate, we can say that positive transfer of learning has taken place. Where it is inappropriate, negative transfer has occurred.

Some examples are video cassette recorders (VCRs) and video cameras. In the case of VCRs, most of the basic controls for functions such as Play, Rewind, Forward and Record are the same as on older devices such as tape recorders and cassette recorders. Most people with experience of these other devices can use the corresponding VCR functions. VCRs also have programmable timers, not found on their more primitive predecessors. It is not surprising that many people never use this function of their VCRs but have no difficulty with the other functions (Norman, 1988).

To complicate matters, the primary functions of many electronic devices are automated and replaced by new secondary functions. For example, in video-cameras, many of the photographic and optical tasks that were a key part of the operation of the parent devices are now automated, but many new functions (such as being able to display the date and time on the recorded sequence) are now available. Much previous knowledge gained from using manual cameras is obsolete when these video cameras are used, but many new 'system' functions have to be mastered if the full potential of the device is to be exploited.

Mobile phones, or cellphones, have positive and negative transfer features when compared to conventional telephones. Very likely, they are not usable by naive users unless the manual is studied. For example, the 'On–Off' function on conventional phones is integrated with the lifting and replacing of the handset: you never have to switch a conventional phone 'On' or 'Off'. With cell phones, a button has to be pressed and a PIN number may need to be entered before the device can be used. Once the phone is activated, a phone number can be dialled as on a conventional phone. Cellphones often offer the user additional features not found on older models of telephone, such as a fax service, automatic redial, phone book and call forwarding. All of these have to be learnt. Thus, there is little positive transfer between older models of telephone and a cellphone. There may be negative transfer initally, in that the naive user will not expect to have to activate and access the phone.

However, many new electronic devices share similar features both within and between products of their own generation. Having learnt to use one make of cellphone, users will have little difficulty in learning to use a different make. Many of the problems of learning to use these new electronic devices are transitory and will diminsh as a new generation of users emerges who have only had experience with the electronic versions.

Mental models

A mental model is an internal representation of a system in which the essential features of the system and their interelationships are stored in memory. N. Moray describes a mental model as a 'many-to-one mapping', as only the essential features of the system are stored in memory. According to Preece (1993), when we imagine ourselves using a device, we are 'running' our mental model of it. Users build up models by experience and refine them with practice. Conversely, system designers begin with a complete model of the system at the earliest stages. The graphical user-interfaces (GUIs) found in most commercial software are built using a desktop as a metaphor for the arrangement and appearance of objects on the screen.

Design models can be powerful tools for enhancing usability and reducing learning times. If a new system is built around an existing model that users already understand, then positive transfer of learning will be expected. Further, if a task has to be learnt by rote, without any possibility of developing a mental model, the user will be unable to cope with novel or unprogrammed events. Alternatively, a user with a fully developed model of the system and the context in which it operates will be able to 'run' the model to predict the outcome of alternative actions.

When the mapping between the mental model and the real system is poor, or incomplete, user difficulties may result. For example, many domestic central heating systems contain a number of key components that make up our mental model. The boiler provides the heat, the thermostat controls the temperature in the rooms and the radiators distribute the heat in each room. Such systems can be controlled in a number of ways. The output of the boiler can be increased or decreased, the setting on the thermostat can be varied and the individul radiators can be turned on and off. In a simple model of this system, these control options are seen as independent. The user might decide to lower the overall temperature in the house and turn off one of the radiators in one of the rooms. The expected result, according to the model is that the overall temperature in the house would be lower than normal and the temperature in one room would be cold. However, in real central heating systems, the control devices are not independent. Turning one radiator off increases the flow of hot water to the remaining radiators. Since the thermostat only detects the temperature in its immediate vicinity, the rest of the house will heat up. A 'higher fidelity' mental model of the system would represent the interdependences between the control devices and might lead to a better control strategy – for example, lowering the setting on the boiler when turning off one of the radiators.

Cognitive systems

Many of the most important ergonomic issues in the design of complex systems are to be found at the abstract, symbolic level and can best be dealt with using the concepts and methods of cognitive psychology. Rasmussen and Rouse (1981) has identified three types of task performance. Skill-based performance depends on the existence of specialised subroutines for the performance of routine tasks and has already been discussed in an earlier part of this chapter. Rule-based performance makes greater demands on conscious processing capacity as explicit rules have to be kept in mind and followed as the operator carries out the task. Knowledge-based performance is essential in complex tasks, particularly where the operator has to engage in problem solving and decision making in novel situations. In systems that are largely automatic, such as nuclear power stations, operators operate primarily in the cognitive mode (Rasmussen and Rouse, 1981). They engage in supervisory control involving goal-setting, monitoring system performance and dealing with non-programmed events. Detection of signals and manual control are relegated to a more secondary role. Car driving is an example of a perceptual–motor control task. Operating a power station is an example of a *cognitive control* task.

Cognitive systems and intelligent action

Hollnagel and Woods (1983) argue that human–machine systems should be analysed, conceived of and designed at a cognitive level – as cognitive systems in which

Table 12.7 Models of the operator and their uses

1. Prediction of operator behaviour
2. Facilitation of task load evaluation
3. Direction of equipment design evaluation procedures (using other ergonomics approaches)
4. Evaluation of the adequacy of operating procedures
5. Evaluation of training programmes
6. Implementation of the model on computer to simulate the behaviour of the operator and find answers to 'what if' type questions (questions not easily answered in other ways)

A cognitive system produces 'intelligent action', that is, its behaviour is goal-oriented, based on symbol manipulation and uses knowledge of the world (heuristic knowledge) for guidance. Furthermore, a cognitive system is adaptive and able to view a problem in more than one way. A cognitive system operates using knowledge about itself and the environment, in the sense that it is able to plan and modify its actions on the basis of knowledge. It is thus not only data driven, but also concept driven. Man is obviously a cognitive system. Machines are potentially, if not actually, cognitive systems. An MMS [man–machine system] regarded as a whole is definitely a cognitive system.

Cognitive models of the human operator

Some of the first cognitive models of human operators were developed in the process control industries. In these industries, the operator monitors the system while the system controls the process (e.g. Bainbridge, 1979; Umbers, 1979). If we want to analyse the task, the best approach is to discover the facts and rules and how they are used to carry out the task. Only then do we assess the design of the workplace and the interfaces.

Operators can be modelled as information processing systems that have to solve a finite set of problems using information from the system (feedback about system behaviour) by the application of programs (usually algorithms or rules that specify how particular events are to be dealt with). The key elements of the model that have to be determined are

1. The rules that the operator uses to control the system.
2. The strategies that determine how the rules are used.
3. The types of system feedback that influence the operator's control strategy.

Table 12.7 shows common uses of models of the operator.

Problem solving

In problem solving there is an interaction between the operation of programs and the movement of data between different memory stores (Newell and Simon, 1972). The memory load may vary depending on the amount of feedback from the system. Excessive load can constrain the operator's choice of control strategies (since the application of rules also takes up space in memory). Memory limitations can affect problem solving by causing the problem solver to return to previous knowledge states. This is known as backtracking.

Table 12.8 The Missionaries and Cannibals problem

Three missionaries and three cannibals seek to cross a river from the left bank to the right bank. A boat is available that holds only two people and that can be operated by any combination of missionaries and cannibals involving not more than two and not less than one person. If the missionaries on either side of the river are outnumbered at any time by cannibals, they will be eaten. When the boat is moored at either bank, the passengers are regarded as being on that bank.

Find the smallest number of crossings that will permit all of the missionaries and cannibals to cross the river safely.

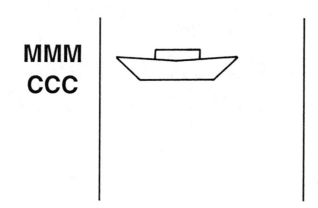

Figure 12.8 A way of representing the starting position of the Missionaries and Cannibals problem.

Models can help specify the conditions under which an operator's ability to control a system will break down. For example, long-term and short-term memory limitations can lead to a breakdown in control behaviour. Long-term memory contains symbolic structures, built up through learning, in which new data can be embedded and represented as a single symbol. Because long-term memory is associative, it enables new data to be represented in the context of past experience. Access to items in long-term memory is fast and does not require search but is blocked in the absence of retrieval cues. Short-term memory contains symbols related to current information processing. Errors associated with short-term memory often involve forgetting important data before it can be consolidated or used.

The well-known 'Missionaries and Cannibals' problem can be used to illustrate some of these points (Table 12.8). Readers unfamiliar with this problem should attempt to solve it before continuing. The problem has first to be approached by finding a way of representing it that will permit rules to be generated and applied. Figure 12.8 depicts such a representation.

According to Bundy (1978), a first step in solving the problem is the realisation that certain types of information can be ignored (the type of boat and the width of the river). The individual missionaries and cannibals can be regarded as interchangeable, since it is only the number of each on either side of the bank that is important. A second step is the realisation that the problem is very difficult to solve without some form of external memory aid. The load on memory is very large when exploring the consequences of moves from a current state if the current state itself has to be remembered rather than recorded. For similar reasons, people normally experience difficulties in anticipating the consequences of a particular crossing (since the alternatives cannot all be held in short-term memory at the same time).

Problems like Missionaries and Cannibals present the solver with choices that have to be made at particular decision points:

1. To select the most appropriate path for exploring the problem.
2. When reaching a new decision point, to decide whether or not to continue.
3. To decide whether the present knowledge state should be remembered for future reference.
4. To abandon a knowledge state and return to a previous one from which further exploration can take place in a different direction.

A state can be reached where there is one missionary and one cannibal on the left bank and two missionaries and two cannibals on the right bank, together with the boat. Since it is impossible to proceed from this point by moving either one cannibal or one missionary back to the left bank (since cannibals will then outnumber missionaries on one of the sides), many people abandon this state and backtrack to the previous one.

Problem representation and information design

The representation of a problem (the form in which it is physically communicated to the problem solver) influences how it is represented at a cognitive level. An appropriate representation of a problem can make its structure more explicit and thus facilitate the selection of appropriate problem solving procedures. The, now classic, experiments of Wason can be used to illustrate this point. Note that Wason did not carry out these experiments in order to generate guidelines for ergonomists. They are none the less very relevant to problems of information design.

Wason and Shapiro (1971), for example, investigated subjects' problem solving ability in a falsification task. The subjects were presented with the following rule:

Every card that has an E on one side has a 4 on the other.

Four cards were placed in front of the subjects who were told that each card had a letter on one side and a number on the other. They were then asked which cards they would have to turn over in order to find out, definitely, if the rule was true or false. Figure 12.9 shows the cards presented to the subjects. Readers unfamiliar with this problem should attempt to solve it before continuing.

A second, and logically equivalent version of the rule was also tested. It stated:

Every time I go to Manchester I travel by train.

The corresponding cards in Figure 12.9 were presented to subjects.

The first version of the rule can be tested by turning over the cards showing E and 7. If the 'E' card does not have a 4 on the other side, the rule is clearly false; if it does, the rule can be provisionally accepted. The card with '7' on it can also disprove the rule and for this reason it must be turned over as well. If there is an 'E' on the

ABSTRACT
REPRESENTATION

CONCRETE
REPRESENTATION

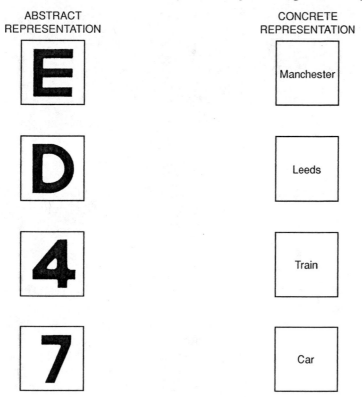

Figure 12.9 Abstract and concrete versions of Wason's 'four card' problem.

reverse of the 7, the rule is clearly false. Because of the way the problem has been structured (we are told in advance that each card has a letter on one side and a number on the other), the 'D' card can be dismissed as irrelevant as to whether or not the rule is true or false (we know that there will be a number on the reverse, not the letter E). Finally, the '4' card is also irrelevant to the testing of the truth of the rule: just because cards with an 'E' on one side are supposed to have a '4' on the other, it does not mean that cards with a '4' on one side must have an 'E' on the other. There is therefore no point in knowing what is on the reverse of the '4' card.

Wason reports that many subjects experience extreme difficulties in solving the problem and in understanding the solution (i.e. why it is essential to turn over the 'E' and '7' cards and not the others) in the alphanumeric format presented above. However, when the problem is recast in a more familiar form, subjects have less difficulty in selecting the 'Manchester' and 'Car' cards to test the truth of the rule 'Every time I go to Manchester I travel by train'. If 'Car' is written on the reverse of the 'Manchester' card the rule is clearly false, as it is if 'Manchester' is written on the reverse of the 'Car' card (i.e. the two cards are the same). It is irrelevant what form of transport is used to travel to Leeds or whether any other cities are visited by train (for a review of these and similar experiments see Wason and Johnson-Laird, 1972).

Wason's experiments illustrate that the way a problem is presented to people can have an influence on the ease with which they can solve it. Generally, the less abstract the form of presentation, the easier a problem is to solve (although one advantage of

Table 12.9 The 'Two Liquids' problem

Imagine two jars, A and B. Jar A contains a liquid called liquid A and jar B contains exactly the same amount of a different liquid, liquid B. Suppose a teaspoonful of liquid A is taken from jar A and mixed thoroughly with the contents of jar B.

When the contents of jar B are thoroughly mixed, a teaspoonful of the liquid in jar B is removed and mixed thoroughly with the contents of jar A.

If, at the beginning, both liquids were completely pure, which is now the most contaminated by the other?

abstraction is that it removes ambiguous or unnecessary information). For practical purposes, abstract or formal terminology is best avoided on products and machines.

Problem solving and cognitive style

Individuals have preferred cognitive styles for conceptualising problems. These depend, at least partly, on prior educational and occupational experiences. The 'Two Liquids' problem (Table 12.9) illustrates some of the different ways in which a problem can be represented in the mind of the problem solver and how this can influence the types of cognitive operations that are bought to bear on its solution. Readers unfamiliar with this problem should attempt to solve it before continuing.

Intuition leads many people to conclude that jar B is the most contaminated because it was mixed with a teaspoonful of pure liquid A, whereas jar A was subsequently contaminated with a teaspoonful of the B and A mixture (less of liquid B was put into liquid A than A was put into B). This is incorrect because it does not account for the reduction in the volume of liquid in jar A after the first teaspoon of A was removed.

Engineers and scientists often approach this problem in a straightforward, although rather stereotyped way. They make some assumptions about the volume of liquid in the two jars (say, 100 ml) and of the teaspoon (5 ml), and from then on all that is needed is a series of simple arithmetical steps to arrive at the solution. The arithmetical solution mode requires a knowledge of addition and subtraction, how to handle percentages or fractions and an external memory aid (such as a notepad) to keep track of the calculations. Mathematicians and statisticians often use their knowledge of algebra to construct a more universal proof using formal notation. This is a more sophisticated version of the arithmetical approach but it entails an essentially similar way of thinking.

An alternative way of representing the problem and of solving it is the elegant non-verbal method of Figure 12.10, based on the use of visual imagery.

A third approach, which depends more on verbal logic begins with and analysis of the problem's structure. Whichever liquid turns out to be the most contaminated after mixing clearly does not depend on the actual volume of liquid in the two jars or on the volume of the teaspoon. What is important is the statement that there is an equal volume of liquid in the two jars initially and that the same volume is transferred from one jar to the other and back again. The structure of the problem remains the same regardless of the actual volumes involved. This realisation enables the problem to be simplified considerably by taking it to its extremes. If the volume of the

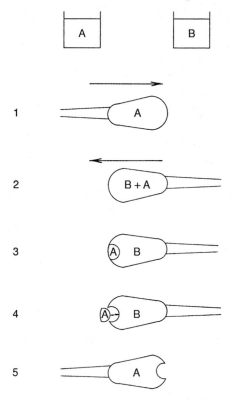

Figure 12.10 A non-verbal, non-numerical solution the two liquids problem. (1) Teaspoon of A transferred to B. (2) Teaspoon of A + B removed from B. (3) Contents of teaspoon are 'mentally unmixed'. (4) The small portion of A is mentally removed from the teaspoon. (5) This discloses that the A remaining in B and the B being transferred to A are the same.

teaspoon is the same as the volume of liquid in the jars, then all of A is first mixed with B. The volume of liquid in jar B (the mixture) is doubled and half of it is returned to jar A. Thus, the two liquids are equally contaminated.

The Two Liquids problem demonstrates that the way a problem is represented determines the types of cognitive operations (verbal, visual, arithmetical, etc.) that can be used to reach the solution. Interestingly, there are many other ways of solving the 'Two Liquids' problem and the reader might find it instructive to explore some of the possibilities.

Effectiveness and cost-effectiveness

In jobs with a large cognitive component, it is not always possible to take a FJM approach through ergonomic design, and organisations rely on selection (FMJ), either of people with the required knowledge and skills or of people who can demonstrate an aptitude to be able to acquire the knowledge and skills. Anastasi (1990) gives examples of the savings acrued due to the use of selection tests for computer programmers in the US Federal Government. Increases in productivity worth over

$30 million were estimated on the basis of using a valid selection test. The reader is referred to Anastasi (1990) for further information.

Research issues

The extent to which psychological theories have really contributed to systems design has been questioned by several authors, including Cox and Walker (1993) as follows:

> Few of the guidelines (for user-interface design) come from traditional scientific experimentation. The psychological laboratories using standard experimental techniques have had a relatively small impact on the engineering of user-interfaces. . . . Progress in the field comes from examination and investigation of inventions rather than application of fundamental principles about the workings of the human mind.

These authors tend to view the growth of information technology in an evolutionary way inasmuch as there is usually an existing system to criticise, improve and elaborate. They argue that the contribution of psychology lies in the evaluation and testing stage of design. Be that as it may, the behavioural sciences do offer principles that can be used to predict the average behaviour of large groups of people. However, these same principles are less successful in predicting the behaviour of any one person (Clement, 1987).

Summary

Operators and users can be considered as processors of information. Psychological theories provide ergonomists with a set of concepts and a language for describing the operator as an information processor. Information flows from the senses, via the perceptual system to working memory or to specific subroutines in which automatic processing takes place, leading to an overt or covert response. Information in working memory decays rapidly unless operated on by mental operations stored in long-term memory. The processing capacity of the system is constrained by channel capacity, the availability of subroutines (level of skill) and the demands made on executive (conscious) resources. The availability of these resources limits the ability to comprehend the state of a system, to diagnose faults and solve problems. Focusing on the strengths of the human information processor and designing systems around these strengths will result in more reliable systems and better products.

Psychological theory has heuristic value in ergonomics. It provides a framework and vocabulary for the structured analysis of design issues in human–machine interaction.

Essays and exercises

1. Describe the elements of the human information processing model presented in this chapter. Illustrate each part of the model with examples from everyday life.
2. Ask colleagues or friends to memorise the recipe below without taking notes. Allow them 10 minutes study time.

Ask them to recall the recipe the next day, after 1 week, and after 1 month. Describe the differences and similarities between the recalled and original versions. In particular, look for errors of omission and sequence and for additions and simplifications. Develop a system for categorising these errors. Does previous culinary experience influence recall?

PEDRO RUIZ'S PAELLA RECIPE

- Obtain 12 large prawns and remove their heads. Simmer the heads in 3 litres of chicken stock for 2 hours together with a large, whole peeled onion. Add 2 teaspoons of salt to the stock. If you have any fresh fish heads, add them to the stock as well.
- Debone, skin and cut into small, lean pieces 1 large, fresh chicken. Remove the fat from 1 kg of tender pork, either chops, leg or steak, and cut into small pieces. Using a large, flat-bottomed frying pan (or a 'paellera' if you have one) add two tablespoons of olive oil and heat. Finely chop three large cloves of garlic and add to the olive oil. Fry the garlic for several minutes.
- Add the diced chicken and pork to the pan and stir making sure that it cooks evenly. When the meat is nearly ready (i.e. almost cooked on the inside) add 2 large red and 2 green peppers that have been previously cut into thin strips. Fry them with the rest of the ingredients until soft on the outside, but still crisp.
- Next add the contents of 2 tins of peeled 'plum tomatoes' and stir. Heat until the mixture simmers and reduce. Remove the stock pan from the burner and strain it to remove the solids. Put 3.5 heaped cups of rice into the frying pan and cover it with 1.5 litres of stock. Simmer gently for 10 minutes.
- Add more stock as needed to prevent the mixture from sticking to the bottom of the pan. Season to taste and add 0.5 g of saffron. Add 500 g of mussels (or clams), 500 g of diced, de-boned hake and 300 g of calmer rings. Make sure there is sufficient stock to cover the ingredients. Finally, arrange the prawns, nicely, around the circumference of the pan. Simmer, gently for 5 minutes or until the prawns are cooked.

3. What is the difference between skilled and unskilled performance? Discuss the ergonomic implications of this.
4. Ask a friend or colleague to solve the problems given earlier in this chapter. In addition, ask them to 'think out loud'. Record the verbal protocol. Analyse it and attempt to construct a representation of the person's problem-solving behaviour.
5. Construct an algorithm that will solve the missionaries and cannibals problem for any equal number of missionaries and cannibals. Give the algorithm to a friend to use in solving the problem and ask them to think out loud while using it (or implement the algorithm in a programming language of your choice).

13 Displays, controls and virtual environments

Visual perception – there's more to it than meets the eye.

(Anon.)

The design of the displays and controls of a machine can either facilitate interaction or increase task difficulty and the probability of error. General principles for the design of displays and controls exist, but task analysis is normally required before these principles can be applied.

Principles for the design of visual displays

The German *Gestalt* psychologists in the first half of the twentieth century described how the way stimuli are structured determines how they are perceived. The structure of a display is superordinate to any particular element within it – hence the Gestalt dictum 'the whole is greater than the sum of its parts'. Conscious perceptions result not just from an analysis of objects in the field of view but from a synthesis of the objects themselves *and* the relations between them.

The Gestalt psychologists identified a number of laws by which the perceptual system was organised. These laws provide a framework for elementary discussion of the design of visual displays.

Figure–ground differentiation

Figure–ground differentiation is a fundamental step in perceptual processing in which

1. The perceived figure has form while the background is formless.
2. The figure appears to stand out against the background.

Since the perceptual system is of limited capacity, figure–ground differentiation can be seen as a way of reducing incoming data to manageable proportions. Although information in the figure receives preferential processing at the expense of ground information, the latter is not entirely lost. Ground information provides a context that influences the way the figure is perceived, demonstrating that ground information is processed beyond the physical level. In advertising, journalism and report writing, the use of special typefaces can influence the way a message is interpreted. Italic or bold letters may be used to convey important points.

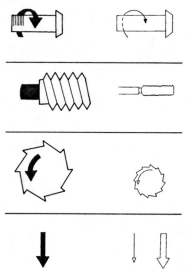

Figure 13.1 Strong contour enhances figure–ground differentiation and the visual impact of a figure. (From Easterby, 1970, with permission of Taylor and Francis.)

Contour, closure and previous experience influence the differentiation of a stimulus array into figure and ground. Contour is a characteristic of a stimulus array that provides cues for figure–ground differentiation (Figure 13.1). In Chapter 10, it was seen that the retina functions as an edge detector, rather than a light meter – sharp contrasts enhance contour detection and produce strong figures.

Sudden changes in brightness or colour provide natural contours between objects. Contour enhancement is a useful method for helping operators differentiate parts of a display panel or recognise a symbol. It is known that well-designed road signs with strong figures are more easily recognised than written messages, giving drivers of all ages almost twice as much time to respond to them (Kline *et al.*, 1990).

Closure can be described as the *rendering of form* – a tendency to complete or 'close' a figure. Closure is the tendency to produce a meaningful percept from incomplete cues and is the mechanism that prevents us from being aware of the effects of the retinal blind spot on vision. The ability to create meaningful percepts from incomplete cues depends on past experience and is a characteristic that distinguishes the skilled from the unskilled operator. When one first views a map, the initial impression may be of a bewildering complexity of unrelated parts. Following study and use, closure may be achieved and the map seen as a whole. Figure 13.2 illustrates the principle of closure in display design.

Grouping

Table 13.1 describes basic perceptual principles that determine how an array of separate elements is 'grouped' to form a complete percept.

According to Rock and Palmer (1990), these laws of grouping are fundamental and have general applicability. Operators seem to group first on the basis of common region and then of proximity. Further grouping occurs if elements in the proximal

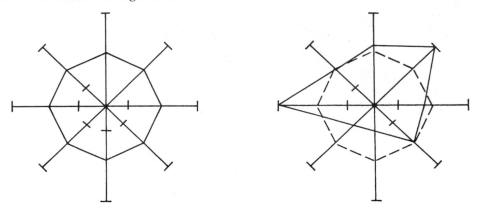

Figure 13.2 Eight linear scales arranged as a polar display. Under normal plant operation the separate readings create a closed figure (an octagon). Departures from normal can be readily perceived as a distortion in the octagon. (Redrawn from Wickens, 1984.)

Table 13.1 Perceptual principles for use in display design

Common region: Elements contained within the same region or area are grouped together.
Proximity: Items close together are grouped together.
Horizontal proximity: groups elements as rows, not as columns.
Similarity: Items similar in colour, size or shape are grouped together.
Symmetry: Elements in an array are grouped to form symmetrical figures.
Continuity: All parts of an array having a common feature are seen as separate from other parts, even if a common point is shared.
Closure: The tendency to see complete figures even when presented with partial information.
Visual apprehension limits: We can 'see by counting' up to 5 or 6 items.

---

cluster differ. Similarity, based on shape or colour can override proximity and provide a common boundary where one is lacking. This is why colour is useful for retrodesigning badly laid out panels. Figure 13.3 shows a display with a variety of elements that has been designed using these grouping principles in mind. Note that the whole display seems to decompose quite naturally into different regions, suggestive of different functions. If the different regions do support different functions, then we can say that the perceptual layout is compatible with the systems structure.

Good mappings between display layouts and system functions can increase the efficiency of visual scanning. Kotval and Goldberg (1998) compared eye movements of subjects scanning the icons shown in Figure 13.4. The same icons were grouped in four different ways as shown. In functional grouping, the icons were arranged on the basis of common function to give three sets of icons dealing with editing, drawing or text. In majority grouping, most of the icons in each group had the same general function, but one differed. In physical grouping, the icons were clustered into groups irrespective of their function. Finally, there was a no-grouping condition in which all the items were laid out equally spaced. Subjects were prompted to find and select an icon and their eye movements were recorded. Not surprisingly, subjects' visual scanning was most efficient when the icons were functionally grouped, as reflected in shorter scanpaths and fewer saccades (ballistic eye movements used in visual search).

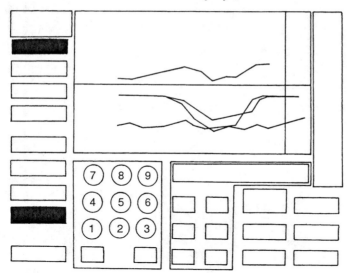

Figure 13.3 Good application of grouping principles. Proximity, similarity, symmetry and continuity have been used to create a coherent display that naturally decomposes into subsystems and different functional areas.

INTERFACE DESIGN

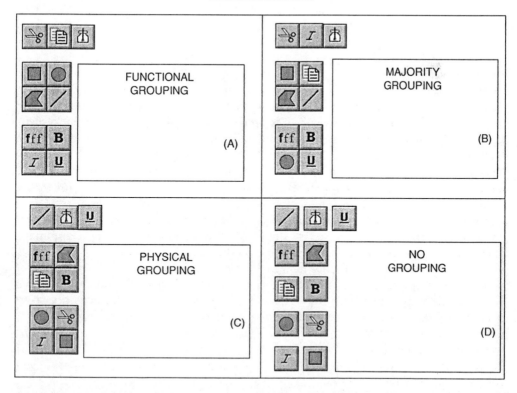

Figure 13.4 Functional groupings support effective visual scanning of screens. (Figure courtesy of Dr X. Kotval.)

Table 13.2 Advantages and disadvantages of the use of colour in display design

Advantages	Disadvantages
Draws attention to specific data	8% of males are colour blind
Faster uptake of information	May cause confusion
Can reduce error	May cause fatigue
Can separate closely spaced items	May cause unwanted groupings
May speed reaction	May cause errors
Adds another dimension	Can cause afterimages
Seems more natural	Appears more frivolous

Interestingly, the no-grouping condition was the next most efficient and the majority and physical groupings were the least efficient. Incorrect grouping seems to be worse than no grouping at all and so grouping should only be used when designers are are certain about functional relationships between different components. A good understanding of the task and feedback from users are essential.

Colour

People have a very strong tendency to perceive similarly coloured objects as belonging together (van Nes, 1986) as long as *no more than three or four colours* are used together. Colour can be used to provide conceptual grouping of text in itineraries, timetables and so on. For example, station names and cities can be presented in one colour, destinations in another and departure and arrival times and on-board facilities in a third. Some of the advantages and disadvantages of the use of colour have been summarised by Filley (1982) and are given in Table 13.2.

Generally, red is used to mean 'Stop' or 'Danger', green to mean 'Go' or that the system is running normally, and orange to signify caution. This is unfortunate because approximately 5–8% of people have colour vision defects (so called 'colour blindness'). The most common deficit is an inability to distinguish red and green! (they both look brown). Approximately 7 million drivers in North America have this problem (University of California, 1993) and some studies suggest that colour-deficient drivers have problems seeing traffic lights. *According to Clement (1987), because 4% of males cannot distinguish green and red, the optimal design for traffic lights is to use red/orange stop indicators and blue/green go indicators.*

The use of unsaturated colours can help to improve the discriminability of coloured signals for colour-blind users. Unsaturated colours emit light at a dominant wavelength (from which the perception of colour arises) but contain light from other parts of the spectrum as well. Differences in the composition of these latter wavelengths may improve discriminability. Alternatively, designers should use shape as well as colour for coding purposes. It is also unfortunate that there does not seem to be a population stereotype for the colour blue because disorders involving the perception of blue are rare.

Resolution of detail

The ability to resolve detail depends on the accommodation of the lens of the eye, the ambient lighting and the visual angle (the angle subtended at the eye by the viewed

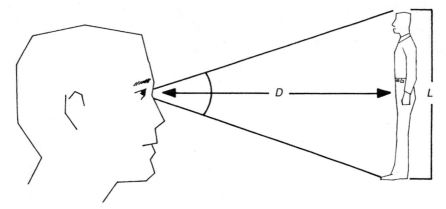

Figure 13.5 The visual angle.

object; Figure 13.5). Visual angle is a more useful concept than absolute object size because it takes into account both the size of the object and its distance from the viewer. Under good lighting, a minimum visual angle of 15 minutes of arc is recommended. Many instruments such as dials and gauges are designed to be read at a reference distance of about 70 cm. If the reference distance is known, the required change in object size to equalise the visual angle when the object has to be viewed at a different distance can be calculated from

$$\text{Size at new distance} = \frac{\text{Size at reference} \times \text{New distance}}{\text{Reference distance}}$$

For example, if the reference distance is 70 cm and the new distance is 140 cm, the size of the object would have to be doubled to maintain the visual angle (see Bailey, 1982, for further information). The ability to resolve detail also depends on human information-processing limitations. The early stages of information processing can handle large amounts of information at a primitive level, but capacity diminishes as soon as more complex processing is required.

Murrell (1971) carried out detailed investigations of the variables influencing dial design. Example designs are given in Figure 13.6.

Colour coding of dials

Colour coded dials do not require the operator to memorise exact scale values that correspond to normal, unsafe or suboptimal states, but exact values can be taken when needed. Colour coding of dials can reduce the memory load of routine check reading tasks (red for danger, green for safe and orange for caution).

Digital displays

Digital displays should be used in situations where highly accurate reading of displayed quantities is required. Digital displays are inferior to analogue displays for conveying information about rates of change, so in those cases where both types of

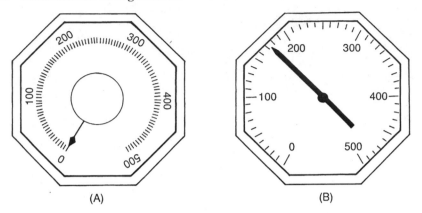

Figure 13.6 Design of scales. Dial (B) has been made easier to read by having fewer, bolder, graduation marks. The scale length has been increased by putting the numerals on the inside. The numerals themselves are larger and level. The whole of the pointer is visible. (Adapted from *Applied Ergonomics Handbook*, 1974, with permission from Butterworth Heinemann.)

information are required a compromise design can be used in which a digital counter is superimposed on the dial face.

Multiple display configurations

Despite the existence of display design guidelines, these are often neglected by designers. In an investigation of nuclear power station control room design, Seminara and Smith (1983, 1984) commented on the complexity of the control rooms and on the massive arrays of control panel elements that were often identical in appearance. There were insufficient visual cues to facilitate the differentiation of subsystems. It was observed that experienced operators became familiar with the panels they operated most frequently, usually through routine operation, and were least familiar with least-used panels – those that were most critical in emergencies.

The Three Mile Island nuclear incident drew attention to the need for ergonomics in the design of control rooms and in the training of operators. Malone *et al.* (1980) evaluated the control room at Three Mile Island. The information required by operators was often found to be non-existent, poorly located or difficult to read. Annunciators were not colour coded and, in one subsystem, 91% of the applicable human engineering criteria for display design were not met. In addition, in the control room, 1900 displays were located on vertical panels, 503 of which could not be read by a short (5th percentile stature) operator. The labelling of panels was also inadequate – it was discovered that the operators themselves had made 800 changes to the existing panels in an effort to improve them.

An example of an ambiguous panel design is given in Figure 13.7a. The same panel, after boundary enhancement, is depicted in Figure 13.7b.

The ability to gather information quickly by shifting focal attention around a complex display depends on several factors. Sanders (1970) demonstrated that as the visual angle between two targets was increased, the reaction time and number of errors needed to make a 'same/different' comparison between them also increased.

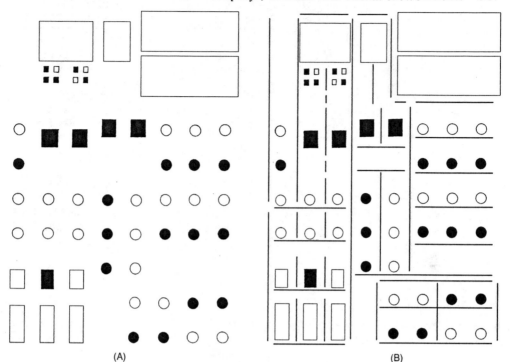

Figure 13.7 Remedial ergonomics to improve panel design. In (A), the original panel design is depicted. In (B), black tape has been stuck onto the panel to make clear the true subsystem boundaries. (Redrawn from Seminara and Smith, 1983, by permission of Butterworth-Heinemann.)

Sanders identified three functional attentional fields. Objects placed within 30 degrees of each other fall into the stationary field and can be detected with the stationary eye. The eye field encompasses a region from 30 to 80 degrees and eye movements are required to compare objects. Beyond 80 degrees head and eye movements are required to detect targets. Focal attention can be likened to a 'spotlight' that illuminates parts of a complex display, making them accessible to higher cognitive processes. Parts of the display within the spotlight can be processed simultaneously, whereas those outside the spotlight require shifting of attention and search.

Data on the extent of the various functional attentional fields are useful in designing complex control panels. If an operator has to monitor several displays simultaneously and make judgements using the data from all of them, they should be placed in the stationary or eye fields with respect to a pre-defined point of observation (such as the operator's desk). Careful analysis of the operator's task and operational priorities including emergency recovery procedures can be used to assist in optimising a panel layout.

Radar screens, switchboards, computer terminals and so on consist of control-display configurations of varying complexity. Operator interaction with these may be constrained or unconstrained. In the former case, the system may permit only a limited number of sequences of specified control actions. In the latter case, many more complex sequences may exist owing to richer task variety (e.g. creating text,

Table 13.3 Common criteria for evaluating the layout of complex panels

1. Eliminate unnecessary or complex movements.
2. Locate controls and displays to reduce postural load.
3. Reduce movement complexity, encourage muscle memory by consistent choice of movements.
4. Control actions should not obscure important display areas.
5. Avoid spatial transformations.
6. Establish spatial proximity of displays and controls used frequently in a sequence of tasks.

programming a computer or playing computer games). In either case, frequent sequences of operator–system interaction should be identified as part of an early attempt to simulate the operators' tasks. Optimal locations for the placement of controls and displays can be arrived at by considering the sequences of hand and eye movements needed to carry out a task. Some examples of criteria for evaluating prototype layouts are given in Table 13.3.

Guiding visual search in complex displays

In large, complex displays, special methods for highlighting potentially hazardous situations and guiding visual search to the appropriate part of the display may be required. The process of interpreting a complex display can be decomposed into a number of stages:

1. Alerting operators to the existence of a signal or target data
2. Orienting the perceptual system to the appropriate part of the display
3. Attending to the data so that it can be transmitted to the central processing centres of the brain

Posner *et al.* (1976) found that *visual stimuli are generally less alerting than non-visual stimuli* and people have to attend to visual stimuli consciously. Non-visual cues are better at alerting operators to hazardous situations, since they demand less attention. Alerting cues appear to improve an operator's readiness to respond and to speed up reaction time, but may do this at the expense of accuracy. Informative cues (which provide information about the type of event to come) seem to assist in the *encoding* of the signal being attended to (the operator is 'primed' to respond in a particular way) as well as having an alerting quality. Improved encoding of a signal would be expected to result in improved response accuracy. It appears that informative, non-visual cues can speed up reaction time and response accuracy and are particularly effective when the positional uncertainty of a target is high. For example, Perrot *et al.* (1990) found that visual search could be aided considerably when the appearance of a visual target was accommpanied by a spatially orientated sound (a 10 Hz click train directly behind the part of the display where the target appeared). Improvements in response latency of up to 300 milliseconds were obtainable using this method. The effectiveness of the auditory signal increased as a function of visual load and the distance of the target from the subject's line of gaze. Auditory spatial information seems to offer benefits in situations where operators have to detect multiple targets in spatially extended displays such as are found on aircraft flight decks and in some process industry control rooms (see Posner (1980) and Wickens (1984) for further discussion of factors influencing visual attention). The information content of auditory cues can be increased by varying pitch, continuity or waveform.

Computer-generated displays

Screen-based displays offer, in addition to increased flexibility, the possibility of using innovative graphical techniques for representing the changing state of a system in a way that is more compatible with operator characteristics. Instead of the values of system variables being displayed on dials, scales and counters, more dynamic, integrated displays such as on-screen histograms, time series and pie charts can be used. These displays can be continually updated as the system state changes. Research is under way to determine criteria for choosing between the various design options. The goal is to ensure that a display supports *direct* perception of a system state. With direct perception the need for interpretive mental operations is minimised, so operator reaction time and mental workload are reduced. Hollands and Spence (1992) compared display options for change and proportion in system variables. They found that the speed and accuracy of detecting change was superior when system data were displayed in the form of time series or histograms (bar charts), whereas perception of proportion was best when in pie chart or divided bar chart presentation mode.

Many computer systems only display information pertinent to the current task. Although this is very economical and avoids some of the problems of visual complexity discussed previously, potential problems are introduced – particularly of users losing track of where they are with respect to the rest of the system. Several design solutions have been proposed to assist users in finding their way around a system. A common approach is to use icons (as in Figure 13.4, for example) to represent system objects, thereby reducing the need for users to learn system terminology and command names. The icon is a symbolic representation of the object (usually a stylised, abbreviated picture of the real object, with strong figure–ground differentiation and closure). A diary can be represented as a book, the calculator facility as a pocket calculator and so on. Icons can be used to create unambiguous, concise abbreviations of system objects but may be less successful when used to represent more abstract parts of the system (such as a file directory). In product design for a global market, the replacement of words with icons to describe system functions would seem to overcome the problem of language differences. There is empirical evidence that well-designed icons can be as effective as English words for describing photocopier functions to English speakers (Howard *et al.*, 1991). However, it is an assumption that icons can overcome language barriers, and replacing English words with poorly designed icons may render a product incomprehensible to English speakers as well as everyone else! Fortunately, international standards on the design of symbols have been developed to overcome this problem (see ISO 7000, 'Graphical Symbols for Use on Equipment–Index and Symbols', for example). *Miniatures* (Nielsen, 1990) are similar to icons but are more flexible. They are created by the system from the information to which they refer. Thus, a page of text and graphics or a letter that a user has recently created can be 'miniaturised' (represented in miniature) and displayed in a window on a separate part of the screen. While carrying out a sequence of tasks, a user can be presented with a symbolic representation of the miniaturised information landscape in which he or she is working. Some modern word processing

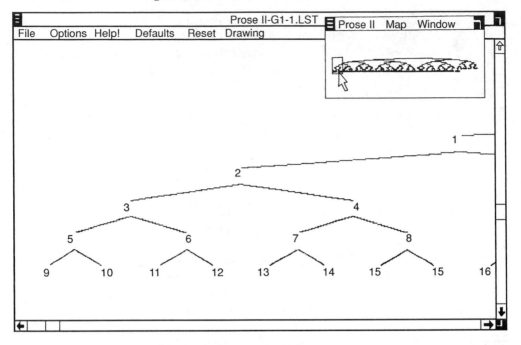

Figure 13.8 Beard and Walker's map window. The arrow indicates the position of the wire frame box used to select part of the database for viewing. (Adapted from Beard and Walker, 1990.)

systems present the user with a miniature of a page to indicate how it will appear when printed. This is particularly useful when different sizes of type are being used together or when text and graphics are combined on one page.

The problem of 'getting lost' in large databases (not knowing the location of the current position in relation to other parts of the database and other system facilities) is exacerbated if the system structure is complicated or ambiguous or if the user's mental map of the system is incorrect or incomplete. For this reason, a number of attempts have been made to present the user with an explicit, graphic representation of the system structure.

Beard and Walker (1990) addressed the problem of user navigation of large data spaces. Such spaces can often be represented as a hierarchical tree structure. Owing to the small size of the VDU screen, only parts of the tree can be displayed at any one time and the user may lose track of his location in the system. Beard and Walker developed a map window in which a miniaturised version of the entire data space is presented. Inside the map window, a box (called the wire frame box) is superimposed over a part of the miniature version of the data space (Figure 13.8). The part of the database within the box is shown on the rest of the screen. Decreasing the size of the box increases the magnification of the data displayed on the screen (the 'zoom' facility). Users can explore the data space by moving the box to different parts of the miniature data space. The map window improved user performance significantly.

With large databases, the usefulness of spatial representations may be limited because the spatial representation becomes visually complex. Other ways of assisting users may be more effective. One alternative is to provide an analogy – an explicit, verbal model that compares interrogating the database with a more familiar activity (Webb and Kramer, 1990). For example, 'It's a bit like shopping in a mall, you first decide which shop to go in, then you enter and choose among the items in the shop, then you leave the shop and choose another shop.' This analogy is appropriate for a hierarchical database because it conveys the idea of there being separate levels of activity – choosing shops and choosing things in a shop – and that you first have to leave a shop (return to a a higher level in the hierarchy) before you can choose something in a different shop (part of the database).

Hypertext systems are databases in which items of information are linked together non-sequentially – posing challenging problems for display designers (Smith and Wilson, 1993). Hypertext spaces contain information at each node and are 'n-dimensional' in the sense that a given node may be the point of departure for many other nodes. A characteristic of user-interaction with hypertext systems is that the development of an optimum search strategy is de-emphasised in favour of a more intuitive approach in which, en route to a desired item, other related items are visited. The user is encouraged to acquire a mental map of the space as a consequence of browsing through it. Failure to develop an adequate mental map can lead to disorientation when using hypertext, which, according to Foss (1989), may have the following results:

1. Users reaching a particular node and then forgetting what was to be done there.
2. Users forgetting to return to the main path after making a digression.
3. Users forgetting to make pre-planned digressions.
4. Users not knowing whether there are other relevant frames in a document.
5. Users forgetting which areas were previously visited.
6. Users having difficulties summarising information examined during a session.

Smith and Wilson discuss some of the issues in the development of navigation aids for hypertext systems. Aids can be either schematic or spatial and they can be displayed in either two or three dimensions. Of particular importance seems to be the correspondence between the type of navigational aid chosen and the semantic content of the hypertext system. Two-dimensional schematic aids, such as those described above, are appropriate for smaller systems. With larger systems, these displays can be confusing because of the 'spaghetti-like' entanglement of links between nodes. Three-dimensional schematic displays can aid user-navigation through complex spaces. In an empirical investigation of the use of maps to assist navigation through hypertext, Stanton *et al.* (1992) found that maps were ineffective (they used two-dimensional maps to represent a small domain with a total of 42 screens). The investigators caution that the use of a map may actually interfere with users' development of their own mental maps of the system and may increase the cognitive demands of learning to use the system. As with any representation of this nature, users have to learn to use it in addition to having to learn the contents of the system itself and its structure.

The representation of complex information spaces seems to require a consideration of the nature of the information in the knowledge domain and the achievement of compatibility between this and the user's typical representations of the knowledge when carrying out representative tasks. It is suggested that certain domains will be best represented in virtual environments (see below).

Maps and navigation aids

Some tasks require an operator to hold a spatial representation of the display in short-term memory. This introduces an added component of mental workload. Wickens (1984) provides a general discussion of the issues involved in navigation and map reading. Map reading is a skill that has to be acquired like any other and does not seem to be well developed in the population at large. Problems can occur when members of the public are expected to navigate complex buildings and facilities. For example, modern shopping centres often have complicated floor plans. Maps of the floor plan ('You are here' maps) are usually provided at intervals to assist shoppers in finding their way around. The difficulty of using these maps depends not only on the map's intrinsic design, but also on its orientation with respect to the shopping centre. We can imagine a compatible situation in which the map is situated horizontally in the middle of the shopping centre with its north end facing the north end of the building. A one-to-one correspondence exists between a route to a destination on the map and a route to a destination in the real building. Alternatively, we can imagine the map situated on a vertical pillar with the north end of the map facing the west of the building. The difference in orientation between the map and the building now has to be taken into account in order to convert the directions of movement required to reach a destination using the map into a route applicable to the real building. This increases the mental workload of the navigation task.

Butler *et al.* (1993) compared the effectiveness of a number of navigation aids (referred to as 'wayfinding aids') in helping newcomers find their way around a complex building. Wayfinding signs (and signposts) were found to be more effective than 'You are here' maps. Signs and signposts seem to result in superior navigation because they place the directional and spatial information required for navigation into the environment itself. The information conveyed by the sign is used when needed and can then be dispensed with. On the other hand, 'You are here' maps impose a considerable mental load on the user. The cues needed to navigate have first to be identified on the map, then held in memory and finally recognised in the environment. Butler *et al.* emphasise that the information displayed on signs must be simple and unambiguous for sign-based navigation systems to be effective. Additional improvements can be obtained by numbering the objects in the space to be navigated and incorporating the numbers into the sign system. Butler *et al.* use the example of hotels, where bedrooms are often numbered and are easy to find, whereas rooms containing other facilities are only named and are often harder to find. The navigability of many shopping centres could no doubt be improved in a similar way by placing more emphasis on the use of numbers, rather than names, as cues to the location of particular shops.

Three-dimensional displays

Our experience of a three-dimensional (3-D) visual world is spontaneous and is dependent on a number of cues, both internal and external to our senses. A knowledge of these cues is essential in the design and analysis of 3-D displays. 3-D displays are becoming of increasing importance. They are used in the design of simulators, in computer systems for the display of 3-D structures such those of as protein molecules and in the design of virtual environments.

Monocular depth cues can be detected using one eye only but are also operative when both eyes are used:

- Accommodation: Proprioceptive feedback from the ciliary muscles provides valuable information about the distance of an object However, this feedback is of most value in judging the distance of objects up to approximately 6 metres.
- Movement parallax: Depth perception is strongly influenced by head movements, which give rise to the relative movement of close objects against an unmoving background. When one is travelling by car, distant objects appear to travel in the same direction of the car and close objects in the opposite direction.
- Interposition: A near object, interposed between a more distant one and the viewer, provides depth cues when it obscures part of the far object.
- Relative size of familiar objects: The size of the retinal image can be used as a depth cue if the size of the object is known. Familiar objects exhibit *size constancy*. A car or a house is perceived as being distant if its image on the retina is small. Young children learn to associate decrease in retinal image size with an increase in viewing distance rather than a change in the size of the object itself (people do not shrink when they walk away from you).
- Texture gradient: Seen in perspective, the surfaces of many objects present a texture gradient, with texture changing from rough to smooth as distance increases.
- Linear perspective: This is an important monocular depth cue that is used by artists to give the third dimension to a painting. Perhaps the most obvious example of perspective is the way parallel railway lines appear to converge as distance increases or the way an object's shape appears to change as it is rotated about an axis. Bemis *et al.* (1988) used perspective to change a 2-D circular graphic display to a more 3-D format by changing to an oblique rather than straight-ahead viewing angle. Response time and errors on a detection task were greatly reduced using the perspective display.
- Casting of shadows: A light source in the direction of a near object will cause it to cast a shadow on objects farther away. This is often used as a cue to depth on computer systems equipped with graphic interfaces.

Figure 13.9 illustrates the effects of some of the monocular depth cues that can be used to add a third dimension to two-dimensional displays.

Binocular depth cues depend on the use of both eyes:

- Retinal disparity: Since the eyes are approximately 6 centimetres apart, the retinal images differ slightly. The two images are 'fused' to create a single percept characterised by depth. This is known as *stereopsis*. Retinal disparity is a powerful cue sufficient to give the perception of depth in the absence of any other cues. It is the basis of stereophotography and '3-D' films. The most common method of producing three-dimensional displays using the disparity principle is to use two cameras a small distance apart to create a stereoscopic pair of images. Special spectacles (differing in colour or plane of polarisation) can be used to view the differently coloured or differently polarised members of the stereo scopic pair such that a different image is presented to each eye in much the same way as when the real object is viewed. As viewing distance increases, the lines of sight of

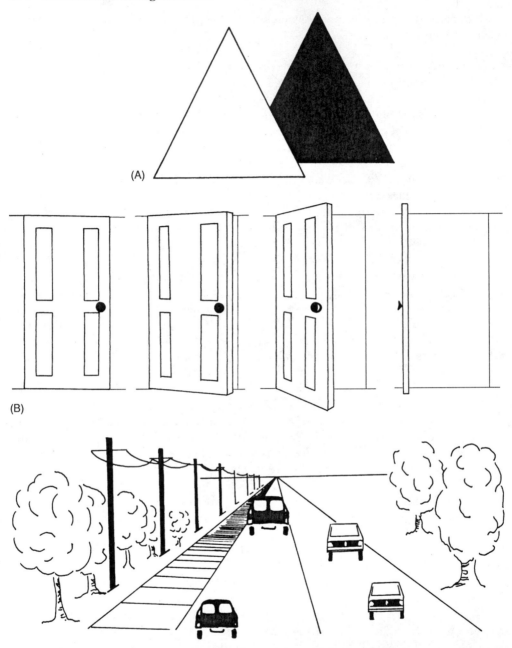

Figure 13.9 Some common depth cues for use in display design. (A) Superposition. (B) Change in shape of a three-dimensional object. (C) Linear perspective: the upper automobile looks bigger because perspective leads us to perceive it as farther away – the object that gave rise to the retinal image must therefore be larger. The types of cues in (A) and (B) are commonly used to give depth in two-dimensional displays and virtual environments.

the two eyes tend to the parallel and retinal disparity diminishes (and is negligible after about 20 metres).

- Convergence: Kinaesthetic feedback from the vergence system is a cue to depth that functions up to about 15 metres.

Auditory displays

Synthetic speech

Voice input to humans (by machines) is possible and is a potentially very powerful display modality. Cowley and Jones (1992) have reviewed some of the ergonomic aspects of the use of computer speech and compared the relative advantages and disadvantages of using digitised versus synthesised speech. Synthesised speech is generated using a set of rules, whereas digitised speech consists of real utterances by humans, which are stored in digitised form and can be called up and manipulated at will. Synthesised speech is less intelligible than digitised speech but may be more flexible.

Marics and Williges (1988) investigated some of the factors influencing the intelligibility of synthetic speech. Subjects transcribed synthetic speech messages via a computer keyboard. Prior knowledge of context reduced transcription error rates by 50%: for example, if a message such as 'rain ending later today' was preceeded by the word 'weather' subjects were less likely to misunderstand the message. Words at the end of an utterance were less likely to be misunderstood than words at the beginning, which suggests that designers might design phrases that begin using common or easily understood words, with low-frequency words being left to the end. Speaking rates of less than 250 words per minute were found to be preferable for novice users, but the authors suggest that user control of speaking rate might be a desirable feature of synthetic speech interfaces, as would a 'would you repeat that?' facility. A final interesting finding that contrasts speech with other types of displays and feedback is that subjects were aware of errors even though the system did not inform them about incorrectly transcribed messages. Speech may therefore be a valuable form of display in situations where it is necessary to block the transmission of errors from one part of a system to another. However, synthetic speech appears to place greater encoding demands on listeners than natural speech and is comprehended more slowly (Ralston *et al.*, 1991). These authors suggest that low-quality synthetic speech should not be used for tasks where rapid response to spoken words is necessary or for linguistically demanding secondary tasks where there is competition for conscious processing resources.

Auditory warnings and cues

Although vision is the dominant sense in humans, auditory warning signals and displays are often extremely valuable. An effective auditory warning should be of sufficient intensity to stand out above background noise and differ in pitch, waveform, etc., so as to be discriminable. The ear is most sensitive to the frequency range 500–3000 Hz, so warnings should ideally be in this range with high intensities at the lower frequencies because these are not deflected by objects and travel farther than high frequencies. Warnings should not be so excessively loud as to increase workers' daily noise dose above acceptable levels (louder is not 'better' beyond a certain point). Certain types of warning signal may be more appropriate for some types of hazards than others. Lazarus and Hoge (1986) found that different types of danger signals were judged by workers to be more compatible with certain situations than others. Sirens were judged to be compatible with danger arising from 'rays' (radioactivity, lasers) and danger areas. Horns were compatible with danger from machines. An alarm similar to that used on police cars was associated with fire or vehicle danger. Impulsed sounds such as bells had low compatibility with both specific and general danger situations and appeared to be more suitable for signifying rest pauses or the end of a shift.

For a sound to serve as an auditory cue it must first be heard above the background noise and it must attract attention. From the theoretical discussion of cochlear function, this suggests that the two key design parameters are cue intensity and frequency. If the frequency of the cue is very different from that of the noise, it can be more easily sensed. Murrell (1971, quoted from Oborne, 1982) suggested that auditory cues should be at least 10 dB louder than the background noise. Miller and Beaton (1994) recommend a signal-to-noise ratio of 8–12 dB(A). ISO 7731 (1986) recommends that the warning sound exceed the the background by 13 dB in at least one third-octave band in the region 300–3000 Hz. Sound attenuates according to an inverse square law, so the farther away the listener is from the source, the greater the attenuation. Less attenuation takes place in confined spaces than in open spaces because of interreflection of sound waves. In practical situations, then, the following considerations are essential in the specification of auditory cues:

1. Intensity of background noise at the point(s) where the cue is to be heard. This requires a noise survey of the premises at times when the background noise is at its loudest to determine background sound levels.
2. Frequency of background noise. A frequency analysis of the background noise can be carried out to assist in the selection of an auditory cue of a different frequency. Much machinery produces complex sounds with many different frequencies (Sorkin, 1987). In this case, the cue must be selected so that it is more intense than the higher-intensity frequencies, particularly if these are of similar frequency to the cue. Sorkin suggests that in the case of very high-intensity background noise (pneumatic machines produce sound levels of over 115 dB(A)), auditory cues should not be used at all.
3. Attenuation of cue intensity from source. If the cue is produced at a different place from where it is intended to be heard, it will be necessary to determine by how much the source intensity of the cue is attenuated at the points where it is to be heard. It is at the listening points that the cue must be distinguished from the noise.

Apart from intensity and frequency, the cue must have high discriminability – it must be different from any other sounds the operator is likely to hear at work. Sirens, for example, can be made to warble or fluctuate in intensity or frequency. Auditory cues should persist for at least 100 milliseconds.

Some of these design considerations are illustrated by Miller and Beaton's (1994) analysis of the problems in detection of emergency vehicle sirens by motorists. In the USA, emergency vehicle sirens have an intensity of 118 dB(A) at a distance of 10 metres from source. This is a maximum limit that was set to prevent ear damage and limit noise pollution. Although pedestrians will usually hear an oncoming emergency vehicle at a great distance, drivers of other vehicles will not. This is because the interiors of many modern cars attenuate sound by 20–30 dB(A) if the windows are closed. Thus, at 10 metres from an emergency vehicle, the intensity of the sound at the driver's ear will be about 98 dB(A). At 50 metres, it will be about 74 dB(A). Since the intensity of sound in a car with the radio turned off is about 70 dB(A), the driver is unlikely to hear the siren of an oncoming emergency vehicle until it is about 25 metres away. If the emergency vehicle is travelling at a speed 50 km/hour faster than the car, the driver will have 1.8 seconds to take evasive action. Clearly this is not long enough and explains why sirens alone are inadequate to prevent emergency vehicles being impeded by traffic (alternative methods of alerting drivers to the presence of emergency vehicles using intelligent vehicle highway system (IVHS) technology are discussed in Miller and Beaton, 1994).

Auditory cueing in visual search

A recent development in the design of auditory warnings is the use of stereo headsets to provide a three-dimensional 'head-up' audio display. Begault (1993) compared a monaural warning with a binaural warning presented over stereo headphones. Pilots had to visually locate a target aircraft while flying a simulator. Auditory collision-avoidance warnings were provided under the two conditions of presentation. Crew members using the three-dimensional display acquired targets approximately 2 seconds faster than those using the monaural warning.

Bolia *et al.* (1999) evaluated the effects of a spatial audio display on the performance of pilots searching a complex visual scene with distractors. The auditory cue was presented spatially via a 'grid' of four loudspeakers. Modulating the intensity of sound from each produces an auditory 'image' of the location of the target in the region of space where the pilot is visually searching for the target. The aiding was found to reduce search time by a factor of 6 or more in complex searches, with no reduction in the accuracy of search.

These findings illustrate the principle that display of targets in several sensory modalities at the same time can aid performance. In particular, the auditory sense seems to be intrinsically directional and largely pre-attentive, which makes it compatible with demands for focal attention in visual search.

Advantages of auditory displays

Advantages of auditory displays include the facts that auditory displays are more alerting than visual displays; they are 'eyes free' and 'hands free', capturing attention during the performance of other tasks; they one suitable when the message is short, when they need not be referred to later, and when the user's job requires movement (for example, among banks of complex control panels). They are also useful to communicate with people working in dark places (such as mines) or at night.

Voice warnings

Wolgalter and Young (1991) investigated behavioural compliance to voice and print warnings. They found that for simple warnings, compliance was greater to voice than to print and greater still when the two were combined. They argue that voice warnings have been relatively unexploited and offer many advantages over other types of warning. Voice is attention-getting and omnidirectional and can orient people to hazards such as 'slippery floors' in complex and visually distracting environments such as malls. Compared to other auditory warnings, voice conveys information directly. However, voice is less useful for conveying complex warnings because of its serial nature, transmission load and fragility. Wolgalter and Young suggest that it may be a useful supplement to complex visual warnings, capturing attention and in providing effective contextual orientation.

Baldwin and Struckman-Jones (2002) point out that auditory warnings are likely to be used in situations where operators are under high mental workload. Demand is placed on the attentional resources required to carry out the task and also to detect and interpret the warning. In conditions of low mental workload and low background noise, warning intensities ranging from 30 to 80 dB give almost 100% correct identifcation of the spoken word. However, in dual-task situations (e.g. detecting a warning while driving a car), warning intensity becomes critical. Subjects in the Baldwin and Struckman-Jones experiment had to judge the veracity of sentences derived from Collins and Quillian (1972) such as 'Birds can fly' and 'Dogs have five legs'. Response time dropped from 1.05 seconds when the warning intensity was 45 dB to 0.9 seconds at 65 dB, while errors dropped by two-thirds. It seems that the detection of 'quiet' warnings does indeed demand additional attentional resources compared to 'noisier' warnings and, in dual task situations, *a minimum intensity of 65 dB is needed.*

Representational warnings and displays

Representational auditory warnings have been termed 'earcons' based on an analogy with icons, their visual counterparts (Blattner *et al.*, 1989). Gaver (in Graham, 1999), states that the maaping between an icon and its referent can be *symbolic* (arbitrary, having to be learnt like the sound of an ambulance siren); *metaphoric,* where the sound shares common features with the referent (a fall in pitch to represent a reduction in quantity); or *nomic,* where the sound the mapping is based on physical causation (the sound of garbage falling into a can when a file is deleted). Nomic icons are the most direct and might be expected to give the quickest reaction for the least learning. According to Mynatt (1994) the requirements for the design of effective icons are as follows:

- The sound must be identifiable (experienced in the work context and sufficiently different from other sounds).
- The mapping between the icon and its referent must be intuitive and close.
- The physical parameters of the sound must be apropriate to the context (e.g. the sound must not be too long).
- The sound quality must be adequate.
- The sound must lead to an appropriate emotional response from the user.

Graham (1999) compared the effectiveness of two conventional and two auditory icon warnings of impeding collision on driving behaviour in a simulator (a buzzer, the word 'ahead' and a car horn and the sound of a car skidding, respectively). All sounds were of 0.7 seconds duration and approximately 60 dB(A). The auditory icons reduced reaction time by 0.1 second but increased the number of false positive responses (braking when there was no imminent danger), suggesting that the icons require less cognitive processing. There were no differences in the number of dangerous situations that were missed. The car horn sound was judged to be less amibiguous than the skidding sound. Belz *et al.* (1999), also compared skidding sounds and horns in car collision-avoidance systems and found a reduction in reaction time of 0.122 seconds when the icons were used instead of conventional auditory alarms. Interestingly, reaction time to the conventional warning was no faster than to no warning at all, suggesting that conventional auditory warnings need time to be interpreted before the driver can respond.

Representational alarms can potentially increase the amount of information that can be communicated, according to Stanton and Edworthy (1994), and may be advantageous in systems or products used by different language speakers. New population stereotypes may emerge among user populations if earcons are implemented in a consistent way in future systems.

Design of controls

Some of the basic considerations in the design of controls have already been discussed in Chapter 5 on musculoskeletal ailments and hand tools. Controls should be designed to be operable in low stress postures and without static loading of body parts, particularly the fingers. Control dimensions should be determined using appropriate hand/foot anthropometry and a knowledge of the mechanical advantage needed to enable the user to easily actuate the control.

Vehicle controls

Steering wheels, joysticks and pedals are commonly used to control vehicles. The usability of controls can be affected by the resistance of the control, which should be operable using forces that are a fraction of the operator's maximum voluntary contraction. However, the control should offer some resistance to movement so that 'bumping' errors and muscle tremor do not cause control errors (Young, 1973).

With position controls, the displacement of the control is in direct relation to the change in the controlled part. Pressure controls (sometimes called 'isometric' controls) are almost immovable and the force on the control stick provides the control signal. Pressure controls are of use in controlling higher-order systems that exhibit lag (see below). Some options for control design are control stick friction, pre-load, control/display ratio and control system hysteresis. Static friction can aid performance in environments in which operators are exposed to vibration. Inertia and viscous damping can be built into controls to assist with the execution of smooth control actions.

Control distinctiveness

In many instances, numerous controls are grouped together on a panel and the designer's task is to ensure that operators can easily distinguish between different controls. Some of the display design guidelines above are relevant here. In addition, manual controls provide the operator with tactile feedback that can be used to give a distinctive identity to a control or related set of controls.

Table 13.4 Recommended resistance values (in newtons) for controls[a]

Control	Optimal resistance to operation (N)
Finger-operated control	2–5
Handles operated forwards and backwards	5–15
Handles operated sideways	5–15
Steering wheel	5–20
Leg-operated pedals	45–90
Ankle-operated pedals	20–30
Buttons operated with finger tips	2

[a] From EN 14386.

McCormick and Sanders (1982) recommend using several dimensions to code different controls so as to enhance their distinctiveness. Designers may choose from shape, colour, texture, size location, operational method, position and labelling. The main tactile cues that may be used to identify controls are texture (e.g. knurled, fluted or smooth surface), shape (e.g. circular, triangular, square) and size. Size differences seem to have to be fairly large to reliably separate different controls (approximately 1.5 cm diameter difference and 1 cm thickness difference). From the research reviewed by these authors, it seems that several combinations each of shape, size and texture can be used to provide a set of easily distinguishable controls. The selection of controls that provide good tactile cues is particularly important when operators have to work in poor lighting or in emergency situations where multiple cues may assist recognition.

Control resistance

Some recommended resistance values for various types of controls are given in Table 13.4 (from EN 14866).

Keyboards

Most keyboards used with visual display terminals have the letters of the alphabet arranged in the QWERTY layout. This layout was designed in the nineteenth century to physically separate the type bars of common letters on mechanical typewriters and so prevent them from jamming. The result was that many common letters were placed at a distance from the 'home' row of keys and this is now thought to slow typists down. The QWERTY layout has costs: it overworks the left hand and some of the fingers and underutilises the home row of keys. Since the arguments in favour of the QWERTY layout no longer apply, several investigators have evaluated alternatives. The best-known are the alphabetic layout, in which the keys are arranged in alphabetical order over three rows and the Dvorak layout (Figure 12.10, with the QWERTY layout for comparison), which was developed in the 1930s using data from time and motion studies.

The Dvorak layout places the most common letters in the middle of three rows. The left hand types the vowels, *a, o, e, u* and *i* and the right hand types the most common consonants, *d, h, t, n,* and *s*. Thus the two hands tend to work in sequence much of the time. Tanebaum (1996) reports that, using the Dvorak layout, typists make 37% fewer finger movements, with the right hand making 56% of the key-

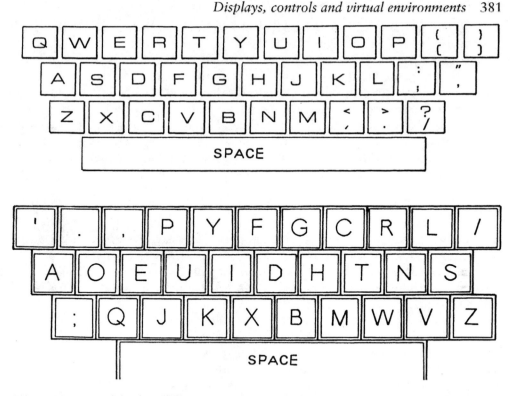

Figure 13.10 Rival keyboard layouts: QWERTY and Dvorak. QWERTY is used because QWERTY arrived first.

strokes (the right hand being stronger in most users). The forefinger and middle finger of both hands type the more common letters *e, u, h* and *t*, whereas with QWERTY these fingers type *d, f, j* and *k*. Although there is evidence that Dvorak users do type faster than QWERTY users (estimates vary from from 25% to 5%), the Dvorak arrangement has never been taken up. Norman (1982) suggests that the Dvorak layout brings only about a 5% improvement in typing speed and this may not be large enough to justify the expense of retraining millions of users. Tanebaum (1996) reports that the Dvorak keyboard was evaluated by the US Navy in 1944. Despite the positive feedback from typists, training on Dvorak required over 50 hours. Although productivity improvements might be expected to justify the expenditure, it is likely that initial training costs act as a barrier to widespread adoption of the layout.

The alphabetic layout is often assumed to be better than the QWERTY layout for naive users ('hunt and peck' typists). Norman (1982) compared several alphabetic keyboards with a random keyboard and with the QWERTY keyboard. Non-typists typed for 10-minute sessions on each of the keyboards. Typing speed was 7.1 words per minute on the random layout, 7.8 words per minute on the alphabetic layout and 13.0 words per minute on the QWERTY layout. Norman suggests that part of the reason for the superiority of QWERTY was that subjects had, at least, some familiarity with it and that its layout is more efficient than that of the alphabetic and random layouts. The alphabetic layout was superior to the random layout by about one word per minute. However, Norman suggested that the alphabetic layout did not yield large benefits because naive subjects do not use their knowledge of the alphabet to type with it. Rather, they visually scan the keyboard looking for the next key in the same way as with the random keyboard. For naive typists, the *mental workload* of reading and typing text leaves little spare capacity for recall and rehearsal of the alphabet.

Modern computer keyboards differ from their mechanical ancestors in another way: lower forces are required to activate the keys and the tactile and auditory feedback from the keys is less distinct. Yoshitake *et al.* (1997) reported 'quiet' keyboards with indistinct tactile feedback gave higher keying error rates than keyboards with a distinct clicking sound and tactile feedback. There may be scope for improving keyboard performance by introducing software-generated auditory feedback in computer keyboards, or at least providing this as an option.

Soft keyboards

Cellphones, personal organisers, etc. are examples of consumer electronics products that require alphanumeric entry for text messaging and storing personal information. One way of supporting these facilities is by means of 'soft' keyboards displayed on screen and accessed by means of a stylus. MacKenzie and Zhang (2001) compared keying peformance on keyboards with large (10 mm by 10 mm) keys and small (6.4 mm by 6.4 mm keys). Subjects entered text using either a random or a QWERTY layout. Although keyboard size had no effect on text entry rate, keyboard layout did, with the QWERTY layout yielding text entry rates of about 20 words per minute compared to 5 words per minute for the random layout. Typing speed on a physical QWERTY keyboard was positively correlated with text entry speed on the soft QWERTY but not with typing speed on the random soft keyboard, indicating positive training transfer. It seems that efforts to improve the efficiency of soft keyboards using new layouts that minimise stylus movement will be hampered by the loss of transfer from 'hard' typing.

Space limitations on products such as cellphones have led to alternative keyboard layouts with non-QWERTY arrangements and sometimes more than one letter per key (Figure 13.11). The FITALY layout is designed for one-handed operation to minimise hand movements. The telephone keypad layout contains three letters per key, as does the JusType layout. Sears *et al.* (2001) compared visual search times in the different layouts with the following conclusions.

- Search times were shortest for layouts that were familiar and placed one letter per key.
- Familiar layouts (such as the telephone keypad) with more than on letter per key produced longer search times.
- Unfamiliar layouts (such as FITALY and Dvorak) produced the longest search times even with one letter per key.

Even for novel and innovative consumer products, transfer of skills, knowledge and experiences cannot be ignored when designing the user-interface. For optimal text entry performance when a keyboard is required, the QWERTY layout seems best. The possible health benefits of split keyboards with separate banks of keys for each hand were described in Chapter 5. If the QWERTY layout is used on these keyboards, only 2 hours of practice is needed for typing performance to reach that of conventional keyboards (Smith *et al.*, 1998). Finally, hardware and touchscreen

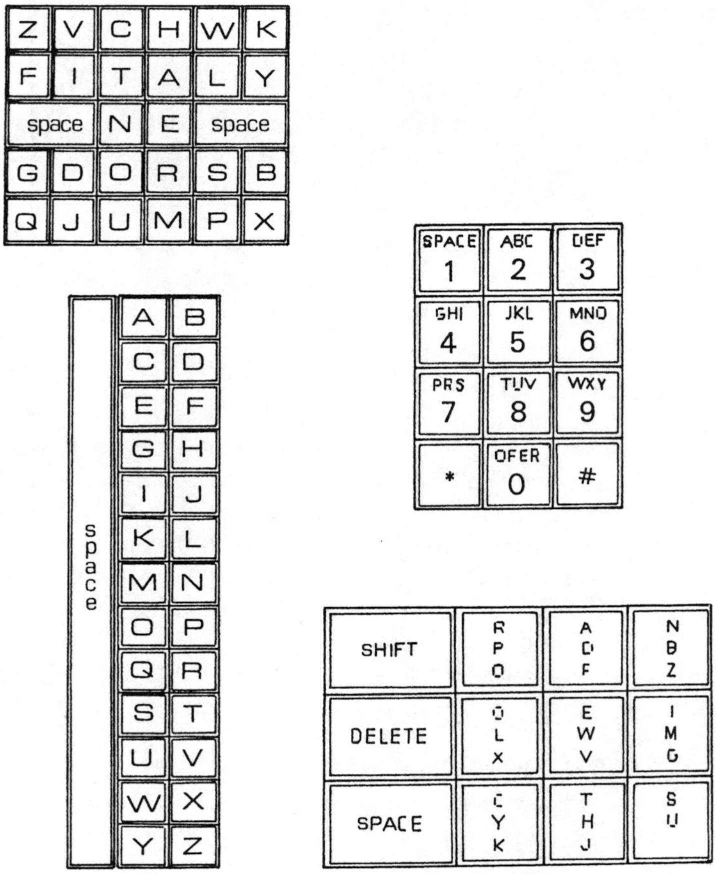

Figure 13.11 Soft layouts for one- and two-handed keyboards.

QWERTY layouts present a problem when used on small hand-held devices because the keyboard is so small. The gap between the keys should be about the same as the width of the tapping device (fingertip or pen). Wright *et al.* (2000) compared the keying performance of older users on hand-held computers with touchscreens or keyboards. Older users, in particular, were slower when using a touchscreen and pen than a keyboard and 88% of them preferred to use the keyboard. It seems that new input devices for palm-size computers would greatly improve their usability. Some possible solutions are text prediction software that attempts to predict the word being entered, voice input, and handwritten input.

Pointing devices

The mouse is the most common pointing device in everyday use. Most of the ergonomic issues are at the physical level. The mouse is designed to be placed at the side of the keyboard and requires abduction of the shoulder and increases the muscular load. Harvey and Peper (1997) found increased EMG activity from the shoulder muscles when a mouse was compared to a centrally positioned trackball. Soft keyboards and miniature input devices may best be operated using a stylus.

Voice control: problems and prospects

The development of speech recognition technology introduces the possibility of using voice as a control device. Voice control has the potential to radically alter the design of human–machine interfaces and is therefore discussed in some detail here. Further considerations are discussed in the Chapter 14 in the section on language. Potential advantages of voice control include the following:

1. It provides an extra communication channel that may take some of the load off more conventional channels.
2. It frees the hands to carry out other activities.
3. Subroutines are already built up for the production of voice commands, so training time should be reduced.

The processing requirements for issuing voice commands would not be expected to conflict with those for manual control and therefore voice is often thought to have the potential to speed up task performance or to increase an operator's information-handling capacity. Some evidence for this comes from Martin and Long (1984), who investigated a simulation of a ship's gunfire control task that involved compensatory tracking as a primary task. They found that a simultaneous pointing task degraded tracking performance, whereas a spoken version of the same task did not. It was concluded that the findings were consistent with a multichannel model of performance in which speech and tracking occupied parallel, but independent, channels with no common capacity limitation.

A potential problem with speech as a medium of control is the time taken to produce it. Visick *et al.* (1984) compared a voice recogniser with a keyboard in the performance of a Post Office parcel sorting task. As each of a number of parcels was sorted, the operators either spoke the name of the destination into a simulated voice recognition machine or pressed a key on a special keyboard where each key was a destination name. It was observed that voice input was slower than keying, despite the bulk of the keyboard that was used, since the average time to produce an utterance was about 600 milliseconds (compared to 50 milliseconds to press a key). In practice, further delays would occur owing to the time taken for the machine to process the utterance. The authors concluded that voice input may be unsuitable for tasks requiring little manual component and may offer inadequate timing feedback for tasks requiring precise sequencing of manual and verbal operations.

Problems faced by the voice recogniser

Speech is a complex acoustic signal containing bands of energy centred around 500, 1500, 2000 and 3000 Hz. These are known as *formants* and they consist of an initial transient segment followed by a steady-state segment. The lower two formants are sufficient for the perception of speech, which enables bandwidth compression to be implemented to reduce the information load of speech transmission systems and speech synthesisers.

Many investigations of speech perception by humans have been carried out by researchers in the field of 'psycholinguistics'. Speech perception is thought to depend on the interplay of both 'top-down' and 'bottom-up' processes. Bottom-up (analytic) processes detect particular features in the speech signal, as it is received. Combinations of features detected in speech form the characteristic signatures of basic speech sounds such as phonemes, syllables or words. When matched with a stored representation of known features, the speech can be said to have been recognised. Top-down (synthetic) processes use higher-order information to synthesize possible items that can be matched against the incoming signal (trying to 'guess what's coming next'). A knowledge of grammar, syntax and context is required to drive these higher-order interpretive processes.

Real speech is a 'messy' signal. It is mixed in with other noises that humans make (e.g. coughs, grunts, sneezes, etc.). The human speech perception system seems to be able to disentangle connected verbal and non-verbal utterances with ease, but it is not clear how this is accomplished or how a machine might best be programmed to do this. People experience *voice fatigue* if they have to speak over the course of a day. This adds to the variability in speech that the voice recognition system has to be able to cope with. Stress, alcohol, colds and flu and other factors also increase the physical variability of a speaker's utterances.

Understanding speech would, at first, appear to be simply a problem of identifying the basic building blocks of speech and then combining them. However, connected speech is characterised by a lack of segmentation. It is as if the building blocks of speech blur into one another in continuous speech. It is not possible to identify a unique set of building blocks at one level.

Many elements of speech are co-articulated – that is, the acoustic form of an utterance differs depending on what precedes and follows it. For example, the spoken form of the word 'bag' is different from that of the word 'beg'. It is not possible to record 'bag' on audiotape and remove the middle 'a' and replace it with 'e' – the whole utterance has to be changed. Similarly, compare the shape of the lips when pronouncing the 'st' in 'stew' and 'stay'. When the limitations of the anatomical structure of the human vocal apparatus are considered, it is easy to see why this occurs. The human mouth, lips and tongue move slowly, have inertia and exhibit hysteresis when speech is produced. Although continuous speech is perceived consistently, it is not produced consistently. Parts of an utterance may sound the same to a human listener but differ acoustically depending on the words that surround them.

Listeners also adjust their perceptions of speech to take account of the way other people produce speech: speech perception can be conditioned by contextual factors. We automatically take account of accents different from our own when listening to someone. It is not clear how this is done or how similar flexibility could be built into an automatic system.

These example demonstrate some of the difficulties of constructing machines to recognise fully connected speech in natural way. More common are 'voice button' systems that recognise isolated words. The ergonomics issues in the design of these systems centre around the choice of words for the 'voice buttons' and the recognition accuracy of the speech recogniser. Recognition accuracy is a crucial factor determining user acceptance; Casali *et al.* (1990) found that improvements in accuracy from 91% to 99% improved task completion time and acceptance in a task that simulated data entry by voice. Given that words may be recognised, not recognised or confused for another word, it is often paramount to design the vocabulary set to minimise the confusability of words rather than simply to ease of recognition. This should be done at the development stage by carrying out user trials.

Applications of voice input

Noyes and Frankish (1989) reviewed the applications of speech recognition technology in the office. A number of applications have been proposed, including 'voice messaging' (the intelligent answering machine that answers back and takes a message), word-processing (using a 'speechwriter' rather than a typewriter) dictation, data entry and environmental control. The limitations of present voice recognition technology seem to pose severe limitations on the usability of these systems. Where speech recognition technology has been successfully implemented, it seems to be because it has clear advantages over other methods and because the task domain is very specific. Voice-input in computer-aided design (CAD) is a good example. When charts and maps are being digitised, a cursor has to held in precise locations on the map while textual and other data are entered by voice. The eyes busy – hands busy task seems to be the paradigmatic one for the successful application of voice-input technology (in principle if not always in practice). Over time, more applications have been found for voice recognition technology. Some banks use it to support telephonic account enquiries. By means of a limited vocabulary and auditory menus, a number of services can be offered using voice recognition and machine-generated speech.

Norman (1993) uses the telephone as an example of a product that can be improved upon using voice control. Manual dialling can be replaced with programable voice dialling (instead of keying in the number, you speak the person's name, having previously entered the name and number into the telephone's memory). This success of this application of voice control depends on its being superior to manual dialling (e.g. being faster, being less error prone, the caller not having to be physically near the phone to dial, and soon). New facilities could also be added – when the phone rings the receiver could answer it or instruct it to return the caller's call in 5 minutes, take a message or ask the caller to phone back some other time, appropriate feedback being given to the caller. The success of these facilities depend on public demand for a more flexible communication device than existing telephones.

Unnatural uses of natural language?

Although interaction with machines by voice is often said to be more 'natural', the limitations of current technology make voice interaction extremely unnatural because the user is restricted to a limited vocabulary and syntax, which has to be spoken in an artificial way. Additionally it can be argued that there is nothing intrinsically 'natural' about talking to a machine.

Combining displays and controls

Approaches to control/display integration

Shepherd (1993) identified a common problem in control/display integration, which he calls a 'breaking the loop' problem. This happens when designers pay insufficient attention to what operators actually do to carry out their work or neglect to carry out a sufficiently detailed task analysis. Shepherd cites as an example a case in which the display was two floors up from the valve where the control action took place. Under these circumstances, it was physically impossible for the operator to control the system on a 'human-in-the-loop' basis even though the designers had deliberately decided not to automate this part of the system.

Three general guidelines can applied to improve the layout of control/display panels:

1. The Gestalt principles can be used to ensure that the functional relationships between system inputs and outputs are correctly represented in the panel layout.
2. Considerations of spatial memory and of operators' limited ability to process mental images can minimise unecessary mental workload.
3. Learned associations between displays, displayed variables and control movements (population stereotypes) should be incorporated to minimise errors.

These can be combined with information obtained from task analyses to optimise the integration of controls and displays.

Use grouping principles in panel design

Proximity and similarity can be used in a consistent way such that controls are perceived to belong to their associated displays (Figure 13.12). One problem noted by Seminara and Smith in their evaluation of a nuclear power plant control room was the lack of correspondence between the proximity and similarity of displays and controls and the actual subsystem boundaries. Under these circumstances, operators may have to work against the tendency to perceptually group together unrelated displays and controls or expend extra effort monitoring groups of apparently unrelated instruments.

Avoid spatial transformations

People are capable of retaining a visual image of a real object in memory and of manipulating the image in a manner analogous to the real object. Shepard and his

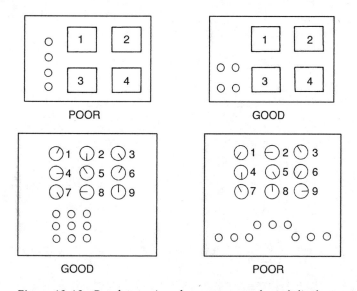

Figure 13.12 Good mappings between controls and displays promote usability.

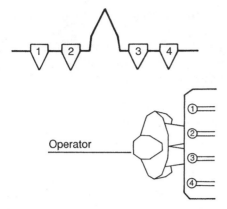

Figure 13.13   If displays and controls cannot be positioned in close proximity, it is important that their layouts be spatially compatible. These controls can be mapped
onto their displays by mentally rotating either through 90 degrees.

colleagues (e.g. Shepard and Cooper, 1982) have shown that this mental rotation of
images takes time depending on the amount of rotation required. The more one has
to rotate a mental image to make some kind of decision about it, the more time it
takes. It seems that correct judgements about displays and their associated controls
can be reached more easily when the spatial correspondence between the two is high
(Figure 13.13).

The domestic stove is a good example of a device in which operational problems
can arise owing to a lack of spatial correspondence between the layout of the switches
and the layout of the burners. This lack of correspondence renders the true control/
display relationships ambiguous and leads to error. Two typical arrangements for a
four-burner cooker are given below (Figure 13.14).

Problems arise because the burners are arranged in a plane (which has two dimensions) whereas the switches are arranged linearly. Since this two-dimensional array
cannot be projected onto one dimension without loss of information, spatial incompatibility is almost inevitable. The loss of the front/rear information makes switch
selection for the front and rear burners arbitrary. Two possible solutions are to either
stagger the layout of the burners to create a two dimensional array that can be

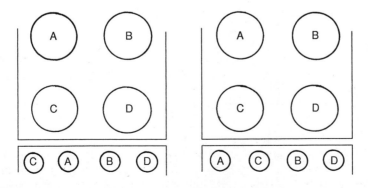

Figure 13.14   Two common arrangements for the controls and burners of domestic stoves.

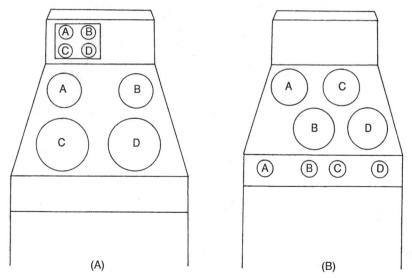

Figure 13.15 Possible solutions to the stove control problem.

projected unambiguously onto one dimension or to arrange the switches themselves in a plane arranged to be spatially compatible with the burner plane (Figure 13.15).

Chapanis and Lindenbaum (1959) found the staggered burner arrangement to be freer of operating error than other arrangements. Many different burner/switch arrangements have been developed by manufacturers over the years and none appears to be optimal. Fisher and Levin (1989) investigated this problem using an unconstrained approach in which subjects were asked which of the switches of a stove burner simulator would turn the burners on. The CABD arrangement was found to be the most consistently chosen and preferred. However, other arrangements were also found to acceptable, although less so. The fundamental incompatibility between the traditional burner/switch layouts probably explains why so many different designs are commercially available and why designers have resorted to techniques such as the use of logos, overlays and mimic displays in an effort to make explicit the functional relationships between the controls and the hobs.

Sensor lines

Sensor lines are used as visual cues to make explicit the relationships between controls and displays in control panels. Chapanis and Yoblick (2001) compared actuation time and errors for the four panels shown in Figure 13.16 both with and without the addition of sensor lines. Panels 1 and 2 have a higher degree of compatibility than panels 3 and 4 because the upper controls are linked to the upper displays. Performance on the compatible panels was superior to that with the incompatible panels (0.7–0.8 seconds to activate the control compared to 1.0–1.2 seconds and about one-third the number of errors). However, the addition of the sensor lines had no effect on performance when the compatible panels were used and impaired performance when the incompatible panels were used. These findings suggest that the closer the spatial mapping between the displays and the controls the better. If a close mapping is not possible, then the layouts in panels 1 and 2 are to be used. Sensor

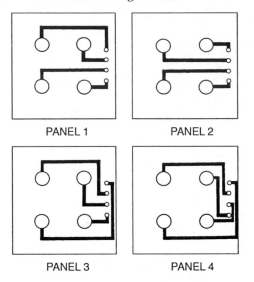

PANEL 1   PANEL 2

PANEL 3   PANEL 4

Figure 13.16 Sensor lines as a palliative intervention for poorly mapped panel elements. (Based on Chapanis and Yoblick, 2001.)

lines *may* be added to these panels. However, sensor lines should not be used as a palliative intervention in the hope of improving performance on existing incompatible layouts (panels 3 and 4).

Population stereotypes

People tend to attribute meaning to the movements of objects and have a tacit understanding of mechanical cause–effect relationships. The famous Swiss psychologist Piaget described how children develop these notions and how their ideas change with age. Adults ascribe meaning to the movement of mechanical objects such as the pointers of dials or scales. Where this is done consistently in a population, it is known as a population stereotype. Displays or controls that behave in a manner consistent with the prevailing stereotype are said to be compatible. One of the most common stereotypes is the 'clockwise-for-anything' stereotype (Barber, 1988). The clockwise movement of a pointer on a dial signifies the passage of time, an increase in the speed of a motor car or an increase in pressure or flow. An upward movement of a pointer on a vertical scale indicates increase as does a rightward movement of a pointer on a linear scale. The development of a repertoire of stereotypes seems to take place over a period of years and is accompanied by *an increasing lack of flexibility and ability to adapt to new situations*. By adulthood, certain stereotypes appear to be well learnt and although new relationships can be acquired, they merely overlay the originals rather than replacing them. In stressful situations, more primitive forms of behaviour may replace even well-learned skills. If the operator lacks the processing capacity to monitor his or her own performance, old subroutines containing the original learning will emerge, leading to reversal errors. It is for this reason that when it is clear that a stereotype exists, designers should ensure that their design is compatible.

A number of exceptions to the clockwise-for-anything rule may be noted. Taps are a good example of an exception to 'clockwise-for-anything' and people appear to be able to learn this exception as a special case. In some circumstances, the ability to correctly interpret the movement of a dial or a pointer will depend on the operator's knowledge of the system or of the displayed variable. For example, the dial on a vacuum guage moves anticlockwise to indicate an 'increase' in vacuum (i.e. a drop in atmospheric pressure). Whether operators will be able to correctly interpret instructions such as 'increase the vacuum' will depend on their understanding of what a vacuum is.

People frequently have strong expectations about how a movement of a control will influence the direction in which a display moves. These expectations are referred to as direction of motion stereotypes. For example, clockwise movements of a rotary knob are associated with upwards or rightwards movements of display pointers. Some of the more straightforward direction of motion stereotypes have been summarised by McCormick and Sanders (1992) and appear to be fairly strong and well-represented in the population. Turning a steering wheel to the right should result in the vehicle moving to the right. When tuning a radio, a clockwise movement of the knob should cause the dial to move to the right or up, and so on.

Operators may not have strong stereotypes about the controls and controlled parts of novel machines; for example, where the controlled part moves in a different plane from the controls or in a complex way. Alternatively, stereotypes may not exist for novel control devices such as palmwheels or voice. Under these circumstances, the designer may have to undertake empirical research to learn more about people's expectations and preferences, ensuring that, if there are any strong expectations, they are not violated. Whatever control logic is then decided upon should be used in a consistent way, throughout the design, to facilitate the learning of any new stereotypes and avoid interference between mutually incompatible ones.

Many activities of daily living involve the chaining together of learnt behaviours and associations. When these are violated, or the environment is altered, increased errors and accidents may occur (see Box).

Britons abroad are the worst drivers in Europe[a]

A recent study by a team at the University of Granada in Spain found that British drivers are more likely to cause an accident when driving on Spanish roads than any other nationality, despite the fact that the UK road accident rate is the lowest in Europe. British drivers in Spain were more than twice as likely to cause an accident than Spanish drivers and more likely than any other nationality.

The relative risk of causing an accident was 2.16 for British drivers, compared with 1.0 for Spanish drivers. Other European countries had relative risks intermediate between these values but most were closer to the Spanish than the British levels.

Of all the groups compared, only the British drivers habitually drove in right hand drive cars on the left hand side of the road. It seems that the increased risk is not due to a difference in driving skill but due to British drivers being unfamiliar with left hand drive cars and driving on the right hand side of the road. During emergencies, they may become confused and revert to visual scanning patterns appropriate to right hand drive vehicles.

[a] From Robert Matthews, *The Sunday Telegraph*, 28/4/02, p. 4

Developing countries

Assumptions about stereotypes made in industrialised countries may not be appropriate in developing countries. Verhagen *et al.* (1975) investigated the strength of direction of movement stereotypes in North and Central Africans. They found that stereotypes were more strongly established in the young, educated people, probably owing to their greater exposure to technology. Fisher and Olivier (1992) tested rural and urban South Arican schoolchildren to compare their acquisition of stereotypes in the operation of everyday items. The tasks given to the children were

1.   Unscrewing and replacing a bottle top
2.   Tuning a clock radio
3.   Turning taps on and off
4.   Dialing a telephone number
5.   Turning a light switch on and off
6.   Operating a four-burner stove

The children's performance was categorised as 'correct', incorrect' or 'don't know'. Significant differences were observed between the two groups, suggesting that correct performance depended on previous exposure. The rural children's performance was more often classified as trial and error (don't know) in the operation of clock radios, telephones and light switches, whereas significantly more urban children failed the tap test (in rural areas of developing countries, a tap, shared by several households, may be a focal point of village life). Significantly more rural children failed to operate the telephone correctly. The authors concluded that there were noteworthy differences in the proportion of rural children unable to manipulate items except on a trial-and-error basis and that technical education was necessary to purposefully uplift them. It would seem that when designing technology for implementation in developing countries, it is essential to know the current experience of technology of the intended user population.

Virtual ('synthetic') environments

Virtual environments (VEs) are more than just 3-D displays. They are interactive models of systems implemented on computer and displayed in three dimensions. The intention is to create the experience of being 'in' the environment, of being able to move around and explore it and of being able to manipulate objects encountered in it. The main uses of VEs are as design tools, as training simulators and as visualisation aids. The degree of fidelity between the model and the real system may vary. The display may be on computer screen or it may viewed through a stereoscopic, head-mounted display (HMD) or wall-mounted screens. In the case of a 'cave', the user is surrounded by screens. Users' subjective reactions to systems can be positive, with the experience of 'presence' or of being 'immersed' in the system, or negative in the case of nausea, disorientation, visual discomfort and postural problems.

The feeling of presence seems to depend on a number of variables including system response time to user actions – reflected in the time taken to update the display – the optic flow as the user moves through the system, and the internal geometry of virtual obbjects ans how well it changes with viewing ditance and angle compared to the real world. According to Kalawsky (1999), the key usability factors for virtual environments are the following:

- Functionality – the interface provides the level of control the user expects to complete a task.
- Natural inputs – the user can interact with the system in a natural way.
- Displays – information is understood, relevant to the task and unambiguous.
- Help – is available online.
- Consistency – system operation is consistent with the user's understanding and with convention.
- Flexibility – users are not artificially constrained by system limitations.
- Model Fidelity – interaction takes place in the context of a model of the system or task.
- Error correction – Errors can be corrected/undone before a permanent change is made.
- Sense of immersion – the user feels part of the system.
- Usability – interaction should be intuitive (consistent with the user's model).

Cybersickness

Interaction with VEs may produce three kinds of unpleasant effects, collectively known as cybersickness (Stanney *et al.*, 1999):

- Nausea and vomiting
- Oculomotor disturbances
- Disorientation ataxia (postural stability problems)

Cybersickness is thought to be due to discrepancies between the optic flow experienced in the VE system and the dynamics of the manual controls, particularly the haptic feedback. When entering the VE for the first time, users may have to learn to 'recalibrate' their control actions in relation to the visual feedback they receive. When leaving the VE, there may be a delay in re-adaptation to the real world. Stanney *et al.* tested subjects before and after performing object manipulation and walking tasks in a VE for 30 minutes. There were significant increase in nausea, oculomotor disturbance and disorientation (self-reported) after the VE session and these persisted for over 30 minutes. Increased errors on a pointing task were also observed, but only when the eyes were closed. This suggests that the VE did induce some recalibration of the processing of proprioceptive information. The practical implications are that users may require a 're-adaptation' period after immersion in VE and before it is safe for them to return to their normal work – particularly if their normal work requires close hand/eye coordination.

The nausea experienced during cybersickness, like motion sickness, is thought to be due to sensory conflict between the feedback from the visual system and the vestibular system. Conflict can occur when the scene changes to indicate movement of the user in the VE but the user's head remains still (in the real world). Thus, the expected vestibular feedback does not match the visual feedback. For users wearing HMDs, the VE can be explored by moving the head. Sensors in the HMD detect this movement and the scene is changed accordingly. However, update lags mean that the scene remains still as the head begins to move and continues to change when the head stops moving. Thus, there are two potential sources of conflict. Howarth and Finch (1999) compared the prevalence and severity of nausea in two VEs; in one the scene was changed by moving a hand control and in the other by moving the head. The results were that both systems were found to be nauseogenic but that the head movement condition was most severe. Of particular interest was that there was an additional sensory mismatch between the movement of the head-mounted display and the movement of the scene displayed – for every 120 degree rotation of the head, the scene rotated 360 degrees.

Vection is another cybersickness symptom that seems to be due to exposure to visual cues in the VE in the absence of any real movement. Vection is the illusion of movement in the opposite direction. Vection seems to be similar to the experience one sometimes gets after looking out of the window of a moving train – when the train comes to a halt, there is the illusion of moving in the opposite direction. Vection itself can be nauseogenic. So *et al.* (2001) investigated nausea and vection in subjects navigating a VE (a city road network). Ratings of nausea and vection increased with the root mean square navigation speed. The range 3–10 m/s showed the greatest increase in the ratings. Above 10 m/s, the ratings levelled off up to 59 m/s. To avoid upsetting virtual tourists, it seems that either very slow, or fairly fast virtual tour buses are needed! In general, the incidence of cybersickness can be reduced by mini- mising the sensory conflict.

Head-mounted displays

HMDs seem to have a number of advantages over CRTs or fixed panels if the update of the display can be closely synchronised temporally and spatially with head move- ments. Disparate views of the VE can be presented to each eye to give the perception of depth. However, the screens that display these images may only be 50–70 mm from the eye with a lens in between and concern has been expressed by both Howarth (1999) and Rushton and Riddell (1999) about undesirable effects on the visual sys- tem. Light from binocular HMD displays passes through a lens before reaching the eye, such that the light rays are collimated (parallel) when the reach the eye, as they would be if the displays were at infinity. Under normal viewing conditions, the ac- commodation and vergence systems are linked, such that accommodation, together with disparity (double images), is a cue for vergence and vergence, together with blur is a cue for accommodation. As a distant object approaches the viewer in the real world, the demand for accommodation and vergence increases. In VE, however, the viewing distance may be infinity but vergence may be required to avoid seeing a double image and to get a stereoscopic view. This disassociation of the accommoda- tion and vergence systems may be a cause of aesthenopia in HMD wearers or of visual aftereffects when the VE session ends. Although the visual system exhibits a certain amount of plasticity, the time scale and the scope for adaptation is not yet known. Individual differences in interpupillary distance do not seem to pose much of a problem, according to Howarth, so adjustable eye pieces may not be needed on HMDs. However, discrepancies between the centre of the lenses used to collimate the rays and the centre of the displays that emit them do seem to produce undesirable visual effects.

Head-mounted displays and space navigation

Intuitively, it would seem that HMDs would offer advantages for the navigation of 3-D information spaces such as those represented in the previous chapter. The kinds of head movements that we use to find our way around the real world would be available in the VE and thus the 'mapping' between the real navigation tasks and VE tasks would be better. The available evidence is equivocal. Westerman *et al.* (2001) compared performance of a search task using either an HMD or a CRT. HMD performance was poorer than CRT performance in terms of time taken to locate target items, navigation efficiency and the number of trials in which subjects ran out of time to complete the task. Subjects' orientation in the space was no different in the conditions, suggesting that subjects were equally able to develop a 'cognitive map' of the space. It may be that the level of development of the current generation of HMDs is inadequate to tap into the kinds of navigational cues that head movements provide in the real world – they simply are not usable or 'natural' enough at present. Westerman suggests that use of these devices may impose additional cognitive demands that tax the capacity of those with already low spatial ability.

Interacting with virtual environments

Virtual environments require control devices that will provide the user with control over movements in six degrees of freedom. There are three translational axes (up and down, left and right, and backwards and forwards) and three rotational axes (roll, pitch and yaw). Figure 13.17 depicts two devices (an isometric spaceball and an elastic 'egg', a kind of three-dimensional mouse) for controlling the movement of virtual objects.

Zhai and Milgran (1997) investigated the performance of subjects using these devices in a pursuit tracking task in which they had to align a tetrahedron-shaped cursor with a tetrahedron-shaped object moving in six degrees of freedom in a virtual environment. Depth was represented using linear perspective, retinal disparity, interposition and partial occlusion. Tracking errors were greatest (35% larger) in the depth dimension (the target moving towards or away from the subject). Subjects varied in their ability to make simultaneous translational and rotational control actions with one hand. When experiencing difficulties, they tended to ignore rotational correction and concentrate moving the cursor to catch the target. Most subjects mastered simultaneous control of translation and rotation after about 40 minutes' practice. It seems that simple one-handed devices are usable for interacting with objects in virtual environments, but provision of good depth cues is essential.

Virtual environment technological limitations are ergonomic issues

The usability and health aspects of VEs have much to do with current technological limitations (Wilson, 1996b). HMD's have a *refresh rate* of 60 Hz (the frequency at which the image on the screen is repeated). Higher refresh rates can improve image quality. The screen update rate is the frequency at which the screen image can change and depends on the capacity and processing speed of the system. A minimum of 12 fps (frames per second) is needed. The faster the update rate, the smoother and more realistic is the representation of user interaction with the virtual world. System

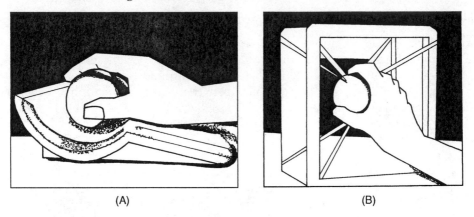

Figure 13.17 (A) An isometric spaceball and (B) an elastic 'egg' (a kind of three-dimensional mouse) for exploring virtual environments.

lag is the delay between the user performing an action and the perception of it in the VE. The shorter the lag, the more likely the user is to experience 'presence'. System lag may induce sensory conflict and cybersickness. According to Wilson, there is a trade-off between field of view and resolution, particularly if the pixel matrix is small (e.g. 360 × 240 pixels). Viewing the VE under such conditions may be equivalent to interacting with the real world without one's spectacles. For enhanced sense of presence and greater comfort, VE systems require:

- High resolution
- Fast refresh rates
- 24 fps or greater
- Minimum system lag

All of these requirements depend of the available computer processing power.

Effectiveness and cost-effectiveness

That ergonomic design of controls and displays to optimise human–machine interaction has sound financial benefits is demonstrated by a research literature going back nearly 50 years. Singleton (1960) redesigned the speed control of an industrial sewing machine and in a one-month comparative trial on the shop floor estimated that the potential production improvement would be 10–15%. Given that some of the increased production would result in an increase in earnings of 5%, it was estimated that the new controls would pay for themselves in 7 years. Similarly, Beevis and Slade (1970) reported performance improvements when sewing machines were fitted with variable-speed as opposed to fixed-speed motors. Trained operators achieved a 9.5% reduction in task time and a 25.6% reduction in errors. For untrained operators, the reductions were 38.8% and 60.4%.

Murrell and Edwards (1963) compared the operation of lathes fitted with traditional vernier-style gauges and digital gauges. The digital indicators resulted in a marked reduction in setting time, from 11.5% to 1.2% of the total time spent processing orders – an increase of machine utilisation time of 11%. Slade (1969) made a

Table 13.5 Industrial comparison of lathes with scale and digital position indicators[a]

	Order quantity	Machine time/order (hours)	Handling time/order (hours)	Setting time/order (hours)	Total time (hours)
Scale and vernier	2400	7644	3860	1500	13 004
Digital	2400	4440	3860	100	8 400
Savings		3204		1400	4 604

[a] From Beevis and Slade (1970).

lengthy comparison of digital versus vernier gauges on lathe operation. Table 13.5 summarises the findings.

Design of brake lights on cars

A large-scale study of the effectiveness of centred, high-mounted brake lights on cars was carried out in the 1970s (Malone, 1986). These lights are placed in the centre of a car at the level of the rear window or higher and they are installed in addition to the existing brake lights that are situated at or around the level of the rear wheels. With centre-high-mounted lights, not only is there a third light warning that the car ahead is braking, but the light appears in the eye field of the driver in the following car. Thus, it would be expected that the driver of the following car would be less likely to miss the signal that the car ahead is braking (since there are now more signals). Furthermore, the reaction time to the centre-mounted light would be faster when the following driver was looking directly ahead. Thus, centre-mounted lights would be expected to reduce the number of rear-end collisions. A field trial was carried out and four groups of 525 cabs each were monitored over a one-year period in Washington DC. Data on rear-end collisions, other collisions and mileage were gathered. At the end of the year, non-rear-end accidents were the same in all groups, whereas a 54% reduction in rear end accidents was found in the experimental group. Cabs fitted with the centre-high-mounted brake light were half as likely to be hit by vehicles following them. Similar reductions in the incidence of rear-end collision have been found in other studies of these brake lights. The cost of installing a light in 1977, when the Washington study was carried out, was $4.00 per vehicle. The average vehicle repair cost for rear-end collisions was $317.00 for vehicles with conventional brake lights and $194 for those with high-centred brake lights, suggesting that when cabs with centre-mounted lights were hit, the damage was less serious. The cost saving for the 3.2 million accidents in 1977 would have been $1.465 billion, not including medical, legal and lost-time costs. All new passenger cars fitted in the USA after 1985 were fitted with these lights with an estimated reduction in property damage of $910 million through preventing 126 000 crashes. The cost of the study and the enforcement efforts totalled $5 million.

Warnings

Clement (1987) illustrates the potential benefits and likely costs of warning signs and instructions. Using an industrial example (warnings in forklift trucks about the dangers of hydraulic failure), the main costs and benefits are as follow.

Benefits

- Avoidance of major injuries or death
- Avoidance of minor injuries
- Enhanced perception that the manufacturer is safety conscious
- Increased attention to maintenance by the user company
- Reduced liability insurance costs

Costs

- Cost of producing instructions
- Cost of fitting warning plates on all trucks
- Perception that the product is dangerous
- Perception that the product may need increased outside maintenance

Assuming that there had been one serious back injury in the company in the previous 10 years, the estimated cost of a further injury would have been $300 000 (the 1987 purchase price of a 30-year policy paying the injured party a total of $1 million). If the probability of the company being found guilty of negligence was 0.3, the cost that would be avoided by proper warnings and instructions would be $90 000 over the next 10 years (this is an underestimate due to not adjusting for inflation). For minor injuries, a cost of $2400 was estimated. The better safety material might result in one more sale per year (with a profit of $500 yielding $5000 over 10 years). Savings in the cost of liability insurance were estimated at $5000 per year for 10 years. The total benefit was estimated by Clement to be $90 000 + $2400 + $5000 + $2500 + $50 000 = $149 900.

The costs of developing the warnings including extra printing costs might be $1000 and for the warning plates $10 000. Assuming a cost of $25 000 over 10 years due to a loss of sales as the product is perceived as being dangerous, the total cost of the warnings would be $25 000 + $10 000 + $1000 = $36 000.

No additional benefits would accrue to the company by deciding not to provide warnings, but the costs of not doing so can be seen to be substantial. According to Clement's estimates, the ratio of benefits to costs is almost 5 to 1. This ratio can of course change, dropping if lower-cost liability insurance becomes available and rising if new legislation is introduced that increases employer liability for health and safety (i.e. if the regulatory authorities increase their propensity to prosecute or if new precedents increase the likelihood of private litigation).

Analyses of this nature provide companies with a rational choice for deciding whether to proceed with the incorporation of safety features and thereby incur extra costs. As soon as the benefit/cost ratio drops below 1, a rational decision can be made to do nothing.

Virtual environments for training

Evidence is mounting that training time and costs can be reduced using virtual environments to simulate critical tasks or tasks that are unpleasant or dangerous to carry out. Stone (2001) has reviewed this evidence. Despite the limitations of many of the existing studies (e.g. small sample sizes, unsophisticated outcome variables to compare VE training and conventional training), VEs do seem to offer economic benefits. Flight simulators have been in existence for many years – without them, it would be extremely expensive to train pilots on real aircraft, not least because more aircraft would be needed just to train the pilots. VEs have been used for more mundane tasks and do appear to be cost-effective. It seems to be the case that as long as the level of realism is large enough for the particular task, learning takes place interactively and transfers positvely to the real world. At the same time, VE costs are dropping and simple environments can be implemented on commercially available, Windows™-based systems. Some examples cited by Stone are as follow.

- Computer-based instruction can reduce training time to reach course objectives by 30%.
- For training in maintenance and repair tasks, VEs can be superior to actual equipment when supported by tutorial guidance, improving performance from the 50th to the 66th percentile.
- Virtual environments for training surgical procedures have been shown to improve surgical efficiency according to criteria such as error reduction, instrument path length and pointing error and are a mandatory part of the training at the European Surgical Institute near Hamburg.
- A flight deck officer training system for the Royal Navy based on six Pentium personal computers brought about significant cost savings because training was no longer dependent on the weather, could take place at any time of the day, and avoided costs of £4000 per day of taking a real helicopter out of service.
- A close-range weapons simulator was developed for the Royal Navy to replace a coastal firing range after it was closed down. The system used HMDs and allowed a number of different threats to be simulated (ships and aircraft). Trainees were able to practise their skills from a position on their own virtual ship in different weather conditions, times of day and sea states. The system recently fired its millionth virtual round of ammunition at a fraction of the real cost of £27 per round.

Research issues

Human–machine interaction and display design are areas subject to considerable 'technology-push' and only a few examples of research issues will be given. There is currently research on the design of virtual environments using novel display modalities such as real-time or near-real-time three-dimensional computer-generated graphics and novel control devices. The possibility that there may be after-effects of using these devices that put users at increased risk of injury is a research issue. Voice input is continuing to receive attention, and voice recognition technology has been implemented in a number of applications.

Summary

What we perceive depends on the operation of processes not accessible to consciousness. These processes organise the stimulus array that is projected from the retina to the occipital cortex of the brain using spatial and other information to maximise its meaningfulness. Visual displays can be designed in accordance with the operation of these processes in order to strengthen the correct interpretation of the data. Lack of consideration of these inherent properties of the perceptual system can result in incompatible designs of less than optimal usability.

Guidelines for the design of control/display combinations have been in existence for many years. Population stereotypes and more general user expectations should be taken into account. Information technology facilitates a more flexible approach as displays can be increasingly implemented via software rather than hardware.

Complementary approaches exist for investigating and evaluating human–machine interaction and in particular the mental workload of tasks. In addition to established physiological and psychological techniques, new theoretical approaches to human behaviour can provide system designers with powerful frameworks for investigating human–machine interaction and human error.

Exercises and essays

1. Analyse the design of examples of the following domestic appliances. Identify and comment on the design of the controls, the displays and the relationship between them.

    • A four-burner stove
    • A video cassette recorder
    • A food mixer
    • A microwave oven

    What feedback do the devices provide users to indicate response adequacy?

2. Describe the design options available for human–machine interfaces. Discuss the advantages and disadvantages, using examples from everyday life.

3. Use the material in this chapter to write a checklist for evaluating the design of complex displays.

4. Mr Jones wants to go on a motoring holiday in the south of France. He lives in London and has only ever driven a right-hand-drive car in England, where they drive on the left-hand side of the road. He can't make up his mind whether to put his own car on the ferry to France (where they drive on the right-hand-side of the road in left-hand-drive cars) or fly to France and hire a left-hand drive car when he gets there.

    (a) What are the ergonomic principles involved in choosing between these two options?
    (b) What would you advise Mr Jones to do?

5. Design a control box for remote-controlled driving of a motorised drilling rig for underground mining. The drills are mounted on columns on a four–wheeled, rectangular carriage. The hydraulic powered carriage can move in either direction and both axles can be steered independently. The vehicle stands between the operator and the rockface and can move left-to-right or right-to-left. It can curve in either direction or 'crab' towards or away from the face. Aim for compatibility and consistency using the minimum number of controls.

14 Human–computer interaction, memory and language

> People forget ideas not consonant with their predispositions but retain without loss or even increment those ideas consonant with their predispositions.
>
> (Hovland *et al.*, 1949)

Analogue representations are usually best for the design of interfaces in perceptual-motor tasks such as car driving. Many computer-based tasks, though, are carried out by manipulating symbols rather than performing skills. Skilled performance is graded – we can talk about the performance of task elements (such as operating the controls, performing hill starts and reacting to traffic signals) in terms of the degree of success or safety with which they were carried out. Performance can be assessed and scored on a continuum or scale. Symbol-based human–computer interaction is much more brittle. Symbols always have a certain arbitrariness about them that increases the requirement for learning and the likelihood of forgetting. When recognising a symbol, we are usually either right or wrong – there are no shades of grey.

Much of the effort in human–computer interaction (HCI) research in the last 20 years has been directed at making interactive systems easier to use. To master any interactive device, users have to learn about

* The characteristics of the device
* The tasks be be performed with the device
* How to perform the tasks using the device

To speed learning, designers aim for positive transfer from everyday life of users by making the device more familiar using quasi-analogue models of the real world – graphic user-interfaces (GUIs) – with the intention of

* Making a new task similar to one that is already known
* Striving for conceptual and operational consistency with known tasks

A number of frameworks have been posited to guide designers through the issues associated with designing user-friendly interactive systems. In *Command Language Grammar* (Moran, 1981), a computer system is represented at several different levels, as described below.

At the *conceptual level*, we distinguish between the user's mental model of the system, the designer's mental model and the real system. Designers should strive to ensure that the design model is compatible with the real system (its essential features can be mapped onto the system, with no mismatches) and with the kinds of mental models the users already have (if any). The model itself can be physical, a verbal description, or a set of rules. Deciding on a design model is an important first step.

At the *sematic level*, the main system objects are specified. Many computer systems nowadays are used for different purposes but use the same objects – files, folders, directories, etc. This is an economical way of getting positive transfer from one system to another, since users will already have experience of creating and deleting files, storing them in folders and subfolders and soon. Thus, the same objects can be used for a wide variety of tasks in different systems.

At the *syntactic level*, decisions are made about how the system objects are presented to the user and how these representations are combined together during task performance. Syntactic rules are developed to govern the interaction – for example, you cannot close a file until you have decided whether or not to save the contents and told the system what to do.

At the *lexical level*, detailed design takes place concerning how the system objects and the syntactic rules are actually presented to the user.

At the *physical device level*, the basic operations needed to interact with the system are allocated to the hardware both in terms of the choice and design of the input devices and the detailed design of the screenfuls of information that are presented to the user. Most of Chapter 13 dealt with physical device issues.

In practice, designers need a good understanding not only of the task requirements but also of the knowledge, expectations and needs of users and *what they already know*. By capitalising on this, the designer can produce new that are usable very quickly. According to Shneiderman (1991), we need to take account of the distinction between syntactic and semantic knowledge.

Users have syntactic knowledge about device-dependent details and semantic knowledge about concepts. Semantic knowledge is in two parts: task objects (the things you need to carry out a task and what you produce) and actions (what you have to do to carry out a task); and computer objects and actions (how task objects and actions are represented on the computer). You may understand how a system is operated, for example, but not how to carry out a particular task. Alternatively, you may understand how to carry out a task in the real world, but not how to do it using a computer.

Acquisition of syntactic knowledge is slow because of the arbitrariness of the material and becomes a problem when syntactic knowledge is system-specific. Now that 'Windows'-based systems are ubiquitous, when learning to use a new system, users should only need to learn the new task because the new system should resemble existing systems (Preece, 1993).

The modern goal is to minimize the knowledge of computer concepts and syntactic knowledge by presenting a visual representation of the task. Thus, an explicit model is used that encourages the user to apply real-world knowledge to the execution of the task (which depends on the fidelity of the mapping between the computer representation of the task domain and the real domain).

Human-centred design processes for interactive systems

ISO 13407 defines usability as 'the extent to which a product can be used by specified users to achieve specified goals with effectiveness, efficiency and satisfaction in a specified context of use'.

To design a usable system, then, designers need to a description of the users them-selves, their goals and the organisation in which the system is to be implemented. Different layers of knowledge are required to design the system through a formal description of the users' goals and the tasks associated with attainment of the goals. These tasks are carried out by performing operations (actions) on system objects in ways that are specified according to a set of rules (syntax). The actions are imple-mented at a physical level using physical devices such as keyboards and screens. Throughout, designers must strive to achieve compatibility and consistency within and between the levels.

According to ISO 13407, user-centred design has to be based on the principles outlined below.

Active involvement of users and a clear understanding of the user and task requirements

Shneiderman (1991) lists the following as advantages and disadvantages of user-involvement:

- There is more accurate information about tasks.
- There is opportunity to discuss design decisions early on.
- A sense of participation fosters involvement and acceptance.

On the other hand, user-involvement brings

- Longer implementation period
- Antagonism of users who are not involved (not everyone can be)
- Antagonism of people whose ideas are rejected
- Compromise of design by designers to satisfy incompetant participants
- Build-up of opposition to implementation
- Possible rejection of novel ideas because users cannot appreciate the potential of the technology

An appropriate allocation of function between users and technology

Many system functions can best be performed by humans or by machines or by both. In practice, functions usually fall into one of several categories:

1. Those that must be carried out by machines (because it is impossible or unaccept-able for humans to do them)
2. Those that must be carried out by humans (because no adequate machines are available or because machine execution of function is not appropriate)
3. Those that might be carried out by either humans or machines or both

The third category is of the most interest to ergonomics. Early attempts to formalise decision making about function allocation used comparisons of the relative strengths and weaknesses of humans and machines. For example, machines can generate larger forces and do not require rest periods but are less flexible than humans and cannot cope with unforeseen (unprogrammed events). This approach has now gone out of fashion for several reasons, including *a lack of trust of machines* (e.g. automatic systems can be used to fly aeroplanes but most passengers would prefer there to be a pilot in the cockpit as well.

In many modern systems, decisions about allocation of function are really decisions about how automated a system should be. System functions are normally interdependent and should be allocated according to some higher-level design goals, with reference to the kinds of jobs people are used to. Function allocation decisions may differ over time, depending on the economic climate and availability of labour. Job design issues, discussed in the final chaper, may need to be taken into account. It is very important that the decisions are *not* simply based on allocating all functions to the computer and expecting the user to handle the left-overs! Users should be left with a congitively and physically coherent set of tasks that make for a meaningful job.

Iteration of design solutions

User-interface design is highly iterative and participatory as designers work from a general design concept to a working prototype. The design ideas are evaluated along the way with focus groups of users or of experts. The process can begin before any software or hardware has been built. Designers may present drawings or sketches of their ideas to user groups, in the form of 'storyboards' (a series of screens from a sample transaction) or other mock-ups, and gain feedback that is used to develop the ideas further. The goal is test out design proposals against real-world scenarios in order to arrive at a robust and effective system.

Multidisciplinary design

It is rarely the case that one person has all of the required skills and knowledge. For many reasons, the involvement of a variety of stakeholders is needed, including

- End-users
- Supervisors
- Support staff
- User-interface designers
- Programmers
- HCI expert
- Technical authors

As the design progresses, trade-offs between, for example, functionality and cost, may be needed and these will be dealt with best by the multidisciplinary team.

Planning the human-centred design process

The design process should be integrated into the overall system development plan (see Chapter 16). In general, the process includes the following steps (Shneiderman, 1991).

- Collect information.
- Define system requirements and semantics (system objects, their properties, etc.)
- Design syntax and support facilities (e.g. online help).
- Specify physical devices (screens, keyboards, pointing devices, etc.).
- Develop software.
- Integrate system and disseminate to users.
- Nurture user community.
- Prepare evolutionary plan for further development.

Human-centred design activities

ISO 13407 specifies four human-centred design activities:

- Understanding and specifying the context of use
- Specifying the user and organisational requirements
- Producing design solutions
- Evaluating the design against requirements

Understanding the context of use This includes details of the organisation, its location and the phyiscal environment and hardware that will house the new system. The system goals are specified together with the tasks needed to accomplish those goals to ensure that the system has the right level of functionality. Inadequate functionality simply means that the system does not provide all the tools needed to carry out the tasks required to achieve the goals. Excess functionality usually results in uneccessary complexity, reflected in excessive learning time, too many ways of carrying out the same task or cluttered screens.

Specifying user and organisational requirements The organisational requirements determine the target peformance of the system. This will be centred on peformance metrics concerning output, capacity, processing times for key tasks and so on. Other requirements will include health and safety considerations, training times, workstation design and implementation and running costs.

Producing design solutions Design solutions are produced iteratively to get feedback from multidiscplinary groups. Concrete proposals using mock-ups, storyboards, sample screens, etc. are presented first. Users are allowed to perform simulated tasks at the early stages of design and real tasks once working prototypes have been developed. The design is modified in accordance with feedback received and the process is managed, often by an ergonomist.

Evaluating the design against requirements Evaluation should begin as early as possible in the design process. In general, changes are less costly to make at the early, rather than the later stages. Although early evaluations using crude mock-ups may lack realism, large savings can be made. Evaluation is centred on how well the proposed design will enable the goals to be achieved in relation to the performance metrics. Some assessment criteria are that commands function as specified, that the database is secure and, with modern systems, that WYSIWYG applies (that 'what you see is what you get').

Design goals for interactive systems

According to Shneiderman (1991) human-centred design has the following goals:

- *Acceptable learning time*. How long should it take for a typical user to master basic system functions? How long to learn new tasks?
- *Speed of performance*. Compare performance on the new system with performance metrics for benchmark tasks
- *Retention over time*. How long can users be expected to retain their system knowledge? This depends very much on how standard the system is, how well the tasks resemble other tasks, the new knowledge needed and acquisition time
- *Subjective satisfaction*. Satisfaction is enhanced when new systems conform with standards for current systems, when the arbitrary or confusing use of terms is avoided, and when the system complies with common expectations built up using other systems (e.g. motion stereotypes).
- *Standardisation* is needed to reduce the training burden. Generic functions such as deleting and creating files should be standardised. Other advantages of standardisation are that the output of the system can be used by other systems, that there are common features across multiple applications, and that action sequences, use of abbreviations, units, layout colour display format, etc. are consistent.

Summary of design guidelines for usability in human–computer interaction

General guidelines applicable to most applications are summarised in Table 14.1.

Dialogues should be designed to yield 'closure'. Subtasks should have a clear beginning, middle and end so that users can track their progress as they carry out a task and are ready to continue with the next task when they get to the end. In practice, support for error handling means that an 'undo' button is provided. Unecessary

Table 14.1 General guidelines for usability[a]

- Know the user population.
- Provide short cuts for experts.
- Make the system flexible and user-driven.
- Do not force users to follow a series of rigid steps.
- Provide alternative paths to each goal.
- Avoid unecessary complexity.
- Minimise memorisation:
    Use menus rather than commands.
    Use names rather than numbers.
    Use meaningful terms.
    Strive for operational similarity with existing systems.
    Use task-related syntax.
- Design for error – serious errors are difficult to make and most errors easy to recover from.
- Develop an explicit design model and check that the system is consistent with the model.
- Minimise worker effort – do not ask users to carry out tasks that the system can do.
- Provide on-line help and meaningful error messages.
- Minimise system response time and provide feedback when there is a delay.
- Maximise worker control by allowing multitasking.

[a] From Preece (1993).

obstacles are avoided by allocating low-level functions to the computer and giving the user control. For example, a badly designed online grant application system for university researchers was developed that promised them the ability to create, edit, update and submit grant applications over the Internet. However, whenever the user logged-on, the system assumed complete control, demanding to know the user's name and password and the date before allowing the transaction to go any further. This example illustrates a number of points of bad practice. Firstly, the transaction is live, so the system already knows what the date is even if the program does not; secondly, the user may only wish to edit the application, not submit it, so the date is not relevant; thirdly, no further dialogue is possible without the date (the user might have prefered to do something else and enter the date later).

Designing information in external memory stores

Graphic user-interfaces should facilitate recognition and retrieval using multidimensional coding, for example using filenames and icons together, within a visual frame of reference. Proponents of GUIs claim that they are more natural than other designs. Although this is undoubtedly true for certain systems and tasks, it is not necessarily universally true – in large systems, the complexity of the visual display itself may become difficult to manage and verbal models may be more appropriate (Webb and Kramer, 1990). It has long been known (Bartlett, 1932) that people have preferred thinking styles – what is appropriate for a visualiser may be inappropriate for a verbaliser.

Database retrieval using keywords

Gomez and Lochbaum (1984) investigated performance on an information retrieval system using the keyword method. Subjects were required to find one of a number of recipes by entering appropriate keywords (descriptive of the dish). A potential problem with systems such as these is that the system designer and users may choose different words to describe the same things. People often use many words to describe the same object and it may be difficult to know, in advance, what their preferred words will be when they interact with a computer.

The size of the set of keywords was varied in a number of experimental trials (from 1.96 to 31.96 keywords per recipe). The larger ('enriched keyword') vocabulary set was found to increase the number of recipes retrieved by the subjects. It had been hypothesised that the enriched keyword vocabulary might incur a performance cost as a result of increased ambiguity, but this was not the case – subjects were able to identify target recipes using fewer total entries with the enriched keyword list. The enriched keyword technique (known as aliasing) seems to work particularly well in semantically rich domains where unambiguous classification of items is not possible.

Keywords should reflect common usage

A university in Cape Town, South Africa, changed the name of its printing department from 'Printing Department' to 'University Document Management Services' with the result that the printing department's phone number was listed under 'U' in the electronic and hard-copy telephone directories. This caused difficulties because customers wanting some printing done were unlikely to begin their search with the letter 'U' (or even under 'D' for document). In fact, all departments in the university could be listed under 'U' in the phone book ('University Mathematics Department', etc.). This illustrates some of the problems caused by using inappropriate keywords and, in this case, a redundant descriptor that might be 'politically correct' but adds no useful search information.

Automatic categorisation of search results

Searching for information on the World Wide Web can be time consuming because of the large number of sites that may be accessed using a single keyword. Chen and Dumais (2000) developed a user-interface for web searches that automatically categorised web searches and presented the results in meaningful categories (standard interfaces normally provide only a numbered list of whatever was found). In searching for information about how to rent a Jaguar car, the sites found would be categorised as, for example, automotive, leisure, business and finance, entertainment and media. Users would then look for Jaguar under likely categories, e.g. 'automotive', to find the site of a company that hired Jaguar cars. The category interface was compared to a standard interface by having subjects search for 15 websites using each interface. The category interface resulted in faster search and fewer aborted searches. Subjects gave up only about 2% of the time using the category interface compared to 5% of the time using the list. Average search time per item was 45% longer (over 80 seconds compared with 55 seconds) using the list interface. Even with complex databases such as the web, presorting search results into meaningful categories improves performance. The category interface was preferred by the subjects.

Personal information storage

Lansdale (1988) reviewed the ways in which people organise the information stored in their offices and the implications of this for the design of the 'paperless office of the future'. Naturalistic observations of pencil-and-paper offices revealed some of the characteristic features of personal information management, which may have profound consequences for office automation and system design.

Neat versus messy offices may reveal more than just the personality of the occupant. Repetitive jobs often have an information content that lends itself to the construction of a straightforward system for classifying information. New data can be classified in an unambiguous way under a clear heading or category name. For example, a general practitioner may classify each patient's data under the patient's name and have several separate systems for different types of data (for example, a system for patient records containing medical histories, a system for accounts and a system for appointments). This type of classification system favours a neat office. However, some jobs have a semantically rich information content and any item of information can be categorised under one of several possible headings. The messy office may really be a symptom of classification problems caused by the information content of an occupant's job rather than a reflection of the occupant's personality.

Similar problems have been found to occur in the design of Videotex systems. Users have been found to make mistakes when interrogating these systems because of the lack of an intrinsically unambiguous classification system for accessing the information on the database.

Files versus piles in the modern office Filing systems may be replaced by 'piling' systems when documents do not fit into any clear category. Users fear that if they file a document, they will never be able to find it again (because of not knowing where to look or looking in the wrong place because the context has changed). In memory theory parlance, the problem does not occur because the item is no longer in storage – rather, retrieval is impeded by a lack of cues or a loss of context. Storing documents in piles is an informal solution to this problem because a pile always reminds the user of its presence. Physical cues indicate how long a document has been stored in a pile (by its distance from the top). Physical appearance (typeface, colour, etc.) enriches the set of retrieval cues (we might not remember the name of a document but we might remember that it had a blue cover and the title was written in startling red letters). According to Lansdale, electronic information systems, if they are to be successful, need to support the types of coping strategies that people typically use to overcome the ambiguity and memory problems associated with large databases. An electronic system that used a neat office as the model for its database query system would be suboptimal for messy applications, because of the categorisation problems described above. Multiple keywords may be a partial solution, but the task of filing itself (choosing categories) and how this should be implemented electronically is also a design issue as is the specification of other attributes (in addition to a name) to enrich the representation of electronic documents on computer.

Some of these issues were tested by Lansdale *et al.* (1990) in a document retrieval task. Enriching information in the form of either words or icons was added to documents either by the subjects themselves or by the system. Recall was found to be higher when enriching information was provided and when subjects made their own selection. There was no evidence that any particular modality of enriching information influenced recall better than any other. Lansdale *et al.* interpreted these findings using the principle of coding diversity. The principle states that encoding an item using a diverse set of modalities enhances its recallability.

Retrieval cues for web pages

The Internet is a vast and highly complex database that users can interrogate in a wide variety of ways. Simple tools have been developed to help users retrieve preferred web sites. *Favourites lists* and *bookmarks* are examples of these, yet how, or indeed whether, users remember their interactions with the web is not well understood. Czerwinski *et al.* (1998) investigated how subjects organised 100 different web pages for storage in their favourites folder and what features of the web page were the best retrieval cues. The features investigated were the title of the page, a short description of the site, and a thumbnail-sized miniature of the whole web page, singly and combined. Most subjects constructed shallow schemes of about 10 categories for organising the websites into categories and very few used subcategories. The kinds of categories used were similar across subjects (e.g. shopping, entertainmment, business, etc.). Title was the most effective retrieval cue with an average retrieval time of about 10 seconds. Retrieval time using the thumbnail was 17.5 seconds. Interestingly, retrieval time and number of incorrect pages selected were greater when all of the retrieval cues were presented together than when the title alone was presented. The additional information may have confused subjects, perhaps because of superficial similarities between different thumbnails that are not reflected in their title.

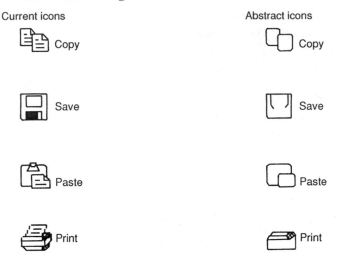

Figure 14.1 Concrete and abstract icons compared by Stotts. (Reprinted from Stotts, 1998. The Human Factors and Ergonomics Society. All rights reserved.)

Multidimensional cueing may assist retrieval from memory but not recognition. It may be that in this study of 100 web pages subjects had no need to use information other than web page title to encode the page in memory. This would explain why title alone was sufficient for recall.

Designing icons for ease of recognition

Icons are ubiquitous nowadays and the design debate has shifted from whether system objects should be represented by icons at all, to how best to design icons. The answer seems to be that icons should be made to resemble their referent objects as much as possible. Stotts (1998) compared search and select times for graphically concrete or graphically abstract icons (Figure 14.1) that were either functionally representative of their referents or were not. Concrete icons were found almost twice as quickly as abstract icons (in about 4 seconds) and the effect was strongest for experienced users. Concreteness was most effective in speeding search and select times when the icons were functionally representative of their referents. It seems, then, that search time depends on the detection of features that can be mapped onto the user's internal representation of the referent and that the more concrete these features are the quicker the icon is recognised.

Sears *et al.* (1998) investigated the quality of the icons used in Microsoft® Word 7.0. Volunteers (384) rated the identifiability and perceived effectiveness of the icons found on Word's toolbar (Figure 14.2). Effectiveness was assessed by presenting subjects with the icon, a verbal description of its function and a 7-point rating scale, and asking them to assess how effectively the icon repesented its referent function. Identifiability was assessed by presenting subjects with each icon and asking them to identify its function. Subjects also indicated how frequently they used each icon. Icons that were perceived as effective and were identified correctly tended to be used to perform frequently performed functions. These included Print, Bold, Italic and Underline. Icons that were rated as ineffective and were not identified tended not to

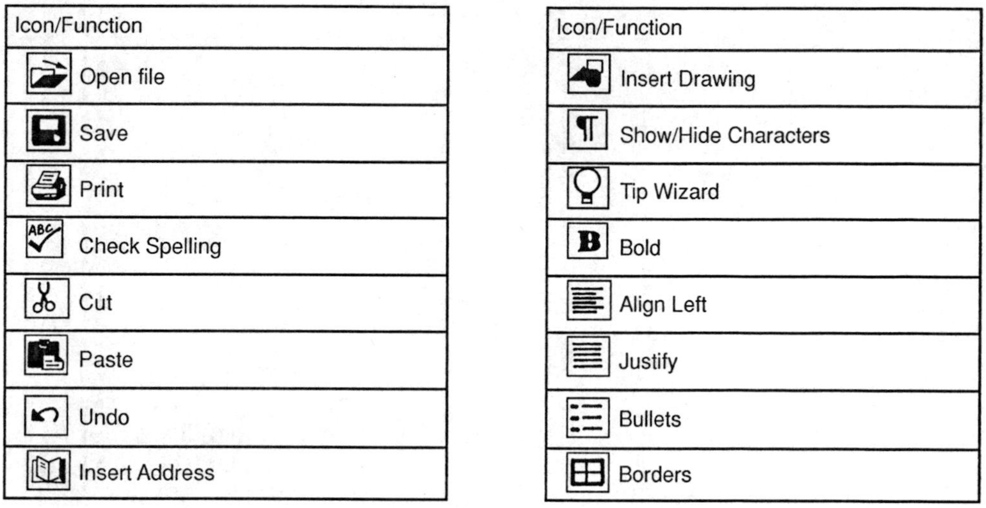

Icon/Function
📂 Open file
💾 Save
🖨 Print
✓ Check Spelling
✂ Cut
📋 Paste
↩ Undo
📖 Insert Address

Icon/Function
Insert Drawing
¶ Show/Hide Characters
💡 Tip Wizard
B Bold
Align Left
Justify
Bullets
Borders

Figure 14.2 Icons of the kind investigated by Sears *et al.* (1998) for effectiveness and identifiability.

be used. These included Auto Format and Insert Address. A number of icons were identified that were used less frequently than the functions they represented (because users employed the pull-down menus or used other techniques to access the functions). Since the purpose of using icons instead of commands is to increase usability, this suggests either that these icons are redundant or that they should be redesigned. Examples of such icons include: File, Save, Print, Preview, Check Spelling, Cut, Copy, Paste, Undo and Insert.

Design of codes

Codes are strings of letters or numbers, or both, that are used to represent system elements. Bailey (1982, 1983) has summarised many of the psychological considerations in the design of codes – memory is one of the most important. The main considerations in code design are

1. Length
2. Sequence
3. Content
4. Chunking
5. Errors

Designers often design codes in accordance with the requirements of the system rather than the strengths and weaknesses of user memory. Codes usually exceed the capacity of short-term memary (STM) (7 items ±2). An alternative is to fix the minimum length of code required to discriminate among the members of the class of objects to be coded and to use this as a basis for development. Bailey (1993) points out that every person in the world could be given a unique code number using only a 10-character code but the New Jersey driver's licence number is 15 characters long,

Table 14.2 Guidelines for the design of abbreviations[a]

1. Determine the required number of letters in the code.
2. Remove suffixes (e.g. -ed, -er, -ing).
3. Select the first letter and the last consonant (e.g. Boulevard → B . . . D).
4. Fill the remaining spaces with consonants (e.g. BLVRD, BLVD).
5. Avoid double consonants.
6. Use vowels only if there are insufficient consonants.

[a] From Bailey (1982, 1983).

even though there are only 4 million drivers. Codes such as these are handwritten and read in many situations and invite unnecessary errors.

The sequencing of items in a code is another important consideration. Research suggests that the middle items of a code have the highest probability of incorrect recall (owing to the primacy and recency effects). In shorter codes, this can lead to the penultimate items being recalled worst. The problem with long codes is that the discriminability of the middle items is low – they are like figures embedded in a background of similar information. In alphanumeric codes for short-term storage, it may be advantageous to sequence the letters and numbers together to increase the 'chunkability' of the code by increasing the discriminability of its component parts (e.g. GA2 as opposed to G2A).

Abbreviations

Abbreviations are also a way of coding information. Some rules for the construction of abbreviations are given in Table 14.2.

Reversal errors

Reversal errors are common in data entry tasks and can be minimised by avoiding the use of character combinations that utilise adjacent keys and, when designing codes for classes with few members, giving each member a unique code whose reversed forms are illegal (and will not be recognised). On one occasion, the author's mortgage repayments, which were automatically deducted from his salary, were being paid to the wrong Building Society. The author, his employer and the two financial institutions involved spent considerable time trying to find out why these incorrect payments were being made. It turned out that the employer's payroll system used the codes '68' and '86' for the two institutions and a reversal error at data entry had caused money destined for institution number '86' to be sent to number '68' – the coding system had reversal errors designed into it rather than designed out.

Code content

There are several important considerations in determining the content of codes. Firstly, an alphanumeric code may be easier to remember than its purely numeric or alphabetic equivalent – people can remember mixed-category lists of words better than single-category lists. Special symbols (!, $, #) should be avoided, particularly if their meaning is incompatible with the application. The use of many special symbols is contraindicated in most keyboard work because it requires the use of the 'shift' key and therefore increases the likelihood of psychomotor errors.

Code content can also be designed to increase meaningfulness by capitalising on already learned associations (e.g. QU versus QV) and by designing 'rehearsable' codes ('BORT' versus 'TBRO'). My student number when I was a student was 'BRDROB003' – a meaningful alphanumeric mix that breaks naturally into three easily retrievable chunks. Chapanis (1990) has investigated STM for numbers. Repetitive triplets such as 999, 112 or 099 are easily recallable as are triplets containing 2 zeros or 2 ones (e.g. 001, 100, 900). They can be used for emergency telephone numbers, for example. Triplets containing the doublets 94, 49, 93 or 64 seem to be difficult to recall.

Coding in errors by design

The design of a coding system determines the probability of particular types of error. Visual (iconic) errors are more likely if physically similar characters are included in the character set used to generate codes (e.g. F,E; C,G; B,3; O,0; and 1,l). This also depends on the quality of print used to display codes on screen or paper. The quality of the various media used to display codes can be evaluated with particular attention paid to the appearance of easily confusable characters. If handwriting as well as print is involved, the following characters are likely to be confused:

Z–2	G–9,6
U–V,W	T–J,7
Y–7,4	Q–O,2
s–5,8	D–0

Confusions can be minimised by removing O, I, Z, Y, S and G from alphanumeric character sets and U, Y, Q, T and J from alphabetic character sets.

Text written in upper-case letters is usually less legible than lower-case text (for most commercial typefaces, at least) because the ascenders and descenders in lower case provide additional visual cues for letter discrimination, which may be particularly beneficial when viewing conditions are poor.

A separate class of errors can occur when codes have to be held in STM. Apart from the forgetting of codes due to excessive code length or trace decay caused by an interruption of rehearsal, auditory confusions can also occur. In this case, items that look very different but sound the same can be confused (e.g. e,D; c,B). This is because storage in short-term memory usually takes place in the acoustic or articulatory modalities. If codes have to be held in STM – while people speak over the phone, for example – similar-sounding letters and numbers should be excluded from the character set.

Human–computer dialogues

Several user typologies have been developed over the years. One of the simplest typologies distinguishes between 'naive' users (those with little knowledge of programming or computer science concepts) and expert users (who normally have professional computing skills). Both types of user may be either 'casual' or 'regular' users of a particular system. Some characteristics of naive users are a lack of insight into

Table 14.3 Human–computer interaction styles[a]

Advantages	Disadvantages
Menu selection	Rapid display rate needed
Little training required	Danger of too many menus
Reduces keystrokes	Skilled users may be slowed down
Structures decision making	
System can 'drive' naive users	
Can be used with minimal keyboard skills	
Form fill-in	
Simplifies data entry	Consumes screen space
Little training needed	Typing skills needed
Shows context for activity	
Command language	
Flexible. User 'drives' the system	Substantial training needed
Appeals to sophisticated users	Language has to be learnt, retained
'Macro' commands increase power	and recalled
	Invites errors due to forgetting
Natural language	
No new computer syntax needs to be learnt	Natural language can be ambiguous
Clarification may be needed	Typing skills needed
Direct manipulation	
What you see is what you get	Additional pointing devices needed
Encourages use of an explicit 'model' of the task	Visually impaired users may have
Good for disabled users who can point but	problems
cannot type	

[a] Adapted from Shneiderman (1988) and reproduced with permission of author and publisher.

the workings of the underlying technology, expectations about computers that differ markedly from those of designers and experts, and an inappropriate 'model' or concept of how the system actually works. Regular naive users' knowledge of a system is likely to be procedural rather than technical (knowing what to do but with limited understanding of how and why it works). Casual naive users can be provided with procedural information in the form of instructions by means of on-line 'help' facilities. They should not need to have technical knowledge in order to operate the system. Regular expert users will have both technical and procedural knowledge and require little assistance. Casual expert users may require on-line help to assist with detailed aspects of executing specific procedures.

Implementation modes for human–computer interaction

Shneiderman (1988, 1991) identified five main interaction styles for dialogue design (Table 14.3).

Menu selection

The main advantage of menus is that the user does not have to remember the options on offer – they are explicitly displayed on the screen. The main ergonomic issues in menu design are

- Mapping the task domain onto a menu structure
- Compatibility of menu structure with user's conception of the task
- Compatibility with the way the task is performed
- Navigation through the menu
- Backtracking
- User's memory load

Highly structured tasks lend themselves to the construction of unambiguous menu trees with simple choices at each level. Care should be taken that the structure of the menu is compatible with the users' conception of the domain (e.g. should the *Financial Times* newspaper be accessed via 'Business' or 'Entertainment'?). It should also be compatible with the task itself (e.g. users should not need to navigate four layers of a heirarchical menu just to turn off the computer, although some systems do come close to this with continual reminders and warnings before shutting down). Thus, menu designers need

- An understanding of the task domain and the task semantics
- An understanding of the users' conception of the domain (hence the need to consult users)
- A description of the task

Several techniques facilitate navigation through menus. These include 'jump-ahead' techniques for experts, backtracking facilities and maps. Memory load can be reduced by optimising menu structure. According to Shneiderman, shallow menu structures are to be preferred because few levels have to be traversed to reach the goal. A disadvantage of this is that there may be too many options at once and users may have forgotten some of the choices by the time they finish reading the list, notwithstanding primacy and recency effects. Deep menu structures have the advantage that the user has fewer choices each time but may get lost in the system and backtracking to previous states may be time-consuming. For complex domains, the compromise position seems to be to have more depth at the root (beginning) of the menu and fewer intermediate choices. Ideally, users should quickly develop a mental model of the menu, particularly if it is compatible with their conception of the domain. Problems occur with domains in which categories are not mutually exclusive and it may then be necessary to provide links with overlapping but separate parts of the menu tree.

Much depends on the ease with which the task domain can be represented hierarchically. There may be categorisation problems of the type described by Lansdale (1988) that complicate menu design. Schwartz and Norman (1986) use as an example of categorisation problems in menu design a main menu that asks users to choose between the following:

1. Services for professionals
2. Home services
3. Business and financial
4. Personal computing

They argue that the menu is unsatisfactory because the *distinction between the first two options is not clear*, firstly because both options offer services and secondly because 'professional' and 'home' are not mutually exclusive categories (many professionals work at home). Options 1 and 3 could overlap completely. In options 2 and 4 the distinction between 'home' and 'personal' is also questionable.

Schwartz and Norman compared the above menu with an alternative that had been modified to be more distinctive:

1.   General home computing information
2.   Specific financial investing information
3.   General leisure information
4.   Specific professional service information

Subjects using the modified menu took less time to find items, they found them using a more direct path through the hierarchical menu structure, and they gave up on fewer occasions than those using the original menu. Thus, categorisation for item discriminability seems to be an important consideration in menu design.

Shneiderman has suggested that it may be better to design a menu structure as a shallow rather than a deep tree. At each level of the menu many items would be presented on the screen at once rather than having a deep hierarchy in which each screen (or level) has a small number of items. This would cut down on the number of different screens and the time taken for users to orient to new screens and facilitate backtracking to previously used screens. STM would presumably limit the maximum number of items per screen to about 7.

STM limitations are also potentially problematic in auditory menus used for telephone access to databases because

1.   The user cannot control the rate of information presentation.
2.   The user does not know the number of items that will be presented.
3.   Presentation mode is sequential and the user cannot scan the list of items.
4.   There are no visual cues to indicate the menu structure.

Another problem with rigid menus is that experienced users cannot interrogate the system much faster or more efficiently than novices. The rigidity of such menu systems is often caused by the constraints involved in traversing the hierarchy. These may cause frustration and annoyance among experienced users who, having learnt the menu structure, are still forced to take several small steps to get to a well-known level. For this reason, many modern menu interfaces are provided with extra facilities for experienced users. For example, Laverson *et al.* (1987) investigated jump-ahead techniques that enable experienced users to take 'short cuts' through the menu structure. A 'direct access' method was found to reduce learning time and lower error rates. The method worked by assigning a unique name to each menu frame. A particular frame could then be accessed directly by entering its name. If the name could not be remembered, the user could fall back on the conventional stepwise method of using the menu.

Form fill-in

According to Shneiderman (1988), there has been little research on this method of interaction. It requires keyboard skills and therefore may not be appropriate for certain types of users. An advantage of form fill-in is the increased density of items that can be displayed on a single screen compared with menus. Many of the design issues would seem to centre around the basic requirement for screen legibility (similar to optimising the layout of any visual display or 'pencil and paper' form).

Greene *et al.* (1992) describe an interesting variation on this theme called 'direct entry with autocompletion'. Subjects performed an airline reservation task in which they had to specify variables in three fields: departure city, destination city and airline. The two main methods of specifying these variables were entry and selection. In the entry conditions, subjects had to type in correctly each variable and received either immediate or delayed feedback as to whether the entry was correct. In the entry with autocompletion condition, the system completed the spelling of the entry once the subject had typed enough letters for it to be recognised. A novel technique, called the 'sponge', absorbed letters typed after autocompletion had been initiated to prevent surplus letters from spilling over into the next field. In the selection condition, menus were placed in the three fields and subjects used the arrow keys to scroll through these menus and select the desired variable.

Autocompletion with a sponge to absorb surplus characters proved to be the fastest and most preferred style of interaction. In situations where the user does not need a prompt about what to type in to the system, this may be a viable mode of interaction and superior to menu selection.

Command languages

Command languages are less used nowadays owing to the emergence of graphic user-interfaces. A discussion of some of the design principles can be found in the previous edition of this volume.

Natural language

It is sometimes said that natural language will ultimately be the best mode of interaction between a user and a computer, but this is debatable. Many written forms of documentation would be easier to understand if they were rewritten using a limited vocabulary program style and using flow diagrams. There is evidence that people perform better at problem-solving tasks when they are presented in flowchart form rather than as paragraphs of fully connected prose (Kaman, 1975). Flowcharts themselves are best designed to be compatible with people's reading patterns and should be oriented in a left-to-right, top-to-bottom pattern (Krohn, 1983).

According to Shneiderman (1988), one of the main challenges of natural language as an interaction mode is posed by the fact that the permissible syntax and system capabilities (the context) may not be as 'visible' as in other, more structured, systems. Thus, the designer has to find some way of providing contextual information that makes clear the scope of the system and the types of verbal inputs it is appropriate for users to make.

Direct manipulation

Much of the software now developed for personal computers is based on direct manipulation platforms with menu selection or form fill-in used to support specific tasks.

A recent experimental study by Rauterberg (1992) suggests that direct manipulation may be a more efficient mode of interaction than menu selection. Novice and expert subjects were given 10 database interrogation tasks to carry out on both types of interface. It was found that both groups needed less time to carry out the task in desktop-interface mode than in menu mode (regardless of differences in the number of keystrokes required to operate the two interfaces). There was no support for the view that direct manipulation using the desktop metaphor is only suitable for beginners and it was concluded that the desktop interface promotes learning.

Memory and language in everyday life

If we think of language as a tool for expressing and communicating ideas, it becomes clear that different languages differ in the 'facilities' they offer for doing this and that the 'facilities' offered by a particular language reflect the distinctions important in the life of its speakers. In English, there is only one form of the word 'we', whereas in other languages there are different forms depending on whether the speaker is referring to you 'you and I', 'you, I and somebody else', 'me and him but not you' and 'me, him and others but not you' (Pinker, 1994). Within a particular language, speakers differ in the compactness and precision with which they express different ideas.

Communication problems

Differences in language use may be found between designers and consumers or users and between operators as opposed to maintenance or design engineers, between line versus staff management and between experts versus laymen. Designers of equipment, manuals and instructions must understand the vocabulary and syntactic habits of the various groups of people who will use their systems.

Jukes (1986) presents a fascinating example of miscommunication during a US Government hearing in the 1970s. Three parties were involved in a debate about whether sugar in food was a health hazard: Senator Schweiker (later US Secretary of Health and Human Services), two nutritionists and an anti-sugar lobbyist (Dufty). The debate seems to go nowhere because of fundamental differences in the way the speakers use certain words and how these words are linked to concepts in their semantic memories. Critically, there is almost no conceptual mapping between the semantic networks of the parties with respect to the issues under debate. Figure 14.3 is a formal representation of the knowledge domain, probably similar to that possesed by the nutritionists. For them, foods contain nutrients and non-nutrients. Nutrients consist of water, carbohydrates, fats proteins and so on. Carbohydrate is a technical term for a class of chemicals and there are different kinds of carbohydrates – starches and different types of sugars. For Dufty, however, sugar is an artifical substance, probably conceived of as a food additive or even a toxin!

At the point we join the hearing, the Senator is exploring the notion that sugar in food is an 'anti-nutrient'. The accusation, made by the 'anti-sugar' lobby, was based on the fact that when glucose is metabolised in the body, B vitamins are consumed in the process, whereas when fat is metabolised, no B vitamins are used.

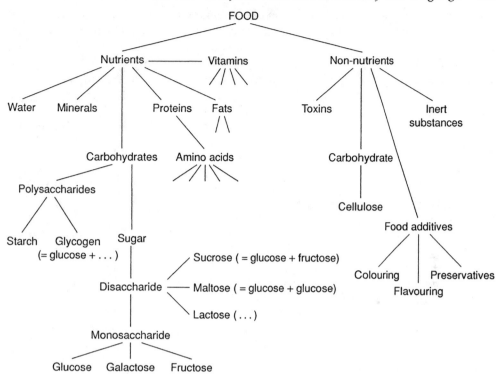

Figure 14.3 Approximate formal structure of the nutrition knowledge domain. Words like 'carbohydrate' are linked to a semantic network when used by scientists and are closely linked to other concepts, in a hierachical manner. Communication problems occur because the mappings between the semantic networks of lay users and experts are poor. In the sugar example, words like 'carbohydrate', when used by lay people, may not to be linked to a semantic network at all!

> **Dr Van Itallie:** *The fact that the requirement for vitamin B_1 and certain other B vitamins will increase when you take more carbohydrates does not justify the statement that carbohydrates – or sugar – is an antinutrient.*
> **Senator Schweiker:** *I am not talking carbohydrate; I'm talking sugar, let's keep it on the right track.*

(The Senator conceives of sugar and carbohydrate as different and does not know that sugars belong to a family of nutrients known as carbohydrates. He is also unaware that carbohydrates, such as starches, are converted to glucose, i.e. sugar, in the digestive system before they enter the bloodstream.)

> **Dr Van Itallie:** *There is no difference between sugar and carbohydrate with respect to vitamin B_1.*

(The anti-sugar lobby then interject, arguing that 'This is inaccurate unless refined carbohydrates are specified', unaware that the word 'carbohydrate' already refers to a refined chemical substance, i.e. a nutrient and not a food such as pasta or potatoes.)

> *Senator Schweiker: Well, we have had a number of dentists just come before us recently and tell us how bad sugar, not carbohydrate, was on dental caries . . .*

(The Senator does not realise that although sugars and starches may affect the teeth differently, they are all broken down into glucose in the intestines and metabolised in the same way.

> *. . . and you are saying that something that increases the need for nutrients in terms of quantity is not an anti-nutrient?*
> *Dr van Itallie: That is correct.*
> *Senator Schweiker: Are you sure we are not getting into a semantic argument here?*

(The Senator does not understand the difference between a nutrient and a food. He seems to think that only vitamins and minerals are 'nutrients' and that sugar is a food. They are, indeed, getting into a 'semantic argument'.)

> *Senator Schweiker: If we market a cereal and say we presweetened the cereal and added sugar, we are working against ourselves. A customer buys a pack of cereal, nutrients added . . .*

(In, fact the nutrients are not 'added' they are intrinsic to the food, what he means is that the cereal is fortified with extra vitamins.)

> *. . . presweetened . . . If we hadn't had the sugar we might not need the nutrients.*
> *Dr Butterworth: Sugar is not an anti-nutrient. Sugar is a nutrient, and sugar is a carbohydrate.*
> *Senator Schweiker: But it does substantially increase the need for nutrients . . . Now the FTC made the sugar associates quit advertising that sugar was an energy builder* and a nutrient. . . .*

(*Sugar is a source of chemical energy rather than an 'energy builder'.)

> *Dr van Itallie: Sugar is a nutrient.*
> *Senator Schweiker: And that is just what the FTC say you can't say because they don't believe it.*

The debate then became coloured by more emotive uses of language such as that Americans now eat '14 megatons' of sugar each year (presumably to conjure-up images of nuclear warfare). Jukes points out that we might equally argue that Americans drink '80 megatons' of water each year (much of the sugar is a naturally occuring nutrient in the foods, such as fruit, that Americans like to eat).

   This example illustrates the futility of engaging in debates about factual matters when the different parties do not share the same concepts, even though they use the same words.

Language comprehension

Language comprehension can be seen as a cognitive process involving both top-down and bottom-up process. According to Greene and Cromer (1978), language

Table 14.4 Levels of knowledge in language comprehension

1. The sound/spelling of words
2. The normal meanings of words
3. The grammatical rules for acceptable utterances
4. Metaphorical/idiomatic usages
5. Contextual knowledge
6. General (world) knowledge

comprehension requires several layers of knowledge (Table 14.4). This applies to all linguistic transactions between speakers. It also applies to the comprehension of verbal exchanges between a user and a computer. Human–computer dialogue is usually limited by the fact that, unlike their users, most systems can only communicate at the lowest levels given in the list.

Sound and spelling of words

Speech utterances are thought to be recognised by an analysis of their constituent features and by synthetic processes that use contextual and other information to construct a meaningful representation from the acoustic features of the stimulus. The parsing of speech is aided by *prosodic* cues (emphasis and rhythm). These are often lacking in synthetic speech, which increases the load on working memory. Recall of synthetic speech is inferior to recall of natural speech (Paris *et al.*, 2000). Synthetic speech, used to provide warnings in vehicles, in aids for the disabled and many other products, can be improved through better design of its prosodic features.

Normal meanings of words

The meaning of a word is understood when the word in question is matched with its stored representation in the mental dictionary and the stored version is linked to a node in the semantic network. When a computer user sees a word or phrase such as 'insufficient disk space' displayed on the screen he must, in order to understand the message, firstly have these words in his vocabulary and secondly be able to link them to the appropriate concepts about the functioning of computer storage media.

Comprehension difficulties can arise because of a lack of the proper concepts, because of the choice of vocabulary or because an excessive number of different terms are being used to describe the same thing. In the first case, the user's knowledge of the word is 'empty' – not linked to a node in the semantic network. The user may simply learn to respond in an appropriate way when the word is presented without ever developing anything more than a superficial understanding of the system. In some applications, this may be appropriate to the requirements of the task. For example, it is not usually necessary to understand how an electric motor works or know the appropriate terminology (what commutators or brushes are or what the field strength of the magnet is) in order to use a vacuum cleaner or a food mixer. However, if the task at hand *does* require a knowledge of the concepts, a user who only has a superficial understanding of the words may experience problems. For example, in knowing what to do when presented with a message such as 'mouse driver not installed'.

Designers must understand how users conceptualise a system and be familiar with the vocabulary they use to describe it.

In the second case, the user may understand the appropriate concepts but the system may use unfamiliar words to describe them and the user may have to infer the meaning of the word from his knowledge of first principles and the current system behaviour. Novices are unlikely either to possess the appropriate concepts or to know the words that refer to them. It is for this reason that they need to be shielded from technical feedback such as 'input buffer overload'.

Lachman (1989) developed an on-line comprehension aid – a window in which definitions of technical words embedded in the text could be displayed. Subjects were found to access 80% of the available definitions and were able to retain a substantial amount of the semantic content of the text. Lachman described his system as a shallow form of hypertext inasmuch as cues can be called onto the screen that are associated with items in the database. He concluded that even simple hypertext systems can enhance learning and comprehension of text.

Grammatical rules for acceptable utterances

A grammar can be defined as a set of rules for generating all grammatical sentences in a language and no ungrammatical ones (Chomsky, 1957). Speakers of English, for example, can tell which of these sentences follows the rules of English grammar:

1. Press the red button in an emergency.
2. In an emergency the red button press.

Syntax is the term used to describe the rules for combining words in an order that will be recognised as 'grammatical' by a native speaker of a language. In the example above, it is the syntax of the two sentences that differs. Note that in the discussion of grammar and syntax, we are not necessarily talking about the meaning of these sentences. It is likely that, in an emergency, the second version would be indeed be understood despite its incorrect syntax.

In order to understand a sentence, the basic syntactic relationships between the words have to be determined. The actual sequence of words (called the surface structure of a sentence) has to be analysed to determine these basic syntactic relationships. Any grammatical sentence can be said to have a deep structure in which the syntactic relationships are represented and it is from the deep structure that semantic analysis of a sentence takes place.

Sentence comprehension requires more than just a knowledge of the meaning of words. In order for a sentence to be understood, a syntactic analysis is required to describe the sentence structure (the sequence of nouns and verbs) This is known as 'parsing'. Further processing is needed to identify the basic syntactic relationships between the words (e.g. subject, verb, object). Semantic processing of the contents of the sentence can then take place. When designing verbal material or analysing comprehension problems, it is important to distinguish between the syntactic and semantic aspects of language comprehension. In ergonomics, language comprehension can be regarded as a mental task, which, like any other mental task, has a workload associated with it. Ergonomists should attempt to design verbal material such as warnings and recovery and emergency procedures that can be comprehended with a minimal amount of mental effort.

Metaphorical/idiomatic usages

One of the remarkable aspects of the use of language by humans is the tendency to find new ways of using words to convey ideas. Everyday speech is also characterised by this type of word usage as in 'John climbed rapidly up the ladder grasping every opportunity with both hands'. In order to comprehend utterances in a language, the listener often has to decide between a literal or a metaphorical approach to interpretation.

Because many of the concepts encountered in information technology are abstract and difficult to express using everyday speech, the area is littered with metaphorical/idiomatic uses of language. Performing a 'screen dump', for example, involves printing the entire contents of the screen onto the printer. Colourful uses of language such as these have high imagery value and are no doubt easy to learn and subsequently recall, but they may cause difficulties on the first encounter because the real meaning of the phrase is nowhere implicit in its constituent words. Special attention needs to be paid, therefore, to explaining such phrases to novice users either in training or in the on-line help facility.

Contextual knowledge

Information about the context in which an utterance takes place can be an important determinant of comprehension. Specific knowledge about people and events is often essential to disambiguate the usage of particular words. For example, Bransford and Johnson's (1973) sentence 'The haystack was important because the cloth ripped' could only be understood when accompanied by the heading 'The Parachutist'.

Document designers should use headings and subheadings to provide local context for readers of extended documents.

General (world) knowledge

Communication using natural language is more than just the interchange of verbal symbols ordered according to a set of grammatical rules. The purpose of language is to communicate meaning. To make sense of a speaker's utterances, the listener decodes the sound pattern and performs a grammatical analysis and also uses knowledge stored in memory of the topic under discussion to assist in interpretation.

The everyday use of a language by its native speakers is underpinned by assumptions about shared beliefs and experiences. Speakers expect listeners to share similar beliefs and to keep them in mind during a conversation (we all expect others to use their 'common sense' when listening to us). This is one of the reasons why individuals who speak English (or any other language) as a second language often experience problems when communicating with native speakers. Of particular importance to the construction of user-friendly language are the assumptions made by the writer about what constitutes common sense on the part of the readers. If these assumptions are invalid, comprehension problems and misunderstandings may result.

Design of visible language

Good design aims for ease of comprehension at the initial encounter with the material and on subsequent readings. Memory for the meaning of a sentence appears to be independent of memory for its wording (Fillenbaum, 1973), so sentences that are difficult to understand at first reading may stay difficult! Since people only remember the 'gist' of things, they may be unaware that their memory of the meaning of a previous instruction is incorrect. If they are unable to retrieve the precise wording of the original material, their initial, incorrect, interpretation will persist.

Complex phrases are often found in official and technical documents: for example,

Pull out the left-hand ignition element retaining screw pin.

It is not readily apparent what is being referred to, the left-hand ignition element or the retaining screw on the left of the element. A more easily understood version is 'Pull out the left-hand pin on the screw that retains the ignition element' (Wright, 1978).

Table 14.5 summarises checkpoints for the design of user-friendly language. Easily understood sentences have four characteristics:

1. They contain few clauses.
2. They utilise the active rather than the passive voice.
3. They are affirmative rather than negative.
4. They are concrete, rather than abstract.

Few clauses   A complex sentence such as 'The man who met the clerk that made the loan that saved the old firm is in prison' (Johnson-Laird, 1988) may cause comprehension difficulties by placing an excessive STM load on the reader. The danger is that STM will overflow before the final verb phrase reaches short term memory – people will forget what the sentence is about before they get to the end.

Problems arise when clauses are embedded within each other in a sentence. For example, 'The man that the girl saw stood up'. The phrase, 'that the girl saw' merely specifies something about the man who stood up, but because of its position in the sentence nothing about the man can be understood until the end of the sentence. The alternative version 'The girl saw the man. The man stood up' involves less memory load and communicates information about its subject more directly. People do seem to be able to cope when two ideas are brought together in a single sentence despite this, although performance breaks down rapidly with increasing complexity. Wright gives an even more complicated version: 'The man that the girl that the dog chased saw stood up'.

Active versus passive voice   Passive sentences rarely occur in everyday spoken or written English and, apart from lacking a certain degree of idiomatic force, can also cause comprehension difficulties, as in 'The notes should be read' as opposed to 'Read the notes'.

Passive constructions have a special place in language – when the speaker wishes to avoid mentioning the agent of a verb or does not know the agent of a verb. They are commonly used by politicians and scientists, as in the example 'Taxes were increased'

Table 14.5 Some checkpoints for evaluating the design of user-friendly language

Level of analysis	Checkpoints
Sound/spelling/appearance of words	Signal/noise ratio of spoken message and background noise.
	Digitised speech, better quality than synthesised.
	Typeface of letters (e.g. upper or lower case).
	Print size (minimum 8-point, preferably 10-point)
	Colour contrast of letters and background.
	Luminance of active displays.
	Illuminance in work environment.
	Visual angle.
Normal meanings of words	Frequency of usage in everyday English.
	Frequency in particular industry.
	Technical jargon: avoid in non-technical context (e.g. substitute 'paid work' for 'gainful employment')
Grammatical rules utterances	Complexity of syntax (embedded clauses, long sentences)
	Number of transformations to convert sentence into SVO (subject–verb–object) format
	Avoid overly officious style (e.g. unnecessary use of complex grammar and words such as 'hereby', 'heretofore').
Metaphorical/idiomatic usages	Replace 'purge files' with 'delete files', for example.
Contextual knowledge	Is the context made clear in opening remarks, instructions or screen design (or the change in context with each new screen in a menu-driven system)?
	Does text have sufficient headings?
General knowledge	Has language designer made unwarranted assumptions about general knowledge of the reader? (e.g. use of words such as 'debentures' assumes some knowledge of investments, business and the workings of the financial services sector).
	Can general knowledge intrude in an inappropriate way?

as opposed to 'The government increased the taxes'. The passive voice usually places a greater comprehension load on the reader but can sometimes be of use in maintaining continuity in a body of text.

Negative versus affirmative sentences   Sentences with negative elements are more difficult to understand than affirmative sentences, as in 'Is the book unavailable?'. To answer this question, many people have to convert it into a positive question ('Is the book available?'), answer it and then re-convert the answer. Negative questions can increase the mental load of language comprehension and, under stress or high primary task workload, comprehension may break down. The negative may be missed altogether through hurried reading or noisy speech (Wickens, 1992).

Double negatives can be even more problematic; for example, 'Do not attempt to adjust this equipment if a technician is not present'. The alternative 'Only adjust this equipment in the presence of a technician' is more easily understood. The value of negative statements is to correct incorrect assumptions held by readers or to make strong statements such as

Do NOT prescribe to children under 5 years of age

To guard against the reader mistaking the negative sentence for a positive one, the word 'not' can be highlighted, printed in bold or upper case or in a different colour. Fillenbaum's 'perverse' sentence illustrates the challenge to comprehension posed by double negatives combined in a conditional sentence,

Don't print that or I won't sue you

which the majority of people do not understand.

Concrete versus abstract words   People seem to cope better with complex syntax when the topic is concrete or when concrete words are used. Unnecessarily abstract words should be avoided in the design of forms or instructions, particularly if complex syntax has to be used. A more concrete, 'imaginable' and perhaps less perverse version of the previous example serves to illustrate this point:

Don't wear that, or I won't take you to dinner

Instructions and warnings

Research has shown that it is easier to search memory for the attributes of a particular object (e.g. to decide whether a particular animal has stripes) than to talk about an attribute and instantiate objects that possess that attribute (e.g. to name stripy animals). As a general rule, then, objects should be followed by attributes. However, this does depend to a certain extent on the context for which the instructions or documents are being prepared. In writing a tutorial in a user manual, the object-attribute approach might be best – telling users what the various controls or commands enable them to do. However, when preparing documentation to be used in fault-finding or recovery from abnormal operating conditions, it may be better to emphasise system functions rather than physical characteristics. Documents suitable for an operator (who is concerned with the system functions and attributes rather than technical details about its components) may not be suitable for the maintenance or design engineer.

There is also evidence that people are inclined to carry out tasks in the order in which they read about them. For example, instructions such as 'Before completing this section, read note 3' would be more compatible if rephrased as 'Read note 3 before completing this section'.

Another consideration in the design in instructions is the likelihood of their being read at all ('If all else fails, read the manual'). Wright (1982) investigated the factors that determined whether instructions would be read using a sample of consumer products. The perceived simplicity of the product was one of the major determinants of willingness to read instructions. Users seem unwilling to read instructions on products, such as packets of ice cream or candy, that appear to be simple. People also seem unwilling to read instructions for frequently used products. In these circumstances, it may be best to do away with instructions altogether and design products that are more 'error-proof'.

Similar considerations apply to the design of warning labels on products. Friedmann (1988) found that there was a steady decline in the percentage of people noticing

(88%), reading (46%) and complying with (27%) warnings on a variety of products. The perceived hazardousness of the product was positively associated with the probability of the warning being read. Friedmann concluded that labels should be designed to increase the perceived hazardousness of the product. Critical warning information can be placed first, together with a well-designed symbol. In addition, warning labels should be placed on a prominent part of the product where they are likely to be seen and should be made of materials that will last the life of the product under fair conditions of use (e.g. paper labels glued onto products that get wet or are regularly washed are unacceptable). Young (1990) has shown that warnings embedded in instruction manuals are better understood when prominent print is used (e.g. bold print against a coloured background) and when accompanied by a warning symbol.

Chapanis (1994) has evaluated some of the perceived hazards associated with warnings. 'DANGER' is associated with the highest level of hazard followed by 'WARNING' then 'CAUTION', which are perceived to be much closer together than to 'DANGER'. Perceived hazard increased through the colours white, yellow, orange and red with 'DANGER' on a red background associated with the greatest perceived hazard. Chapanis suggests that this research needs to be repeated on different groups so as to establish international standards for warnings. In view of the perceived proximity of 'CAUTION' and 'WARNING', he also suggests that many products currently labelled with 'WARNING' should be relabelled with 'DANGER'.

Composite signs work better

Wogalter *et al.* (1997) compared elevator users' compliance to four signs, three from the classic paper by Chapanis (1965). The signs were intended to improve the efficiency of elevator usage by exhorting people to use the stairs for short trips (one floor up or two floors down). The signs are shown in Figure 14.4.

Subjects rated the signs according to how understandable they were and their willingness to comply. Sign (a) was rated lowest on both criteria and sign (d) was rated highest. A field study confirmed that sign (d) was the most effective of all the signs – the proportion of people complying was 0.91 when sign (d) was placed outside all the elevators in a building compared to 0.67–0.77 for the other signs. So, when badly designed signs are used, up to one-third of users may not comply compared to only 1 in 10 when a better design is used.

Effectiveness and cost-effectiveness

Design of forms

Forms are a primary interface between information workers and information processing systems. Fisher and Sless (1990) describe an information design project carried out in a large Australian insurance company in the 1980s. It was decided to redesign the proposal forms used for new business. The forms were completed by salesmen in the field but were designed by people who had not sold insurance policies. Any errors made by the salesmen were amplified as the forms moved through different departments on their way to the data entry operators.

Changes were made to forms in a piecemeal fashion, by their 'owners' in a particular division or department in the light of changes in the computer system, new legislation

Figure 14.4 Composite sign evaluated by Wogalter (1997). (Redrawn from Wogalter *et al.*, 1997.)

or new insurance products. Of critical importance was the fact that the design of the forms was not always compatible with the design of the computer system interfaces that received the data. Before being entered into the computer, data had to be transcribed onto other forms, thus increasing the time taken to enter the data and the probability of error along the way.

Because changes to forms were made in one division according to its needs (normally the needs of the computer systems specialists), the needs of users in other divisions (such as the salesmen) were often neglected.

Costs of badly designed-forms   The direct cost of having to correct the errors was over A$500 000 per annum. This estimate is conservative because it does not take into account the costs of having to hire extra people to correct the errors, or telephone, postage or stationary costs, nor does it include the cost of salesmen having to pay return visits to clients. Opportunity costs due to lost business are also not included.

The main problems with the system were that there were too many different forms, they had to be signed by too many different people, and there were too many errors. Information-processing staff tended to blame the sales force for filling in the wrong forms or for missing questions. As is often the case with problems at the user-interface level of systems, it is the behaviour of users that tends to get the blame

Table 14.6 Analysis of errors in the completion of 200 proposal forms for new business

Percentage of proposal forms containing errors	100%
Total number of errors	1560
Range of errors per form	1–27
Number of error types	5
Turn-around time to acceptance of proposal	1–167 days
Number of proposals taking >100 days to be accepted	6

Table 14.7 Costs of error correction

Total number of completed proposal forms per year	41 105
Average no. errors per form	7.8
Total no. errors per year	320 619
Average error repair time	10 minutes
Total repair time	53 436.5 hours
Average hourly salary of repair staff	A$10.32
Total cost	A$551 464

Table 14.8 Performance of old and new forms

	Old form	*New form*
Percentage of proposal forms containing errors	100%	15%
Total number of errors	1560	44
Range of errors per form	1–27	1–4
Turn-around time to acceptance of proposal days	1–167 days	2–46
Number of proposals taking >100 days to be accepted	6	0

rather than the design of the interface itself. Table 14.6 summarises the error data for the forms at the beginning of the study. The cost of repairing the errors in these forms is presented in Table 14.7.

New forms were designed using an ergonomic approach, after consulting all stakeholders. After the new forms had been in use for 6 months, a detailed evaluation was carried out. The data are presented in Table 14.8 together with data from the old form. The reduction in errors turned out to be 97.2% (i.e. only 2.8% of the new forms had errors, compared with the old forms. Conservatively, the cost saving was estimated at A$536 023.

Clearly, attempts to increase productivity by upgrading the hardware and software will not realise their potential unless the interfaces, such as forms, that connect the users with the information system are also optimised.

Usability: Psychological costs and benefits in system design

Professor K. D. Eason of Loughborough University carried out a study of computer system usability. It highlights some fundamental characteristics of human behaviour that can strongly influence the efficiency with which computer systems are used in practice.

The study involved bank employees' use of an online inquiry system. Users were provided with a set of 36 different codes. Each code enabled a different type of information to be retrieved and displayed. To deal with an enquiry about a customer's account, the employee entered the code followed by the account number. There were codes for retrieving the account balance, the last 16 items on the account, and so on. Close analysis revealed that employees did not use the different codes efficiently. Instead of using the appropriate code for the query (i.e. the code that would elicit the correct information, and only the correct information, and would display it in the best way) they used a subset of codes (75% of the time, only four codes were used). They then adapted the information displayed to the needs of current query.

The real power of the system was not exploited. For many transactions, excessive or inappropriate information was elicited with employees recalculating or inferring the needed information from the information displayed. Presumably, this resulted in longer transaction times.

Eason suggested that there are diminishing returns in computer system design. Each incremental increase in the functionality offered to the user results in a decrease in usability. A major limitation in delivering more power to the user occurs when the point of diminishing returns is reached. Users ignore new functions to avoid getting lost in the complexity. There is a trade-off between the costs of learning to access greater functionality and the benefits it offers. In the banking system studied by Eason, a small set of query codes were 'bent' to handle a much wider range of queries.

Professor Herbert A. Simon, in a different context, has suggested that people behave not so as to maximise their returns but as 'satisficers' who maximise returns in relation to the effort involved. In the banking example, users saved themselves the cost of learning how to use new codes by 'bootstrapping' themselves to solve new problems using what they already knew. This seems to be an important characteristic of human behaviour. In problem solving, it was long ago demonstrated that people try to make problems fit their preconceived ideas of what the solution 'must' be than conceive novel approaches.

One of the current trends in the software industry is to produce systems with graphic user-interfaces 'to encourage exploration' – the idea being that users will learn more about the system because it is designed to enable them to explore it. This view may be unduly optimistic, at least in occupational settings where people are using the system to carry out work tasks. Users are more likely to learn the minimum necessary to carry out their work tasks with minimum effort rather than the minimum needed to maximise their use of the system's power.

Whether current software packages really are easy to explore and do deliver real benefits in return for the exploration effort is debatable. Perhaps Pareto's law also applies to users: if a system is designed to allow users to carry out 80% of their work tasks using 20% of the functionality, do not expect the remaining 80% to be used (or do not automate this part of the job in the first place).

Returning to the bank example, Eason did suggest ways of improving the usage of the system. Firstly, the list of codes was reformatted so that they were *clustered* in a way that was relevant to real work tasks (to reduce the psychological costs of learning new codes). Secondly, it was ensured that each branch had a knowledgeable 'local expert' who could advise and encourage employees (faced with a problem, people seem to prefer to seek help from colleagues rather than manuals or 'on-line' help).

Table 14.9 Costs of user-centred approach to system design[a]

Item	Cost ($)[b]
Market analyses	
Focus groups (×3)	6 285
Product	
Focus groups (×3)	6 285
Product acceptance	
Product mock-up	3 760
Product analysis	
User survey	7 200
Task analysis	
User study	7 320
Lab construction	17 600
Prototype	
Initial design	6 400
Prototype construction	
Design change (20@$320)	6 400
User interface system	16 080
User testing and evaluation	
User studies (4@$7320)	29 280
User survey	7 200
User testing	
User study	7 320
Product survey	
User survey	7 200
Total	128 330

[a] From Mantey and Teorey (1988).
[b] Costs are in 1988 dollars.

Costs and benefits of ergonomics in the software development life cycle

The costs of applying ergonomics through the software development life cycle are (according to Mantey and Teorey, 1988),

- The cost of running focus groups
- The cost of building mock-ups
- The cost of developing prototypes
- The cost of purchasing prototyping software
- The cost of running user trials
- The costs of usability laboratory space
- The costs of any user surveys

Table 14.9 gives these costs in 1988 dollar prices. The costs of equipment, professional time and premises for the development of a system with 32 000 lines of code to be used by 250 employees are included.

The benefits of the inclusion of ergonomics in the design process are

- Reduced learning times
- Fewer user errors
- Lower maintenance costs

Projected cost savings in the first year were estimated to be $48 000 for the avoidance of late design changes, about $180 000 for error reduction and $55 000 in reduced training costs. As a rule of thumb, Mantey and Teorey recommend that the prototyping approach should be used whenever its cost is less than 25% of the total project cost.

Costs and benefits of interface standardisation

There is evidence for the cost-effectiveness of standardisation of products both at the corporate level and beyond (Rosenberg, 1989). Standardisation at the user-interface means fixing the way actions such as 'delete', 'exit' and 'cancel' are performed and standardising other aspects of the interface such as the physical layout, the graphics and the icons used to represent system objects. Such standardisation will lead to shorter learning times, fewer errors and more rapid implementation of new programs onto the existing platform.

Although standardisation may have benefits, there are possible drawbacks, or at least, more controversial implications. For example, standardisation may inhibit creativity and the implementation of fundamentally new ways of doing things. Rosenberg points out that it is possible both to patent the routines or algorithms that generate the interface 'feel' (how it works) and to copyright the 'look', which essentially locks-out competitors or forces smaller companies to pay a licensing fee to adopt another company's standard. Many aspects of user-interface design have become increasingly standardised and users seem to be developing 'population stereotypes' or expectations about the user-interface that are stored deeply in long-term memory and cannot be easily expunged. Thus, portions of interface development are now generic. This has speeded up development time and lowered costs. Standardisation has also facilitated quantitative predictions (or what 'time and motion study' experts call 'standard times') for carrying out interactive tasks. Thus, at the earliest prototyping stages, designers and ergonomists can predict how long it will take to perform key tasks with the new system and thus develop performance standards for the evaluation of later prototypes.

Rosenberg describes several ways in which standardisation can reduce software development costs:

- There can be savings in the time needed to specify the system and its user-interface.
- Major portions of user-interface code can be recycled from one project to another (standard functions such as 'edit' 'save' 'delete', etc. can be used on different systems from a wide range of application areas).
- Test procedures can be standardised.

User-interface costs increase as the system becomes more interactive or 'user-friendly'. Standardisation is one way of reducing interface development costs. According to Rosenberg, if standardisation reduces development costs by 20%, it will pay for itself in five products or fewer (allowing for inflation of labour costs over the development time for the five products).

International developments for interface design have progressed with respect to both the Internet and the ISO.

Costs and benefits of iterative usability testing

Clare-Marie Karat of the IBM Thomas J. Watson Research Center presented a 'dollars and cents' case study of the costs and benefits of iterative usability testing (Karat, 1990). A data entry and enquiry mainframe system was being upgraded to simplify access across transactions in an application. Prior to this, users had to remember each of a series of security codes or mnemonics per transaction. With the new system, only one master mnemonic was required. Usability testing was done firstly in branch offices and a usability laboratory using foil (overhead projector transparancy) prototypes. The last stage was done using a prototype on-line system. The foil panels that mimicked real screens were developed at 20% of the cost of the online prototype.

The sign-on dialogue was simulated by an experimenter who placed a foil in front of the subject. The subject then wrote his or her response on it with an erasable pen. The experimenter then removed the foil and replaced it with the next foil, corresponding to that which would appear on the system according to the subject's response. This continued unbtil the task was completed. In this way, the sign-on task was simulated at 10% of the cost of a real interactive simulation.

Performance measures were taken, including the time taken to complete the task, the error rate, the number of attempts needed before 95% of the participants could sign on error-free and the percentage of participants who could complete each task. At the end of the experiment, participants were invited to give suggestions for improvements in the design of the interface. Usability staff then recommended improvements that were implemented and tested in the final trial.

The costs of the usability trial were estimated at \$20 700. The benefits of the trial were estimated on the basis of the time savings for the completion of the interactive tasks (time saved × personnel costs × productivity ratio) and amounted to \$41 700 (Karat, 1990).

Two important conclusions of the study are that

- Usability trials need not be expensive.
- If usability trials are carried out early on in the design process, then low-tech methods have cost advantages. They obviate the need to implement prototype interfaces before there is a real system on which to test people.

Research issues

The real benefits of design features such as thumbprints and voice in multimodal displays have yet to be evaluated, and in some applications it seems to be true that these 'enriched' displays are as likely to confuse users as they are to help them. There is plenty of scope for empirical evaluations to establish best practice. Some of the issues in the design of 'learnable' interactive systems have been discussed in this chapter. Apart from the choice and design of codes, symbols and language, there are more general issues such as the selection of an appropriate level of system complexity, the degree to which certain functions are executed automatically or by the user, and the degree to which previous knowledge can be transferred from one system to another.

Table 14.10 General principles of software ergonomics

Appropriate software:	
1. Is suitable to the task	Presents users with no unnecessary obstacles or problems. Is not unnecessarily complex or elaborate
2. Is easy to use and adaptable	It is easy to master. Uses dialogue appropriate to user's ability. Has an interface adaptable to different skill levels. Minimises consequences of error: Lost data are recoverable. 'Undo button' is provided.
3. Provides feedback on system performance	Timely error messages. Appropriate level of information. Help on request.
4. Works at the user's own pace	User 'drives' the system. System displays all keystrokes when they are made. System response time and response time variability are minimised.
5. Does not 'spy' on the user	No covert monitoring of user performance. Users are informed of any recording of their input. Emphasis on quality rather than quantity.

Summary

Both the strengths of human memory (its multidimensional, associative nature) and weaknesses (the limited capacity of STM and the semantic nature of long-term memory, where only the 'gist of things' is remembered) have to be accounted for in HCI design.

The need to develop more usable, error-resistant systems introduces the study of communication into ergonomics. The study of language provides a framework for approaching the design of artificial means of communication – between users and interactive systems – in a systematic way. It can be seen that for successful human–machine communication to take place, the symbols being exchanged have to be filtered through several layers of knowledge. At a physical level, the letters and words have to be recognised. Rules then have to be applied to disclose the syntactic structure according to which individual words are ordered. Knowledge about the meaning of words and pragmatic (common-sense) knowledge have also to be applied. Communication can fail if either of the parties involved in the symbolic exchange lacks knowledge at any particular level or if the contents of this layered knowledge structure differ between individuals.

Recent European Union regulations about work with display screen equipment (Health and Safety Commission, 1992) presents general principles of software ergonomics. (Table 14.10 summarises this and other general principles. See also ISO/DTS 16982:2000, Ergonomics of Human System Interaction – Usability Methods Supporting Human Centred Design.)

Essays and exercises

1. Evaluate the operating instructions of some common consumer products. Use the language comprehension framework described in the chapter as a guide. Describe the knowledge required to understand the instructions.

2. Evaluate a menu-based user-interface of your choice. Comment on the number of levels, the hierarchical structure and representation of the domain, the provision of backtracking and any other facilities for navigation.

3. Refer to the icons in Figure 14.2. Using the Help facility on your computer, define each function and describe the conceptual mapping between the function and its corresponding icon. Do you agree with the subjects in Sears' study?

4. Carry out a survey of product warning labels on commonly used consumer products.

15 Human–machine interaction, human error and safety

A new born baby died in a hospital intensive care unit after a member of staff had entered a decimal point in the wrong place in a drug prescription . . . The (decimal) point had been written-up in the wrong place and so the baby received too much of a particular drug.

(*The Times*, 11 October 2000)

People interact with machines via interfaces. Although the interface is not the task, its design is a representation of the task. The interface is the platform for task performance. If users' conceptions of the interface are compatible with their conceptions of the task, then the interface will be 'transparent' to the user when the task is performed and the need to master secondary tasks associated with the operation of interface will be minimised, even if the task itself is difficult. Human errors resulting in accidents and injuries always take place in the context of a task, embedded in an organised system of work. Badly designed systems invite users to make errors and then do nothing to prevent the error from having disastrous consequences. Well-designed systems make it unlikely that users will make errors in the first place. When errors do occur, they block, ignore or feed back to users the errors that they make; and when they do fail, they either

- *Failsafe*: revert to a state that is non-threatening or dangerous.
- *Failsoft*: slowly cease to function, giving plenty of indication that something is wrong and plenty of time for it to be put right.

Human error and equipment design

Interface features and operator error

Mosenkis (1994) reviewed some of the problems associated with medical device design that can increase the likelihood of error. Some examples follow.

- *Unusual or inconsistent operation.* Users may assume that if they know how to use one type of infusion pump, they know how to use all types. Problems occur when different makes have different control logic. Good practice standardises the procedures for operating different types of products to enhance positive transfer of learning. Dixon (1991) demonstrated that even if there is a large conceptual difference between two systems, procedural similarity enhances the learning of the other (e.g. landing an aeroplane and shutting down a nuclear power station).

- *Lack of protective incompatibility.* Some devices can be misconnected owing to the design of their leads and plugs. An apnoea monitor had an electrode at one end (placed on an infant's chest) and a plug at the other that could be attached to the AC mains electricity supply. Several infants were electrocuted because people saw unconnected leads close to power connectors and assumed they should be connected. Similar problems have occurred with enteral feeding tubes. Liquid feed is slowly pumped into the the patient's stomach, via a gastric tube. These pumps have inadvertently been connected to the cuff inflators on endotrachreal tubes (providing an airway to the lungs). When the cuff inflator bursts, fluid is pumped into the lungs. Unconscious patients have been known to drown in their beds when this happens. Such errors happen because of the *affordances* offered by unrelated components. The psychologist J. Gibson coined the term 'affordance' to describe the cues inherent in objects that indicate what they can be used for. In both these examples, different parts of different systems shared the affordance that they could be connected and so sometimes they were connected – with disastrous results.

- *Unclear/incompatible control/display relationships.* Traditionally, medical equipment has been fitted with analogue displays and controls that conform to existing stereotypes (e.g. turn the knob to the right to increase the voltage through the electrothermal unit, or to the left to increase the flow of anaesthetic gas). A defibrillatory/monitor displayed only alphanumeric codes whenever a problem arose. Explanations for the codes were only to be found in the user manual. Staff had cut out the codes and taped them to the device. The provision of such a table – on the device – was really the designer's job, given the questionable decision to use such codes in the first place.

- *Defeatable or ignorable safety features.* LCDs and other low-power displays are less effective than flickering lights. Smell and vibration are particularly good non-directional warnings. However, some medical devices have auditory warnings with volume control. It is possible to turn down the volume of the alarm (e.g. in quiet periods, when there is only one person and one patient in the room) such that the warning would not be heard at other times. This is particularly hazardous when people work shifts and workload and staffing vary between shifts.

- *Lack of cues to aid discrimination.* Mosenkis gives the example of a contact lens package that contained two, almost identical, bottles, one containing a caustic lens cleaning fluid and the other containing saline to wash the fluid off when the lens had been cleaned. Small, difficult-to-read labels were the only clues as to the contents of the bottles. Eye damage occurred when users cleaned the lenses with saline and then washed them with cleaning fluid. The example illustrates the failure to take into account the users' capacity throughout the task – at the critical point, the user's vision is at its poorest. Shape and colour could be used to distinguish the two bottles (e.g. a red, hexagonal bottle for the cleaner and a blue, smooth one for the saline).

Modes

Modes are nothing more than particular states of systems, but states in which the system behaves differently in response to user actions. Modes are all around us.

Johnson (1990) gives examples of everyday objects and their different modal states. Modes can cause people to make errors for a number of reasons – when we don't expect the system to have modes, for example (e.g. an electric toothbrush) or when the mode is not explicit. As an example, a slide projector had a hand-held remote control with two buttons, one labelled 'advance' and another labelled 'forward/recall'. The effect of the advance button depended whether the projector had first been put into the 'forward' or the 'recall' mode. In forward mode, advance presented the next slide, whereas in recall mode the previous slide was displayed. The only cue to which mode it was in was the setting of the 'advance/recall' button itself, which users tend not to look at when they are presenting slides. Clearly, this design invites a particular kind of error when the user forgets which mode the device is in. A better design might have one mode ('on' or 'off') and separate buttons for next slide or last slide.

Modes do have the advantage of simplifying console and control layouts. For example, radios have different frequency bands. One small dial can be used to support several different bands as the user switches between modes (FM, AM, longwave or shortwave, etc.). Modes can also boost the efficiency of a device. Cars have a 'disengage' mode when the clutch is depressed that automatically reverts to 'engaged' mode when the clutch is released. To drive a car with a manual transmission requires operation through a fairly strict sequence of modes, with the 'disengage' (clutch depressed) mode alternating between each drive mode (gear). Telephones also have modes. You pick up the receiver and the dial tone prompts you to dial a number. Note that the act of picking up the phone automatically initiates the transition from 'off' mode to 'dial' mode – the transition is embedded in the task. Cellphones are slightly different, however, and the transition from 'off' to 'dial' may have to be made explicitly by pressing the 'on' button. These examples may seem trivial until it is remembered that the usability of many everyday items depends on users' knowledge and expectations about them. Johnson points out that the usability of many everyday objects can be degraded as they become 'computerised' and modes start to appear in previously 'modeless' devices.

Most computer users are familiar with using modes on their computers. Word processors have a text editing mode, file management modes, and so on. Until recently, many non-computer devices such as domestic appliances did not have modes and if modern 'moded' versions of these devices do not display their modes explicitly, usability problems may result. Setting the alarm time on an old-fashioned alarm clock used to be a simple matter of moving the 'alarm' dial to the time on the clockface and flipping a lever to put the clock into 'alarm' mode. Modern radio alarms are rarely this simple and the alarm can only be set once the system is already in 'alarm' mode. User difficulties and errors are likely to be further compunded if the user's model of the system is incompatible with the system's design model. This is never more true than when modes exist where they are not expected.

Mental models can reduce the likelihood of error

Learning to use products such as word processors, electronic calculators or video recorders can be seen as a problem-solving exercise. The way the user-interface is designed can influence how the operations required to use it are represented in the minds of users and the types of cognitive processing required for successful operation.

As a user learns how to operate a new product or system, he or she constructs an internal representation of what the system is and how it works. This is known as the user's model of the system.

Models of interactive devices guide users' control actions and assist in the interpretation of the device's behaviour. Umbers (1979), in a detailed review of research on operator mental models, concluded that the ability to predict was an underlying feature of human controllers. In a similar vein, an experiment by Flowers (1978) showed that even in a simple tracking task subjects made use of predictive control, beyond the explicit data, which enabled them to continue accurately controlling a system despite gaps in the sensory feedback. Subjects had developed a mental model of the task that enabled them to predict the behaviour of the system.

Young (1981) investigated the mental models implicit in the design of a variety of pocket calculators. Of interest here is the differences in control logic of 'reverse Polish notation' (RPN) calculators and the more familiar 'algebraic' (ALG) calculators. In both cases, the model is a metaphor or cover story for thinking about the calculators, how they work and how they can be used to solve problems. RPN calculators have a very particular and explicit design model. Error-free operation depends on there being a very close mapping between the user's model and the design model.

Pocket calculators

There is a fundamental difference in the operation of RPN and ALG calculators that arises from the differences in control logic. To evaluate the sum of 2 and 3 using an ALG calculator the sequence of key presses is

'2' '+' '3' '='

and the number 5 appears on the calculator's display. When the number 2 is first entered into the calculator, it is immediately displayed. However, when '+' is entered, nothing is displayed. This implies that there is a memory, or register, for storing arithmetic functions such as '+', '−', '×' etc. When the number 3 is entered, it replaces the '2' in the display (the implication being that the '2' has not been lost but is being stored in some kind of memory for later use). Pressing '=' causes the '3' to disappear to be replaced by '5' (the implication being that the '2' and '+' have been retrieved from somewhere and that the expression has been evaluated. The calculator's behaviour implies that it has a memory (or a number of 'registers') and that if something disappears from the display, it can still be 'in' the machine.

According to Young, the ALG calculator's behaviour when handling more complex expressions leads to a very complex model if the concept of inner registers is adhered to. Most people are unlikely to do this and will abandon the implied register way of thinking in favour of a verbal rule such as

> *You type in an expression. The machine examines it, analyses it and calculates the answer according to the rules of arithmetic.*

This is a rule-based model of the calculator as 'an engine for evaluating written formulas' (Young, 1981). It focuses attention on what has to be done rather than on how the machine works. However, the model does nott enable the user to infer what to do if a mistake is made in entering expressions for evaluation. If a mistake is made during complex calculations, it is difficult to determine what the error was and how to rectify it, and it advisable to start again from the beginning and concentrate on keying in the formula correctly. You are OK as long as you follow the rules of arithmetic.

To evaluate the sum of 2 and 3 using an RPN calculator, the sequence of key presses is

'2' 'ENTER' '3' '+'

and the number 5 appears on the display. The RPN calculator's behaviour also implies that it has registers for storing numbers, but the register are organised in a very particular way, as a 'stack' memory.

The stack is made up of four registers X, Y, Z and T. Each register can store one number. The number in X (the top of the stack) is displayed to the user at all times. Pressing the 'enter' key 'pushes' the stack. The number in X goes into Y, that in Y into Z, that in Z into T and the number that was in T is lost. To evaluate the sum of 2 and 3, the number 2 is pressed (2 goes into the X register), followed by 'ENTER', which 'pushes' 2 into the Y register. At this point, a copy of the original 2 remains in X but it is overwritten when the '3' key is pressed. Pressing '+' at the end 'pops' the stack. The operation is carried out and the numbers in the Y and X registers are added. The result is written into the X register (the contents of X are always displayed) and the contents of the Z and T registers move up one place.

To evaluate an expression such as

$2 \times (3 + 4)$

the RPN calculator requires the following sequence of key presses:

'2' 'ENTER' '3' 'ENTER' '4' '+' '×'

'2' is first pushed into the stack (into register Y) when 'ENTER' is pressed followed by the number 3. The stack is pushed again, 2 moves into Z and 3 into Y. The copy of 3 in X is overwritten when the number 4 is entered. When the '+' key is pressed, the contents of Y are added to those of X (3 and 4 are added). The result is written into X and the contents of Z (the number 2) move up to Y. Pressing '×' causes the contents of Y (the number 2) to be multiplied by the contents of X (the number 7) and the result (14) is written into X.

The stack model is very explicit. It draws attention to how the calculator works rather than what has to be done in order to use it. The machine's behaviour is incomprehensible without the stack memory model. The calculator is a good example of a device whose efficient usage depends on the user possessing an explicit conceptual model of its operation. Once a user understands the model, the calculator's behaviour for any sequence of key presses (whether or not they are arithmetically meaningful) can be predicted.

Predictability of system behaviour, rather than performance, is the advantage of using a good conceptual model in design. In fact, Hoffman *et al.* (1994) compared subject performance in solving mathematical problems using RPN and ALG calculators. The limiting factor for solving relatively straightforward problems was the time to physically enter the numbers and the operators. RPN was faster than ALG for most problems because less keying was needed. When the number of keying operations for both calculators was controlled, there was no difference in solution time.

User models and product usability

Explicit design models can facilitate learning. Complex computer operations can be made to appear simple. For example, in text editing on a word processor, text is moved around using the 'cut' and 'paste' functions. The model is based directly on the real world – text on real paper is cut using scissors and pasted into a new position. Thimbleby (1991), in a review of the concept of usability, argues that user-interfaces must be designed to support algorithms for their use. An algorithm is a set of procedures for solving a problem or achieving a result or, less formally, a step-by-step method for solving a problem. In the cut-and-paste example, the algorithm for moving electronic text around maps directly onto the algorithm for editing real text on real paper. Text editing using this algorithm is easy to learn and easy to do once learnt. Not all devices are that simple, however, and the lack of an explicit and clear design model is often the reason. Thimbleby describes some of the design deficiencies of video cassette recorders (VCRs).

1. *Complexity.* A VCR and television have over 100 keys between them – more than found in much more complex devices such as computers.
2. *Inconsistency.* Differences in the logic used to specify the functions of the remote controls and the controls on the VCR front panel.
3. *Lack of meaningful feedback.* Sometimes no feedback is given; at other times feedback is given even if nothing happens. Frequent tasks are difficult to perform and infrequent tasks are too easy to perform accidentally.

Thimbelby's main point is that well-designed user-interfaces are not just a collection of 'ergonomically clustered' buttons for activating the machine functions (such as STOP, RECORD, TIMER, etc. on a VCR). A well-designed user-interface is based on a more formal representation of what the system is (such as an explicit conceptual model). It is the designer's task to develop such a model and only then to implement the user-interface (in a way that is consistent with the model). For example, one approach to the design of the REWIND function would regard the task of rewinding a tape as a subgoal of a more complex task (e.g. ejecting the tape to end a session, replaying the tape, etc.) and not an end in itself. In this case, it should be permissible to develop simple programs by combining REWIND with EJECT or with PLAY or RECORD. Similar programs could be developed for other functions as well. An alternative model would represent the VCR as a system without a memory – all tasks would have to be broken down into subgoals (directly mappable onto the control keys) and executed sequentially with appropriate feedback after the execution of each subgoal. Thimbleby's argument is that the design of user-interfaces should be based on more formal models of the set of tasks involved in using products.

That VCRs are difficult to use or have too many unnecessary functions is supported by the results of a manufacturer's survey of 1000 families: 30% of people never used the timer on their VCRs. Norman (1988) has made similar observations about the use of other products such as word processors; many of the facilities they offer are never used because of the extra learning required to master them. It is often easy for manufacturers to add more functions to their products (in the hope of adding value and gaining competitive edge), but a point may soon be reached where the perceived usefulness of the new functions is outweighed by the increased cognitive complexity of the system. If new products offer new functions that are only accesible by switching modes, users may avoid them if the previous products they were accustomed to were modeless.

Mental workload in human–machine interaction

There are many reasons for assessing the mental workload of a task, including the following.

1.   To compare alternative methods or designs.
2.   To evaluate the usability of a prototype.
3.   To identify the high stress aspects of a job.
4.   To evaluate operator performance and the likelihood of error.

Physical workload can be measured objectively at the level of both physics and physiology. The problem with measuring mental workload is that the operator's thought processes are not directly observable. The most direct and obvious approach would be to ask an operator or user how difficult it is to carry out a task. Some potential limitations of this approach are that the operator might not want to say or may confuse the important aspects of a task with those that are difficult, or there may be other sources of bias.

   To overcome these problems, investigators have developed a number of techniques to measure mental workload (further details can be found in Rohmert, 1987; Ogden *et al.*, 1979; Wierwille, 1979; and Williges and Wierwille, 1979):

1.   Physiological measures of mental workload
2.   Psychological measures of mental workload
3.   Evaluation of the performance of the main task (including error)

Physiological measures of mental workload

These methods typically assume that high workload causes an increase in the state of arousal of the central nervous system. This is hypothesised to cause involuntary changes in certain physiological variables that can be objectively measured. Some of the main physiological variables used are described in Table 15.1. All of the physiological methods require specialised equipment and can be influenced by factors other than workload. However, they can be profitably used in conjunction with other methods.

Psychological measures of mental workload

The main psychological methods of measuring mental workload are

Table 15.1 Physiological measures of mental workload

GSR (galvanic skin response). The electrical resistance of the skin changes when sweating increases owing to changes in the concentration of ions in the skin cells. GSR is often measured on the hand, foot, wrist or forehead. Changes in GSR have been shown to correlate with anxiety. In studies of driving, GSR has been shown to correlate with roadway events that increase workload. Although stressful situations do reduce the GSR, temperature, humidity and physical workload can have the same effect and need to be controlled when GSR measurements are made.

Heart rate. Heart rate does not necessarily change with mental workload although it can increase under stress. Consumption of tea or coffee and cigarette smoking and heat can also increase the heart rate. Heart rate variability measured on the basis of the instantaneous heart rate or of the interbeat intervals is sometimes used. Heart rate variability may decrease with mental workload but it is also affected by changes in respiration rate that can confound the results.

EMG (electromyography). Increases in forearm and forehead EMG have been correlated with high task loading in a variety of aircrew tasks.

EEG (electroencephalogram). The electrical activity of the brain can be measured using scalp electrodes. Semi-periodic activity in the EEG correlates with states of alertness (4–10 Hz, drowsiness or sleeping, 20–30 Hz, alertness). The EEG is an extremely complex signal that requires expert interpretation and its use as an instrument for mental workload measurement is limited.

Eye movements and blinking. Fixation time, lookaway time and sequencing of eye movements can be used as indices of workload. Alertness and directed attention may reduce the blink rate. Overarousal or emotional stress may increase the blink rate.

Speech pattern analysis. Changes in speech are detectable when a person is under stress (the modulation of speech may change, for example).

---

1. Subjective opinions
2. Measures of spare mental capacity by means of secondary tasks

Operator opinions about mental stress can be obtained using rating scales or by questionnaire and interview. In aviation, rating scales such as the 'Cooper–Harper' scale have been developed for assessing the handling quality of aircraft. The validity of these scales depends on the items included and how they were selected and the size of the sample on which the scale was validated.

Interviews and questionnaires are primarily supplementary sources of information on workload. In aviation and astronautics, debriefing of pilots is commonly used, as is self-reported logging of stressful activities. Data from these sources can correlate well with those from other sources.

Measurement of spare mental capacity is based on the rationale that the operator has a limited capacity to process information. As task difficulty increases, this capacity is used up and less is available to be allocated to other tasks if they have to be carried out at the same time. If the operator is given a secondary task to do concurrently with the main task, the level of performance on the secondary task will reflect the difficulty of the main (primary) task. If the main task is not too difficult, processing capacity will be available for performing the secondary task as well. If the difficulty of the main task increases, less capacity will be left for the secondary task and performance on the secondary task will suffer. Thus, measures of secondary task performance can be used as an index of primary task load. For this reason, secondary tasks are designed so that performance can be measured objectively by means of performance time and/or number of errors.

Brown carried out a number of classic experiments on driving using the secondary task technique (Brown and Poulton, 1961; Brown, 1966). Brown (1966) found that a secondary task could predict success and failure in trainee bus drivers. Trainees were given a subsidiary task to do while driving (checking 8-digit strings for changes). The trainees' scores on the subsidiary task while driving were expressed as a percentage of their scores while doing only the subsidiary task. The percentage score was taken as an index of the amount of 'spare mental capacity' while driving. Previous driving experience was found to be a good predictor of whether the trainees were successful in passing a final driving test. Secondary task performance while driving was also found to be a good predictor of future success, but only if performance on the secondary task was investigated at the beginning of the training programme. These findings were interpreted as follows.

1. Those trainees who initially experienced the smallest decrements in secondary task performance when driving already had driving skill (either naturally or through previous experience). This meant they had spare processing capacity to carry out the secondary task even during the early stages of learning to drive a bus.
2. Those whose secondary task performance suffered while driving had little skill (i.e. few subroutines) and had to use conscious mental capacity to drive at the early stages of training. This left less mental capacity for the subsidiary task.

Some problems with secondary tasks are that they may not be as sensitive as required and they may disrupt performance of the primary task. This, of course, violates the logic of the procedure and renders it invalid. However, the secondary task procedure is intuitively appealing in the sense that, at certain times, when the primary task load is high, operators will be too busy to perform the secondary task well. Wetherall (1981) concluded that the technique has merit if the secondary task is selected carefully with respect to the primary task (so that it does not interfere with primary task performance) and is used in conjunction with other measures.

Performance of the main task

Performance of the main task is a source of information about the rate of work from which the mental workload of the operator may be inferred. For tasks such as driving, specially instrumented vehicles may be used in which control actions (e.g. steering, braking, accelerating) can be recorded. Although these variables objectively quantify task performance, they do not necessarily have a one-to-one relationship with mental workload. A skilled operator may be able to cope with a wide range of task loadings with minimal demands on conscious processing capacity. Subjectively, these may all be perceived as being equally easy to perform. In conjunction with users, the investigator may carry out an analysis of critical points in task performance and assess workload. Gawron (2000) gives the *task difficulty index* in Table 15.2 as an example of a simple measure of workload.

Table 15.2 Task Difficulty Index (sum of scores on four dimensions)[a]

Task	Score
Familiarity of stimuli	0 = Familiar (e.g. letters)
	1 = Unfamiliar (e.g. spatial patterns, unfamiliar map)
Number of concurrent tasks	0 = Single task
	1 = Dual task
Task difficulty	0 = Memory set size 2 or less
	1 = Memory set size 4 or more, delayed recall
Resource competition	0 = No competition
	1 = Competition for either modality of stimulus (e.g. visual, auditory or central capacity)

[a] Gawron (2000).

Workload is assumed to be higher when stimuli are unfamiliar, when tasks are performed concurrently, when a load is placed on STM and when there is competition for central capacity. The higher the value of the index, the more difficult the task. For example, looking for weeds (depressingly often) in a familiar place (your garden) among familiar stimuli (cabbages) and doing nothing else is a low-difficulty task. Looking for weeds among unfamiliar and exotic plants in a strange country while attempting to communicate in a foreign language is a highly difficulty task.

Psychological aspects of human error

Human–machine interaction can be conceived of as a stream of behaviour that is governed by the user's intentions, expressed as goals and subgoals and executed using 'pockets' of well-learnt behaviours known as plans. Norman (1981, 1984) calls these sequences of actions 'action schemas' (a kind of 'memory–motor program' that can be triggered by particular events to enable a person to carry out a sequence of actions). Action shemas are usually well-learnt and are carried out without conscious attention, once initiated. Many of the errors that people make in everyday life can be explained once it is understood that intentions are usually disassociated from the behaviours that are executed to satisfy them. Errors can occur when the intention to act is correct but the wrong action schema is activated. Very well-learnt schemas often turn-up in unexpected places.

The intention to act involves a hierarchy of action schemas. A general intention gets broken down into more detailed components that must be executed in order. A particular kind of 'action slip' can occur when part of a sequence becomes disassociated by interference or by overload of working memory. The higher-level goal being lost, the person no longer knows why they are doing something (as in searching through the contents of a draw but not knowing what you are looking for).

Error categorisation

Several categorisation schemes have been developed for handling error data. The power of these schemes lies in their making explicit the psychological processes underlying the error and the scope for prevention by ergonomic interventions. A distinction is commonly made between errors of omission, ervors of commission and psychomotor or 'bumping' errors. Errors of omission involve an operator not doing something they were supposed to have done (e.g. forgetting to save an updated file on a computer system before calling up a new file to edit). Errors of commission involve executing a correct action at the wrong time or performing an incorrect action. Psychomotor errors involve accidently operating a control or executing a sequence of actions in the wrong order. Error can also be categorised according to the level of human information processing involved:

- Errors at the skill-based level include *slips* (failure of attention to the task or goal) and *lapses* (failure of memory).
- Errors at the rule and knowledge-based levels include *mistakes* (using the wrong rule or deciding on a course of action that it incorrect) and *violations* (deliberately doing something wrong).

Error may also be viewed as a process that develops over time as follows:

- Error production
- Error detection, particularly self-detection
- Error identification
- Error recovery
- Error prevention

Error production

1. Errors in the formation of the intention to act (e.g. choosing the wrong goal)
2. Faulty activation of the action schema (e.g. looking in the wrong place)
3. Faulty triggering of the schema (e.g. looking in the right place at the wrong time)

Norman (1981) describes the following kinds of 'slips' associated with forming goals and executing actions.

Errors in the formation of the intention

Mode errors   Mode errors involve misclassifying the situation. For example, trying to set the alarm on a radio alarm clock when it is in 'time' mode, with the result that the clock displays the incorrect time and therefore the alarm will sound at the incorrect time. When mode errors are made, the actions are correct but they are applied in the wrong situation. Mode errors may be difficult to detect.

Description errors   Description errors occur when the intention is insufficiently specified, perhaps owing to memory limitations or high workload. Putting the lid of the sugar container onto the coffee cup that has a similar shape is an example given by Norman. The writer Douglas Adams described how he bought a packet of biscuits and a newspaper at a railway station and sat down at a table to eat the biscuits. A man sitting opposite him started to eat one of the biscuits. Adams was horrified but said nothing and took a biscuit from the packet. The man did the same. When the packet was empty, the man went away. Adams folded his newspaper only to find his unopened packet of bisuits on the table in front of him.

Faulty activation of the schema

Unintentional activation Unintentional activation occurs, for example, when a familiar habit is invoked at an inappropriate time. The psychologist William James described going to the bedroom to change for dinner and ending up in bed! Well-learnt behaviours will often be triggered by environmental cues when attention is diverted elsewhere (for example, to internally generated stimuli). When daydreaming, or when distracted by sudden events, we are particularly susceptible to these errors.

External activation External activation of action shema occurs when stimuli in the environment trigger unintended actions. For example, many people have to give their business telephone number during routine calls at work. When, exceptionally, a business call is received at home and the work number is required, the person gives their home number. Norman gives an example of someone, who, after one failed attempt to buy cigarettes, tried again after getting more coins but put the coins into the adjacent coffee vending machine.

Associative activation Associative activation of action schemas is similar to external activation and occurs when the stimulus is strongly associated with the desired response. Associative activation often occurs in speech when people misname a person or thing but the name used is strongly associated with the one intended.

Loss of activation

The classic example of loss of activation is when when finds onself standing in a room or looking in a cupboard without knowing why one is there or what one is looking for. The cause is often due to interference with memory either by external events or by internally generated thoughts. Loss of activation, when it happens in the middle of a stream of closely linked behaviours, may result in one or more steps being missed. For example, pouring boiling water into an empty teapot or putting the car in gear and pulling away from the kerb before all the pasengers are in. Loss of activation may result in steps being repeated if the achievement of the goal has not been registered (for example, brushing one's teeth twice in the morning).

Faulty triggering

Faulty triggering may involve selection of the wrong schema or failure to execute actions that are appropriate to the situation.

Detection of error

Sellen (1994) carried out a study in which people kept diaries of the errors they made in everyday life and then described the mechanisms by which they detected their own errors. She found three mechanisms by which people detected their own errors and also identified situations in which the errors were detected by others. The main mechanisms of self-detection are *action-based*, *outcome-based* and *process-based* (in which the constraints in the external world prevent further action).

Action-based detection   This involves 'catching oneself in the act', according to Sellen. The incorrect actions are detected as they are being performed. In skilled activities, such as typing, the person may detect the error by means of tactile feedback. In speech, the error may be detected immediately when the word uttered does not match the word in mind. Action-based detection depends for its immediacy on the person paying conscious attention to the task or to feedback from the task. If, for example, the wrong action schema has been initiated, this will only be detected if the executive processor is monitoring the task and the intention (or goal state) is in working memory. Sellen gives an example in which a person intended to turn on the radio to listen to music but found themselves with their hand on the telephone. The absence of any goal for using the telephone triggered recall of the original goal and the person then turned on the radio. Thus, action-based detection can work in two ways. Feedback/monitoring of the actions detects a mismatch between the actions being performed and those intended. Evaluation of actions with respect to the goal state leads to the realisation that an error is being performed, but this only works if the conscious attention is being paid to the desired goal or if the goal can be retrieved from memory.

Outcome-based detection   This is the next barrier to error and comes into play if the actions were not monitored sufficiently to enable mismatches to be detected. Sellen gives an example in which a person placed a teabag into a cup, intending to pour boiling water over it, and poured coffee over it instead. The error was only detected when the teabag floated to the surface of the coffee. If the intention itself is wrong and if the action is appropriate to the intention, then detection of the error on the basis of the action is not possible. Sellen gives the example of a person walking down some stairs who thought there was only one more stair to go, when there were really two more. The feeling of the descending leg being in the air for too long and the painful recovery of balance was the only cue that an error had occurred. Outcome-based detection may be inadequate to prevent an accident or injury from occurring. There are several requirements for successful detection of error on the basis of outcome. Firstly, the person must have expectations about what the outcome of the behaviour is likely to be; secondly, the effects of the actions must be perceptible; thirdly, the behaviour and the system must be monitored throughout the action sequence; and fourthly, the person must attribute any mismatches between desired and actual outcomes to their own actions and not to the hardware or to extraneous factors. Outcome-based detection works in two ways. The person detects a mismatch between the desired outcome and the expected outcome. This depends on the mismatch being visible or indeed being alerting, in the case of a person who is not expecting anything to go wrong. If the person *is conscious of the possibility of error*, they may actively evaluate the outcome when it occurs. In January of each year, many people are conscious of the fact that they may write or enter the date incorrectly, using the number of the previous year instead of the new year, and actively check to see whether this has happened.

Process-based detection   This occurs when constraints in the world 'short-circuit' the action schema, preventing further activity. Norman calls these constraints 'forcing functions' and gives, as an example, the practice of putting a barrier on the ground floor landing of stairwells in tall buildings. In emergencies, when occupants evacuate these buildings using the stairs, they have been known to become disorientated and continued descending into the basement. Unable to escape, they may suffocate if the building is on fire. The erection of a barrier on the ground floor landing breaks the action sequence and acts as a cue for exit on the ground floor. Another example, this time in HCI, is the practice of interrupting the user when the user wishes to close a file that has not been saved. The system interrupts the action and switches to a warning mode in which the only legal key presses are those that will answer a prompt such as 'Do you want to save the changes to Chapter 15'. The key difference between forcing functions and other methods is that the error detection is cued by the design of the environment, not by user expectation or evaluation of feedback.

Human error and higher cognitive processes

So far, the discussion of human error has been confined to 'slips' or 'lapses' in the context of well-learnt behaviours. Mistakes are another source of error and include poor judgement, bias and error due to the inability to solve problems.

People frequently have to make decisions under conditions of uncertainty: for example, whether to play golf or squash depending on how likely it is to rain that day, or whether to buy a car now or in 6 months time, depending on how likely it is that prices will go up or come down. Experts such as medical doctors, financial managers, etc. are continually faced with difficult decisions and make them using specialised knowledge and techniques.

There are two main (and complementary) approaches to the study of decision making. Normative models attempt to specify the optimum way of making a decision under a particular set of conditions. They specify how a decision should best be made and are often couched in mathematical terms. Descriptive models attempt to characterise the behaviour of human decision makers. They attempt to describe what people actually do when making a decision. The first approach is a branch of applied mathematics known as decision theory. The second approach is a branch of applied psychology known as behavioural decision theory. Clearly, both approaches are essential if decision making in complex human–machine systems is to be optimised. If the behaviour of human decision makers departs from the optimum, and situations in which suboptimum decisions might occur can be predicted, preventive steps can be taken (such as the provision of decision-support systems, as described in a later section). For a more advanced treatment of decision making, see Edwards (1987).

Cognitive aspects of decision making

General theories of how humans make decisions have an important place in ergonomics, as do specific models of the decision-making behaviour of experts in a particular knowledge domain. Jacob *et al.* (1986) define a decision as a choice between alternative courses of action. In reviewing strategies and biases in human decision making, they divide the process up into four stages.

Information acquisition This stage involves information search and storage to specify the alternatives to be decided between. Selective perception and memory can have an impact on this process; for example, people may choose to ignore certain information because it is unpleasant or they may be unable to recall information that is crucial for making distinctions between possible courses of action.

Evaluation of alternative courses of action   Theorists have developed many models of how people choose between various courses of action. For example, in the 'elimination by aspects' strategy each alternative is selected and alternatives that do not include a key aspect or dimension are eliminated. The process continues until only one alternative remains. A number of models have also been developed in which decision making is based on the utility of the various outcomes and the probability that a particular outcome will occur. If the decision maker does not know the probabilities of the various outcomes, he or she may use subjective estimates of probability as in subjective expected utility theory, in which the choice of a particular alternative depends on its utility and the subjective probability of its occurrence. For example, the chances of winning the UK National Lottery are so low that it is completely irrational for anyone to bet on it. However, the fact that people do buy tickets can be understood when it is remembered that it is the subjective utility of winning as well as the probability that determines behaviour.

Execution   Once an alternative has been chosen, a course of action can be taken. The decision maker may choose not to implement an action if inaction involves less responsibility or is emotionally less demanding than action.

Feedback   Feedback can be essential to the learning and improvement of decision-making skills by enabling incorrect decisions to be identified and by making the consequences of the chosen course of action explicit. However, the provision of feedback does not, in itself lead to learning and improvement of decision making skill if the decision maker does not know that there is something to be learnt or if what is to be learnt is unclear. The value of feedback can be degraded if it is ambiguous or cannot be mapped onto the behaviour that produced it (owing to excessive delays, for example).

Heuristics and biases in human decision making

Tversky and Kahneman (1974, 1981) have attempted to characterise some of the ways humans make decisions. Of particular interest to this discussion is the use of heuristics ('rules of thumb') to simplify the mental operations required to make a decision. Three main classes of heuristic technique are

1. Judgement by representativeness
2. Judgement by availability
3. Judgement by adjustment and anchoring

Representativeness   People often have to make judgements that really depend on the probability that an event or object belongs to a particular class or is the result of certain process. The representativeness heuristic replaces judgements about probability with judgements about whether the event is representative of the class or process. For example, in trying to decide whether an individual is a farmer or a librarian, the similarity of the individual to the stereotype of a farmer or librarian is used.

Although judgements about similarity and probability are often the same, use of the rule leads to systematic and predictable errors. In order to come to a decision about someone's occupation, as in the above example, it would be useful to know about the relative proportions of librarians and farmers in the population (the prior probability of a person being a librarian or a farmer), but judgements about similarity do not take data of this type into account. In the countryside, farmers probably outnumber librarians, so it is more likely that someone living there will be a farmer irrespective of any other considerations such as whether they look like a farmer. In the middle of a large city, the opposite is likely to be the case.

To use an example more relevant to ergonomics, if a manual worker complains of arm pain at work, it is often concluded that the pain is caused by the work on the basis of its similarity to occupational arm pain. However, to make a correct decision, it is important to know the probability of someone experiencing the pain both at work and outside of work. Failure to do this causes the role of occupational factors to be overestimated. Representativeness also leads to misconceptions of chance. People expect sequences of random events to be similar to the process that gave rise to them even if the sequence is short. In tossing a coin, for example, people will say that a sequence such as

H T H T T H H T

is more likely than

H H H H T T T T

or

H H H H H H H T

Although it is true that when a fair coin is tossed there will be an equal number of heads and tails *over the long term* (because the probability of either outcome is 0.5), short sequences of tosses need not resemble long sequences. This misconception underpins the well-known Gambler's Fallacy or 'Law of Averages' in which random events are seen as self-correcting processes. The misconception states, for example, that after a long run of heads, a tail is now 'due' to restore the appearance of the process. Unfortunately, a fair coin does not 'know' that the sequence of prior tosses 'looks wrong' and the probability of a head or tail remains the same each time the coin is tossed.

The representativeness heuristic can lead to systematic biases when people accept that a small sample is representative of the process that gave rise to it. Even engineers and scientists are prone to this over-reliance on small samples and make unwarranted extrapolations from them. Many complex processes, in economics or meteorology, are affected by overlapping trends with varying periodicities. For example, a person may decide after experiencing five increasingly long and gloomy winters, that the weather is getting colder. However, this conclusion is an artefact of extrapolation from a short time series that may not be representative of long-term trends (in the case of Britain, current weather may be slightly cooler than 15 years ago, but is definitely warmer than 200 years ago; definitely cooler than 1000 years ago and definitely much warmer than 10 000 years ago at the end of the last ice age). In process industries (and other complex situations) where historical plant data can be presented using computer graphics, designers must use a knowledge of system dynamics to ensure that variables are sampled, presented and displayed so that real trends, rather than random fluctuations or noise, are presented to operators.

Availability   When people use the availability heuristic, they judge the likelihood of events according to the ease with which instances or occurrences of the event can be brought to mind. Things that are difficult to imagine are deemed to be unlikely. Sometimes this is a useful cue for making probabilistic judgements because instances of large classes of objects can indeed be imagined more quickly and more frequently than instances of small classes. However, availability – the ease with which events can be imagined or recalled – is influenced by variables that do not influence probability. For example, people may judge one activity to be more dangerous than another because it conjures up more vivid images (e.g. working in an explosives factory as opposed to working in one's own kitchen) or because more potential causes of failure can be imagined.

Judgements based on availability may be biased by the effectiveness of a search strategy. If people are asked whether more English words have the letter *r* as their first or third letter, many will choose the former even though it is incorrect. It is easier to retrieve from memory words beginning with *r* than with *r* as the third letter.

Adjustment and anchoring   When people are required to make quantitative estimates or evaluations, they often begin at an initial value and adjust it to arrive at a final value. As Tversky and Kahneman point out, the amount of adjustment and the final value can be influenced by the size of the initial value (the 'anchor'). Salesmen and advertisers often use this as a technique to influence people's perception of the fair price of goods (e.g. 'How much would you expect to pay for this packet of washing powder, $5.00, $4.50? Why not take advantage of our special offer of $2.50?).

In one experiment, subjects were first asked whether the percentage of African countries in the United Nations was greater or less than a particular value (the anchor). They were then asked to estimate the actual percentage. With an anchor of 10%, an estimate of 25% was obtained. With an anchor of 65% and estimate of 45% was obtained. In another demonstration, subjects asked to estimate 8! (eight factorial) written in two ways:

$$8 \times 7 \times 6 \times 5 \times 4 \times 3 \times 2 \times 1 \quad \text{or} \quad 1 \times 2 \times 3 \times 4 \times 5 \times 6 \times 7 \times 8$$

and gave estimates of 2250 in the first case and 520 in the second (the correct answer is 40 320).

The section on anchoring and adjustment highlights the need for careful design of training programmes for operators of systems since it is at these early stages of learning that anchor and adjustment biases can be inadvertently set.

Many other extraneous factors can cause biases in human decision making (Table 15.3). For much of the time, though, heuristic principles can produce essentially correct decisions and have considerable power in enabling complex problems to be dealt with quickly.

Table 15.3 Breakdown of problem solving behaviour

1. *The one solution fixation.* The problem solver may develop a 'one solution fixation'. After failure to solve the problem, a return is made to the same starting point and the same premises or approach are used again. The problem solver is unable to think of alternative approaches. This is sometimes accompanied by the irrational belief that it is the problem, rather than the solution, that is incorrect.

2. *Getting 'stuck in a loop'.* The problem solver may get 'stuck in a loop' – repeating a set of moves that lead nowhere except back to the starting point. When this occurs it can be particularly difficult to break out.

3. *Inability to think ahead.* This is often due to memory limitations. In the Missionaries and Cannibals problem and in games such as chess and draughts, the number of alternative configurations soon becomes unmanageable when the problem solver tries to consider all of the possibilities even two or three steps ahead.

4. *Unwillingness to consider counterintuitive actions.* In many problems, there are certain essential subgoals that have to be executed in order for the main goal to be attained. Sometimes these may appear to be in direct conflict with the main goal. For example, in the Missionaries and Cannibals problem, the main goal is to transfer the missionaries and cannibals from the left to the right banks in as few moves as possible. Intuitively, the best way to do this would appear to be to transfer two people at a time from the left to the right bank and send only one person back each time. However, the problem cannot be solved if this approach is rigidly adhered to.

   The minimum number of one-way crossings to move everyone across the river is 11, the solution to the problem, for those who still require it, can be found in Bundy (1978).

Skill-based, rule-based and knowledge-based errors

Rasmussen proposed three levels of human performance in human–machine interaction and Reason (1990) has summarised the different causes of error at each of these levels. Selection of the most appropriate method for error prevention depends on the level at which the operator is performing.

Breakdown of problem-solving behaviour

Problem solving behaviour can break down for a number of reasons (Table 15.3). Wason has shown that one of the main weaknesses of human problem solvers is the preference for confirmatory evidence over evidence that will refute their ideas (Wason and Johnson-Laird, 1972). This can lead to rigid, stereotyped behaviour that is very difficult for the problem solver to break out of. For example, a novice computer programmer attempting to debug a program may continually search for syntax errors as the cause of a fault, unable to shift attention to search for faults in the structure of the program itself. Preference for confirmatory evidence can have serious consequences for fault diagnosis and correction in process control tasks. One solution is to train operators to recognise this type of behaviour and provide them with better specific problem-solving skills, including how to test their hypotheses by searching for evidence that will refute rather than confirm them (Wickens, 1987). A related problem is known as *bounded rationality*, which is essentially a failure to appreciate the 'big picture'. Bounded rationality restricts the problem solver's efforts to a limited part of the problem space and reduces the chances of finding evidence to refute incorrect hypotheses.

Characterising human–machine interaction

Before ergonomic principles can be applied to the design of interactive systems, a sound understanding of the task is needed. There are many ways of capturing data on task performance, the simplest being observation, either in usability laboratories or in the field. Typically, the interaction is recorded and any errors are noted. Users may be debriefed after the interaction so that further insights into the usability of the device can be obtained. Of the more formal methods of characterising human–machine interaction, task analysis is probably the most used.

Task analysis

Task analysis was defined by Snyder (1991) as

> an ordered sequence of tasks and subtasks, which identifies the performer or user; the action, activities or operations; the environment; the starting state, the goal state; the requirements to complete the task such as hardware, software or information.

Task analysis drives human-centred design by providing a system-specific context for the application of the fundamental ergonomic principles. In broad terms, the procedure for carrying out a task analysis is as follows (Rasmussen and Ouse, 1981).

1. Identify a prototypical task by collecting detailed descriptions from expert users of what different people do.
2. Identify all the processes that comprise the activity.
3. Analyse the descriptions in terms of the various options for action and the criteria used to select between them.
4. Generate a prototypical task specification by selecting a characteristic set of tasks and specifying the work sequences common to them:

    *   Who must be involved?
    *   What are the subcomponents of the activity?
    *   How are participants involved in the various tasks?
    *   What information is required at each stage in the task?
    *   Where does the information come from?
    *   How is the information exchanged?
    *   How might any of the above be improved?

Ergonomists have developed more formal ways of analysing tasks. These include hierarchical representation of task behaviours, observational techniques for obtaining data about the behaviour involved in carrying out a task and methods for representing the dynamic aspects of human–machine interaction. The outcome of a task analysis consists of the following:

1. A description of the behaviours required to carry out the task
2. A description of the system states that occur when the task is carried out
3. A mapping of the task behaviours onto the system states

Table 15.4 Levels of description in task analysis: an example[a]

Job title: fuel pump station operator

Assignments

A1	Inspection of environment, equipment and machinery
A2	Execution of start-up procedures for pumping
A3	Monitoring of system condition when running
A4	Execution of close-down procedures
A5	General maintenance

Segments

(A1) S1	General inspection; safety, lighting, housekeeping
(A2) S2	Start-up filtration of fuel in reservoir tank
(A3) S3	Periodic inspections of pumps
(A4) S4	Close down pump motors
(A5) S5	Changing the filter elements

Tasks

(A1S1) T1	Replace broken lamps in station
(A2S2) T2	Start-up transfer pump
(A3S3) T3	Check torque settings on pump retaining clamps
(A4S4) T4	Close down pump motor 1
(A5S5) T5	Remove filter element 1

Operations

(A1S1T1) O1	Switch off power to lighting unit at wall switch
(A2S2T2) O2	Press green button on transfer pump housing
(A3S3T3) O3	Using a torque wrench, tighten nuts in clockwise direction until torque setting is displayed
(A4S4T4) O4	Press red button on pump motor 1
(A5S5T5) O5	With an 8 mm spanner, loosen nuts retaining filter element housing

[a] There are many more segments, tasks and operations. Those that are shown are included to illustrate the detail required at the different hierarchical levels.

This information can be used for a variety of purposes:

1. Evaluation or the design of the human–machine interface
2. Identification of the skills needed by an operator of the system
3. Design of training materials and operating instructions
4. Identification of critical elements of the task to predict or evaluate the reliability of the system

Table 15.4 gives an example of the stages involved in the hierarchical decomposition of a job from a job title to specific operations. This type of analysis can produce enormous quantities of documentation, as can be imagined. However, in complex systems, it is often essential to describe tasks at this level of detail because even small changes in equipment or procedures can effect the mapping between required human behaviours and system states. Table 15.5 gives examples of the types of questions the systems designer must ask when looking at the mapping between individual task behaviours and system states. A basic question is 'How are operator task behaviours to be invoked and how will the system respond to their execution?'

Table 15.5 Example mapping of task behaviour and system operation in system design and evaluation

---

Task: To transfer fuel from mobile tanker to on-board storage unit

Mapping elements

1. Indications and when to do the task	When requested by Captain When indicated by fuel guage With engines switched off
2. Control objects and operation	Transfer valves Transfer hose Hose retaining clamps Transfer pump controls Slide hose into transfer valve orifices on tanker and storage unit (outlet and inlet) Use retaining clamps to secure hose to transfer couplings on tanker and storage unit Open tranfer valves on tanker and storage unit Press green button on transfer pump
3. Precautions	Valves in open position and hose secured Fuel level in mobile tanker above red line Tanker on level surface and brakes in 'on' position
4. Feedback modality and Indication of response adequacy	
• Visual/kinaesthetic	Valves 'click' into open position
• Auditory/kinaesthetic	Pump vibrates/'whines' if flow is impeded
• Visual	Transfer pump pressure gauge reads 150–250 kPa. Absence of leaks Fuel level in storage unit gauge rises
5. Fault diagnosis and maintenance	If pump 'whines' or flowmeter indicates blockage, press red button to stop pump
• Check	Transfer valves open Hose not twisted or crushed Fuel level in tanker Pressure gauge on transfer pump

---

Cognitive task analysis

The purpose of cognitive task analysis is to identify the cognitive processes involved in the performance of tasks. The focus is on the demands made on memory, decision making, attention and so on in order to ensure that operator capability is not exceeded and to propose cognitive aids, where needed. Morphew *et al.* (1998) present the following framework for a cognitive task analysis (Table 15.6). In the example given, the task of landing aircraft onto the deck of an aircraft carrier was analysed from the point of view of the Landing Signals Officer on board the vessel. Cognitively demanding aspects of the task were identified and a monocular display was developed to support the cognitive processes (the display was worn on a monocle). One demanding cogntive task was changed to a perceptual task when the monocle was worn and another task was simplified by presenting objective data on deck motion to lessen the burden on judgement.

Cognitive task analysis is an example of the new knowledge-based methods for systems design. Sturrock *et al.* (1997) provide a good example of the use of knowledge

Table 15.6 Framework for a cognitive task analysis[a]

What is the decision?	Why is it difficult?	Uncertainty involved	Potential errors	Critical cues	What is the HCI aid and how will it help?
Describe the decision that has to be made	What factors influence the ease of decision making?	What is the nature of the process in terms of predictability	What can go wrong?	What information is needed to reach the right decision?	How can the interface be designed to lessen cognitive load?
Example[a] Is the deck clear? Can it be cleared in time, if not?	LSO is looking ahead, at plane deck is behind him	What is the obstruction? Status of attempts to clear deck	Divided attention. LSO attends to deck rather than approach. Plane cannot land	Foul deck light on. Verbal cues from colleagues	Move foul deck light to monocular display Perception replaces cognition
Is deck motion safe for landing plane?	Anticipation needed – both future motion and pilot responses	Where will the deck be when the plane lands? How will the pilot react?	Pilot has to 'chase' the deck. Pilot ability overestimated in relation to conditions Failure to switch to automated system	Horizon Ramp motion alarms	Create ship turning indicator and cross winds indicator on monocular display Provides objective measure of critical parameters

[a] Adapted from Morphew *et al.* (1998). Task is that of a landing signals officer (LSO) on an aircraft carrier. Subtask is to decide whether an incoming plane can land on the deck.

requirements analysis to design the user-interface. Baber and Stanton (1996) describe how task analysis can be used for error identification.

Verbal methods of data capture

There are several verbal methods of capturing data for modelling the human operator. Some basic 'on-line' and 'off-line' methods are noted below. System state/action state diagrams can be used off-line. The operator is presented with a matrix of possibilities concerning a controlled variable (such as temperature). At each cell in the matrix the operator is asked what action (if any) he or she would take if the system behaved in the way described in the matrix. Although the method provides a structured approach to data capture, it has a number of limitations, such as the validity and reliability of the answers and whether the operator really comprehends or can imagine the occurrence of all the states presented.

Questionnaires can be also be used to gather data off-line about operator control strategy. The general problems of questionnaire design need to be considered – the wording of questions to avoid ambiguity and avoidance of leading questions that suggest there is a correct answer. With off-line methods there is also the problem of operators responding by saying what should be done under a particular set of circumstances (if they followed the procedures) rather than what they would actually do.

Structured interviews can be used to obtain data on the rules and general principles that operators utilise and can be accompanied by the use of static simulation in which operators are given a number of hypothetical situations that they must talk the interviewer through.

These methods have the advantage that the most interesting aspects can be isolated. The disadvantages include an absence of realism and a lack of time pressure. A wide sample of situations must be investigated to characterise the operator's knowledge.

Verbal protocol collection is an on-line method of capturing data on operator control strategy. Operators are asked to 'think out loud' while carrying out the job and what they say and do is recorded. The method attempts to 'get inside the operator's head' and is intuitively appealing. For example, it may be difficult to determine the difference between a good car driver and a highly skilled car driver using objective measurements alone. Unless they are exposed to extremely demanding situations, most of what they do will seem the same. However, the running commentary of the two drivers might differ considerably, with the highly skilled driver showing more evidence of higher cognitive activity such as thinking ahead, detecting potentially dangerous situations well in advance, and so on.

The verbal protocol technique has a number of advantages over other verbal methods. Objective data about system functioning can be recorded simultaneously with the protocol. The protocol itself can be used to design follow-up questions for use in off-line interview. However, the technique has been questioned, particularly in relation to its validity (e.g. Umbers, 1979):

1. It makes assumptions about the operator's ability to 'put into words' their control strategy.
2. Operators may neglect to mention things that seem obvious to them but are not obvious to anyone else.
3. Most people can think more quickly than they can talk.

4. The process of thinking aloud may distract the operator – it can be seen as a secondary task that, depending on the modality and processing requirements of the primary task, may compete for common processing resources (car drivers often break off in the middle of a conversation when carrying out a difficult manoeuvre).

GOMS

GOMS stands for goals, operators, methods and selection rules (Card *et al.*, 1983). It is a method of representing human–machine interaction over time as a sequence of stages that can take place in series or in parallel. For HCI applications, there is also a keystroke level model that enables the analyst to predict the time taken by keying actions and system response. GOMS analyses are primarily concerned with how users achieve goals and subgoals using the methods and procedures ('operators') that the system makes available. Operators are rather like words in a language and are the elementary motor and cognitive acts whose execution is needed either to change the user's mental state (e.g. choosing an item on a menu) or the system's state (e.g. selecting the item). For example, the goal 'edit chapter 15 title' might consist of a number of subgoals, 'Open file', 'Find heading', 'Delete heading'. Selection rules are used to choose among the alternative ways of accomplishing the goal (e.g. using either the mouse or the cursor control keys to move the cursor; highlighting the text to be deleted then pressing 'DEL' or using ←; the rule might be to use the mouse if the cursor is more than two lines from the text to be deleted).

The GOMS analysis can be used to evaluate ease of task performance and to identify any task demands that conflict with one another. The keystroke-level model can be used to predict the time taken for the user to accomplish the task. Together, two methods can be used to compare alternative systems. This can be done before either has been built.

In 'Project Ernestine', Gray *et al.* (1993) used GOMS to compare an existing and a proposed interface for telephone toll assistance operators (TOAs). The time taken to deal with calls from customers was the key criterion for acceptance of the new interface. In dealing with telephone enquiries from customers, TOAs had to answer three questions – who should pay for the call, at what rate should the call be charged, and when to terminate the interaction. In GOMS terms, these are the three goals of the interaction with the customer. The operators needed to achieve the goals included listening to the customer, talking to the customer, keying, waiting for the computer system to respond and the cognitive activities needed to make a decision. GOMS breaks the call handling task into a functional stages ('Receive Information', 'Request Information', 'Enter Information' and 'Release Call'). Each of these is broken down into a series of activities involving basic operators (e.g. 'Read Info from Screen', 'Listen to Customer', 'Enter Billing Rate', etc.). The GOMS model is built by establishing the correct sequence in which the operators are applied to achieve the goal and the time taken to perform each operator. The model also specifies which operators are carried out in series and which are carried out in parallel. The serial operators form a *critical path* and the sum of the durations of each operator along the path determines the time taken to reach the goal. These times were estimated in two ways: from empirical data on TOA performance (average operator times gleaned from videotapes of TOAs at work) and from normative estimates in the GOMS literature. Gray used the following normative operator times for his model:

- Attend to visual information   50 ms
- Attend auditory information   50 ms
- Initiate speech, keystroke, eye movement   50 ms
- Move eye to screen location   180 ms
- Detect visual signal   100 ms
- Perceive visual signal (alphanumeric codes, small words)   290 ms
- Perceive auditory (no signal)   300 ms
- Perceive and encode short word   340 ms
- Interspeaker pause duration   400 ms

The GOMS model for the existing workstation was found to be accurate and under-predicted average call times by only 3% (0.6 seconds). For the proposed workstation, the GOMS model was built using manufacturer's data on system response time, the new sequence of operators forming the task and the normative data above. Figure 15.1 shows GOMS models for the two workstations. As can be seen, the new workstation removed seven motor and three cognitive operators, but since these were not on the critical path, there was no reduction in call time. At the end of the call, the new workstation added one keystroke to the critical path and this required the addition of one cognitive and three motor operators. The GOMS modelling enabled Gray to predict that the new workstation would be slower than the old one and this was confirmed in a 4-month trial.

Prevention of error in human–machine interaction

The ergonomic approach to *error prevention* focuses on the design of the environment in which the task takes place and the design of the interface. More deeply, ergonomists are becoming increasingly interested in the design of tasks themselves. The second line of defence is *error detection*, which can implemented by having the system detect error or by improving users' ability to detect their own errors.

Interface design considerations

The main considerations are summarised below. For usability and error prevention provide:

- A good conceptual model with consistent system image
- Good mappings between system states and task stages
- Good mappings between task actions and their effects
- Continuous feedback
- Informative error messages
- Explicit cues for distinguishing between system states
- Explicit cues to indicate change of state
- Minimal use of modes
- Coherent, consistent system image
- Forcing functions to block error
- External memory aids in the interface
- An undo facility to make actions reversible

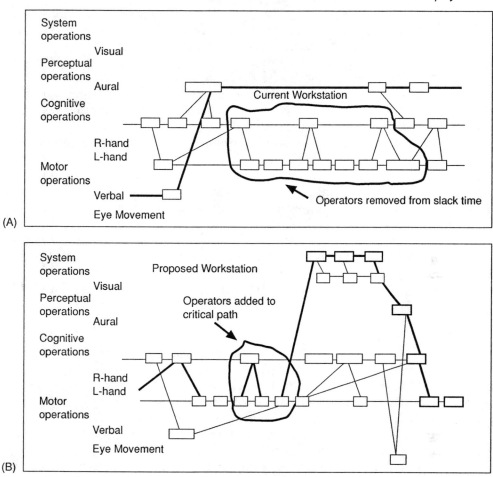

Figure 15.1 GOMS models of VDU task performance. The redesign removed operations that were not on the critical path and added new operations in series with existing operations. Transaction times increased accordingly. (Lawrence Erlbaum Associates, with permission.)

Make system states visible

System state should be as visible as possible to prompt the correct user behaviour. For example, some vending machines have a back-lit plastic cover displaying the product logo. This informs users what the machine sells and that the machine is switched on and ready. A glass window may enable buyers to see the product itself or to choose between a variety of options. That the system is 'On, ready to sell, and with stock available' can be deduced just looking at the machine. A display by the coin slot may inform that buyer that the machine gives change and what coins to use.

The author has a kettle made of a blue thermoplastic that becomes pink when the water boils. There is a level indicator and a red light that goes on when the kettle is switched on. The 'on' switch slides from the 'off' to the 'on' position with a clicking sound and switches off automatically. The kettle, displays whether it has water in it and even whether it has just boiled and been refilled with cold water (it turns blue up to the cold water level when refilled after boiling). Practically all of this kettle's states are visible and different from each other. The mapping between the dynamics of the task and the representation of the system state is excellent. The kettle turns pink very slowly and from the bottom-up because the heating element is in the bottom. The spout only turns fully pink when heated by steam from the boiling water. Contrary to the old saying 'A watched pot never boils', with this kettle the user quickly learns to guess how long is left for the kettle to boil.

In HCI, menus serve a mapping function, prompting users to choose between the options on offer. The item selected then becomes the title page or context marker of the next screen. Good menus provide good mappings between user actions and task stages.

It is tempting to speculate that some of the deaths due to drug overdose are caused, in part, because the user forgets how much of a drug has already been ingested. In the case of mind-altering drugs used recreationally, the drug itself may impair memory and lead to accidental overindulgence. Some medicines in tablet form are prescribed in 'blister' packs that may or may not have a date or number by each tablet. In any case, the user is presented with a record of how many tablets have been taken in the form of the empty blisters.

Forcing functions

Perhaps the most extreme example of human error is suicide. Whether it is preventable is debatable, but some researchers have suggested that its incidence can be diminshed by placing constraints in the environment (Gunnel and Frankel, 1994). Examples of such constraints are catalytic converters in cars to reduce the toxicity of exhaust fumes, and redesign of exhaust pipes to make it more difficult to attach pipes (to prevent people deliberately gassing themselves in cars); providing off the shelf and prescription drugs in small amounts to deter overdosing; addition of other drugs to paracetamol to reduce its toxicity in overdose; better design of barriers around the top of tall buildings or provision of nets. These interventions are designed to prevent suicide long enough for the suicidal urge to subside.

Forcing functions vary in the degree of 'force' applied. The system may prevent any further progress until the user performs a required action; for example, the keyboard goes dead until the user responds to a request to save a file. Alternatively, the system may simply ignore an illegal entry. For example, a common psychomotor error in data entry is to reverse adjacent symbols. This can be blocked by designing a coding system in which reversed codes are illegal and are met with no response when entered. As a third possibility, the system may warn the user that an error has been made – some cars sound a warning if the driver leaves the lights on after turning the ignition off and opening the door.

Error detection

Sellen argues that those human errors that are easiest for systems to detect are the same ones that are easiest for users to self-detect. Task analysis for error identification (Baber and Stanton, 1996) could be used in advance to design an error-detection facility into a system (by delineating the space of possible errors at different stages of the task). If the system were 'self-aware' (i.e. knew what state it was in at any time), the set of errors at a given state would be reduced and correct identification would be more likely. Self-detection can be enhanced by providing explicit feedback throughout the task. 'Rumble' strips at the side of roads warn drivers of a tracking error by virtue of the noise and vibration they generate. This is a highly alerting and discriminable cue whose meaning is easily learnt. Reduction of high mental workload is another strategy, since the conscious detection of error requires a degree of self-monitoring that is lost under high workload. Outcome-based detection depends on the user identifying a discrepancy between the real and expected outputs. This can be enhanced by making the system outputs as explicit as possible, in time and space. If the expected outcome is unclear and its arrival uncertain, self-detection of error will be prevented.

Knowledge in the head and knowledge in the world

Norman (1988) discusses the trade-offs involved in making systems 'transparent' through good design (embedding the knowledge needed to use them in the design of the device) and making use of prior knowledge and training (embedding the knowledge in the user's mind).

Knowledge in the world Knowledge in the world is always available to the user when the system is visible or audible. The need for learning is minimised if natural mappings and consistent design are used. Although the system is usable at the first encounter, the user has to find and interpret the necessary information. The need to embed the knowledge in the device places constraints on the design, particularly the design of the interface. There may be space problems and other problems with adjacent devices.

Knowledge in the head If the knowledge to use a system is in the user's head, there are fewer constraints on the designer. However, learning and formal training may be needed, which may have cost implications. Usage once learnt may be vulnerable to error due to memory limitations, but very efficient performance is possible.

Alarms

In complex systems and in everyday life, alarms *can* help people detect problems. Two main drawbacks of alarms are habituation and denial. Alarms sometimes activate when there is no problem and, like 'the boy who cried wolf', lose the trust of the user:

> A motoring magazine got two men to pose as thieves and load cars onto a lorry in broad daylight, dozens of passers-by ignored the blaring alarms and deliberately suspicious behaviour of the men . . . The experiments confirm the widely-held belief that so many alarms go-off by accident that anyone in earshot is more likely to turn up the radio than call the police.
>
> (From the *Daily Mail*, 1 March 2000)

In fact, the only call the police did receive was a complaint that the lorry was blocking the street! Apart from saturation, another problem with alarms might be termed *denial*. The author was woken one morning at 4:30 a.m. by an alarm bell, while staying in a hotel. The sequence of reactions went something like this:

- Why is that bell ringing?
- It sounds like the fire alarm, why are they testing the fire alarm at 4:30 a.m.?
- Maybe the alarm is malfunctioning.
- This is a ridiculous time to hold a fire drill.
- I'd better get up. Sounds like there's a fire.

Whether the same sequence of reactions would have occurred if a verbal alarm (FIRE!) had sounded is debatable and highlights the importance of *salience* in alarm design.

Automation

Highly complex systems such as nuclear power stations are also highly automated. Most of the time, these systems can run themselves and Reason (1990) suggests that operators are only retained to handle non-design emergencies (emergencies that could not have been foreseen by the designers). Since these emergencies could not have been foreseen by the operators, either, they have the problem of having to solve a novel problem under pressure. Bainbridge (1982) describes several more ironies of automation:

1. The automated system is supposed to perform the task better than the operator. It is ironic that the operator is supposed to determine whether the machine is performing optimally and to take over when the machine fails.
2. If operators assume control only when the machine breaks down or in emergencies, they will be less skilled than if they always controlled the system. In automated systems then, operators are required to assume control at the most difficult times and with suboptimal control skills.
3. Second-generation operators (who have never controlled the system) will have no skills (subroutines), only knowledge.
4. Operators of poorly designed automated systems will have better control and recovery skills than those of well-designed systems.

   Some possible solutions to these dilemmas are to build manual control practice into the organisation of the work or to use simulators to train operators in the necessary manual control skills.

Provide decision support systems

1. Signal detection support consists of computer-generated probability that a signal appearing on a complex array is a target rather than noise.
2. 'Cascading' of alarms and warnings can overload the operator. Decision support systems use a model of the process to process large amounts of fault data rapidly.
3. Diagnostic problem-solving aids (see Shortliffe *et al.*, 1979; Shortliffe, 1984; and Kleinmuntz, 1984, for further information) Multiple faults can have a single cause (e.g. poor roadholding and fuel consumption due to too-low tyre pressures) and single faults can have multiple causes (e.g. uneven engine performance due to worn-out spark plugs, incorrect ignition timing, a dirty air filter and leaky gaskets on the inlet and outlet manifolds). Until all faults are fixed, the problem

remains – each remedial action, although correct in itself, is not followed by the expected improvement in performance.

- Decision trees consist of a hierarchy of questions with a limited number of responses (sometimes only 'yes' or 'no'). The user is prompted by a series of questions and gleans data from the system to arrive at the answer which then leads to a question at the next level down in the hierarchy.
- Statistical pattern recognition techniques are based on survey data about the prevalence of a problem and the variables associated with its manifestation (the 'signs and symptoms' in medicine). Survey data are subjected to statistical analysis using regression and other multivariate techniques. This produces a mathematical model of the disease pattern. The model gives the probability that a patient has a particular disease, given that certain signs or symptoms are present.
- Bayesian decision aids help decision makers handle conditional probabilities, which are not easy for human decision makers to deal with intuitively. Judgements about the likelihood of events that depend on one or more co-conditions are difficult in the absence of other information to assist the decision maker. Bayesian decision aids deal with the computational complexity of assessing conditional probabilities.
- Expert systems attempt to mimic the decision-making behaviour of the expert. In diagnostic tasks, judgement is made on the basis of gross chunks of knowledge rather than detailed facts. Heuristic rules are applied to these knowledge chunks to arrive at a decision. The rule-based approach directs attention to pertinent areas and obviates the need for detailed analysis of the problem. Knowledge is in the form of production rules – simple conditional statements that relate observations to inferences (such as 'If a bacterium is a Gram-positive coccus growing in chains, then it is likely to be streptococcus'). The inference engine decides that rules to use and how they should be chained together to make decisions. The system's knowledge structure can be modified by changing the rules or adding new ones and the rules themselves can be used to explain how the system came to a particular decision.

These aids help to prevent poor decision making because they are free of the biases that affect human decision makers.

Training

In diagnostic and other problem-solving tasks, training in fundamental principles alone is usually inadequate (Morris and Rouse, 1985). According to Schaafstal (1993), theoretical instruction has to be combined with training in the use of the knowledge to solve problems in the context of the actual equipment that will be used. Morrison and Duncan (1988) have described some of the characteristics of easy and difficult faults. Easily diagnosed faults can be predicted and are associated with an explicit algorithm that will elucidate them. Difficult faults are intermittent or novel, they cannot be predicted from presently available information, and there is no predetermined algorithm for their elucidation. It is with these types of problems that knowledge of fundamental principles is useful and sometimes the only means of solving the problem. The opportunity for practice is an important component of training and industries such as aerospace and nuclear power spend considerable sums on simulators to enable off-line practice to be carried out.

Support problem solving

People's ability to solve problems can be improved by providing them with information about the underlying structure of the problem. There are several ways to facilitate the disclosure of a problem's structure:

1. Remove unnecessary and irrelevant information.
2. Choose a representation of the problem that operators are familiar with.
3. Identify the critical information operators need and ensure that it is displayed appropriately.

Shepherd (1993) gives an interesting example of a problem in the design of a computer-based control system for a process control industry that affected operators' ability to handle plant disturbances efficiently. The three control panels in the control room were organised on the basis of geographical area of the plant. Two VDUs displayed information about the plant that could be retrieved via a hierarchical menu also structured according to geographical area. When a plant disturbance occurred, operators had to decide whether to close the plant down or whether it could be set at some intermediate state while the disturbance was rectified. Although it was the case that disturbances in plant functioning did tend to be confined to one geographical area, information needed to restore functioning had to be collected from all geographical areas of the plant. The result was that, with the control system configured geographically, operators were placed under extreme memory load as they moved from panel to panel and interrogated different areas of the geographically configured menu, while having to retain in short-term memory the information gained to date and the intermediate decisions they had made.

Shepherd describes this class of design problem as one of information fragmentation. He emphasises that it can be avoided by carrying out, at an early stage, a particular type of task analysis – an information requirements analysis – that emphasises the information operators need to solve the problems they are confronted with at work. Woods and Roth (1988) give a similar example, a menu-driven database designed to computerise paper-based procedures for nuclear power plant emergency operation. Although the user-interface had been designed according to ergonomic recommendations, operators were unable to complete recovery from emergencies by following the procedures interactively. They were unable to keep procedure steps in line with plant behaviour and got 'lost' in the system. The problem occurred because the real cognitive demands of emergency of recovery had not been considered in the design. In real emergencies, operators carry out several activities in parallel and shift from one part of a procedure to a different procedure and back again depending on plant conditions. Paper-based documents support this type of activity well because they permit reading ahead, scanning several documents at once and leaving open documents in one place while doing something else. They provide a physical trace of the sequences of actions in concurrent activities. However, standard menu systems provide fewer cues as to the sequence of activities. The authors redesigned the

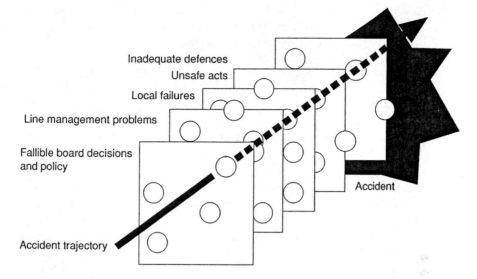

Figure 15.2 'Swiss cheese' model of accidents. (Redrawn from Noyes and Stanton, 1997.)

interface so as to better support the cognitive demands of the task. The interface was made more 'booklike'; two onscreen windows were provided to permit more parallel display of information; steps across procedure boundaries were made easier; and incomplete steps were tracked by the system and flagged with 'electronic bookmarks' to ease resumption of previously interrupted activities.

Accidents and safety

Accidents can happen for a variety of reasons, including technical failure. Accidents are often attributed to human error. Bailey (1982) states that there is general tendency to equate poor system performance with poor human performance, yet detailed analyses of accidents and near accidents reveal that human error is almost never the sole cause. Real systems are multilayered and hierarchical and errors can only lead to accidents if they have consequences at other levels. Reason's 'Swiss cheese' model (Figure 15.2) uses an analogy in which an initial fault gets into a system, progressing along an 'accident trajectory' that depends on a weakness at the next level of system organisation. If all the weaknesses, at all levels, coincide, like slices of Swiss cheese momentarily aligned so that you can see through the holes, an *accident trajectory* exists and an accident will happen.

Medical error and Swiss cheese

Iatrogenics refers to medical problems caused by the system of medical care, including clinicians, managers and technology. Iatrogenic complications can have ergonomic causes that illustrate the Swiss cheese model of accident causation. Many different factors can combine to cause a problem; some examples are as follow. Much medical equipment in hospitals has no 'owner', being on wheels and handled daily by different people, sometimes in emergencies. Medical equipment is often mechanically 'traumatized' – out of calibration or dysfunctional. Disposable equipment is sometimes re-used but does not survive autoclaving, for which it was not designed. Hospitals may lack formal systems of quality control and formal maintenance standards. Feedback about equipment performance and reliability may be delayed or non-existent. Shortcomings in treatment or bad practice may persist owing to a lack of negative feedback. Professor I. Levine of Yale University Medical School reported that about 10 000 patients die each year in the USA as a result of errors in anaesthesia. A study of 1800 diagnoses revealed an error rate of 20%; half of these mistakes may have resulted in patient death. Almost two-thirds of tests ordered on patients were found to be inappropriate or unecessary. Blood pressure cuffs and X-ray machines are frequently out of calibration, damaged or leaking. Bridger and Poluta (1998) concluded that many of the iatrogenic problems occur owing to the complex relationships between owners of technology, users and patients.

Scope of accident investigation

To understand why accidents, errors or any unexpected system behaviours occur, it is necessary to look beyond human behaviour to the rest of the system (Kirwan, 1994). The following important factors need to be investigated:

1. Design of system components, particularly human–machine interfaces
2. The state of the system leading up to the incident (stable/unstable, quiet/busy, on course/off course, etc.)
3. Operators' mental and physical workload
4. Work organisation (e.g. shift system and time during shift, supervision, design of work groups)
5. External factors (e.g. weather)

Human error and system inefficency are often inadvertently programmed into systems by design deficiencies that do not take account of the characteristics and limitations of humans.

Risk homeostasis theory

Risk homeostasis theory (RHT) was developed during research into road safety by Wilde (1995). It challenges the assumption that when we make one part of a system safer, the number of accidents will drop. A model of RHT is given in Figure 15.3.

RHT states that, when allowed to do so, individuals will modify their behaviour so that the level of risk they are exposed to is compatible with their target (desired) level of risk. If there is a discrepancy between the target level of risk and the actual level, people modify their behaviour to bring the actual level in line with the target level. For example, a driver on a long journey may increase the cruising speed when arriving at a motorway only to slow down later on when it starts to rain. Changes in the environment bring about discrepancies between target risk and real risk that trigger changes in behaviour. This is known as *risk compensation*. According to Wilde, the way to reduce accidents is to lower the target level of risk in a population of users. Behaviour will then change accordingly to reduce the actual level of risk to which people *expose themselves*. Note that RHT is a behavioural theory and only applies to

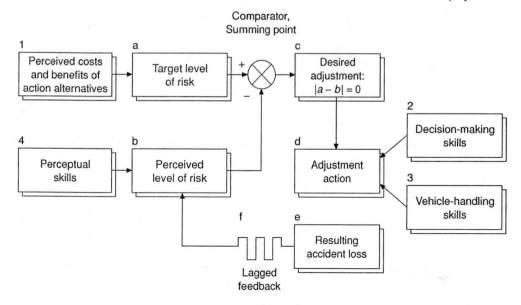

Figure 15.3 Wilde's model of risk homeostasis. (Redrawn from Wilde, 1994.)

situations where people have sufficient control over their behaviour to influence the real level of risk. It only applies to situations where people get sufficient feedback from the system to enable them to estimate the real risk. Thus, it can be applied to car driving but not to paced industrial jobs where the worker has no control over the work rate or the work environment.

Wilde (1994) and Adams (1998) cite statistical data on road accidents in different countries to argue that the provision of many safety features does not influence road fatality rates. According to Wilde, the introduction of compulsory seat belt wearing has not reduced fatalities, probably because seat belts cause risk compensation (belt wearers feel safer and therefore drive faster or take more risks). There is support for the concept of risk compensation, if not for RHT itself. Janssen (1994) reports a study of driver behaviour in Holland. Car drivers drove an instrumented car along a 105 km route. People who normally wore a seat belt, wore a belt as usual in the study. Habitual non-wearers drove under two conditions – with and without wearing a seatbelt. It was found that non-wearers drove faster than belt wearers and that their speed increased even more when they wore a belt. This effect was found to be stable over a one-year period. RHT would attribute the difference in driving speed between habitual belt wearers and non-wearers as evidence for a lower target level of risk in the former. The increase in driving speed when non-wearers wore belts is evidence for risk compensation (the level of risk was perceived to be lower when the belt was worn, so they took more risks). Petersen and Hoffer (1995) analysed Virginia State Police accident reports and insurance company claims data to compare accident and injury rates for drivers of cars with and without airbags. Drivers of cars newly equipped with airbags had significantly larger relative losses under their personal injury protection cover than drivers of cars without airbags. This increase was stable over several years, controlling for other changes in car design such as re-styling. Furthermore, the accident data revealed that there was no evidence that airbags reduced the probability of a driver dying in a single vehicle accident. For multiple-vehicle accidents, drivers of airbag-equipped cars were found to initiate an unusually large number of accidents involving other vehicles. The authors concluded that risk compensation resulting from the installation of airbags causes more aggressive driving that negates the protection offered by the airbag and increases the risk of death to others.

Wilde cites many other examples of risk compensation having unexpected and unwanted effects. For example, in California in the 1970s, the introduction of compulsory child-proof containers for storing medicines led to an increase in the number of child poisonings in the following year (presumably because parents were less vigilant in storing medicines in 'child-proof' bottles and the degree of safety was relative, rather than absolute). Rabinowitz *et al.* (1998), in a laboratory study, found that subjects perceived themselves to be safer when lifting loads using a squat technique and wearing an abdominal belt than in other conditions, even though there was no physical evidence that they were safer. Bridger and Freidberg (1999) investigated managers' estimates of safe loads for manual handling in a number of light manufacturing industries and found that the estimates were higher when the managers were told that the workers would be wearing a 'back-support' belt.

What determines the target level of risk?

The target level of risk is determined by the aggregate value of the costs and benefits associated with risky and safe behaviour and can be represented as a utility matrix (Table 15.7).

According to Wilde, in all situations where people have a degree of control over the level of risk to which they expose themselves, the best way to reduce accidents is to lower the target level of risk. This can be done in many ways, for example:

*   Where risk perception is faulty, e.g. due to ignorance, by training people to improve their perception of risk
*   By providing incentives to increase the benefits associated with safe behaviour (e.g. free driving licences for accident-free drivers)

Table 15.7 Utility matrix of the costs and benefits of engaging in safe and unsafe acts: Unloading weekly shopping out of boot of car and carrying it indoors.

	Safe method[a]	*Unsafe method*[a]
Costs	Task takes longer Physiological cost higher Boredom More chance of forgetting something	Muscle and joint strain Difficult to negotiate obstacles Less attention available for other stimuli More chance of dropping/damaging goods Less manoeuvrability
Benefits	Biomechanical load lower Less chance of injury Less chance of dropping anything Less localised muscle fatigue	Task completed more quickly (free to move on to next goal) Sense of achievement Demonstration of prowess Less chance of leaving anything behind

[a] The 'Safe' method would be to make several journeys carrying a small number of bags each time. The 'Unsafe' method is to attempt to carry everything in one trip.

• By increasing the costs associated with the unsafe acts (e.g. having dangerous drivers interviewed by a psychiatrist and recording this fact on their medical records).

Anecdotally, Wilde reports that in Sweden, for example, the road accident rate drops in very bad weather and that it dropped dramatically in 1967 in the month after the Swedes changed from driving on the left-hand side of the road to the right-hand side (despite predictions that the opposite would happen). These are examples of risk compensation in response to real increases in accident risk. In the 1970s, road death rates dropped in many countries at the time of the oil crisis following the massive increase in the petrol price. People drove less, and when they did drive they drove more slowly to conserve fuel. This is an example of how increasing the costs of unsafe acts lowers the target level of risk.

Risk homeostasis theory is controversial but it has heuristic value in challenging the often unwarranted assumption that when one part of a system is changed to improve safety, the rest of the system will stay the same. It reminds us that users may 'consume' the additional safety as a performance benefit and so the accident rate may remain unchanged. According to RHT, then, better roads and safer cars will result in more miles travelled more quickly for the same number of road deaths.

Effectiveness and cost-effectiveness

Automatic teller machine dialogues

Zimmerman and Bridger (2000) compared transaction times for cash withdrawals and the prevalence of card loss between two kinds of automatic teller machine (ATM). The ATMs differed little in terms of their external appearance but greatly in the dialogue needed to withdraw cash (Figure 15.4). Critically, dialogue A (found on all ATMs of one particular bank) dispensed cash first and then reverted to the main menu. To get the card back, users had to choose the 'no more transactions' option on the menu.

Dialogue B was found on many ATMs from several other banks. Cash withdrawal required fewer actions by the user and, critically, did not require the user to actively terminate the transaction. Rather, once cash had been requested and the transaction confirmed, the card was automatically returned to the user, followed by the cash. The transaction then ended automatically. In both cases, the cash withdrawal dialogue was initiated once the user had entered the card into the ATM. For dialogue A, the focus on commencement of the task was on all the functions offered by the ATM, including cash withdrawal. Dialogue B focused on cash withdrawal at the beginning, placing all other functions in a separate menu. Thus, subsidiary tasks and the functions to achieve them were explicit throughout the task under dialogue A, but were hidden for most of dialogue B. It was hypothesised that transaction time for cash withdrawal and card loss prevalence would be greater when cash was withdrawn from ATMs using A. Having attained the goal of withdrawing cash from an ATM under dialogue A, users would be more likely to leave in pursuit of other goals (e.g. spending the cash) rather than perform the remaining operations needed to secure the return of their cards.

Over 200 people were unobtrusively observed withdrawing cash from ATMs. Transaction times were recorded and short interviews were conducted to assess user

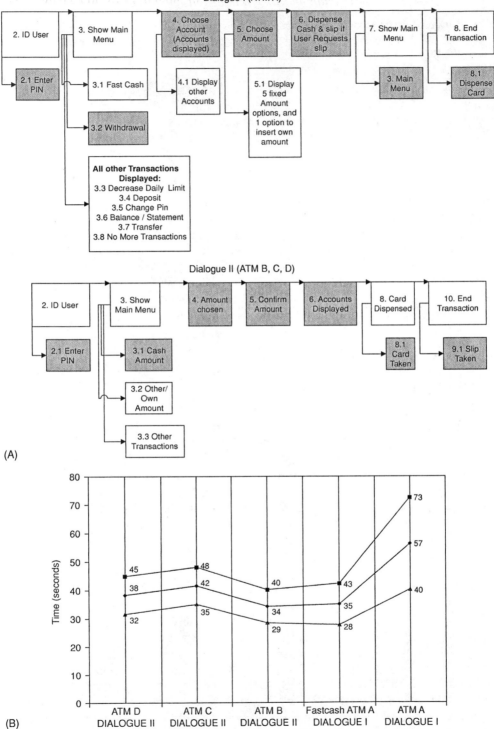

Figure 15.4 (A) ATM dialogues compared by Zimmerman and Bridger (2000). In dialogue I, the transaction ends with the dispensing of the card, after the cash has been dispensed. In dialogue II, the transaction ends wih the dispensing of the cash. (B) Mean and standard deviation cash withdrawal times (seconds) for different ATMs. Dialogue I is significantly slower. (From Zimmerman and Bridger, 2000.)

difficulties and errors associated with the use of the ATMs. As hypothesised, transaction times for cash withdrawals were longer when ATMs with dialogue A were used (a mean of 57 seconds, compared with means of 34, 38 and 42 seconds for dialogue B at three other banks' ATMs). The difference, about 30% longer, was statistically significant and was not due to differences in computer response time.

About half of all habitual users of ATMs with dialogue A had lost their card compared to about a third of other users (users of ATM A were about twice as likely to have lost their card *at least once*). Thirty-one per cent of users of ATMs with dialogue A could recall forgetting their card and leaving it in the ATM. Only 4.5% of other banks' users could recall having done this. *Norman's 1989 prediction that forcing people take their card before their cash will reduce card losses is correct.*

Zimmerman and Bridger estimated the cost implications of these differences in dialogue. Since all the banks imposed service charges for cash withdrawals, it is in their interest to process customers as quickly as possible at peak times. Long queues might have several effects that would reduce service charge revenue. For example, customers might be deterred by long queues and visit another bank's ATM or they might withdraw larger amounts of cash and so reduce the number of times they visit the ATM. Assuming a peak business period of 5 hours per day, dialogue A can process a maximum of 52 customers/hour compared to 105 customers/hour for the fastest ATM with dialogue B. The estimated increase in service charge revenue by changing dialogue A to achieve the faster transaction time was $17 840.35 per ATM per year. The bank using dialogue A in its ATMs had 1600 machines, so the potential for increased revenue through faster transaction times is 1600 × $17 840.35 or $28 544 560!

The cost to the bank of dealing with lost cards consists of the cost of the card itself ($0.25–0.50 for magnetic strip cards and $2–10 for smart cards at 1997 prices; Puri, 1997) and the administrative cost of replacing it (estimated as the cost of 20 minutes of clerical worker time). The total number of customers forgetting their cards was estimated from the prevalence data of the study and the bank's estimated number of customers to give an annual card replacement cost of $2 069 072. The use of a dialogue that included a forcing function at the cash dispensing stage might reduce this cost by about one-third.

A new workstation for telephone operators

Gray *et al.* (1993) compared an existing workstation at a telephone company in New England, USA, with a new workstation proposed as a replacement. The new workstation was expected to reduce the time spent by telephone operators in handling calls. Both workstations were computerised and the operators carried out a dialogue with the computer, via a keyboard and screen, and a related dialogue, by voice, with customers. The new workstation was predicted to reduce call handling time for a variety of reasons – it had a high-resolution display with a 1200 baud rate compared to a baud rate of 300 on the old workstation. The new workstation had a graphic user-interface with icons and windows and a keyboard layout designed to minimise travel distance between frequently used keys. Keystroke number per transaction was, in general, reduced by replacing two-key sequences on the old keyboard with one key press on the new keyboard. Further, the whole of the new screen could be updated quickly whereas the old screen was slower and new information was displayed only as it became available (usually in the form of alphanumeric codes).

Because the telephone operators tended to handle a high number of calls falling into a small number of different categories, the task was fairly low-level in terms of skills and knowledge and was easily learnt and practised. It was expected that the new workstation would pay for itself quickly and save the company money by reducing the transaction time per call. The main time savers were expected to be the reduced number of keystrokes per transaction and the faster screen update rate. Average transaction time was expected to drop by 20%. The company had predicted that a reduction of 1 second per call would save $3 million per year.

Gray *et al.* carried out two assessments to compare the performance of the two workstations. The first was a modelling exercise using GOMS (goals, operators, methods and selection rules, see earlier) and the second was a field trial in which performance with the new station was recorded and compared with performance of the old workstation. The GOMS modelling was a simulation of the task and it revealed that most of the transactions carried out by the operators could be modelled using existing data on human–machine interaction. Some of the subgoals could be carried out wholly or partially in parallel, whereas others formed a critical path of sequential activities that were interdependent (one had to be completed before the next could be initiated).

The field trial was carried out for 4 months to allow subjects to become familiar with the new system. Surprisingly, in an average month, the average call took 6% longer on the new workstation (1.3 seconds longer). This difference was stable across different call categories and persisted after the subjects had learnt the new system. Referring to the GOMS model, it became clear that there were several reasons for the unexpected result and these all had to do with the nature of information flow during call handling. For example, the faster screen update time of the new system turned out *not* to be an advantage because, at that point in the transaction, operators only needed to see the first line of the new screen to proceed with the next subgoal. Thus, the faster screen update time had no benefit. In fact, the old screen displayed the first line of information more quickly than the new screen displayed a whole page. The new workstation required fewer keystrokes than the old one. However, the keystrokes that it did away with were *not* on the critical path – that is, other activities were critical to progress and there was no effect on transaction time by keying or not keying. In fact, the time taken to complete a transaction depended almost entirely on the time taken to complete a conversation with a caller and system response time. Few of the operators' interactions with the workstation were on the critical path. It made no difference that the new keyboard made better use of functional grouping to reduce travel time. In fact, the old keyboard had one keystroking action that saved time – the keystroke needed at the end of a call was on the left of the old keyboard, whereas the remaining keys were on the right. Thus, both hands could work in parallel to end a call quickly on the old keyboard, whereas the functional grouping of the new keyboard forced the right hand to work sequentially.

In practice, it was estimated that the increase in transaction time bought about by the adoption of the new system would have cost the company $2 million/year, not including the cost of purchasing the new system and training operators to use it and training others to maintain it. This study illustrates why it is so important to understand the nature of real tasks before implementing ergonomic recommendations. The new workstation delivered less than it promised because most of the benefits it bought occurred during 'slack' time in the information flow and a few of the changes increased the length of the critical path.

Advanced driver training

The occupational health service of the UK Post Office had a car fleet of 48 vehicles travelling an average of 13 500 miles/year. The ratio of insurance claims to vehicle numbers was 80%. The average accident claim cost was £2000 and, in a large proportion of claims the driver was to blame. The total direct claims cost was £75 000 in 1994/5. An intervention programme was initiated consisting of

- Raising driver awareness
- Providing advance driver training to high-risk drivers
- Providing support to drivers after an accident

Raising awareness consisted of feedback about fleet performance compared to other postal fleets and sharing information thought to be useful for prevention ('lessons learned'). Accident costs and a league table of performance were published on a 6-month basis.

Driver training consisted of a seminar highlighting accident risk and then a driving day spent in the company of an instructor who taught key road skills. At the end of the day, drivers were given a report highlighting their strengths and weaknesses and given simple 'games' to practise new driving techniques. When accidents did happen, drivers received support from their managers and were offered further training.

A steady reduction in accidents was observed in the occupational health fleet in the months following training. No reduction was observed in a control fleet that received no training, feedback or support. Later, the control fleet was also trained and a simular reduction in accidents occurred. Claims as a percentage of fleet size dropped from 80% to less than 40% and the average net cost per claim fell from £2000 to below £500. The number of blameworthy accidents dropped to below 20%. The total accident cost fell from over £70 000 per annum to below £10 000. The total cost of the training was less than £7000 (£125 per driver per day, with no opportunity cost as the training was carried out on days already set aside for health and safety training). A number of factors may underlie the effectiveness of the programme, including better perception of risk through increased awareness of the aggregate fleet accident loss (better feedback), better identification of risky situations when driving and improved recovery skills. The reduction in accidents for which the Post Office drivers were liable suggests a change in driving behaviour indicative of better risk perception (Boorman, 1999).

Research issues

Research continues into human error and how to prevent it, particularly in complex systems where the use of information technology is intensive.

Summary

People make mistakes all the time. The only way to deal sensibly with these mistakes is to view them in the context of the system in which they occur. Many mistakes can be predicted in advance and systems can be designed to block/negate or feedback errors to the user.

Essays and exercises

1. Observe users operating a vending machine. Time how long it takes them to perform the task and note any errors and difficulties. Try to relate the errors to the design of the machine and its dialogue.
2. Carry out a task analysis of the task of paying for a parking ticket using a machine in a multistorey car park. Use such a machine yourself and identify all its possible states. List all the syntactic and semantic errors that *could* occur on the basis of the task analysis. Observe people performing the task. Time how long the operations take and note any errors. Compare the actual distribution of errors with all the errors that could occur and relate any differences to the ergonomics of the machine.
3. Observe people carrying out a variety of everyday tasks. Compile a task difficulty index for each one. Ask people to rate the difficulty of each task and plot the subjective ratings against the task difficulty indices. Discuss your findings.

16 System design: organisational and social aspects

> We will make people before we make products.
>
> (Attributed to a Japanese car manufacturer)

Systems ergonomics is the term for the set of techniques used to apply the knowledge base of ergonomics to the design of systems. Successful design requires that the expertise of different specialists is carefully managed and utilised at appropriate stages in the design process so that ergonomics can be integrated with the engineering and personnel functions.

Systems design methods for ergonomics

According to the recently drafted ISO/DIS 6385 (Ergonomic Principles in the Design of Work Systems), the steps needed for an ergonomic approach to design are

- Formulation of the system's goals
- Analysis and allocation of function
- Design concept
- Detailed design
- Realisation, implementation and validation
- Evaluation

These are described below.

Formulation of goals and requirements analysis

Deciding what the system has to do is the first stage in design. It takes the form of a specification of the requirements. The focus is on what is needed, not how it is to be achieved. When existing systems are being upgraded, the requirements may already be known and only the parameter values may need to be changed (e.g. the new requirement is to deliver 100 tonnes of product per hour, instead of 50). With new systems, information about the performance requirements of the work process may need to be gathered, including information about the role of human operators. In both cases, design should be regarded as an opportunity to solve any existing ergonomics problems and to avoid generating new ones.

Table 16.1 Data capture methods for requirements specification

1. Unstructured interviews
2. Structured interviews
3. Unobtrusive observation of the current system
4. Brainstorming sessions
5. Questionnaire surveys
6. Market research

Specification of requirements   Two central issues at this stage concern the choice of techniques used to specify the requirements and the choice of people to sit on the design team. A common mistake is to leave requirements specification to technical managers, engineers or business experts and ignore the potential contributions of operators or users. This can be avoided using a participatory approach. The involvement of people closer to the actual work situation provides more information because it takes into account the informal structure and actual working arrangements of the organisation (how work is really carried out as opposed to how it is supposed to be carried out). Critical information regarding health and safety may be fed into the requirements at this stage.

Systems design is an open-ended process and open-ended methods can be used to gather information for requirements specification. Some examples are given in Table 16.1.

The goal of the requirements specification stage is not to describe how the system is to work so much as its purpose – what it must do. In the design of a new aeroplane or motor car, the requirements specification is not that the aeroplane flies or that the car's wheels stay on but the characteristics of the product with respect to a defined section of the market. For example, the number of passengers the aeroplane should carry, the routes it would operate, its performance characteristics and fuel consumption, etc. Thus, the requirements for an aeroplane would be designed through an analysis of the functioning of the parent system – the air transport system.

Analysis and allocation of function

A system's functions are the activities that enable it to achieve its goals. In the case of new systems, they may first have to be identified before they can be allocated to different operators or machines.

Identification and description of functions   Information on functions can be obtained in several ways. Panel discussions, interviews with users and experts and observation can provide function-related information. System design methods often stress that initially functions be identified and described in abstract terms. Some advantages of abstract functional thinking (according to Singleton, 1972) are that

1   It encourages the search for new methods.
2   It provides a common language across disciplines.
3   The range of solutions is not limited by particular components or methods.

For example, ignition timing is an essential function in the operation of a petrol engine. It can be implemented electronically or mechanically. Contact breakers are not a function, they are an example of a particular type of component that implements the function mechanically.

Huchingson (1981) suggests that functions be stated in terms of a verb, a noun and modifiers: e.g. 'Make a mid-course correction'. The use of noun–verb pairs to describe functions is thought to assist the design team in clearly identifying the important functions around which the search for creative design solutions can take place.

Allocation   Allocation of function between operators and machines determines the level of automation in a system. In practice, functions usually fall into one of several categories:

1. Those that *must* be carried out by machines (because it is impossible or unacceptable for humans to do them)
2. Those that must be carried out by humans (because no adequate machines are available or machine execution of function is not appropriate)
3. Those that might be carried out by either humans or machines or both.

According to Dearden *et al.* (2000), 'The goal of allocation of function is to design a system for which the performance is high; the tasks of the operator are achievable and appropriate to the operator's role; and the development of the system is technically and economically feasible.' When allocating functions, then, designers need to consider safety, reliability and psychosocial factors to arrive at a set of tasks for the operator that will be coherent and consistent and provide the basis for a satisfying job. Candidate functions for total automation can be selected on the basis of the feasibility and cost of developing automated means of implementing them. The relation of the function to the operator's role also needs to be considered. According to Dearden *et al.*, functions should be automated if they are separable from the operator's role and do not interact with it. Functions that are central to the operator's role, or provide information that critically affects the performance of the role, should not be automated.

Workload analysis is a critical part of the function allocation process. Dearden *et al.* recommend that scenario analysis be used to assess the effects of different allocation strategies on the operator's workload, awareness of the situation and performance (including safety and reliability). Under normal conditions of operation, a set of functions allocated to the operator may appear to provide optimal workloading and a coherent and compatible set of tasks. However, under foreseeable emergencies or hazardous states, the workload may be too high or the tasks incompatible. In such cases, partial or flexible allocation of function may be needed so that the operator can delegate control of a chosen function to the system for the duration of the problem. Adaptive aiding (Rouse, 1988) is an example of this.

Piecemeal allocation of several human functions to one operator can create potentially stressful jobs if there is a conflict between the requirements and responsibilities of the different functions. Role ambiguity is an example of a type of job stress in which workers are unclear about where their real responsibilities lie and what they are really rewarded for by the organisation (Sell, 1980). *Role conflict* occurs when the successful achievement of one function negates the achievement of another.

Function allocation decisions may differ over time, depending on the economic climate and availability of labour, and across cultures. In developed countries, many functions previously carried out by operators can now be automated, sometimes leading to the 'ironies of automation' described in Chapter 15.

Design concept

The translation of the requirements specification and the function allocation into a design concept may be facilitated using brainstroming techniques or structured methods of concept generation such as value engineering. The concept itself should be detailed enough to describe the structure of the work system and the interactions between the components, including the human–machine interactions described in Chapter 1. Those functions allocated to workers are organised into tasks, roles and jobs as part of the emerging design concept. The concept is reified, iteratively, using methods such as task description and analysis, the construction of scale models and mock-ups and discussions with users and consultants.

Detailed design

The key to modern systems engineering is that the human and machine components are designed in parallel to save development time. Traditional ergonomics acts as a link between these activities. It is not necessary for the system to be complete before training and job design can take place. The Apollo space programme is an example where all training and development took place off-line using simulators and prototypes.

It is during the detailed design phase that jobs are designed, either implicitly or explicitly. Singleton has described how workers have sometimes been thought of as the 'elastic glue' that holds the system together. The operator's function is simply to do all those things that the machines cannot do for themselves. Occupational psychologists, however, have proposed alternative ways of designing jobs. Job-centred approaches attempted to optimise the content of jobs so that they would be perceived as satisfying to do. Person-centred approaches are based on the idea that people are motivated to work well if the work satisfies their needs. Thus, it was thought that successful job design required some consideration of people's needs. The sociotechnical systems approach attempted to optimise the design of a social system of work to support the technical system. These approaches are described in more detail below.

Draft ISO 6385 suggests that detailed design should include all of the following:

- Design of work organisation
- Design of work tasks
- Design of jobs
- Design of the work environment
- Design of the work equipment, hardware and software
- Design of workspaces and workstations

Ergonomic principles for the design of the above are described in the preceding chapters.

Realisation, implementation and validation

Realisation involves the procurement and installation of the new system on site. Implementation involves management of the changeover from the old system and introduction and training of operators. If necessary, a temporary back-up system may be installed or there may be a handing-over period in which both systems run in parallel. Validation requires that the system be shown to function according to the requirements specified earlier on – that it achieves its goals as intended.

Evaluation

Evaluation occurs after the system is up and running. The integration and implementation phase is often characterised by 'teething troubles', which are temporary phenomena, not characteristic of the final design. It is for this reason that in activities such as facilities design, a delay is introduced between implementation and final evaluation. This enables teething troubles to be overcome so that true pros and cons of the facility can be more easily identified. For example, post-occupancy evaluations of new buildings are usually only carried out after one year of occupation.

Evaluation seeks to determine how well the system achieves its goals in practice. Many criteria can be used in evaluation and these are well summarised by simple human–machine models such as those given in Chapter 1. For example, evaluations of noise, lighting, temperature, the health and safety of the workforce including the prevalence of any medical conditions and the occurrence of errors and accidents may all be carried out at this stage.

Quality control and quality assurance

A system is of good quality if it meets the standards that were set at the requirements specification stage. A high-quality system is simply one that meets high standards. There is a growing body of standards, such as those of the ISO, quoted throughout this book, that can be used to design systems with high-quality ergonomics. Quality control takes place throughout the design stage to ensure that all relevant standards are being used as the design progresses. Once the system has been built and is functioning, assessments are made to check compliance. If the system does comply with the ergonomic standards, then we are assured that it is of the appropriate ergonomic quality. Quality control is not possible without reference to some kind of standard.

Organisational aspects: some macroergonomic examples

Macroergonomics is the term used to describe ergonomics at the level of the organisation (most of this book deals with 'microergonomics' – ergonomics at the human–machine interface). Macroergonomic considerations can have a big impact on how ergonomics is applied in practice and can dictate the key issues in a design. For example, demographic change in industrialised countries is driving a renewed interest in ageing and design for the older worker. It is also driving the development of 'lean-manned' systems with fewer workers and more automation as the size of the workforce shrinks.

Sociotechnical systems theory

Most human–machine systems are 'sociotechnical systems'. There is a technical sub-system (which integrates the machines themselves and the interconnections between them, the infrastructure and power supply, etc.) and a social subsystem (which integrates the activities of the people involved in operating and maintaining the machines and the relationships between them). Work organisation is the link between machines and the social organisation of the individuals who operate them. The optimal utilisation of technology depends on an appropriate system of work organisation that itself determines the social organisation of the workforce and the relations and inter-dependencies between individuals. For example, in a classical production-line system of manufacture, workers form a homogeneous group in terms of status and skill. The relationship between them is one of linear dependency in one direction (the direction of the production line). Control is exercised by supervisors and managers. This can be contrasted with a patient care team in a hospital. Here, the members of the team differ in status; individuals perform different functions, have different expertise and are mutually interdependent. Individuals can use their own initiative about matters falling within their own area of expertise and can contribute to decision making at the group level.

Sociotechnical systems theory was developed by members of the Tavistock Insti-tute of Human Relations in the years following the Second World War. Cherns (1976) has summarised the design principles derived from sociotechnical systems theory (Table 16.2).

Legislative aspects and management of ergonomics at work

Ever since the nineteenth century, government bodies in the developed nations have attempted, for social as well as economic reasons, to influence the way business is conducted. This trend has continued and firms now have to comply with regulations that limit workers' exposure to the health-threatening aspects of their jobs.

OSHA: ergonomic safety and health management rules

In the USA, OSHA (the Occupational Safety and Health Administration), which monitors about 6 million workplaces with about 93 million workers, has attempted to implement standards on workplace ergonomics in response to the rapid increase in claims for workers' compensation for incapacitating musculoskeletal injury. According to the Bureau of Labor Statistics, the number of disorders associated with repeated trauma has more than tripled since 1984. The increase is apparently due to changes in production processes and technologies, increased overtime and piece-work and a lack of integration of ergonomics into production processes. Many tasks are now more specialised, assembly line speeds are higher and cycles times lower (Department of Labor, 1992).

OSHA's approach to ergonomic safety and health management aims to 'prevent, eliminate and reduce ergonomic hazards in the workplace'. The ergonomic hazards and disorders covered appear mainly to be of a musculoskeletal nature (other ergo-nomic hazards such as noise are already regulated; less tangible hazards relating to mental stress, error, information processing, etc. are probably outside the scope or are covered by other agencies).

Table 16.2 Principles of sociotechnical job design[a]

1. *Compatibility.* The way sociotechnical design is carried out must be compatible with the changes that are made. For example, if the one of the reasons for change is to make the most of the cognitive skills of the workforce, these skills must be utilised in the process of change itself.
2. *Minimal critical specification.* In the design of new systems, no more should be specified than is absolutely essential. This leaves individuals with freedom to determine the precise details of how a task is to be carried out. It also leaves open options for change and improvement and enhances adaptability to unanticipated events. Strictly specified rules and procedures can inhibit an organisation's ability to adapt.
3. *The sociotechnical criterion.* This states that variance in a system (the occurrence of unspecified and unprogrammed events) should be reduced by controlling it as close to its source as possible. Much of the system of supervision, inspection and maintenance in industry is an attempt to reduce variance from a distance – to correct the consequences of variance rather than control it. If operators carry out their own inspection, the size of the variance control 'feedback loop' is reduced.
4. *The multifunction principle.* This is aimed at increasing the adaptability of the organisation by allowing workers to fulfil more than one role, but designing jobs so as to reduce the interchangeability of people.
5. *Boundary location.* In all organisations, boundaries between different departments have to be drawn up. This is usually on the basis of function, technology, territory or time. For example, engineering products often pass through several different departments such as milling or grinding shops in the manufacturing process. Of the total time taken to manufacture a product, only a portion is spent with the item in contact with machines. The rest is taken up by transport, storage, etc. An alternative is the 'group' or 'unit' production method in which each department makes a complete product. This approach has been tried in automobile assembly, for example.
6. *Information flow.* This principle states that information should go where it is needed in the first instance rather than via senior management to subordinates. The tendency for information to 'filter down' from above can have disadvantages as well as being inefficient. Managers may become preoccupied with matters for which their subordinates should be responsible. An information system designed according to sociotechnical principles should direct information efficiently to those parts of the organisation where it is needed and acted upon. It should also support two-way communication.
7. *Support congruence.* Administrative and management systems should be designed to reinforce those behaviours that the organisation wishes to encourage. If an aim is to encourage employees to take a more responsible attitude top their jobs, then supervision and payment systems should be designed to be congruent with this.
8. *Design and human values.* This principle emphasises that systems must be designed to provide high-quality jobs. This is a difficult principle to apply – individuals may respond differently to changes in job design and have different needs.
9. *Incompletion.* Design never ends.

[a] After Cherns (1976).

In addition to general guidelines, OSHA has attempted to to implement specific guidelines for high-risk industries such as construction, where there is a high risk of falls, and meat packing, where there is a high level of musculoskeletal stress. The general thrust is to reduce the prevalence of avoidable ill-health by providing employers with specific ergonomic guidelines with which they must comply. Those failing to comply with the guidelines are liable to prosecution.

In 1992, OSHA carried out over 42 000 inspections covering more than 2 million employees and imposed over $116 million worth of fines (National Safety Council,

1993). These data refer to a wide range of industries and violations of all OSHA health and safety rules. In January 1987, OSHA inspected Iowa Beef Producers and fined them $2.5 million for record-keeping violations, many of which were work-related musculoskeletal disorders (WMSDs) of the upper body. John Morel and Co. were fined $4.3 million in April 1988 for ergonomic and record-keeping violations (Brogmus *et al.*, 1996).

The requirement for good working conditions is not a new one. The UK Occupational Safety and Health Act (1970) requires all employers to 'provide their employees with a workplace free from recognised serious hazards' irrespective of whether these hazards are covered by specific standards. If poor ergonomics constitutes a hazard, then employers are required to act. In the last decade, OSHA has attempted to draft standards that specify what constitutes an 'ergonomic hazard' and what action to take to remove the hazard. Importantly, these standards offer guidance on *how to implement these efforts at an organisational level.*

OSHA's Ergonomics Program Management Guide for Meatpacking Plants (OSHA 3123, 1993) is a good example of how to implement ergonomics in large organisation, who to involve and what practices and procedures to put in place. The guide takes a top-down approach to the management of ergonomic hazards, beginning with the need to get the support of top management and to involve employees (Table 16.3). Health and safety is given a high priority and key people are made responsible for appropriate parts of the programme and given the resources and authority to intervene. Everyone knows what is expected them, who they are accountable to and what resources they have to carry out their responsibilities. Deadlines for the attainment of objectives are set and a written plan is made available to all concerned.

Employees are involved in the monitoring of injuries, the investigation of problem areas and the recommendation of solutions. A committee is established to receive information on problem areas and empowered to act on this information. The programme is reviewed regularly and a report is issued detailing current progress and any revisions of goals.

Worksite analysis begins with an analysis of accident and sickness records to evaluate the severity of any existing ergonomic problems and prioritise areas for intervention. Worksite evaluations utilising checklists to characterise tasks in terms of force, posture and repetition are then carried out. Worksite analysis is carried out on an ongoing basis and repeated at least annually. Hazard prevention utilises a combination of engineering, work practice and work organisation controls as described in previous chapters. Adequate resources for the medical management of patients must be provided. The emphasis is placed on prevention, early diagnosis, conservative treatment (rather than surgical intervention, if possible) and on-the-job rehabilitation. In the case of back injury, it has been shown (Lusted, 1993) that the probability of an injured employee returning to work is greater the earlier after the injury the patient is referred for rehabilitation. Early referral for rehabilitation is more likely under a well-integrated system of medical management. Finally, employers must provide workers with education and training by qualified persons. The training must be geared towards the needs of particular groups of workers and must include a general component (about ergonomic disorders, their cause, recognition and prevention) as well as job-specific training (e.g. workers who use knives would receive training on the correct way to hold and cut with knives and proper knife care to maintain sharpness).

Table 16.3   Key features of OSHA's programme guidelines for meatpacking plants[a]

Management Commitment and Employee Involvement	
Commitment by top management	*Written programme*
Eliminating hazards given top priority	Outlines corporate goals and plans
Health and safety as important as production	Advocated at the highest level
Delegation of accountability and responsibility	Communicated to all employees
Provision of resources and authority	Objectives set out with deadlines
Employee involvement	*Review and evaluation*
Complaint/suggestion procedure implemented to provide feedback	Analysis of trends in injury/illness
Prompt reporting of disorders encouraged	Employee surveys
Health and Safety Committee established	Pre- and post-change comparisons
Ergonomic intervention teams established	Record of job improvements implemented

Programme Elements	
Worksite analysis	*Hazard prevention and control*
Examination of injury and illness records	Engineering controls
Identification of problem areas and trends	• Workspaces
Visit worksites to identify risks	• Work methods
Follow-up when operations change	• Tools
	Workpractices
	• Work techniques
	• Equipment maintenance
	• Employee conditioning
	• Monitoring
	Personal protective equipment
	• Proper fit
	• Protection against cold
	Administrative controls
	• Increase cycle time
	• Provide rest pauses
	• Use more people
	• Rotate jobs
	• Train 'standby' workers
	• Use preventive maintenance
Medical management	*Training and education*
Injury and illness recording	General training
Early recognition and reporting	Job-specific training
Systematic evaluation and referral	Training of supervisors
Conservative treatment	Training for managers
Rehabilitation	Training for engineers and maintenance personnel
Adequate staffing and facilities	

[a] From OSHA 3123 (1993).

As can be seen, the OHSA approach to ergonomics operates at all key levels of the work system and includes technical, medical, cognitive, environmental and organisational components.

Legislation in the European Union

Six sets of directives required each member state to introduce its own legislation covering the following:

- Management of health and safety
- Manual handling of loads
- Work with display screen equipment (i.e. VDUs)
- Use of personal protective equipment
- Use of work equipment
- Workplace design (health, safety and welfare)

The regulations concerning management of health and safety are particularly interesting in that the onus of responsibility for health and safety is placed on the employer, who is required to

- Carry out risk assessments
- Make any arrangements necessary for health and safety
- Provide health surveillance
- Develop proper procedures for dealing with emergencies
- Provide employees with specific information about health and safety at their work
- Assess employee capacity for work
- Provide training
- Use competent people

Legislation of this nature has enormous ramifications for the way work is carried out and for employers' and employees' expectations. For example, under the directives, employees using VDUs on a regular basis as part of their jobs become entitled to eye and eyesight tests and corrective appliances. This is a radical departure from the traditional expectation of employers that it is the responsibility of the employee to ensure his own capability for work in terms of physical characteristics (Health and Safety Commission, 1992).

Organisational aspects of safety

Further evidence of the need for a systems approach comes from a study of industrial accidents by Dwyer and Raftery (1991). Traditionally, accidents have been characterised at the level of the human–machine interface and concepts such as the 'accident-proneness' of workers or unsafe working conditions have been used to explain why an accident happened. However, Dwyer and Raftery point out that organisational factors have been shown to be related to accident frequency at one time or another. Examples includes:

1 Working on a piece rate basis
2 Undernutrition of employees
3 Extended work hours
4 Absence of integration of different work groups

In their own investigation of industrial accidents, these authors concluded that accidents could be prevented by a system of work in which workers exercised greater *auto-control* of their activities and by management if, in the absence of conditions favourable to auto-control, a proper safety management programme was initiated.

Modern approaches to safety can be contrasted with traditional approaches. Traditional approaches regarded safety as the responsibility of the individual. Safety propoganda was directed at workers in an attempt to change their attitudes towards a safety issue, such as the wearing of personal protective equipment. Research in the area of social psychology highlights the weakness of approaches based on attitude change. Firstly, attitudes often bear only a tenuous link to behaviour – many smokers, for example, know that smoking is undesirable, unhealthy, a waste of money and antisocial when carried out in public, but they still smoke. People's attitudes are usually 'general' and 'principled' and are a projection of core beliefs onto an idealised view of the world. Improved safety, however, depends on specific behaviours taking place in specific situations. Modern approaches to safety focus on the establishment of safe systems rather than on the individual. These systems focus on specific behaviours rather than attitudes – safe behaviours are made explicit, monitored and fed back to individuals.

Incentive schemes for health and safety promotion

Incentive schemes for safety and health make use of behaviour modification techniques. Usually, the approach is to reward desired behaivours (positive reinforcement). Undesirable behaviours are usually ignored rather than punished (negative reinforcement) to avoid causing anxiety, denial and maladaptive coping strategies (such as non-reporting of accidents and near misses). Incentives for safety are usually accompanied by feedback – employees are kept informed of accident rates or other variables that reflect good practice. The incentives can take many forms – a cash bonus, awarding of points or tokens such as trading stamps that can be redeemed for benefits later – or they may include a social component as in 'employee of the month' awards. McAfee and Winn (1989) reviewed reports of 24 evaluations of safety incentive schemes. All of the studies brought about short-term improvements in measures associated with safety such as the percentage of employees following safety procedures, the number of accidents, etc. There was evidence that these schemes do not rely merely on their novelty and that they can be kept in place for many years.

Zohar and Fussfeld (1981) used a token economy system to increase ear plug usage among workers in a textile factory where noise levels were around 106 dB(A). Tours of the factory were conducted three times per week and each worker wearing ear plugs was issued with one token, on each occasion. Tokens could be exchanged for consumer products varying in token 'value'. During the period in which the token system was designed (using a participatory approach), the baseline measure of ear plug usage was 35%. Following introduction of the token system, usage rose to 90%. There was no change in usage in a control department where no token economy system was operating. The new level was found to be stable, at around 90%, 9 months after termination of the token economy system. It seems that ear plug use is inherently reinforcing after workers become accustomed to the plugs. The tokens provide an incentive that enables workers to get beyond the initial adjustment phase. Once they are comfortable wearing the plugs, plug use is self-reinforcing (the plugs reduce the noise to a comfortable level). In the case of behaviours that are not self-reinforcing (when the link between the behaviour and the outcome is more tenuous), explicit feedback and continuous incentivisation may be needed.

Organisational aspects of the implementation of new technology: the productivity paradox of IT

Productivity refers to a company's ability to produce goods or services, the amount produced in relation to the time taken and the economic resources consumed during production. There is a growing literature on the supposed 'productivity paradox' of information technology (IT). Despite 30 years of investment in information technology in the white-collar sector, productivity gains up to the year 2000 have been modest. This is in stark contrast to the gains achieved in manufacturing industry through automation.

New technologies take time to mature   One of the most compelling arguments for IT investment is that the technology is still immature and that its architecture has not yet stabilised in a form that fully exploits the potential. In time, the real benefits of IT will emerge and they will emerge rapidly. History suggests that new technologies do take time to develop before their true potential can be exploited. When they are first adopted, they are implemented using design concepts (or 'architectures') appropriate to a previous technology. For example, the ancient Egyptians, having mastered stone cutting and with a large workforce and a formal, written arithmetical system, were able to build pyramids. These impressive structures are essentially hills made of stone. They are highly stable but are also highly inefficient (in the Great Pyramid, over 2 million blocks of stone weighing 10–15 tonnes each were needed to enclose the Pharaoh's tomb).

Early steam engines resembled ox wagons with a steam engine attached to them and early cars did, indeed, resemble 'horseless carriages'. As designers became accustomed to working with a new technology, new design ideas emerged that enable the true potential of the technology to materialise. After an initial period of experimentation, the new designs undergo 'architectural stabilisation', designers stop experimenting with new system architectures and the design effort is focused on optimising the new, standard design.

There are parallels with the development of information technology. Hopwood (1983) described how a motor manufacturer upgraded its production system using automated control processes and robotics only to find that the new system had built into it all of the rigidities of the old production line system. A competitor used the same technologies to redesign its production system in a more fundamental way. The improvement in flexibility meant that it was better placed to deal with changing market demands.

Constraints on IT productivity   New systems may introduce extra tasks in the process of automating an old one. These secondary tasks can consume human and material resources to negate the benefits of new technologies. By focusing too much on what it is possible to do with the new technology, it is easy to overlook the administrative costs and other overheads associated with owning, operating and maintaining the new system. Some of the biggest costs of new systems occur during development and implementation phases. These costs can be recouped if the system is owned and used for years or if it brings large increases in productivity. According to W. Gibbs of *Scientfic American* magazine, US companies spent $213 billion on computer hardware alone in 1996 – more than they spent on any other type of equipment. Much of this went on replacing obsolete information technology rather than on new systems. Compared with machine tools, buildings and other forms of capital expenditure, constant innovation and rapid obsolescence of systems may prevent IT from becoming a long-term investment for companies.

Productivity paradoxes   There are several explanations for the 'productivity paradox' of IT and these are explored below.

- *Quality.* The real benefit of computerised systems at the higher end of the value chain is an increase in service quality and convenience, rather than productivity. ATMs permit access to the banking system 24 hours per day, 365 days per year. Productivity is hard to measure and perhaps traditional methods of productivity measurement are inappropriate.
- *Lack of technology.* Most companies do not possess sufficient computer resources to make a real difference to overall productivity.
- *Time lags.* The first electric dynamos were installed in 1881, but it took 40 years for the technology to become used widely and even longer before companies organised their production lines to optimise their use of electricity. Similarly, more time may be needed before the productivity improvements of IT really become apparent.
- *Inappropriate expectations for productivity enhancement.* High productivity growth during the 1960s may have been due to factors unrelated to technology, such as changing demographics due to the post-war baby boom resulting in a large number of well-educated young people entering the labour market.
- There is evidence that the computerisation of administrative functions in large organisations has led to an increase in the number of administrative staff rather than a reduction, and that information-processing functions can take more as opposed to less time. In 1968, according Paul Strassman of Method Software, US hospitals employed 435 000 administrative workers to serve 1.4 million patients. By 1992, 1.2 million workers handled 835 000 patients. We may speculate that as ever more computer memory and processing power becomes available, the complexity of administrative systems expands to consume it.

Although personal computers themselves are inexpensive, they must be linked to a network and installed with appropriate software and they require a network administrator and technicians to maintain the system, so the unit costs are several times higher. Non-productive uses of IT include

- Waiting for programs to run or for help to arrive.
- Double checking print-outs for accuracy.
- Rearranging files.
- Computer games.
- *The Internet.* Despite its popularity, the Internet itself consumes time because of its large response time and response-time variability, even when used for productive work tasks.
- *Overemphasis on quality.* Many systems come with packages of software including games, Internet browsers, 'paintbrush' artist toolboxes and software for making professional presentations. These packages may be used for mundane tasks such that the enhanced quality is excessive in relation to the material being presented and the time taken to develop it.
- Employees coping with ever more complex programs. Much modern interactive software is 'generic' rather than specific. New packages contain ever more features, many of which will only ever be used by a small percentage of users. There is lack of a good mapping between the software interface and the user's work task. Generic software interfaces may not map well onto specific jobs, causing unanticipated rigidities and constraints.
- User attention becomes focused on the software (the tool) instead of the business (the work task). Inappropriate system complexity wastes human mental resources as more time is spent mastering the technology than achieving work targets.

Organisational braking factors on IT productivity   Hopwood (1983) described some of the organisational factors that impede IT productivity enhancement. Decision makers may carry out cost-benefit analyses of proposed systems that focus too strongly on costs and cost savings rather than benefits such as increasing the organisation's responsiveness to market demands. This may happen simply because costs and cost savings (e.g. automating a function and reducing the workforce) are easier to quantify than more subtle changes introduced by new systems. For example, a motor manufacturer implemented an inventory recording system in its dealer network to improve management of its inventory. An unexpected consequence was that dealers used the improved access to the information to reduce their own stocks. The manufacturer then had to increase its own stocks at extra cost. Hopwood suggests that information systems can make organisations less flexible and less able to adapt to external change. In white-collar organisations, automation of the information system may institutionalise information management and foster an inward-looking perspective at the expense of the more externally focused marketing and strategic planning functions.

All organisations have an informal as well as a formal structure and much important work may take place using informal channels. Computerised information systems that ignore these processes may render the organisation less able to deal with change. For example, a British company that survived difficult market conditions found that the key to its survival had been the close physical proximity of key employees, which facilitated informal communication. Information could permeate quickly through the organisation because of this. Plans to relocate 66% of head office staff to a suburban site were quickly shelved.

Incremental alternatives to organisational development   Sundstrom (1986) pointed out that the time taken for organisations to modify their infrastructure to take advantage of new technologies introduces enormous lags. Electric lights were commercially available around 1880 but did not become common in offices and factories until the 1930s. This was due to convergence with the development of electric air conditioning. Air conditioning made new building designs possible, but, with workers seated far from windows, electric illumination became mandatory. The tasks carried out in these electrified buildings were essentially unchanged by the technology and the same may be said for many of the tasks carried out in electronic offices today. Perhaps new work roles and new organisational structures are needed to fully exploit the benefits of information technology.

When open-plan offices were introduced in the 1960s they were supposed to herald the arrival of a new 'democratisation' of work, but the evidence suggests that, despite the advantages of better space utilisation and control of the office environment, organisations themselves changed little. In fact, open plan may facilitate authoritarian rather than democratic styles of management because of the way it facilitates supervision.

Sundstrom argues that new technologies have an evolutionary rather than a revolutionary effect and that there is a tendency to overestimate the pace of change. In fact, we might question whether information technology is a force for change in organisations or whether, in fact, organisations can ever change fast enough to keep up with technological developments. Many companies, far from being 'computerised organisations' merely use computers for data processing in the same way as they use electricity for illumination and air conditioning. The productivity paradox of IT according to this argument is nothing more than a reflection of organisations neglecting to use the potential of IT to restructure themselves more efficiently.

Finally, Norman has suggested that the way forward for information technology is via the development of simple tools ('information appliances') to help people carry out specific functions better. These tools would work in a stand-alone manner – presumably like washing machines, vacuum cleaners and telephones – but could be linked into larger networks when necessary.

Successful interventions at the macro-level: case study on containerisation

Macroergonomic interventions seek to bring improvements by redesigning the organisation of work, rather than by redesigning user-interfaces or workspaces. They can bring enormous benefits in terms of both cost and safety. The mechanisation of the transport of goods is an example.

The world economy has been expanding at over 3% per year, yet the volume of goods traded internationally has expanded at over double this rate. Many different factors underlie this trend of increasing world trade. Deregulation of markets, world trade agreements and other factors such as the emergence of regional trading communities are obviously important. One important factor, which is often overlooked, is the way traded goods are transported.

Freight costs have dropped significantly over the years. Thanks largely to the introduction of containerisation, many of the difficulties as well as the high costs of shipping goods from one country to another have disappeared. World trade is now founded on a *standardised system* for handling goods. Irrespective of the value or nature of the commodity traded, it is shipped in standard containers and handled with the same machines all over the world.

Early and modern trading patterns (from *The Economist*, Nov 15, 1997, p. 123)   At the turn of the century, most international trade consisted of low-value-added products. Commodities such as wheat, coal, iron ore and wood formed the bulk of what was traded. US exports, 41% were 'raw' (either materials or food). These goods are either bulky or heavy, or both, so the transport costs are high in relation to the unit price per tonne of material traded.

Nowadays, much of world trade consists of goods that have high added-value. This means that their value is unrelated to their size and weight (e.g. compact discs, disk drives, roses and wine) so the freight costs are low in relation to the value of the goods themselves. Thus, there is little price advantage in buying locally manufactured as opposed to imported goods unless governments impose import tariffs or take other anti-competitive measures to protect local industries. Some goods can be shipped by air, such is their value in relation to their weight or size. Software can even be shipped electronically.

Cargo handling productivity before mechanisation   Prior to the introduction of containerisation, goods were handled in a complex and laborious way from their factory of origin to their destination in another country. For example, a vineyard in South Africa would have to packs its wine into crates, then load manually onto a truck, which would deliver them to the docks or to the railway station (where the goods would be loaded onto the train). At the docks, the goods would be offloaded, case by case, and stored in a bonded warehouse. Dockyard workers would then haul the cases into the ship's hold where they would be stacked in place. At the other end, the process would be reversed and there might be further stages – transporting the goods to a city by train, offloading onto trucks for transport to a warehouse and finally distribution to retail outlets.

The system was very labour intensive, consisting of many different stages at which workers could either injure themselves or damage or steal the goods. At each stage, delays were possible and an extra layer of management was needed to coordinate transportation and the workers.

Containerisation   In 1955, Malcom McLean, owner of a transport business in the USA, came up with the idea of loading the entire back section of his trucks, including the wheels, onto the ship, instead of unloading the cargo. The cab would then drive back to the depot and another empty cab would meet the load at the destination. This system eliminated the need for dockworkers to handle the cargo itself. Nowadays, container ships carry thousands of containers that are handled at each end by special cranes and fork-lift trucks. The use of containers kills many birds with one stone – it is no longer necessary to build wooden crates to protect fragile items, the containers can be locked and provide secure storage both on deck and at the quayside, fewer workers are required to move them, and a truck delivering a container at the dockyard can collect another to transport back, irrespective of what load is in the container. Similarly, transporting goods using a combination of transport modalities along the way (train, ship, truck, etc.) is greatly simplified because of the speed and ease of transferring a container from one modality to the other.

The container revolution caused a massive reduction in transport costs. Although manual handling is still required to load and offload the containers, the total amount of handling needed to transport goods from a factory in one country to a retail outlet in another has been greatly reduced.

Shiftwork

It has long been known that many physiological processes exhibit a 24-hour or circadian rhythm. Body temperature is a representative example. It is low in the early morning, rises throughout the day before peaking in the later afternoon, and becomes lower again at night. Just after midday, many physiological variables, which have been rising steadily since early morning, suddenly dip to a lower level. This is known as the 'post-lunch dip'. Subjectively, we often feel overcome by feelings of lassitude in the early afternoon, which can be mistakenly attributed to lunchtime overindulgence. In fact, the post-lunch dip occurs even if lunch is missed.

Research on circadian rhythms Experimental subjects have been put into bunkers in which all time-of-day cues have been removed as well as social contact. Under these circumstances, body rhythms change to a slightly longer cycle. Physiological rhythms are thought to be genetically programmed but become entrained to the environment by external cues ('zeitgeber'). People living on the equator, where there is a 12-hour day/night cycle throughout the year, have pronounced circadian rhythms. Social cues are also important in entraining circadian rhythms in humans, probably more so than light. People living in the Arctic under total daylight in summer and total night in winter also have identifiable rhythms due to the powerful effect of social cues.

There is evidence that task performance varies over the course of the day – errors and accidents in industry are more common at certain times than others. Time-of-day effects have been found in the speed of reading gas meters or answering a switchboard, the frequency of car drivers falling asleep behind the wheel and of train drivers missing warning signals. These effects roughly track body rhythms – they are more common at night and during the post-lunch dip. Problems can occur when workers adapted to a particular cycle of light/dark and sleep/wakefulness have to adapt to a new cycle, as when flying to a different time zone and particularly when the difference is greater than 4 hours.

Factors influencing adaptation to shiftwork Social factors may inhibit physiological adaptation to shiftwork: many individuals prefer to revert to the normal cycle on rest days and at weekends. Incomplete adaptation to shiftwork may be a cause of increased sickness absence. Colligan *et al.* (1979) examined the records of 1219 nurses on permanent day, afternoon, night and rotating shifts. They found that those on permanent shifts had a lower rate of clinic visits due to ill-health than those on rotating shifts. Although sickness absenteeism was similar for the two groups, those on rotating shifts had more sick days for serious illnesses – that is, rotators were in poorer health. Smith *et al.* (1982) compared day workers with night workers and people on rotating shifts. The latter two groups had poorer sleep, altered eating habits, greater alcohol consumption, greater sickness absence and more on-the-job injuries. It appears from the work of Smith *et al.* that working a permanent day shift is the least stressful and working a rotating shift system is the most stressful.

Some considerations in the design of shift systems One solution to the shiftwork design problem may be to identify individuals who prefer permanent night work, thereby avoiding the need to rotate workers between the night and day shifts and thus reduce the need for adaptation. Some individuals seem to prefer night work and choose jobs that can only be done at night (such as night watchman), whereas others may have a clinical intolerance to shiftwork (Motohashi, 1992). There is still considerable debate about how best to design shift systems to minimise unpleasant health and social effects. Wilkinson (1992) reviewed the literature on the rotation of shift systems and concluded that fixed night shift systems were superior on most counts. For 24-hour operation of a work system, the permanent night shift can be supplemented by two rapidly rotating day shifts. Wedderburn (1992) disagreed with this conclusion and proposed the opposite view – that rapidly rotating shifts provide the best compromise solution, providing for both the physical and social needs of workers. Finally, Folkard (1992) has argued that Wilkinson probably overestimated the problems associated with rapidly rotating shift systems and underestimated the problems associated with implementing permanent night shift systems. Folkard has concluded that there is no such thing as a good shift system and that in designing any such system there is a trade-off between accommodating workers' social needs (using a rapidly rotating system) and health and safety considerations (with workers adapted to permanent night work). Although there is still controversy about how best to design a shift system for a particular industry, the literature suggests the following.

1   Adaptation to a new shift system takes place over at least a week and must be allowed for. Workers may never adapt properly to rapidly rotating systems.
2   Where possible, minimise the change in working hours. If it is 4 hours or less, workers will be able to accommodate to the change with few problems.
3   If a voluntary, permanent night shift cannot be found, a rapidly rotating three-shift system may be the best compromise. It is pointless to try to adapt workers to night work if they revert to daytime wakefulness in their time off and a rapidly rotating system may cause less disruption to circadian rhythms under these circumstances. Folkard (1987) suggests rotating shifts in the order morning–evening–night–morning, etc.
4   The permanent night shift with rapidly rotating day shifts may be the best compromise if enough people can be found to work the permanent night shift. Workers on the rotating day shift will be able to accommodate the change since their hours of sleep and wakefulness are unlikely to be disrupted by more than 4 hours.
5   Permanent night shift systems may not work, even with a willing workforce, if night workers live in noisy inner city areas or in crowded shanty towns in Third World countries. Disturbances of daytime sleep may result in chronic sleep deprivation of night workers that cancels out any health or performance benefits of their being adapted to the night shift.
6   Shift designers must also consider the social customs and circumstances of their employees in order to arrive at the best compromise system. If night workers do not have personal means of transport, special arrangements may have to be made. Social, recreational and entertainment facilities geared to the needs of night workers may assist them in maintaining their adaptation to night work during their free time.
7   It is sometimes suggested that the working week be compressed from five, 8-hour days out of every seven to two, 12-hour days out of every four. Rosa and Bonnet (1993) investigated performance and alertness on 8-hour and 12-hour rotating shifts. Performance and alertness decrements were found on the 12-hour system

as were reductions in sleep. Greater fatigue was apparent over the course of the 12-hour shift than on the 8-hour shift and the investigators suggested that critical activities be scheduled for the early part of 12-hour shifts, particularly at night. However, not all investigations have found there to be more problems with 12-hour rather than 8-hour shifts and the former seem to be popular among workers because of the increased number of days off. Williamson *et al.* (1994) evaluated a 12-hour shift system that involved two 12-hour day shifts followed by two 12-hour night shifts followed by 4 days off work. The workers were computer operators and the 12-hour system replaced an irregular 8-hour system. Data on mood state, sleeping and eating patterns, work quality, productivity, staff turnover and sickness were collected when the 8-hour system was in place and when the 12-hour system had been worked for 7 months. The 12-hour system produced positive changes in mental state and physical symptoms and improved sleep quality and quantity. There were no significant effects on absenteeism, staff turnover or productivity. The 12-hour system may be appropriate in certain jobs, particularly when arranged so as to provide 4 days off for every 4 worked.

Rosa and Bonnet caution that, although workers may gladly tolerate increased fatigue in exchange for more free time, further research into both work-related and non-work-related accidents and injuries as a function of extended workshifts is needed as is illustrated by a newspaper (see box):

Death Crash Driver Fined £200

A motorist who killed two teenage boys after falling asleep at the wheel of his car has been fined £200.

The motorist had been working 12-hour shifts at a factory before his car mounted a kerb and hit Gareth Horton, 14 and Kevin Parsons, 17.

But he was charged only with driving without due care and attention because he was not speeding.

Andrew Hancocks, chairman of the bench said: 'We are here today to adjudicate on a charge of driving without due care and attention. Our penalties are based only on the charge before us'.

(*Weekly Telegraph*, 14–20 November 1995)

Psychosocial factors

Much of the emphasis of ergonomics is on human capabilities and limitations and the need to design systems that take these into account. Although this is clearly necessary, it can be debated whether it is sufficient to optimise the design and performance of human–machine systems. Moores (1972) commented that one of the similarities between ergonomics and work study was the tendency to ignore the behaviour patterns of workers either as individuals or groups. Definitions of ergonomics of the fitting the job to the worker variety were condemned by Moores for being too mechanistic, conjuring up images of matching at the level of an animal rather than a human being. Sell (1980) has taken a similar line and suggests that a major problem with ergonomics is that it neglects to take account of a person's social and psychological needs. Ergonomic solutions can remove barriers to effective productivity but cannot compel people to work productively or safely (or to find work a satisfying experience). These factors depend, to some extent at least, on the level of motivation and motivation depends on higher level factors such as work organisation.

Motivation

Maslow produced one of the first theories of motivation to be applied to the investigation of work situations. According to Maslow (1950), motivation is the driving force that directs behaviour – the mechanism for the reduction of needs. For an understanding of what motivates people to work, a consideration of their needs is essential. Maslow postulated five classes of needs, which he arranged hierarchically. Low-level needs include the basic physiological requirements – food, oxygen, etc. Next came the need for security and avoidance of danger, followed by social needs – companionship, social interaction, etc. A fourth need was for self-respect and the esteem of others including social feedback about performance. At the top of the hierarchy was the need for 'self-actualisation', which can be seen as an ideal – how we would like to see ourselves in some ideal world. A key point of Maslow's theory is that higher-level needs do not come into play until lower level needs are satisfied (e.g. there is no point in giving a hungry person a bunch of flowers). Although the theory can be questioned, it provides the ergonomist with a useful framework for understanding employees' emotional responses to job design.

Job enlargement and job enrichment

Several early investigators made the observation that many industrial jobs consisted solely of simple, repetitive tasks and considered this to be unsatisfactory for a variety of reasons (these jobs waste human potential, lead to boredom and cause physical and mental fatigue). Attempts were made to make industrial jobs more interesting. Job enlargement is an example of such an approach. According to Barnes (1963):

> The job should be redesigned so that it consists of a complete piece of work with at least one unit or component being completed by the operator. Thus, in placing a latch on a vacuum cleaner, one operator could assemble the latch, assemble the plate over the two bolts, start the two nuts and tighten them with a power wrench. If one person were to assemble the latch, another assemble the plate and start the nuts and the third person tighten them, each would be doing but a fragment of the operation. It is not the length of the cycle that is important but the fact that the task must be a complete unit, having a beginning, a duration and an ending.

Job enlargement entails a horizontal extension of the operator's tasks – it gives the operator more tasks at the same level of system operation. Although it increases the physical variety of the job (and may reduce the risk of fatigue or injury), it is only superficially more meaningful and varied. Hertzberg (whose theory of job satis-faction is described below) criticised job enlargement on the grounds that it only increases the number of meaningless tasks given to the operator.

Job enrichment

Job enrichment programmes attempt to extend a worker's duties vertically as well as horizontally. One approach is to make workers responsible for quality control of their own work. This does increase the meaningfulness and responsibility of a job since this function is normally carried out by a more senior person – the supervisor. As such, a part of the management function is delegated to the worker on the factory floor. However, enrichment of an unskilled worker's job can result in a correspond-ing impoverishment of the supervisor's job and introduces other complex issues: for example, whether the increased responsibility should be rewarded with a correspond-ing increase in pay.

Job-centred approaches to improving labour effectiveness have been criticised for a number of reasons: for example, that they are piecemeal and that it is not always clear whether they will fulfil any of an employee's higher-order needs.

Job satisfaction

According to Maslow's theory, employment in industrialised countries caters well for people's lower-order needs. In order to motivate people at work, higher-level needs such as self-esteem and the opportunity for self-actualisation should be considered. Hertzberg (1966) developed a theory of job satisfaction that stressed the importance of those aspects of a job that determine whether or not it is intrinsically satisfying. Hertzberg regarded satisfaction and dissatisfaction as separate dimensions rather than opposite ends of a continuum. Job satisfaction was determined by a set of factors known as Motivators, which were intrinsic to the work itself. Dissatisfac-tion, on the other hand, was said to be influenced by Hygiene factors, extrinsic to the work (Table 16.4.)

The term 'hygiene' was chosen using an analogy with medicine. Unhygienic condi-tions can cause ill-health, but increasing the level of hygiene, beyond a certain point, cannot lead to positive increases in health. Because satisfaction and dissatisfaction are regarded as separate factors in Hertzberg's theory, the absence of job satisfaction is no satisfaction and the absence of dissatisfaction is no dissatisfaction. Poor work-ing conditions (an absence of hygiene factors) will cause dissatisfaction but a good

Table 16.4 Hygiene factors and motivators in job design

Hygiene factors	Motivators
Company policy and administration	Achievement
Supervision	The work itself
Relationship with supervisor	Responsibility
Work conditions	Advancement
Salary	Personal growth
Security	

Table 16.5 How to improve job design (after Hertzberg)

Do:
1. Encourage responsibility by removing low-level supervision while retaining accountability for performance.
2. Encourage achievement and coherence in the job by giving the worker a natural unit of work (e.g. testing a complete car rather than just the brakes).
3. Provide recognition by giving direct feedback to workers about their performance.
4. Introduce new and more difficult tasks to encourage learning.
5. Assign individuals to specialised tasks to enable them to build up unique skills.

Don't:
1. Add additional tasks to an already fragmented one in an attempt to make a job more meaningful.
2. Rotate low-level tasks in the hope of enriching a job.
3. Remove the most difficult parts of a job, since these are often the most interesting.

---

working environment cannot make people like the work that they do. According to Hertzberg, liking a job depends on factors that are intrinsic to the job itself.

Hertzberg proposed a number of principles for job design based on his theory of job satisfaction. Improvements in labour effectiveness and employee motivation can be achieved, it was argued, by changes in the organisation of work through better design of the jobs that people do (Table 16.5).

Criticisms of Hertzberg's theory   Hertzberg's theory can be criticised in several ways; for example, because of the lack of evidence that satisfaction and dissatisfaction really are separate and because of its job-centred nature. The job itself is seen as the source of satisfaction or lack of satisfaction without considering what it means to the worker, his or her expectations or other social and cultural factors. It can be argued, for example, that some people do not expect to have interesting jobs and will not be dissatisfied if they do not get them. Others may be more interested in their families or in hobbies or other outside interests and see work as merely a means of achieving other, more important, goals. Satisfaction would then be determined by the extent to which the job enables them to achieve these external goals, rather than the design of the job itself.

Both Maslow's and Hertzberg's theories were developed in a Western industrial milieu, where much emphasis is placed on the individual and his or her achievements. Whether these theories are applicable in societies where people customarily view themselves and their work as part of a community endeavour, rather than an exercise in personal growth, is debatable (see Meshkati and Robertson, 1986, for further discussion of this point). Finally, for many people all over the world, having any job at all is a major goal in life and a salary that raises them above a basic subsistence level is likely to be a major source of satisfaction.

Evidence for the effectiveness of job design   Kelly (1992) reviewed 31 studies where job redesign had taken place using the principles described above. Changes in job design that had an impact on how job content was perceived resulted in increased job satisfaction. Increases in job performance, on the other hand, were bought about by pay rises and job losses. There was scant evidence that increased job satisfaction resulted in increased motivation and improved job performance. Kelly proposed a 'twin track' model in which the determinants of job satisfaction and job perform-ance are different. What satisfies employees about their jobs and what makes them produce more are not the same, and there is no guarantee that a satisfied worker will perform better or that what makes an employee perform better will result in increased satisfaction with the job. It seems, then, that the linkage between the psychosocial work environment and the physical work environment is not a strong as previous researchers have suggested. Unproductive jobs can, in principle, be better engineered while having no impact on job satisfaction. Jobs can be redesigned to make them more satisfying and have no impact on output. There are many reasons for making jobs more satisyfing – to recruit and retain skilled workers and to im-prove industrial relations. However, contrary to the predictions of the Human Rela-tions Movement (derived from the earlier interpretations of the findings of the Hawthorne experiments), it should not be assumed that people will be more product-ive just because they like their jobs. Increased pay does seem to result in improved performance, though.

Aggregate levels of job satisfaction   More recently, West and Patterson (1998) com-pared the job satisfaction of the workforce and company performance in 42 UK manufacturing companies. The aggregate level of job satisfaction of all employees in a company *was* related to the company's overall level of performance one year later. The linkage between the overall level of satisfaction and company performance may lie in better teamwork. If everyone is satisfied, the argument goes, they are more likely to cooperate and form teams, both formally and informally, to deal with the issues of the day. Increased satisfaction leads to increased responsiveness of workers to each other and better cooperation. Job satisfaction seems to work, but only at the macro-level of the organisation and not at the individual level.

Psychosocial factors and physical stressors

Psychosocial factors have been found to account for significant amounts of the vari-ability in the prevalence of musculoskeletal disorders. Modern studies of WMSD prevalence take into account not just the prevalence of a disorder, but also the behaviours associated with it. These *disability behaviours* include the frequency with which medical treatment is sought, absenteeism, interference with work activities (including job change due to the disorder), interference with activities of daily living (ADLs) and fear-avoidance behaviour. Psychocial factors such as self-reported stress, unhappiness and a sense of hopelessness seem to amplify the deleterious effects of the disorder itself. Psychosocial factors can *mediate* the prevalence of a disorder or the behaviours associated with it (mediating factors increase or reduce the size of the outcome). They can also *moderate* the effects of the exposure, reducing the severity of the outcome. Moderating factors include the individual's coping style and the presence of social support.

Lindon (2000) reviewed 37 studies on back and neck pain for evidence of the role of psychosocial factors. The findings are summarised in Table 16.6.

Burton (1997) pointed out that the lifetime prevalence of back pain is about 60% and that it occurs most often in people of working age. Although ergonomic factors may precipitate a first back injury, the transition to chronic back pain and back pain disability depends on a number of psychosocial factors, including the presence of

Table 16.6 Psychosocial risk factors for back and neck pain[a]

1. Psychosocial factors influence the transition from acute pain to chronic pain disability and have predictive value.
2. Psychosocial factors are associated the onset of pain.
3. Back pain disability depends more on psychosocial factors than on biomedical factors or biomechanical exposures.
4. There is no simple 'pain prone' personality and the role of personality traits is unclear.
5. Cognitive factors are related to pain development and disability:
   • Fear avoidance facilitates pain development.
   • Passive, rather than active, coping strategies facilitate pain development.
   • Over-reaction (catastrophising) of back pain enhances pain disability.
   • Depression, anxiety and stress increase pain disability.
   • Self-perceived poor health increases pain disabilty.

[a] Lindon (2000).

fatalistic beliefs about back problems, a lack of job satisfaction and social support at work and a mentally stressful job. Much of the back pain experienced at work by people with these beliefs may be erroneously attributed to the work. In fact, back pain is part of life and is exacerbated by strenuous activities both at work and outside of work. Workers may develop inappropriate beliefs about their back pain – that certain work activities should therefore be avoided and that they need special treatment. These beliefs may reinforce the very behaviours that prevent proper recovery and re-integration at work. Fear-avoidance prevents the kinds of muscular stimulation that facilitates restoration of function. Rest and the avoidance of exertion, together with restricted work duties, may re-inforce disability behaviours and actually smooth the transition to chronic disability. Burton has argued that back problems at work should be dealt with by stressing return to normal duties as soon as possible, countering negative beliefs with information booklets and advice and fostering active coping strategies instead of coping strategies based on the avoidance of normal activities. Boos *et al.* (2000) found that medical consultation for low back trouble in individuals with intervertebral disc herniation was predicted, over a 5-year period, by listlessness, job dissatisfaction and shiftwork. Work incapacity was predicted by physical job characteristics (a combined score including lifting and carrying, working in a forward flexed posture, exposure to vibration and sedentary work), job dissatisfaction and shiftwork. It seems that psychosocial factors mediate withdrawal by back pain sufferers, causing them to avoid physically stressful work.

Kompier (2002) has reviewed the current status of research in this area. There is evidence that psychosocial factors are related to WMSDs, but further research is needed. In particular, longitudinal studies are needed to explain how exposure to risk factors influences the development of WMSDs and how the development is mediated by psychosocial factors. Intervention studies are needed to determine optimum methods of dealing with the problem.

Psychosocial factors in the design of experiments and field trials

Subject reactivity is the term used to describe changes in behaviour bought about by participation in experiments or field trials. Social psychologists have done a great deal of research on the behaviour of subjects during experiments and the factors that can bias the findings. In general, participants in experiments or trials are conscious of the fact that they are being observed and often feel that they themselves are being evaluated (even if they are not). The result is that they are usually compliant and make an effort to perform well. When new equipment or methods are being tested, the benefits may be overestimated because of this. Subject reactivity can take several forms, such as the following manifestations.

A response to the 'demand characteristics' of the experiment Demand characteristics are all the cues from the experimental situation that influence subjects' behaviour. Orne (1962) coined the term after conducting a classic experiment in which subjects were asked to carry out a tiresome and meangingless task. Orne was impressed by the subjects' diligence and how readily they accepted the task. Subjects may try to guess the experimental hypothesis and to confirm it by behaving in what they perceive to be the 'correct' way. The demand characteristics of experiments can have serious implications for the assessment of the effects of mild stressors on task performance. This is probably one of the reasons why research concerning the effects of noise on performance is inconclusive (experimental subjects work harder to overcome the distracting effects of the noise). In real work situations, mild stressors can have a large impact on performance of work tasks because people are not so readily motivated to continue working in unpleasant conditions. They may even have the right to stop work or go home if work conditions exceed certain limits.

Novelty and disruption effects Novel interventions may have short-term effects on performance because they are novel, not because they work better. The novelty creates interest, diverts attention from problems and may even heighten arousal and therefore boost performance. New office furniture, for example, may lead to reductions in musculoskeletal complaints in the short term as everyone experiments with it and gets used to the change. In situations where change is common, novelty effects may be less likely than in situations where change is rare.

Pretest sensitisation Administration of questionnaires or interviews before a trial may 'prime' people to respond in ways that differ from how they would respond had the pre-test not been administered.

Experimenter effects The experimenter may unintentionally bias the trial. Subjects may infer, from the experimenter's behaviour, what is expected of them or what the hypothesis might be. The experimenter may unconsciously reinforce certain behaviours by means of body language or attention. Subjects may then behave in such a way as to confirm the hypothesis. It is also possible that subjects may try to sabotage the experiment if they respond in a regative way to the experimenter or the procedure.

Placebo effect A placebo is something that is not a medicine but is made to appear like one and is given to patients who are told that it is a medicine. Patients receiving placebos who believe they are being given an effective treatment often improve. Up to 40% of the benefits of some medicines are thought to be due to the placebo effect. The placebo effect is good for consultants, but not for researchers, or as J. N. Blau put it:

The doctor who fails to have a placebo effect on his patients should either become an anaesthetist or a pathologist.

Placebos are nearly always used in clinical trials of new treatments: the treatment is only judged to be efficacious if it brings about a larger improvement than that due to the placebo.

Hawthorne effect   The Hawthorne effect is a controversy that is often mistaken for a fact. There is no standard definition. Sundstrom defined it as 'the motivating effect of being observed' and Kroemer as the fact that people behave differently when they know they are in an experiment. According to Wickstrom and Bendix (2000), the term was coined by French (1953), referring to the Hawthorne experiments:

> From a methodological point of view, the most interesting finding was what one may call the 'Hawthorne effect' – a marked increase in production related only to special social position and social treatment.

In the most frequently quoted of the Hawthorne experiments, lighting levels were gradually reduced while workers assembled relays or wound coils. Lighting was kept constant in a control group. Productivity increased in both groups and the usual intepretation was that the workers were responding to the social situation – any change that was noticed by the workers increased output. Many have challenged such interpretations of the Hawthorne experiments and called into question the existence of any 'Hawthorne effect'. Parsons (1974) reviewed the experimental procedures at Hawthorne and proposed that the increase in productivity, although still an artefact of the experimental situation, was due to changes in the way bonuses were paid and the provision of performance feedback. Instead of receiving a bonus on the basis of the output of a department of 100 workers, the 5 workers in the experimental test room received a bonus on the basis of the output of their own group. Furthermore, feedback on output was collected and displayed continuously, giving the workers accurate information on their performance at any time. Parsons interpreted the productivity increase as nothing more than operant conditioning of the workers. The Hawthorne researchers had, unwittingly, created the ideal conditions for skill acquisition. Parsons redefined the Hawthorne effect as, 'the confounding that occurs if experimenters fail to realize how the consequences of subjects' performance affect what subjects do'.

Franke and Kaul (1978) re-analysed the data fom the Hawthorne experiments using multiple regression analysis. Contrary to the Hawthorne experimenter's view that the increase in output was difficult to explain without considering the 'human relations' between supervisors, workers and the experimenters themselves, they found that 90% of the variance in the output data could indeed be accounted for. Most of the variance was accounted for by the imposition of managerial discipline, economic adversity (the work took place during the economic depression of the 1930s and workers were naturally concerned to keep their jobs) and improvements in the quality of raw materials. There was also evidence that fatigue reduction, through the design of rest periods and the economic incentives of the new bonus scheme also increased output.

The Hawthorne experiments have always been controverisal. The paradox of Hawthorne is that the improvements in output were attributed, at the time, to better 'human relations', athough the researchers made no attempt to measure these 'human relations'. More conventional explanations such as better work organisation, better economic incentives and good industrial engineering, were discounted and, even though these variables were measured, the appropriate analyses of the data were not done. Re-analysis by Franke and Kaul, 40 years later, demonstrated that very little of the variance in output could not be explained by the more mundane, some would say 'Tayloristic', changes to work practices that the researchers made when setting up the experiment.

Researchers have had difficulty replicating the Hawthorne effect. It seems *not* to occur in intervention studies carried out in the real world over the long term. Rubeck (1975) reports on some field trials of teaching in schools that were carried out to determine whether students' performance would improve if they knew they were in an experiment. The results were negative – there was no evidence that telling students that they were subjects in an experiment on teaching methods had any effect on their performance in reading or mathematics.

In their review of the Hawthorne experiments, Wickstrom and Bendix (2000) concluded that 'Vaguely referring to the Hawthorne effect should be avoided as it adds more to confusion than to clarity'. Given that the very existence of the effect is in question, that there is no agreed definition of it and that it has been shown not to occur in long-term field trials in the real world, it is recommended that the term be abandoned.

Reduction of bias in field trials

There are many ways to reduce bias in field research. The double-blind, randomised controlled trial is probably the 'gold standard'. A real treatment and a control or placebo treatment are randomly allocated to subjects. Neither experimenters nor subjects are aware of which one is being administered. An alternative to the placebo is to compare a new treatment with the best conventional method. In the evaluation of seats and other artefacts, 'blinding' of subjects and experimenters is not possible. Experimenter bias can sometimes be avoided by using naive experimenters, who are not aware of the hypothesis. There are many good references on this topic and the reader is referred to Cook and Campbell (1976).

Litigation

The last few decades have seen an increase in the number of cases in which injured employees seek compensation from their employers in the courts. Frequently, the injuries experienced may have been caused or exacerbated by ergonomic factors in the workplace and an ergonomist is summoned as an expert witness by lawyers representing the plaintiff or the defendant. A medical doctor usually carries out an examination to determine the nature and severity of the injury and give an expert opinion as to the cause. In those cases where there is a real injury and there are grounds for believing it to be work-related, an ergonomist is called in to visit the workplace, analyse the task requirements and determine the following:

- Whether there were risk factors in the work environment
- Whether the employer could reasonably have been expected to know about the risk factors at the time
- Whether, if there were risk factors and the employer should have known about them, the employer was negligent in not taking all reasonable steps to safeguard the employee's health.

This report then forms part of the evidence given in court for its consideration. Further discussion of legal aspects (in the context of carpal tunnel syndrome) can be found in Owen (1994).

Cross-cultural considerations

The environment in which people grow up and the formal and informal educational processes to which they are exposed have a major influence on the cognitive structures they develop. Merely to state that people from different cultures think differently is trivial – much of human behaviour (and physiology) is clearly an adaptation to the particular surroundings. Beliefs that may be appropriate in one culture may be inappropriate in another.

Cognitive differences between cultures are important when people whose cognitive structures have developed in a particular sociotechnical milieu are exposed to new technologies or industries. This can happen when technologies or work systems from industrially developed countries are transferred to industrially developing countries without taking into account the knowledge and beliefs of workers in the recipient country.

Different thinking styles and the beliefs associated with them can become an important consideration during times of change. There may be mismatches between the knowledge and cognitive styles of the users and the operational requirements of the technology being introduced, which will result in severe cognitive incompatibilities. In industrially developing countries, people's formal exposure to technology as well as to the infrastructure that makes it possible to own and use technology may be lacking. They therefore do not have the opportunity to internalise many key concepts about how technology, and more generally technological society, actually works. A common coping strategy for the operator or user when faced with a mismatch of this nature is to learn by rote, developing only a minimal form of 'mental model' or high-level representation of technology, the context in which it operates and the environment that supports it. This is a workable strategy that enables the individual to function and to interact with machines on a routine basis. It breaks down, however, when the individual is presented with novel situations or unprogrammed events beyond the scope of the learnt behaviours.

This is one of the reasons why technology transfered from developed to industrially developing countries may fail – often spectacularly and as a result of behaviour that may seem bizarre or inconceivable to someone with appropriately internalised concepts. The cognitive mismatch between the user and the system can be profound. In cases of such severe mutual incompatibility there may be a need to 'fit the worker to the job' by means of appropriate training and upliftment via exposure to enriching technological experiences. Alternatively, designers must attempt to select intermediate technology that will fit the cognitive structures of the workforce in the recipient country. This may be extremely difficult as it nearly always involves making assumptions about the operator's knowledge and the boundaries within which 'common sense' can be expected to prevail, as is aptly illustrated below.

Social beliefs and ergonomic controls

People in technologically advanced societies are not fundamentally different from those in existing or past pre-industrial societies and are only rational in an objective sense some of the time and in particular contexts. However, irrational religious and superstitious beliefs that influence many aspects of our daily behaviour (such as the purchase of lottery tickets and the belief in 'luck' in the western industrial milieu) are supressed in the workplace by formal education and training. This may not be the case in pre-industrial societies where an animistic religious outlook endows everyday objects with religious as well as practical significance. The distinction between religious and superstitious thinking and objective nature may be blurred as custom, ritual and taboo infiltrate working life and impede correct understanding of phenoma unembellished by mystical conceptions.

Sanwo (1996) has described positive and negative ergonomic consequences of the social beliefs of the Yoruba, an indigenous people in Nigeria. These beliefs consist of normative beliefs, superstitious beliefs, taboos, proverbial beliefs and religious beliefs. An understanding of these beliefs can have practical utility in designing safety slogans and other propaganda and practices for work design. Irrespective of differences in education or social status, belief in the supernatural is endemic among the Yoruba and the animistic outlook pervades all aspects of life. Sanwo argues that ergonomics can be promoted in developing countries, such as those found in Africa, by building on exisiting beliefs and their associated controls. Normative beliefs, which are the most binding, can be built on although new normative beliefs are difficult to create. Superstitions can be exploited to support guidelines for efficient and safe work practices by means of appropriate slogans displayed in local languages (e.g. 'spirits do not control work, where common sense does'). Religious beliefs can also be used to convey ergonomic advice (as in 'God protects a careful worker'). Religious beliefs can have negative consequences when expressed in a fatalistic way to suggest that 'God protects all' and the worker may depend on good luck charms, prayers, etc. to ensure safety rather than correct work practices and the use of protective clothing. The view that nothing unpleasant will happen if God is on our side is pervasive in many cultures and is illustrated by the habit of appending phrases such as 'God willing' or 'Si Dios quiere' to statements or predictions. In a Mexican factory investigated by Lazcano (1996), an altar to the the Virgin of Guadalupe is found at the entrance to all production areas and is used daily by workers at the beginning of their shift in order to feel protected. Lazcano emphasises the need to acheive compatibility between the cultural conceptualisations of workers and the needs of industry.

Effectiveness and cost-effectiveness

From a macroergonomic perspective, there are two main approaches to cost-effective ergonomic interventions. The first, most dramatic, is to to identify high-risk areas and high-risk groups of individuals where the loss associated with accidents and injuries is high. Even expensive interventions will be cost-effective if they bring about reductions in risk. The second, more mundane, approach is to identify problems that are extremely common in society or in a company and direct population-based interventions in the hope of making a small improvement. Even a small reduction in the incidence of problems will be effective because the absolute numbers of people involved is large. Ergonomics offers both tightly focused interventions aimed at specific groups or occupations and more general interventions of wide applicability.

A third approach, one perhaps where ergonomists have been less successful, is in designing new systems that work by using people and technology more effectively, by improving the efficiency of the system and, at the same time, providing higher-quality work for people and saving money. The example below illustrates how cost-beneficial this can be.

Containerisation: a worked example

Freight costs for containerised transport are low even in relation to fairly unsophisticated products such as table wine. Consider a vineyard owner in Stellenbosch, South Africa, wanting to ship a container of wine to a wine merchant in the United Kingdom. The cost of delivering the container to the vineyard, loading it and returning it to the dock is approximately $250, handling charges included (1998 prices). The sea freight charge is $600. If we assume that the cost of offloading the container at the other end and transporting it to the customer's warehouse is double the local charge, i.e. $500, then the total cost is $1350.

A 6-metre container holds 1200 cases of wine (14 400 bottles). The transport cost is therefore 9.3 cents per bottle. If the retail price of the wine is $6.00 per bottle, the transport cost per bottle is approximately 1.5% of the price of the bottle. In relation to other factors that influence the cost of the wine to the consumer (and therefore the competitiveness of the product) such as import duties, value added tax, etc., the transport cost is trivial. Even if the above estimate leaves out significant incidental costs, it is clear that the transport costs are small in relation to the likely profit margins of the producers, wholesalers and retailers. It is easy to see why high-value-added products such as compact discs or computers can be made in one part of the world and shipped to other countries and why containerisation is one of the key drivers in globalisation of the economy.

Productivity trends

Productivity growth in the richer countries of the world has slowed down from 4.5% in the 1960s to about 1.5% today. IT has bought productivity gains to some industries – particularly those that have automated low-value clerical tasks such as clearing of cheques. Chapanis (1991) cites a study carried out among data entry operators in the Pacific Gas and Electric Company in the USA. A computerised method of depositing cheques was optimised by redesigning the workstations, instituting appropriate rest periods for employees and the introduction of a piece-rate incentive plan of remuneration. The average backlog of $415 million waiting to be deposited was reduced to $169 million. The interest earned by the $246 million deposited a day or two earlier amounted to approximately $40 000 per day. Clearly, well-designed computer systems can produce large productivity gains and savings if targeted correctly.

A word of caution

Hopkins (1999) discussed some of the limitations of cost arguments for the improvement of safety. These limitations apply equally well to ergonomics. It is not sufficient to argue that ergonomics will pay for itself or save money. The arguments for cost-saving must specify how ergonomics will save money for relevant decision makers. Many middle managers, for example, are only concerned about their own budgets and may be unimpressed by the claim that an intervention will save the *company* money. Hopkins gives some examples of major accidents that *appear* to be very damaging to the companies concerned. However, closer scrutiny of the aftermath of these disasters reveals that many of the losses were covered by insurance, that the resulting reorganisation of the company may have been beneficial in the medium term and that top management was not personally affected by the incident. Hopkins argues that there are good reasons for reminding decision makers of their ethical and moral responsbilities by couching the arguments in social and emotional terms.

Research directions

Psychosocial factors have come to the fore in much ergonomic research – in particular, their role as factors that mediate the success of ergonomic interventions. Some other general trends include the effects of demographic change and the ageing population and the design of increasingly complex, lean-manned systems.

Summary

Ergonomics is focused on the human–machine interface but is applied in larger systems that are often complex. Ergonomists must integrate their efforts with those of designers, engineers, users, medical staff and architects, to name but a few. Structured approaches exist to marry ergonomics with system design and systems engineering. These approaches call for ergonomic input at all stages in the design process – from requirements specification and concept generation to post-implementation evaluation. At all stages, practitioners must be aware of the macroergonomic issues that influence design decisions.

Gardell (1981) summarised the requirements for the psychosocial work environment as follows:

1   Work must be arranged to allow workers to influence their own working situation.
2   Work must be arranged to provide the worker with an overview and an understanding of the whole process.
3   Work must be arranged to give workers the chance to use and develop their human resources.
4   Work must be arranged to allow for human contact and cooperation.
5   Work must be arranged to allow workers to satisfy other demands on their time – e.g. from family or social/political commitments.

Essays and exercises

1. What factors determine the implementation of ergonomics in practice?
2. Discuss possible future technological developments of your choice and the ergonomic problems that will have to be solved before they can be successfully implemented as systems or products.
3. Discuss the macroergonomic implications of the following global issues. Discuss how these issues might influence the design of work practices, the management of organisations and the training of future generations of ergonomists:
    *   Global warming
    *   Massive enlargement of the hole in the ozone layer
    *   Continued demographic ageing of populations on a global scale
    *   The replacement of an oil-based energy supply with a solar/hydrogen-powered energy base

Further reading in and around ergonomics

Books

Astrand PO, Rodahl K. (1977) *Textbook of Work Physiology*. McGraw-Hill.

Guyton AC (1991) *Textbook of Medical Physiology*, 8th edition, WB Saunders.

Kirwan B (1994) *A Guide to Practical Human Reliability Assessment*. Taylor and Francis.

International Labour Organisation (1996) *Ergonomic Checkpoints*. ILO. ISBN 92-2-10-9442-1.

Kapandji JA (1974) *The Physiology of the Joints*, vols 1, 2, 3. Churchill Livingstone.

Karwowski W (ed) (2001) *International Encyclopaedia of Ergonomics and Human Factors*. Taylor and Francis.

Kuorinka I, Forcier L (1994) *Work Related Musculoskeletal Disorders – a Reference Manual for Prevention*. Taylor and Francis.

Mascie-Taylor CGW, Bogin B (1995) *Human Variability and Plasticity*. Cambridge University Press.

Mital A, Nichholson AS, Ayoub M (1997) *A Guide to Manual Materials Handling*, 2nd edition. Taylor and Francis.

Niebel NW, Frievalds A (1993) *Methods, Standards and Work Design*, 10th edition. WCB McGraw-Hill.

Norman DA (1988) *The Psychology of Everyday Things*. Basic Books.

Peebles L, Norris B (1998) *Adult Data. The Handbook of Anthropometry and Strength Measurements*. Department of Trade and Industry, UK, DTI/Pub 2917/3K/6/98/NP.URN.98/736.

Pheasant ST (1995) *Bodyspace*, 2nd edition. Taylor and Francis.

Preece J (1993) *A Guide to Usability*. The Open University Press, Addison Wesley.

Reason J (1990) *Human Error*. Cambridge University Press.

Ridley J, Channing J (1999) *Safety at Work*, 5th edition. Butterworth Heinemann.

Ringelberg JA, Koukoulaki Th (2002) *Rish Estimation for Musculoskeletal Disorders in Machinery Design – Integrating a User Perspective*. European Trade Union Technical Bureau for Health and Safety.

Sundstrom ED (1986) *Workplaces*. Cambridge University Press.

Wald PM, Stave GM (1994) *Physical and Biological Hazards in the Workplace*. Van Nostrand-Reinhold.

Wickens CD (1992) *Engineering Psychology and Human Performance*, 2nd edition, Pearson Education.

Wilde GJS (1994) *Target Risk*. PDE Publications.

Woodson WE (1981) *Human Factors Design Handbook*. McGraw-Hill Inc.

Journals
Human Factors
Ergonomics
International Journal of Industrial Ergonomics

Applied Ergonomics
Human–Computer Interaction
Work and Stress
Ergonomics in Design
Journal of Applied Psychology
Occupational Medicine
Scandinavian Journal of Work Environment and Health
International Journal of Cognitive Ergonomics

References

Aaras A (1994) The impact of ergonomic intervention on individual health and corporate prosperity in a telecoms environment. *Ergonomics*, 37: 1679–1696.

Abeysekera JDA, Shahnavaz H (1989) Body size variability between people in developed and developing countries and its impact on the use of imported goods. *International Journal of Industrial Ergonomics*, 4: 139–149.

Adams J (1998) *Risk*. UCL Press.

Adams MA, Dolan P (1995) Recent advances in lumbar spinal mechanics and their clinical significance. *Clinical Biomechanics*, 10: 3–19.

Adams MA, Hutton WC (1980) The effect of posture on the role of the apophyseal joints in resisting intervertebral compressive forces. *Journal of Bone and Joint Surgery*, 2B: 358–362.

Adams MA, Hutton WC (1983) The effect of posture on the fluid content of the intervertebral discs. *Spine*, 8: 665–671.

Adams MA, Hutton WC (1985) The effect of posture on the lumbar spine. *Journal of Bone and Joint Surgery*, 67B: 625–629.

Aghazadeh F, Mital A (1987) Injuries due to handtools. *Applied Ergonomics*, 18: 273–278.

Agnew J, Suruda AJ (1993) Age and fatal work-related falls. *Human Factors*, 35: 731–736.

Akerstedt T, Torsvall L (1981) Shift work: shift dependent well-being and individual differences. *Ergonomics*, 24: 265–273.

Allen AA, Fischer GJ (1978) Ambient temperature effects on paired associate learning. *Ergonomics*, 21: 95–101.

Allnutt MF, Allen JR The effects of core temperature elevation and thermal sensation on performance. *Ergonomics*, 16: 189–196.

Allport DA, Antonis B, Reynolds P (1972) On the division of attention: a disproof of the single channel hypothesis. *Quarterly Journal of Experimental Psychology*, 24: 225–235.

Allread WG, Marras WS, Burr DL (2000) Measuring trunk motions in industry: variability due to task factors, individual differences and the amount of data collected. *Ergonomics*, 43: 691–701.

Altman I (1975) *The Environment and Personal Space*. Wadsworth Publishing Co.

American Industrial Hygiene Association. 1975. *Industrial Noise Manual*.

Anastasi A (1990) *Psychological Testing*. Macmillan Publishing Co.

Anderson JR (1983) *The Architecture of Cognition*. Harvard University Press.

Andersson GBJ (1986) Loads on the spine during sitting. In *The Ergonomics of Working Postures*, edited by EN Corlett, J Wilson, I Mannenica. Taylor and Francis.

Argyle M (1975) *Bodily Communication*. Methuen.

Ariens GAM, van Mechelen WV, Bongers PM, Boutier LM, van der Waal G (2000) Physical risk factors for neck pain. *Scandinavian Journal of Work, Environment and Health*, 26: 7–19.

Armstrong TJ, Buckle PD, Fine LJ, Hagberg M, Jonsson B, Kilbom A, Kuorinka I, Silverstein BA, Sjogaard G, Viikari-Juntura ERA (1993) A conceptual model for work-related neck and upper limb musculoskeletal disorders. *Scandinavian Journal of Work, Environment and Health*, 19: 73–84.

Ashby P (1979) *Ergonomics Handbook 1: Body Size and Strength.* SA Design Institute, Private Bag X191, Pretoria 0001.

Ashdown SP (1998) An investigation of the structure of sizing systems. *International Journal of Clothing Science and Textile Technology.* 10(5): 324–341.

Aspden RM (1987) Intra-abdominal pressure and its role in spinal mechanics. *Clinical Biomechanics,* 2: 168–174.

Aspden RM (1989) The spine as an arch. A new mathematical model. *Spine,* 14: 266–274.

Astrand PO, Rodahl K (1977) *Textbook of Work Physiology.* McGraw-Hill.

Ayoub MA (1982) Control of manual lifting hazards: II Job redesign. *Journal of Occupational Medicine,* 24: 668–676.

Ayoub MA, Mital A (1997) *Manual Materials Handling.* Taylor and Francis.

Azer NZ, McNall PE, Leung HC (1972) Effects of heat stress on performance. *Ergonomics,* 15: 681–691.

Baber C, Stanton NA (1996) Human error identification tachniques applied to public technology. *Applied Ergonomics,* 27: 119–131.

Baggett P (1983) Four principles for designing instructions. *IEEE Transactions on Professional Communication,* PC-26: 99–105.

Bailey RW (1982) *Human Performance Engineering: A Guide For System Designers.* Prentice Hall.

Bailey RW (1983) *Human Error in Computer Systems.* Prentice-Hall.

Bainbridge L (1979) Verbal reports as evidence of the process operator's knowledge. *International Journal of Man–Machine Studies,* 11: 411–436.

Bainbridge L (1982) Ironies of automation. Analysis, design and evaluation of man–machine systems. *Proceedings of IFAC/IFIP/IFORS/IEA Conference,* Baden-Baden, Germany, pp. 151–157.

Baldwin CL, Struckman-Johnson D (2002) Impact of speech presentation level on cognitive task performance. *Ergonomics,* 45: 61–74.

Ballal MA, Fentem PH, MacDonald IA, Sukkar MY, Patrick JM (1982) Physical condition in young adult Sudanese. A field-study using a self-paced walking test. *Ergonomics,* 25: 1185–1196.

Barber P (1988) *Applied Cognitive Psychology.* Methuen.

Barnes RM (1963) *Motion and Time Study: Design and Measurement of Work,* 5th edn. Wiley.

Bartholomae RC, Kovac JG (1979) USBM develops a low-noise percussion drill. *Coal Age Conference and Expo V,* Louisville, Kentucky, Oct. 23–25 1979.

Bartlett FC (1932) *Remembering.* Cambridge University Press.

Barton NJ, Hooper G, Noble J, Steel WM (1992) Occupational causes of disorders in the upper limb. *British Medical Journal,* 304: 309–311.

Battie MC, Bigos SJ, Fisher LD, Spengler DM, Hansson TH, Nachemson AL, Wortley MD (1990) The role of spinal flexibility in back pain complaints within industry. *Spine,* 15: 768–773.

Bauk DA (1992) *Saude and Seguranca E Terminais de Video.* Departmento de Saude Ocupacional IBM, Brazil, Rio de Janeiro.

Beard DV, Walker JQ II (1990) Navigational techniques to improve the display of large two-dimensional spaces. *Behaviour and Information Technology,* 9: 451–466.

Becker WGE, Ellis H, Goldsmith R, Kaye AM (1983) Heart rates of surgeons in theatre. *Ergonomics,* 26: 803–807.

Beevis D, Slade IM (1970) Ergonomics – costs and benefits. *Applied Ergonomics,* 1: 79–84.

Begault DR (1993) Head-up auditory displays for traffic collission avoidance system advisories: a preliminary investigation. *Human Factors,* 4: 707–717.

Beiers JL (1966) A study of noise sources in pneumatic rockdrills. *Journal of Sound and Vibration,* 3: 166–194.

Beiring-Sorensen F (1984) Physical measurements as risk indicators for low-back trouble over a 1 year period. *Spine*, 16: 1179–1184.

Bejjani FJ, Gross, CM Pugh JW (1984) Model for static lifting; relationship of loads on the spine and the knee. *Journal of Biomechanics*, 17: 281–286.

Belbin RM, Stammers D (1972) Pacing stress, human adaptation and training in car production. *Applied Ergonomics*, 3: 142–146.

Bell CR, Crowder MJ, Walters JD (1971) Durations of safe exposure for men at work in high temperature environments. *Ergonomics*, 14: 733–757.

Belz SM, Robinson GS, Casali JG (1999) A new class of auditory warning signals for complex systems: auditory icons. *Human Factors*, 41(4): 608–617.

Bemis SV, Leeds JL, Winner EA (1988) Operator performance as a function of type of display: conventional versus perspective. *Human Factors*, 30: 163–169.

Bendix T, Beiring-Sorensen F (1983) Posture of the trunk when sitting on forward inclining seats. *Scandinavian Journal of Rehabilitation Medicine*, 15: 197–203.

Bendix T, Hagberg M (1984) Trunk posture and load on the trapezius muscle whilst sitting at sloping desks. *Ergonomics*, 27: 873–882.

Bendix T, Jessen F (1986) Wrist support during typing – a controlled, electroymographic study. *Applied Ergonomics*, 17: 162–168.

Bendix T, Jessen F, Winkel J (1986) An evaluation of a tiltable office chair with respects to seat height, backrest position and task. *European Journal of Applied Physiology*, 55: 30–36.

Bendix T, Winkel J, Jessen F (1985) Comparison of office chairs with fixed forwards or backwards inclining, or tiltable seats. *European Journal of Applied Physiology*, 54: 378–385.

Benson JD (1986) Control of low back pain in industry through ergonomic redesign of manual materials handling tasks. In *Trends in Ergonomics/Human Factors III*, edited by W Karwowski. Elsevier Science Publishers BV (North Holland).

Benson JD (1987) Application of manual handling task redesign in the control of low back pain. In *Trends in Human Factors/Ergonomics IV*, edited by SS Asfour. Elsevier Science Publishers BV (North Holland).

Benway JP (1998) Banner blindness: the irony of attention getting on the world wide web. In *Proceedings of the 42nd Annual Meeting of the Human Factors and Ergonomics Society*, pp. 463–467.

Bernoux P (1994) Participation: a review of the literature. In *P+, European Participation Monitor, The Economics of Participation*, edited by M Gold. European Foundation for the Improvement of Living and Working Conditions. ISSN 1017–6713.

Bhagwati J (1993) The case for free trade. *Scientific American*, November: 18–23.

Bhatnager V, Drury CG, Schiro SG (1982) Posture, postural discomfort and performance. *Human Factors*, 27: 189–199.

Biederman HJ, Shanks GL, Forrest WJ, Inglis J (1991) Power spectrum analysis of electromyographic activity. *Spine*, 16: 1179–1184.

Bilzon E, Chilcott P, Bridger RS (2000) *Investigation of Anthropometric Criteria for Escape Through a Kidney Hatch, An Escape Hatch and for Passage through a Bulkhead Door.* INM Report No. 2000.037, July 2000.

Birch L, Juul-Kristensen B, Jensen C, Finsen L, Christensen H (2000) Acute response to precision, time pressure and mental demands during simulated computer work. *Scandinavian Journal of Work, Environment and Health*, 26: 299–305.

Bisdorff AR, Bronstein AM, Wolsley C, Gresty MA, Young DA (1999) EMG responses to free fall in elederly subjects and akinetic rigid patients. *Journal of Neurological and Neurosurgical Psychiatry*, 66: 447–455.

Bittner AC, Guignard JC (1985) Human factors engineering principles for minimising adverse motion effects: theiry and practice. *Naval Engineers Journal*, 97: 205–213.

Blattner MM, Sumikawa DA, Greenberg RM (1989) Earcons and icons: their structure and common design principles. *Human Computer Interaction*, 4: 11–44.

Boas F (1910) Changes in body form of descendents of immigrants. Senate document 208, 61st Congress, second session. Washington DC.

Bobjer O, Johansson SE, Piguet S (1993) Friction between the hand and handle: effects of oil and lard on textured and non-textured surfaces; perception of dicomfort. *Applied Ergonomics*, 24: 190–202.

Bogin B (1995) Plasticity and the growth of Maya refugee children living in the United States. In *Human Variability and Plasticity*, edited by CGN Mascie-Taylor, B Bogin. Cambridge University Press.

Boldsen JL (1995) The place of plasticity in the secular trend for male stature: an analysis of Danish population history. In *Human Variability and Plasticity*, edited by CGN Mascie-Taylor, B Bogin. Cambridge University Press.

Bolia RS, D'Angelo WR, McKinley L (1999) Aurally-aided visual search in three-dimensional space. *Human Factors*, 41: 664–669.

Bone J (1993) Chisel out a hand-tool Ergonomics plan. *Safety and Health*, May: 64–68.

Boorman S (1999) Reviewing car fleet performance after advanced driver training. *Occupational Medicine*, 559–561.

Boos N *et al.* (2000) Natural history of individuals with asymptomatic disc abnormalities in magnetic resonance imaging. *Spine*, 25: 1484–1492.

Boreham P (1992) The myth of post-Fordist management: work organisation and employee discretion in seven countries. *Employee Relations*, 14: 13–24.

Borg GAV (1982) Psychophysical bases of perceived exertion. *Medicine and Science in Sports and Exercise*, 14: 377–381.

Boshuizen HC, Bongers PM, Hulshof CT (1992) Self-reported back pain in fork lift truck and freight container tractor drivers exposed to whole body vibration. *Spine*, 17: 59–65.

Botha WE, Bridger RS (1998) Anthropometric variability, equipment usability and musculoskeletal pain in a group of nurses in the western Cape. *Applied Ergonomics*, 29: 481–490.

Bough B, Thakore J, Davies M, Dowling F (1990) Degeneration of the lumbar facet joints. *Journal of Bone and Joint Surgery*, 72-B: 275–276.

Bower GH (1972) Mental imagery and associative learning. In *Cognition in Learning and Memory*, edited by LW Gregg. Wiley.

Bowler JN (1982) Welbeck colliery noise level survey. *The Mining Engineer*, September: 159–162.

Boyce PR (1982) Vision, light and colour. In *The Body at Work*, edited by WT Singleton. Cambridge University Press.

Brand JL, Judd KW (1993) Angle of hard copy and text editing performance. *Human Factors*, 35: 57–69.

Brandfonbrener A (1990) The epidemiology and prevention of hand and wrist injuries in performing artists. *Hand Clinics of North America*, 6: 365–376.

Bransford JD, Johnson MK (1973) Consideration of some problems of comprehension. In *Visual Information Processing*, edited by W Chase. Academic Press.

Branton P (1969) Behaviour, body mechanics and discomfort. *Ergonomics*, 12: 316–327.

Bridger RS (1988) Postural adaptations to a sloping chair and worksurface. *Human Factors*, 30: 237–247.

Bridger RS, Freidberg S (1999) Managers' estimates of safe loads for manual handling; evidence for risk compensation? *Safety Science*, 32: 103–111.

Bridger RS, Orkin D (1992) Effect of a footrest on standing posture. *Ergonomics SA*, 4: 42–48.

Bridger RS, Poluta M (1998) Ergonomics: introducing the human factor into the clinical setting. *Journal of Clinical Engineering*, May/June: 181–188.

Bridger RS, Ossey S, Fourie G (1990) The effect of lumbar traction on stature. *Spine*, 15: 522–525.

Bridger RS, Orkin D, Henneberg M (1992) A quantitative investigation of lumbar and pelvic postures in standing and sitting: interrelationships with body position and hip muscle length. *International Journal of Industrial Ergonomics*, 9: 235–244.

Bridger RS, Verweckken B, Whistance RS, Adams LP (1994) A prototype standing workspace. In *Contemporary Ergonomics 1994*, edited by SA Robertson. Taylor and Francis.

Bridger RS, Caborn N, Goedecke J, Rickard S, Schabort E, Westgarth-Taylor C, Lambert MI (1997) 'Physiological and subjective measures of workload when shovelling with conventional and two-handled ("levered") shovels'. *Ergonomics*, 40: 1212–1219.

Bridger RS, Sparto P, Marras WS (1998) Spade design, lumbar motions, risk of low back injury and digging posture. *Occupational Ergonomics*, 1: 157–172.

Brisson C, Montruil S, Punnett L (1999) Effects of an ergonomic training programme on workers with video display units. *Scandinavian Journal of Work, Environment and Health*, 25: 255–263.

Broadbent C (1989) Gasp! Wheeze! Sniff! Air conditioning and you. In *Ergonomics, Technology and Productivity*, Proceedings of the 25th Annual Conference of the Ergonomics Society of Australia. Ergonomics Society of Australia, Fortitude Valley, Queensland, Australia.

Broadbent DE, Little EA (1960) Effects of noise reduction in a work situation. *Occupational Psychology*, 34(1): 133–140.

Brogmus GE, Sorock GS, Webster BS (1996) Recent trends in work-related cumulative trauma disorders of the upper extremities in the United States: an evaluation of possible reasons. *Journal of Occupational and Environmental Medicine*, 38: 401–411.

Brouha L (1960) *Physiology in Industry*. Pergamon Press.

Brown CD, Nolan BM, Faithful DK (1984) Occupational repetition strain injuries: guidelines for diagnosis and management. *The Medical Journal of Australia*, March: 329–332.

Brown ID (1966) Subjective and objective comparisons of successful and unsuccessful trainee drivers. *Ergonomics*, 9: 50–56.

Brown ID, Poulton EC (1961) Measuring the spare mental capacity of car drivers by a subsidiary task. *Ergonomics*, 4: 35–40.

Brown P, Rothwell JC, Thompson PD, Britton TC, Day BL, Marsden CD (1991) New observations on the normal auditory startle response. *Brain*, 114, 1891–1902.

Brun TA, Geissler C, Kennedy E (1991) The impact of agricultural projects on food, nutrition and health. In *Impacts on Nutrition and Health*, edited by AR Simopoulos, World Reviews of Nutrition and Dietetics 65. Karger, pp. 99–123.

Brunswic M (1984) Ergonomics of seat design. *Physiotherapy*, 70: 39–43.

Bundy A (1978) *Computational Models for Problem Solving*. The Open University Press.

Burdorf A, Naaktgeboren B, deGroot HCWM (1993) Occupational risk factors for low back pain among sedentary workers. *Journal of Occupational Medicine*, 35: 1213–1220.

Burgess R, Neal RJ (1989) Document holder usage when reading and writing. *Clinical Biomechanics*, 4: 151–154

Burton K (1991) Measuring flexibility. *Applied Ergonomics*, 22: 303–307.

Burton K (1997) Back injury and work loss. *Spine*, 22, 2575–2580.

Butler DL, Acquino AL, Hissong AA, Scott PA (1993) Wayfinding by newcomers in a complex building. *Human Factors*, 35: 159–173.

Butler DS (1991) *Mobilisation of the Nervous System*. Churchill Livingstone.

Cady LD, Thomas PC, Karwasky RJ (1985) Program for increasing the health and physical fitness of fire fighters. *Journal of Occupational Medicine*, 27: 110–114.

Card SK, Moran TP, Newell A (1983) *The Psychology of Human Computer Interaction*. Lawrence Erlbaum and Associates.

Cartas O, Nordin M, Frankel VH, Malgady R, Sheikhzadeh A (1993) Quantification of trunk muscle performance in standing, semi-standing and sitting postures in healthy men. *Spine*, 18: 603–609.

Carter S (1994) Baxi Partnership: critical success factors. In *P+, European Participation Monitor, The Economics of Participation*, edited by M Gold. European Foundation for the Improvement of Living and Working Conditions. ISSN 1017–6713.

Casali JG, Park MY (1990) Attenuation performance of four hearing protectors under dynamic movement and different user fitting conditions. *Human Factors*, 32: 9–25.

Casali SP, Williges BH, Dryden RD (1990) Effects of recognition accuracy and vocabulary size of a speech recognition system on task performance and user acceptance. *Human Factors*, 32: 183–196.

Cavanagh PR, Rodgers MM, Iboshi A (1987) Pressure distribution under symptom-free feet during barefoot standing. *Foot and Ankle*, 7: 262–276.

Ceesay SM, Prentice AM, Day KC, Murgatroyd PR, Goldberg GR, Spurr GB (1989) The use of heart rate in the estimation of energy expenditure: a validation study using indirect whole-body calorimetry. *British Journal of Nutrition*, 61: 175–186.

Chaffin DB (1987) Biomechanical aspects of workplace design. In *Handbook of Human Factors*, edited by G Salvendy. Wiley.

Chaffin DB, Andersson GBJ (1984) *Occupational Biomechanics*. Wiley.

Chapanis A (1990) Short term memory for numbers. *Human Factors*, 32: 123–137.

Chapanis A (1991) The business case for human factors in informatics. In *Human Factors for Informatics Usability*, edited by B Shackel, S Richardson. Cambridge University Press.

Chapanis A (1994) Hazards associated with three signal words and four colours on warning signs. *Ergonomics*, 37: 265–275.

Chapanis A, Lindenbaum LE (1959) A reaction time study of four control display linkages. *Human Factors*, 1: 1–7.

Chapanis A, Yoblick DA (2001) Another test of sensor lines on control panels. *Ergonomics*, 44: 1302–1311.

Chase WG, Simon HA (1973) Cognitive psychology. In *Visual Information Processing*, edited by WG Chase. Academic Press.

Chatterjee DS (1992) Workplace upper limb disorders: a prospective study with intervention. *Occupational Medicine*, 42: 129–136.

Chavalitsakulchai P, Shahnavaz H (1990) *Woman Workers and Technological Change in Industrially Developing Countries from an Ergonomic Perspective*. Center for Ergonomics of Developing Countries CEDC, Dept of Human Work Sciences, Lulea University Sweden TULEA 1990: 01.

Chen H, Dumais S (2000) Bringing order to the web: automatically categorising search results. *CHI*, April: 145–153.

Cherns A (1976) The principles of sociotechnical design. *Human Relations*, 29: 783–792.

Cholewicki J, Manohar M, Panjabi M, Khachatryan A (1997) Stabilising function of trunk flexor-extensor muscles around a neutral spine posture. *Spine*, 22(19): 2207–2212.

Chomsky N (1957) *Syntactic Structures*. Mouton.

Chong I, McDonough A (1984) A new angle on sports equipment. *Brief*, 1: 66–68.

Christ RE (1975) Review and analysis of colour coding research for visual displays. *Human Factors*, 17: 542–570.

Ciriello VM, Snook SH (1995) The effect of back belts on lumbar muscle fatigue. *Spine*, 20: 1271–1278.

Ciriello VM, Snook SH, Hughes GH (1993) Further studies of psychophysically determined maximum acceptable weights and forces. *Human Factors*, 35: 175–186.

Citron N (1985) Femoral neck fractures: are some preventable? *Ergonomics*, 28: 993–997.

Clark C, Haswell MR (1964) *The Economics of Subsistence Agriculture*. Macmillan and Co.

Clarke TS, Corlett EN (1984) *The Ergonomics of Workspaces and Machines: A Design Manual*. Taylor and Francis.

Clement DE (1987) Human factors, instructions and warnings, and products liability. *IEEE Transactions on Professional Communications*, 30(3): 149–156.

Coch L, French J (1948) Overcoming resistance to change. *Human Relations*, 34, 512–536.

Cole S (1982) Vibration and linear acceleration. In *The Body at Work*, edited by WT Singleton. Cambridge University Press.

Colligan MJ, Frock IJ, Tasto DL (1979) Frequency of sickness absence and worksite clinic visits among nurses as a function of shift. *Applied Ergonomics*, 10: 79–85.

Collins AM, Quillian MR (1972) Experiments on semantic memory and language comprehension. In *Cognition in Learning and Memory*, edited by LW Gregg. Wiley.

Collins M, Davis B, Goode A (1994) Steady state accommodation and VDU screen conditions. *Applied Ergonomics*, 25: 334–338.

Collins MJ, Brown B, Bowman KJ, Caird D (1991) Task variables and visual discomfort associated with the use of VDUs. *Optometry and Vision Science*, 68: 27–33.

Colquhoun WP, Goldman RF (1972) Vigilance under induced hyperthermia. *Ergonomics*, 15: 621–632.

Cook TM, Campbell DT (1976) The design and conduct of quasi-experiments and true experiments in field settings. In *Handbook of Industrial and Organisational Psychology*, edited by M. Dunnette. Rand McNally Publishing Co.

Cook TM, Neumann DA (1987) The effects of load placement on the activity of the low back muscles during load carrying by men and women. *Ergonomics*, 30: 1413–1423.

Corlett EN, Eklund JAE (1986) How does a backrest work? *Applied Ergonomics*, 15: 111–114.

Cote P, Cassidy JD, Carroll L (2000) The factors associated with neck pain and its related disability in the Sasketchewan population. *Spine*, 25: 1109–1117.

Cowley KC, Jones DM (1992) Synthesised or digitised? A guide to the use of computer speech. *Applied Ergonomics*, 23: 172–176.

Cox K, Walker D (1993) *User-Interface Design*. Prentice-Hall.

Cox R, Shephard J, Corey P (1981) Influence of an employee fitness programme upon fitness, productivity and absenteeism. *Ergonomics*, 24: 795–806.

Craik FIM, Lockhart RS (1972) Levels of processing: a framework for memory research. *Journal of Verbal Learning and Verbal Behaviour*, 11: 71–84.

Craik M (1943) *The Nature of Explanation*. Cambridge University Press.

Crockford GW (1962) Air fed permeable clothing for work in hot conditions. *Industrial Safety*, 8(9).

Crockford GW (1999) Protective clothing and heat stress: Introduction. *Annals of Occupational Hygiene*, 43: 287–288.

Croney J (1980) *Anthropometry for Designers*. Batsford Academic and Educational.

Culver CC, Viano DC (1990) Anthropometry of seated women during pregnancy: defining a fetal region for crash protection research. *Human Factors*, 32: 625–636.

Cushman WH (1985) Data entry performance and operator preferences for various keyboard heights. In *Ergonomics and Health in Modern Offices*, edited by Egrandjean. Taylor and Francis, pp. 495–505.

Cushman WH, Crist B (1987) Illumination. In *Handbook of Human Factors*, edited by G Salvendy, Wiley.

Cutlip RG, Marras WS (2000) Soft tissue pathomechanics and its application to ergonomics. In *Ergonomics for the New Millenium, International Ergonomics Association XIVth Triennial Congress*, San Diego, USA.

Czerwinski M, Larson K, Robbins D (1998) Designing for navigating personal web information: retrieval cues. *Proceedings of the 42nd Annual Meeting of the Human Factors and Ergonomics Society*, pp. 458–462.

Daanen HAM, Water GJ (1998) Whole body scanners. *Displays*, 19: 111–120.

Dabbs JM Jr (1971) Physical closeness and negative feelings. *Psychonomic Science*, 23: 141–143.

Dainoff MJ, Dainoff MH (1986) *People and Productivity: a Manager's Guide to Ergonomics in the Electronic Office*. Holt, Rinehart and Winston of Canada.

Daltroy LH, Iverson MD, Larson MG, Lew R, Wright E, Ryan J, Zwerling C, Fossel AH, Liang MH (1997) A controlled trial of an educational programme to prevent back injuries. *New England Journal of Medicine*, July 31: 322–328.

Datta SR, Ramanathan NL (1971) Ergonomic comparison of seven modes of carrying loads on the horizontal plane. *Ergonomics*, 14: 269–278.

Datta SR, Chatterjee BB, Roy BN (1983) The energy cost of pulling handcarts ('thela'). *Ergonomics*, 26: 461–464.

Davies DR, Jones DM (1982) Hearing and noise. In *The Body At Work*, edited by WT Singleton. Cambridge University Press.

Davies NV, Teasdale P, 1994, *The Costs to the British Economy of Work-Related Accidents and Work-Related Ill-health*. HSE Books.

de Puky P (1935) Physiological oscillation of the length of the body. *Acta Orthopaedica Scandinavica*, 6: 338–347.

de Wall M, van Riel MPJM, Snijders CJ (1991) The effect on sitting posture of a desk with a 10 degree inclination for reading and writing. *Ergonomics*, 34: 575–584.

Dearden A, Harrison M, Wright P (2000) Allocation of function: scenarios, context and the economics of effort. *International Journal of Human–Computer Studies*, 52: 289–318.

DeLaura D, Konz S (1990) Toespace. In *Advances in Industrial Ergonomics and Safety II*, edited by B Das. Taylor and Francis.

Dellemanm N, Berndsen MB (2002) Touch typing VDU operation. *Ergonomics*, 45: 514–535.

Dempster WT (1955) Anthropometry of body action. *Annals of the New York Academy of Sciences*, 63: 559–585.

Denniston HD (1935) Physical treatment in postural defects. *Archives of Physical Therapy, X-Ray, Radium*, 16: 525–527.

Department of Labor (1992) Ergonomic Safety and Health Management: proposed Rule, 20 CFR Part (1910) Federal Register, 57(149): 34 192–34 200.

Diamond J (1991) Pearl harbour and the Emperor's physiologists. *Natural History*, 12: 5–7.

Diaz E, Goldberg GR, Taylor M, Savage JM, Sellen D, Coward WA, Prentice A (1989) Effects of dietary supplementation on work performance in Gambian labourers. *Proceedings of the Nutrition Society*, 49(44-A): 45.

Dimberg L (1987) The prevalence and causation of tennis elbow (lateral humeral epicondylitis) in a population of workers in an engineering industry. *Ergonomics*, 30: 573–580.

Dixon P (1991) Learning to operate complex devices: effects of conceptual and operational similarity. *Human Factors*, 33: 103–120.

DonTigny RL (1985) Function and pathomechanics of the sacroiliac joint. *Physical Therapy*, 65: 35–44.

Dowell J, Long J (1989) Towards a conception for an engineering discipline of human factors. *Ergonomics*, 32: 1513–1535.

Dressel DL, Francis J (1987) Office productivity: contribution of the workstation. *Behaviour and Information Technology*, 6: 279–284.

Drury CG, Addison JL (1973) An industrial study of the effects of feedback and fault density on inspection performance. *Ergonomics*, 16: 159–169.

Drury CG, Francher M (1985) Evaluation of a forward sloping chair. *Applied Ergonomics*, 16: 41–47.

Drury CG, Roberts DP, Hansgen R, Bayman JR (1983) Evaluation of a palletising aid. *Applied Ergonomics*, 14: 242–246.

Ducharme RE (1977) Women workers rate 'male tools' inadequate. *Human Factors Society Bulletin*, 20: 1–2.

Duncan PW, Studenski S, Chandler J, Bloomfield R, LaPointe LK (1990) Electromyographic analysis of postural adjustments in two methods of balance testing. *Physical Therapy*, 70: 36–44.

Durrett J, Trezona J (1982) How to use colour displays effectively. *Byte*, April: 50–53.

Dwyer T, Raftery AE (1991) Industrial accidents are caused by the social relations of work: a sociological theory of industrial accidents. *Applied Ergonomics*, 22: 17–178.

Eason KD (1986) Job design and VDU operation. In *Health Hazards of VDUs?*, edited by B Pearce. Wiley.

Eason KD (1984) Towards the experimental study of usability. *Behaviour and Information Technology*, 3: 133–143.

Edwards W (1987) Decision making. In *Handbook of Human Factors*, edited by G Salvendy. Wiley.

Ekdahl C, Jarnlo GB, Andersson SI (1989) Standing balance in healthy subjects. *Scandinavian Journal of Rehabilitation Medicine*, 21: 187–195.

Eklund JAE (1995) Relationships between ergonomics and quality in assembly work. *Applied Ergonomics*, 26(1): 15–20.

Eklund JAE, Corlett EN (1984) Shrinkage as a measure of the effect of load on the spine. *Spine*, 9: 189–194.

English W (1994) The validation of slipmeters. In *Contemporary Ergonomics 1994*, edited by SA Robertson. Taylor and Francis, pp. 347–352.

Evanoff BA, Button JH, Wolf LD (2000) Effects of an ergonomics intervention programme on hospital billing dept employees. *Proc. IEA 2000/HFES 2000 Congress*, Human Factors and Ergonomics Society, San Diego CA, pp. 1-700–1-703.

Evans GW, Johnsson D (2000) Stress and open-office noise. *Journal of Applied Psychology*, 85: 779–783.

Feinstein J, Crawley JE (1968) The ergonomic design of a slab shear pulpit at British Steel. BISRA Report OR/HF/8/68.

Feldstein A-L (1993) Five-year follow-up study of hearing loss at several locations within a large automobile company. *American Journal of Industrial Medicine*, 24: 41–54.

Feng Y, Grooten W, Wretenberg P, Arborelius UP (1999) Effects of arm suspension in simulated assembly line work. *Applied Ergonomics*, 30: 247–253.

Fernstrom E, Ericson MO, Malker H (1994) Electromyographic activity during typewriter and keyboard use. *Ergonomics*, 37: 477–484.

Feuerstein M, Armstrong T, Hickey P, Lincoln A (1997) Computer keyboard force and upper extremity pain. *Journal of Occupational and Environmental Medicine*, 39: 1144–1153.

Fillenbaum S (1973) *Syntactic Factors in Memory?*. Mouton.

Filley RD (1982) Opening the door to communication through graphics. *IEEE Transactions on Professional Communication*, PC-25: 91–94.

Fisher GH (1973) Current levels of noise in an urban environment. *Applied Ergonomics*, 4: 211–218.

Fisher J, Levin N (1989) Display control compatibility in the design of domestic cookers for the South African population. *Ergonomics SA*, 1: 29–41.

Fisher J, Olivier JA (1992) Display-control stereotypes among. South African children in urban and rural settings. *Ergonomics SA*, 4: 20–29.

Fisher P, Sless D (1990) Information design methods and productivity in the insurance industry. *Information Design Journal*, 6: 103–129.

Flowers KA (1978) The predictive control of behaviour: appropriate and inappropriate actions beyond the input in a tracking task. *Ergonomics*, 21: 109–122.

Folkard S (1987) Circadian rhythms and hours of work. In *Psychology at Work*. Penguin Books.

Folkard S (1992) Is there a 'best' compromise shift system? *Ergonomics*, 35: 1453–1463.

Forouzandeh B, Wright RA (1998) Do back support belts cause gastroesophageal reflux? *Journal of Clinical Gastroenterology*, 27: 47–49.

Foss CL (1989) Tools for reading and improving hypertext. *Information Processing Management*, 25: 407–418.

Fox JG, Jones JM (1967) Occupational stress in dental practice. *British Dental Journal*, 123: 465–473.

Fox RH, Bradbury PA, Hampton IF (1967) Time judgement and body temperature. *Journal of Experimental Psychology*, 75: 88–98.

Franke RH, Kaul JD (1978) The Hawthorne experiments: first statistical interpretation. *American Sociological Review*, 43: 623–643.

French J (1953) Experiments in field settings. In *Research Methods in Behavioural Sciences*. 53, edited by L Festinger, D Katz. Holt Rheinhart and Winston, pp. 98–135.

Frey JK, Tecklin JS (1986) Comparison of lumbar curves when sitting on the Westnofa Balans Multichair, sitting on a conventional chair, and standing. *Physical Therapy*, 66: 1365–(1369).

Friedmann K (1988) The effect of adding symbols to written warning labels on user behaviour and recall. *Human Factors*, 30: 507–515.

Gagnon M, Plamondon A, Gravel D (1993) Pivoting with the load. *Spine*, 18(11): 1515–1524.

Galinsky TL, Swanson NG, Saurt SL, Hurrell JT, Schleiffer LM (2000) A field study of supplementary rest breaks for data entry operators. *Ergonomics*, 43(5): 622–638.

Gallagher S, Hamrick CA (1991) The kyphotic lumbar spine: Issues in the analysis of the stresses in stooped lifting. *International Journal of Industrial Ergonomics*, 8: 33–47.

Gardell B (1981) Psychosocial aspects of industrial production methods. In *Society, Stress and Disease*, 4, edited by L Levi. Oxford University Press, pp. 65–75.

Garg A, Owen B (1992) Reducing back stress to nursing personnel: an ergonomic intervention in a nursing home. *Ergonomics*, 35: 1353–1375.

Garrone B (2001) Retail falls: the whole story. *Occupational Health and Safety*, Sept.

Gawron VJ (2000) Guide to measuring workload and situational awareness. *Morning-Only Workshop, Human Factors and Ergonomics Society 44th Annual Meeting*, San Diego, CA, USA.

Genaidy AM, Waly SM, Khali TM, Hidalgo J (1993) Spinal compression tolerance limits for the design of manual material handling operations in the workplace. *Ergonomics*, 36: 415–434.

Gerard MJ, Armstrong TJ, Rempel DA, Wooley C (2002) Short term and long term effects of enhanced auditory feedback on typing force EMG and comfort while typing. *Applied Ergonomics*, 33: 129–138.

Geras DT, Pepper CD, Rodgers SH (1989) An integrated ergonomics programme at the Goodyear Tire and Rubber Company. The forcing strategy. In *Advances in Industrial Ergonomics and Safety*, I, edited by A. Mital. Taylor and Francis.

Gibbs WW (1997) Taking computers to task. *Scientific American*, July: 64–71.

Gibson D, Norton M (1981) The economics of industrial noiose control. *Noise Control Engineering*, 16(3): 126–133.

Gilad I, Kirschenbaum A (1988) Rates of back pain incidence associated with job attitudes and worker characteristics. *International Journal of Industrial Ergonomics*, 5: 267–272.

Gillespie R (1981) *Manufacturing Knowledge: A History of the Hawthorne Experiments*. Cambridge University Press.

Gilson M (1946) Review of 'Management and the Worker'. *The American Journal of Sociology*, 98–101.

Goetzel R, Sepulveda M, Knight K, Eisen M, Wade S, Wong J, Fielding J (1994) Association of IBM's 'A Plan for Life' health promotion program with changes in employees' health risk status. *Journal of Occupational Medicine*, 36: 1005–1009.

Gold M (1994) Editorial. In *P+, European Participation Monitor, The Economics of Participation*, edited by M Gold. European Foundation for the Improvement of Living and Working Conditions. ISSN 1017–6713.

Goldhaber MK, Polen MR, Hiatt RA (1988) The risk of miscarriage and birth defects among women who use visual display unit during pregnancy. *American Journal of Industrial Medicine*, 13: 695–706.

Gomez LM, Lochbaum CC (1984) People can retrieve more objects with enriched keyword vocabularies. But is there a performance cost? In *Interact '84, First IFIP Conference on Human Computer Interaction*, vol. 1, pp. 1429–1433.

Goodman TJ, Spence R (1981) The effect of computer system response time variability on interactive graphical problem solving. *IEEE Transactions on Systems, Man and Cybernetics*, SMC-11: 207–217.

Gordon AM, Julian FJ, Huxley AF (1966) The variation in isometric tension with sarcomere length in vertebrate muscle fibres. *Journal of Physiology*, 184: 170–192.

Gorsche R, Wiley JP, Renger R, Brant R, Gemer TY, Sasyniuk TM (1998) Prevalence and incidence of stenosing flexor tenosynovitis (trigger finger) in a meatpacking plant. *Journal of Occupational and Environmental Medicine*, 40: 556–560.

Graham R (1999) Use of auditory icons as emergency warnings. *Ergonomics*, 42: 1233–1248.

Grahame R, Jenkins JM (1972) Joint hypermobility – asset or liability? *Annals of Rheumatic Disease*, 31: 109–111.

Granata KP, Marras WS (2000) Cost–benefit of muscle cocontraction in protecting against spinal instability. *Spine*, 25(11): 1398–1404.

Grandjean E (1973) *Ergonomics of the Home*. Taylor and Francis.

Grandjean E (1980) *Fitting the Task to the Man: An Ergonomic Approach*. Taylor and Francis.

Grandjean E (1987) *Ergonomics in Computerised Offices*. Taylor and Francis.

Gray WD, John BE, Atwood ME (1993) Project Ernestine: Validating a GOMS analysis for predicting and explaining real-world task peformance. *Human Computer Interaction*, 8: 237–309.

Greene J, Cromer R (1978) *Language as a Cognitive Process*. The Open University Press.

Greene SL, Gould JD, Boies SJ, Rasamny M, Meluson A (1992) Entry and selection-based methods of human–computer interaction. *Human Factors*, 34: 97–113.

Gregg HD, Corlett EN (1988) Developments in the design and evaluation of industrial seating. In *Designing a Better World*, Proceedings of the 10th Triennial Conference of the International Ergonomics Association, edited by AS Adams, RR Hall, BJ McPhee, MS Oxenburgh. Ergonomics Society of Australia.

Grieco A (1986) Sitting posture: an old problem and a new one. *Ergonomics*, 29: 345–32.

Grieve D, Pheasant S (1982) Biomechanics. In *The Body at Work*, edited by WT Singleton, Cambridge University Press.

Griffin MJ (1990) *Handbook of Human Vibration*. Academic Press.

Grignolo FM, Bari A, Bellan B, Camerino L, Maina G (1998) Long-term refractive and phoric changes in visual display unit operators. *European Journal of Opthalmology*, 8: 76–80.

Gunnarson E, Soderberg I (1983) Eye strain resulting from VDU work at the Swedish Telecommunications Administration. *Applied Ergonomics*, 14: 61–69.

Gunnel D, Frankel S (1994) Prevention of suicide: aspirations and evidence. *British Medical Journal*, 308: 1227–1233.

Guo L, Genaidy AM, Warm J, Karwowski W, Hidalgo J (1992) Effects of job-simulated flexibility and strength/flexibility training protocols on maintenance employees engaged in manual handling operations. *Ergonomics*, 35: 1103–1117.

Guyton AC (1991) *Textbook of Medical Physiology*, 8th edn. WB Saunders.

Hadler NM (1996) A keyboard for 'Daubert'. *Journal of Occupational and Environmental Medicine*, 5: 469–475.

Hagberg M (1987) *Occupational Shoulder and Neck Disorders*. The Swedish Work Environment Fund, Box 1122, S-11181, Stockholm, Sweden.

Hales T (1994) Ergonomic hazards and upper-extremity musculoskeletal disorders. In *Physical and Biological Hazards of the Workplace*, edited by PH Ward, GM Stave. Van Nostrand Rheinhold.

Handbook of Noise and Vibration Control (1979) Trade and Technical Press, Morden, Surrey.

Hanna JM, Brown DA (1979) Human heat tolerance. Biological and cultural adaptations. *Yearbook of Physical Anthropology*, 22: 163–186.

Hansson JE, Eklund L, Kihlberg S, Ostergren CE (1987) Vibration in car repair work. *Applied Ergonomics*, 18: 57–63.

Hardman AE, Hudson A, Jones PRM, Norgan NG (1989) Brisk walking and plasma high cholesterol concentration in previously sedentary women. *British Medical Journal*, 299: 1204–1205.

Harris JD, Kerivan JE, Russotti JS (1979) The necessity of high frequency hearing in passive sonar listening. *Military Medicine*, 144: 333–336.

Harrison MH, Brown GA, Belyavin AJ (1982) The 'Oxylog': an evaluation. *Ergonomics*, 25: 809–820.

Hartvigsen J, Lebeuf-Yde C, Lings S, *et al.* (2000) Is sitting-while-at-work bad for your low back? A systematic, critical literature review. *Scandinavian Journal of Public Health*, 28: 230–239.

Harvey R, Peper E (1997) Surface electromyography and mouse position. *Ergonomics*, 40(8): 781–789.

Haslegrave CM (1990) Auditory environment and noise assessment. In *Evaluation of Human Work: A Practical Ergonomics Methodology*, edited by JR Wilson and EN Corlett. Taylor and Francis.

Hawkins LH (1984) The possible benefits of negative ion generators. In *Health Hazards of VDUs*, edited by B Pierce. Wiley.

Hayne CR (1981) Manual transport of loads by women. *Physiotherapy*, 67: 226–231.

Haynes (1999) Can it work? Does it work? Is it worth it? *British Medical Journal*, 319: 652–653.

Health and Safety Commission (1991) *Manual Handling of Loads. Proposals for Regulation and Guidance*. Health and Safety Executive, 1 Chepstow Place, Westbourne Grove, London, W2 4TF, UK.

Health and Safety Commission (1992) *Work with Display Screen Equipment. Proposals for Regulation and Guidance*. Health and Safety Executive, 1 Chepstow Place, Westbourne Grove, London W2 4TF, UK.

Heaney CA, Inglish P (1995) Are employees who are at risk for cardiovascular disease joining worksite fitness centers? *Journal of Occupational Medicine*, 37: 718–724.

Hellebrandt FA (1938) Standing as a geotropic reflex. *American Journal of Physiology*, 121: 471–474.

Hermans V, Hautekiet M, Haex B, Spaepen AJ, Van der Perre G (1999) Lipoatrophia semicircularis and the relation with office work. *Applied Ergonomics*, 30: 319–324.

Hertzberg F (1966) *Work and the Nature of Man*. World Publishing Company.

Heuer H, Owens DA (1989) Vertical gaze direction and the resting posture of the eyes. *Perception*, 18: 363–377.

Heus R, Daanen HAM, Havenith G (1995) Physiological criteria for functioning of hands in the cold. *Applied Ergonomics*, 26: 5–13.

Hewes GW (1957) The anthropology of posture. *Scientific American*, 196: 122–132.

Hill SG, Kroemer KHE (1986) Preferred declination of the line of sight. *Human Factors*, 28: 127–134.

Hillman DJ (1985) Artificial intelligence. *Human Factors*, 27: 21–31.

Hitchcock D (1992) A convert's primer to socio-tech. *Journal for Quality and Participation*, June.

Hocking B, Savage C (1989) Cost–benefit models of hearing conservation. *Journal of Occupational Health and Safety, Australia and New Zealand*, 5(6): 525–518.

Hodges PW, Richardson CA (1996) Inefficient muscular stabilisation of the lumbar spine associated with low back pain. *Spine*, 21: 2640–2650.

Hoffman E, Ma P, See J, Yong C-K, Brand J, Poulton M (1994) Calculator logic: when and why is RPN superior to algebraic? *Applied Ergonomics*, 25: 327–333.

Holdo J (1958) Energy consumed by rockdrill noise. *The Mining Magazine*, August.

Hollands JG, Spence I (1992) Judgments of change and proportion in graphical perception. *Human Factors*, **34**: 313–334.

Hollnagel E, Woods DD (1983) Cognitive systems engineering: new wine in new bottles. *International Journal of Man–Machine Studies*, **18**: 583–600.

Hoogendoorn W, Bongers PM, de Vet HCW, Douwes M, Koes BW, Miedema MC, Ariens GAM, Bouter LM (2000) Flexion and rotation of the trunk and lifting at work are risk factors for low back pain. *Spine*, **25**: 3087–3092.

Hopkins A (1999) For whom does safety pay? The case of major accidents. *Safety Science*, **32**: 143–153.

Hopwood AG (1983) Evaluating the real benefits. In *New Office Technology: Human and Organisational Aspects*, edited by HJ Otway, M Peltu. Commission of the European Communities.

House JR, Holmes C, Allsopp AJ (1997) Prevention of heat strain by immersing the hands and forearms in water. *Journal of the Royal Naval Medical Service*, **83**: 26–30.

Hovland CI, Lunisdaire AA, Sheffiled FD (1949) *Experiments on Mass Communication*. Princeton University Press.

Howard C, Boyle MW, Eastamn V, Andre T, Motoyama T (1991) The relative effectiveness of symbols and words to convey photocopier functions. *Applied Ergonomics*, **22**: 218–224.

Howarth PA (1990) Assessment of the visual environment. In *Evaluation of Human Work: A Practical Ergonomics Methodology*, edited by JR Wilson, EN Corlett. Taylor and Francis.

Howarth PA (1999) Occulomotor changes within virtual environments. *Applied Ergonomics*, **30**: 59–68.

Howarth PA, Finch M (1999) The nauseogenicity of two methods of navigating within a virtual environment. *Applied Ergonomics*, **30**: 39–45.

Huchingson RD (1981) *New Horizons for Human Factors in Design*. McGraw-Hill.

Hughes PC (1976) Lighting in the office. *The Office*, **84**(3): 127.

Hughes PC, Meer RM (1981) Lighting for the elderly: a psychobiological approach to lighting. *Human Factors*, **23**: 65–85.

Hunter GR, McGuirk J, Mitrano N, Pearman P, Thomas B, Arrington R (1989) The effects of a weight training belt on blood pressure during exercise. *Journal of Applied Sports Science Research*, **3**: 13–18.

Hunting W, Laubli T, Grandjean E (1981) Postural loads at VDU workplaces: constrained postures. *Ergonomics*, **24**: 917–931.

International Classification of Diseases Codes (9th revision). ??????????

International Organization for Standardization (ISO) (1985) *Evaluation of Human Exposure to Whole Body Vibration – Part-1: General Requirements*. Geneva (ISO 2631-1).

Irvine CH, Snook SH, Sparshatt JH (1990) Stair way risers and treads: acceptable and preferred dimensions. *Applied Ergonomics*, **21**: 215–225.

Jacob, V, Gaultney LD, Salvendy G (1986) Strategies and biases in human decsision making and their implications for expert systems. *Behaviour and Information Technology*, **5**: 119–140.

Jacobsen FM, Wehr TA, Sack DA, James SP, Rosenthal NE (1987) Seasonal affective disorder: a review of the syndrome and its public health implications. *American Journal of Public Health*, **77**: 57–60.

Jampel RS, Shi DX (1992) The primary position of the eyes, the resetting saccade and the transverse visual head plane. *Investigative Ophthalmology and Visual Science*, **33**: 2501–2510.

Janssen W (1994) Seat-belt wearing and driving behaviour: an instrumented vehicle study. *Accident Analysis and Prevention*, **26**: 249–261.

Jaschinski W (1982) Conditions of emergency lighting. *Ergonomics*, **25**: 363–372.

Jaschinski-Kruza W (1991) Eyestrain in VDU users: viewing distance and the resting position of ocular muscles. *Human Factors*, **33**: 69–83.

Jastrzbebowski W (1857) *An Outline of Ergonomics or the Science of Work*. Published by the Central Instititute for Labour Protection, Warsaw, Poland, 2000.

Jayson MIV (1997) Presidential Address. Why does acute back pain become chronic? *Spine*, **22**(10): 1053–1056.

Jensen LK, Eenberg W (1996) Occupational as a risk factor for knee disorders. *Scandinavian Journal of Work, Environment and Health*, **22**(3): 165–175.

Johnson J (1990) Modes in non-computer devices. *International Journal of Man–Machine Studies*, **32**: 423–438.

Johnson SL (1988) Evaluation of powered screwdriver design characteristics. *Human Factors*, **30**: 61–69.

Johnson-Laird PN (1988) *The Computer and the Mind: Introduction to Cognitive Science*. Harvard University Press.

Jonsson BG, Persson J, Kilbom A (1988) Disorders of the cervicobrachial region among female workers in the electronics industry. A two year follow-up. *International Journal of Industrial Ergonomics*, **5**: 1–12.

Jorna PGA (1993) Heart rate and heart rate variations in actual and simulated flight. *Ergonomics*, **36**: 1043–1054.

Jukes TH (1986) Sugar and health. *World Reviews of Nutrition and Dietetics*, **48**: 137–194.

Kagan A, Levi L (1971) Adaptations of the psychosocial environment to man's abilities and needs. In *Society, Stress and Disease*, vol. 1, edited by L Levi. Oxford University Press.

Kalawsky RS (1999) VRUSE – a computerised diagnostic tool. *Applied Ergonomics*, **30**: 11–25.

Kaman R (1975) The comprehensibility of printed instructions and the flowchart alternative. *Human Factors*, **17**: 183–191.

Kapandji IA (1974) *The Physiology of the Joints*, vols 1, 2, 3. Churchill Livingstone.

Karat C-M (1990) Cost–benefit analysis of iterative usability testing. In *INTERACT '90*, Proceedings of the 3rd International Conference on Human-Computer Interaction, edited by Diaper *et al.* Elsevier Science Publishers (North Holland).

Karwowski W, Marek T, Noworol C (1994). The complexity–compatibility principle in the science of ergonomics. In *Advances in Industrial Ergonomics and Safety VI*, edited by F Aghazadeh, Taylor and Francis.

Kasra M, Shirazi-Adl A, Drouin G (1992) Dynamics of human intervertebral joints. *Spine*, **17**: 93–102.

Kates RW, 1994, Sustaining life on earth. *Scientific American*, October: 114–117.

Keegan JJ (1953) Alterations of the lumbar curve related to posture and seating. *Journal of Bone and Joint Surgery*, **35A**: 589–603.

Kelly J (1992) Does job re-design theory explain job redesign outcomes? *Human Relations*, **45**: 753–773.

Kemp M (1985) Repetition strain injuries: a need for proper management. *Work and People*, **10**: 25–28.

Kerslake DMcK (1982) Effects of climate. In *The Body at Work*, edited by WT Singleton. Cambridge University Press.

Keyserling WM, Punnett L, Fine L (1988) Trunk posture and back pain: identification and control of occupational risk factors. Applied Industrial Hygiene, 3: 87–92.

Keyserling WM, Stetson DS, Silverstein BA, Brouwer ML (1993) A checklist for evaluating ergonomic risk factors associated with upper extremity cumulative trauma disorders. *Ergonomics*, **36**: 807–831.

Khaleque A (1981) Job satisfaction, perceived effort and heart rate in light industrial work. *Ergonomics*, **24**: 735–742.

Kildeso J, Wyon D, Skov T, Schneider T (1999) Visual analogue scales for detecting changes in symptoms of sick building syndrome in an intervention study. *Scandinavian Journal of Work, Environment and Health*, 25: 361–367.

Kim JY, Stuart-Buttle C, Marras W (1994) The effects of mats on back and leg fatigue. *Applied Ergonomics*, 25: 29–34.

Kingery PM, Ellsworth CG, Corbett BS, Bowden RG, Brizzolara JA (1994) High cost analysis. *Journal of Occupational Medicine*, 36: 1341–1347.

Kirkesov L, Eenberg W (1996) Occupation as a risk factor for knee injuries. *Scandinavian Journal of Work, Environment and Health*, 22: 165–175.

Kirwan B (1994) *A Guide to Practical Human Reliability Assessment*. Taylor and Francis.

Kjellstrom T, Dirks KN (2001) Heat stress and work ability in the context of climate change. *Proceedings of the Australian Physiological and Pharmacological Society*, 32(2): Supplement 1.

Klausen K (1965) The form and function of the loaded human spine. *Acta Physiologica Scandinavica*, 65: 176–190.

Klausen K, Rasmussen B (1968) On the location of the line of gravity in relation to L5 in standing. *Acta Physiologica Scandinavica*, 75: 45–52.

Kleeman I (1991) *Interior Design of the Electronic Office*. Van Nostrand Rheinhold.

Klein AB, Synder-Mackler L, Roy SH, Deluca CJ (1991) Comparison of spinal mobility and isometric trunk extensor forces with EMG spectral analysis in identifying LBP. *Physical Therapy*, 71: 445–454.

Kleinmuntz B (1984) Diagnostic problem solving by computer: a historical review and the current state of the art. *Computers in Biology and Medicine*, 14: 255–270.

Kline TJ, Ghali LM, Kline DW, Brown S (1990) Visibility distance of highway signs among young, middle-aged and older observers. Icons are better than text. *Human Factors*, 30: 609–619.

Klingenstierna U, Pope MH (1987) Body height changes from vibration. *Spine*, 12(6): 566–568.

Knight CA, Caldwell GE (2000) Muscular and metabolic costs of uphill backpacking: are hiking poles beneficial? *Medicine and Science in Sports and Exercise*, 2093–2100.

Knight KK, Goetzel RZ, Fielding JE, Eisen M, Jackson GW, Kahr TY, Kenny GM, Wade SW, Duann S (1994) An evaluation of Duke University's LIVE FOR LIFE health promotion program on changes in worker absenteeism. *Journal of Occupational Medicine*, 36: 533–536.

Knowlton RG, Gilbert JC (1983) Ulnar deviation and short term strength reductions as affected by a curve-handled ripping hammer and a conventional claw hammer. *Ergonomics*, 26: 173–179.

Kochan TA (1988) On the human side of technology. *ICL Technical Journal*, November: 391–401.

Kompier M (2002) The psychosocial work environment and health. *Scandinavian Journal of Work, Environment and Health*, 28: 1–4.

Konig HL, Kreuger AP, Lang S, Sonning W (1980) *Biologic Effects of Enviromental Electromagnetism*. Springer Verlag.

Konz S (1986) Bent hammer handles. *Human Factors*, 28: 317–323.

Konz S, Bandla V, Rys M, Sambasivan J (1990) Standing on concrete versus floor mats. In *Advances in Industrial Ergonomics and Safety II*, edited by B Das. Taylor and Francis, pp. 991–998.

Kotval X, Goldberg JH (1998) Eye movements and interface component grouping. In *Proceedings of the Human Factors and Ergonomics Society 42nd Annual Meeting*, 486–490.

Krijnen A, Boer EM, Ader HJ, Osinga DSC, Bruynzeel DP (1997) Compression stockings and rubber floor mats: do they benefit workers with chronic venous insufficiency and a standing profession. *Journal of Occupational Medicine*, 39: 889–894.

Kroemer KHE (1970) Human strength, terminology, measurement and interpretation of data. *Human Factors*, 20: 481–497.

Kroemer KHE (1991) Working strenuously in heat, cold, polluted air and at altitude. *Applied Ergonomics*, 22: 385–389.

Krohn GS (1983) Flowcharts used for procedural instructions. *Human Factors*, 25: 573–581.

Kukkonen-Harjula K, Rauramaa R (1984) Oxygen consumption of lumberjacks in logging with a power saw. *Ergonomics*, 27: 59–65.

Kuller R, Lake T (1998) The impact of flicker from fluorescent lighting on well-being, performance and physiological arousal. *Ergonomics*, 41: 433–447.

Kumar S (1990) Cumulative load as a risk factor for back pain. *Spine*, 15: 1311–1316.

Kuorinka I, Forcier L (1995) *Work-Related Musculoskeletal Disorders (WMSDs): A Reference Manual for Prevention*. Taylor and Francis.

Lachman R (1989) Comprehension aids for on-line reading of expository text. *Human Factors*, 31: 1–15.

Lamble D, Kaurenen T, Laakso M, Summala H (1999) Cognitive load and detection thresholds in car following situations. *Accident Analysis and Prevention*, 31: 617–623.

Langham M, Hole G, Edwards J, O'Neil, C (2002) An analysis of 'looked but failed to see' accidents involving parked police vehicles. *Ergonomics*, 45: 167–185.

Lansdale M (1988) The psychology of personal information management. *Applied Ergonomics*, 19: 55–66.

Lansdale M, Simpson M, Stroud TRM (1990) A comparison of words and icons as external memory aids in an information retrieval task. *Behaviour and Information Technology*, 9: 111–131.

Laverson A, Norman K, Shneiderman B (1987) An evaluation of jump-ahead techniques in menu selection. *Behaviour and Information Technology*, 6: 97–108.

Lawrence JS (1969) Disc degeneration: its frequency and relation to symptoms. *Annals of Rheumatic Disease*, 28: 121–138.

Lazarus H, Hoge H (1986) Industrial safety: accoustic signals for danger situations in factories. *Applied Ergonomics*, 17: 41–46.

Lazcano O (1996) Ergonomic analysis from a social point of view: industrial processes in Mexico using the workers of Nathua ethnic origin. *Cyberg Conference*, Curtin University, Australia, September, 1996.

Leclerc A, Landre M-F, Chastang J-F, Niedhamer I, Roquelare Y (2001) Upper limb disorders in repetitive work. *Scandinavian Journal of Work, Environment and Health*, 27: 268–278.

Lee K, Swanson N, Sauter S, Wickstrom R, Waiker A, Magnum M (1992) A review of physical exercises recommended for VDU workers. *Applied Ergonomics*, 23: 387–408.

Lee KS, Chaffin DB, Herrin GD, Walker AM (1991) Effect of handle height and lower-back loading in cart pushing and pulling. *Applied Ergonomics*, 22: 117–123.

Lee YH, Her LL, Tsuang YH (1999) A comparison of sitting posture adaptations of pregnant and non-pregnant females. *International Journal of Industrial Ergonomics*, 23: 391–396.

Leigh JP, Miller TR (1997) Ranking occupations based upon the costs of job-related injuries and diseases. *Journal of Occupational and Environmental Medicine*, 39: 1170–1182.

Leithead C, Lind A (1964) *Heat Stress and Heat Disorder*. C Davis.

Leutwyler K (1995) The price of prevention. *Scientific American*, April: 124–129.

Levine F, Simone de L (1991) *Journal of the International Association for the Study of Pain*, 44: 69–72.

Liberty Mutual Research Center for Safety and Health (1996) *From Research to Reality*. Hopkinton, MA, USA.

Lie I, Watten R (1987) Oculomotor factors in the aetiology of occupational cervicobrachial diseases (OCD). *European Journal of Applied Physiology and Occupational Physiology*, 56: 151–156.

Lie I, Watten R (1994) VDU work, oculomotor strain and subjective complaints: an experimental study. *Ergonomics*, 37(8): 1149–1160.

Lim J, Hoffman E (1997) Appreciation of the zone of convenient reach by naive operators performing an assembly task. *International Journal of Industrial Ergonomics*, 19: 187–199.

Lindner H, Kropf S (1993) Asthenopic complaints associated with fluorescent lamp illumination. *International Journal of Lighting Research*, 25: 59–69.

Lindon SJ (2000) A review of psychological risk factors in back and neck pain. *Spine*, 25: 1148–1156.

Link CS, Nicholson GG, Shaddeau SA, Birch RB, Gossman MR (1990) Lumbar curvature in standing and sitting in two types of chairs: relationship of hamstring and hip flexor muscle length. *Physical Therapy*, 70: 611–618.

Linton SJ, van Tulder MW (2001) Preventative interventions for back and neck problems. *Spine*, 26: 778–787.

Liublinskaya AA (1957) The development of children's speech and thought. In *Psychology in the Soviet Union*, edited by B Simon. Routledge and Keegan Paul.

Long GM, Garvey PM (1988) The effects of target wavelength on dynamic visual acuity under photopic and scotopic viewing. *Human Factors*, 30: 3–13.

Lord MJ, Small JM, Dinsay JM, Watkins RG (1997) Lumbar lordosis: effects of standing and sitting. *Spine*, 22: 2571–2574.

Loslever P, Ranaivosa A (1993) Biomechanical and epidemiological investigation of carpal tunnel syndrome at workplaces with high risk factors. *Ergonomics*, 36: 537–554.

Lotz JC, Chin JR (2000) Intervertebral disc cell death is dependent on the magnitude and duration of static loading. *Spine*, 25(12): 1477–1483.

Louhevaara V, Smoleander J, Korhonen O, Tuomi T (1986) Maximal working times with a self-containing breathing apparatus. *Ergonomics*, 29: 77–85.

Lowe BD, Frievalds A (1998) Effect of carpal tunnel syndrome on grip force coordination on hand tools. *Ergonomics*, 41: 851–855.

Lundberg U (1995) Methods and applications of stress research. *Technology and Health Care*, 3: 3–9.

Lusted M (1993) Predicting return to work after rehabilitation for low back injury. *Australian Journal of Physiotherapy*, 39: 203–210.

Lutman ME (2000) What is the risk of noise-induced hearing loss at 80, 85, 90 dB(A) and above? *Occupational Medicine*, 50: 274–275.

Mackenzie IS, Zhang SX (2001) An empirical investigation of the novice experience with soft keyboards. *Behaviour and Information Technology*, 6: 411–418.

Mackworth NH (1948) The breakdown of vigilance during prolonged visual search. *Quarterly Journal of Experimental Psychology*, 1: 6–12.

Magnusson M, Pope MH, Hansson T (1996) Does a back support have a positive biomechanical effect? *Applied Ergonomics*, 27: 201–205.

Magora A (1972) Investigation of the relation between low back pain and occupation. *Industrial Medicine*, 39: 504–510.

Maida JC (1993) Review of Jack 5.4. *Ergonomics in Design*, July: 35–36. Human Factors and Ergonomics Society.

Malone TB, Kirkpatrick M, Mallory K, Eike D, Johnson JH, Walker RW (1980) *Human Factors Evaluation of Control Room Design and Operator Performance at Three Mile Island: Part 2*. National Technical Information Service USA Department of Commerce, Springfield VA 22161.

Malone RB (1986) The centred high-mounted brake light: a human factors succes story. *Human Factors Society Bulletin*, 29: 1–3.

Mandal AC (1981) The Seated Man (Homo sedens): the seated work position, theory and practice. *Applied Ergonomics*, 12: 19–26.

Mandal AC (1991) Investigation of the lumbar flexion of the seated man. *International Journal of Industrial Ergonomics*, 8: 75–87.

Manenica I, Corlett EN (1977) A study of a light repetitive task. *Applied Ergonomics*, 8: 103–109.

Mantey MM, Teorey TJ (1988) Cost/benefit analysis for incorporating human factors in the software lifecycle. *Communications of the ACM*, 31: 428–439.

Margulies F (1981) Evaluating man–computer systems. In *Man–Computer Interaction. Human Factors Aspects of Computers and People*, edited by B Shackel. Sijthoff and Noordhof.

Marics MA, Williges BH (1988) The intelligibility of synthesised speech in data inquiry systems. *Human Factors*, 30: 719–732.

Marras WS, Kim JY (1993) Anthropometry of industrial populations. *Ergonomics*, 36: 371–378.

Marras WS, Mirka GA (1992) A comprehensive evaluation of assymetric trunk motions. *Spine*, 17: 318–326.

Marras WS, Schoenmarklin RW (1993) Wrist motions in industry. *Ergonomics*, 4: 341–351.

Marras WS, Rangarajulu SL, Lavender SA (1987) Trunk loading and expectation. *Ergonomics*, 30: 551–562.

Marras WS, Lavender SA, Leurgans SE, Rajulu SL, Allread WG, Fathallas FA, Ferguson SA (1993) The effects of dynamic three dimensional trunk position and trunk motion characteristics on risk of injury. *Spine*, 18: 617–628.

Marras WS, Davis KG, Kirking BC, Bertsche PK (1999) A comprehensive analysis of low-back disorder risk and spinal loading during the transfer and repositioning of patients using different techniques. *Ergonomics*, 42: 904–926.

Marras WS, Allread WG, Burr DL, Fathallah F (2000) Prospective validation of a low-back disorder risk model and assessment of ergonomic interventions associated with manual materials handling. *Ergonomics*, 43: 1866–1886.

Martin J, Long J (1984) The division of attention between a primary tracking task and secondary tasks of pointing with a stylus or speaking in a simulated ship's gunfire control task. *Ergonomics*, 27: 397–408.

Martin JP (1967) *The Basal Ganglia and Posture*. Pitman Medical Publishing.

Mascie-Taylor CGN, Bogin B (1995) Human variability and plasticity. Cambridge University Press.

Matsumoto J, Hallet M (1994) Startle syndromes. In *Movement Disorders*, edited by CD Marsden, S Frans. Butterworth-Heineman.

Maslow AH (1950) *Motivation and Personality*. Harper and Row.

McAfee RB, Winn AR (1989) The use of incentives/feedback to enhance work safety: a critique of the literature. *Journal of Safety Research*, 20: 7–19.

McClelland L, Cook SW (1979) Energy conservation effects of continuous in-home feedback in all-electric houses. *Journal of Enviromental Systems*, 9: 169–173.

McCook A (2001) Napolean's revenge. *Scientific American*, July: 16.

McCormick EJ, Sanders MS (1982) *Human Factors in Engineering and Design*. McGraw-Hill.

McCoy MA, Congleton JJ, Johnston WL, Jiang BC (1988) The role of lifting belts in manual lifting. *International Journal of Industrial Ergonomics*, 2: 259–266.

McCullogh CE, Paal B, Ashdown SP (1998) An optimisation approach to apparel sizing. *Journal of the Operational Research Society*, 49: 492–499.

McFarlane GJ, Thomas E, Papageorgiou AC, Croft PR, Jayson MV (1997) Employment and physical work activities as predictors of future low back pain. *Spine*, 22(10): 1143–1149.

McGill S, Norman RW, Sharratt (1990) The effect of an abdominal belt on trunk muscle activity and intra-abdominal pressure during squat lifts. *Ergonomics*, 33: 146–160.

McGill SM, Norman RW (1992) Low back biomechanics in industry. In *Current Advances in Biomechanics*, edited by M Grabiner. Human Kinetics Publishers, Longman.

McKenzie F, Storment J, van Hook P, Armstrong TJ (1985) A program for the control of repetitive trauma disorders associated with hand tool operation in a telecommunications manufacturing facility. *American Industrial Hygiene Journal*, 46(11): 674–678.

Melamed S, Brouhis S (1996) The effects of chronic industrial noise exposure on urinary cortisol, fatigue and irritability. *Journal of Occupational and Environmental Medicine*, 38: 252–256.

Melhorn JM, Wilkinson L, Riggs JD (2001) Management of musculoskeletal pain in the workplace. *Journal of Occupational Medicine*, 43: 83–93.

Meshkati N, Robertson MM (1986) The effects of human factors on the success of technology transfer projects to industrially developing countries: a review of representative case studies. In *Human Factors in Organisational Design and Management – II*, edited by O Brown Jr, HW Hendrick. Elsevier Science Publishers (North Holland).

Middlemist RD, Knowles ES, Matter CF (1976) Personal space invasions in the lavatory: suggestive evidence for arousal. *Journal of Personality and Social Psychology*, 33: 541–546.

Miedema HS, Chorus AMJ, Wever CW, van der Linden S (1998) Chronicity of back problems during working life. *Spine*, 23: 2021–2029.

Milerad E, Ekenvall L (1990) Symptoms of the neck and upper extremities in dentists. *Scandinavian Journal of Work, Environment and Health*, 16: 129–134.

Miller GA (1956) The magical number seven, plus or minus two: some limits on our capacity for processing information. *Psychological Review*, 63: 81–97.

Miller ME, Beaton RJ (1994) The alarming sounds of silence. *Ergonomics in Design*, February: 21–23.

Miller RJ (1990) Pitfalls in the conception, manipulation and measurement of visual accommodation. *Human Factors*, 30: 27–44.

Ministry of Science and Technology (1988) *Pesquisa Antropometrica e Biomecanica dos Operarios da Industria de Tranformacao RJ*. Instituto Nacional de Tecnologia, Rio de Janeiro, Brazil.

Mitchell RV, Lawler FH, Bowen D, Mote W, Asundi P, Purswell J (1994) Effectiveness and cost-effectiveness of employer-issued back belts in areas of high risk for back injury. *Journal of Occupational Medicine*, 36: 90–94.

Miyamoto K, Iinuma N, Maeda M, Wada E, Shimizu K (1999) Effects of abdominal belts on intra-abdominal pressure, intramuscular pressure in the erectores spinae muscles and myoelectric activitiy of the trunk muscles. *Clinical Biomechanics*, 14: 79–87.

Moore JS (1997) De Quervain's tenosynovitis. *Journal of Occupational and Environmental Medicine*, 39: 990–1002.

Moores B (1972) Ergonomics or work study? *Applied Ergonomics*, 3: 147–154.

Moray N, Haudegond S, Delange M (1999) An absence of vigilance decrement for complex signals in fault detection. In *Contemporary Ergonomics 99*, edited by MA Hanson EJ Lovesy, SA Robertson. Taylor and Francis, pp. 188–192.

Morphew ME, Thordsen ML, Klein G (1998) The development of cognitive analysis methods to aid interface design. *Proceedings of the Human Factors and Ergonomics Society Conference 42nd Annual Meeting*.

Morris JM, Lucas DB, Bresler MS (1961) Role of the trunk in the stability of the spine. *Journal of Bone and Joint Surgery*, 43A: 327–351.

Morris JN, Chave SPW, Adam C, Sirey L, Epstein D, Sheehan DJ (1973) Vigorous exercise in leisure time and the incidence of coronary heart disease. *The Lancet*, 333–339.

Morris JN, Heady JA, Raffle PA, Roberts CG, Parks JW (1953) Coronary heart disease and physical activity of work. *The Lancet*, 2: 1053–1057, 1111–1120.

Morris N, Jones DM (1991) Impaired transcription from VDUs in noisy environments. In *Contemporary Ergonomics*. Taylor and Francis, pp. 184–189.

Morris NM, Rouse WB (1985) Review and evaluation of empirical research in troubleshooting. *Human Factors*, 27: 503–530.

Morrison DL, Duncan KD (1988) Strategies and tactics in fault diagnosis. *Ergonomics*, **31**: 761–784.

Mosenkis R (1994) Human factors in design. In *International Perspectives in Health and Safety*, edited by CWD van Gruting. Elsevier Science BV.

Motohashi Y (1992) Alteration of circadian rhythm on shift-working ambulance personnel. Monitoring of salivary cortisol rhythm. *Ergonomics*, **11**: 1331–1340.

Murdock BB (1962) The serial position effect in free recall. *Journal of Experimental Psychology*, **64**: 482–488.

Murrell KFH (1971) *Ergonomics: Man and His Working Environment*. Chapman and Hall.

Murrell KFH, Edwards E (1963) Field studies of an indicator of machine tool travel with special reference to the ageing worker. *Occupational Psychology*, **37**: 267–275.

Mynatt ED (1994) Auditory presentation of graphical user-interfaces. In *Proceedings of the 1st International Conference on Auditory Displays, New York*, edited by G Kramer. Addison-Wesley, pp. 533–555.

Nachemson A (1966) The load on the lumbar discs in different positions of the body. *Clinical Orthopaedics*, **45**: 107–122.

Nachemson A (1968) The possible importance of the psoas muscle for stabilisation of the lumbar spine. *Acta Orthopaedica Scandinavica*, **39**: 47–57.

Nachreiner F (1977) Experiments on the validity of vigilance experiments. In *Vigilance: Theory, Operational Performance and Physiological Correlates*, edited by RR Mackie. Plenum Press, pp. 665–678.

Nathan PA, Meadows KD, Doyle LS (1988a) Occupation as a risk factor for impaired sensory conduction of the median nerve at the carpal tunnel. *The Journal of Hand Surgery*, **13-B**: 167–170.

Nathan PA, Meadows KD, Doyle LS (1988b) Relationship of age and sex to sensory conduction of the median nerve at the carpal tunnel and association of slowed conduction with symptoms. *Muscle and Nerve*, November: 1149–1153.

Nathan PA, Keniston RC, Myers LD, Meadows KD (1992a) Obesity as a risk factor for slowing of sensory conduction of the median nerve in industry. *Journal of Occupational Medicine*, **34**: 379–383.

Nathan PA, Keniston RC, Myers LD, Meadows KD (1992b) Longitudinal study of median nerve sensory conduction in industry: relationship to age, gender, hand dominance, occupational hand use and clinical diagnosis. *The Journal of Hand Surgery*, **17-A**: 850–857.

National Institute of Safety and Health (1997) Musculoskeletal disorders and workplace factors. a critical review of epidemiological evidence for work-related musculoskeletal disorders of the neck, upper limbs and back. Centers for Disease Control, Cincinnati, OK.

National Safety Council (1993) OSHA update. *Safety and Health*, May: 10.

Nelson CM, Griffin MJ (1989) *Vibration White Finger in Dockyard Employees*. ISVR Technical Report, No, 170, University of Southampton.

Nemecek J, Grandjean E (1973) Noise in landscaped offices. *Applied Ergonomics*, **4**: 19–22.

Nesse RM (1995) *Technology Review*, May–June: 34–37.

Neumann DA, Cook TM (1985) Effect of load and carrying position on the electromyograhic activity of the gluteus medius muscle during walking. *Physical Therapy*, **65**: 305–311.

Newell A, Simon HA (1972) *Human Problem Solving*. Prentice Hall.

Newsholme EA, Blomstrand E, Ekblom B (1992) Physical and mental fatigue: metabolic mechanisms and importance of amino acids. *British Medical Bulletin*, **48**: 477–495.

Nielsen J (1990) Miniatures versus icons as a visual cache for videotex browsing. *Behaviour and Information Technology*, **9**: 441–449.

NIOSH (1981) *Work Practices Guide to Manual Handling*. NIOSH Technical Report No. 81-122. US Department of Health and Human Services, National Institute for Occupational Safety and Health, Cincinnati, OH.

Noakes T (1988) Implications of exercise testing for prediction of athletic performance: a contemporary perspective. *Medicine and Science in Sports and Exercise*, 20: 319–329.

Noakes T (1992) *The Lore of Running*. Oxford University Press.

Norman DA (1981) Categorisation of action slips. *Psychological Review*, 88: 1–15.

Norman DA (1982) Why alphabetic keyboards are not easy to use: keyboard layout doesn't matter much. *Human Factors*, 24: 509–519.

Norman DA (1984) Stages and levels in human–machine interaction. *International Journal of Man–Machine Studies*, 21: 365–375.

Norman DA (1988) *The Psychology of Everyday Things*. Basic Books.

Norman DA (1993) Toward human-centred design. *Technology Review*, July: 47–53.

Noyes JM, Frankish CR (1989) A review of speech recognition technology in the office. *Behaviour and Information Technology*, 6: 475–486.

Oakley EHN (2000) Heat and cold. In *Hunter's Diseases of Occupations*, edited by PJ Baxter *et al*. Oxford University Press.

Oberoi K, Dhillon MJ, Miglani SS (1983) A study of energy expenditure during manual and machine washing of clothes. *Ergonomics*, 26: 375–378.

Oborne DJ (1982) *Ergonomics at Work*. Wiley.

Oborne DJ (1983) Cognitive effects of passive smoking. *Ergonomics*, 26: 1163–1171.

Oborne DJ, Boarer PA (1982) Subjective responses to whole-body vibration. The effects of posture. *Ergonomics*, 25: 673–681.

O'Brien S (1994) Autonomous working at Galtee Foods Dairygold. In *P+, European Participation Monitor, The Economics of Participation*, edited by M Gold. European Foundation for the Improvement of Living and Working Conditions. ISSN 1017–6713.

Ogden GD, Levine JM, Eisner EJ (1979) Measurement of workload by secondary tasks. *Human Factors*, 21: 529–548.

Ohnaka T, Tochihara Y, Murumatsu T (1993) Physiological strains in hot-humid conditions while wearing disposable protective clothing commonly used by the asbestos removal industry. *Ergonomics*, 36: 1241–1250.

Oksa J, Ducharme MB, Rintamaki H (2001) The combined effect of repetitive work and cold on muscle strength and fatigue. *Proceedings of the Australian Physiological and Pharmacological Society*, 32(2): Supplement 1.

Olsen PL (1989) Motorcycle conspicuity revisited. *Human Factors*, 31: 141–146.

Ong CN (1984) VDU workplace design and physical fatigue: a case study in Singapore. In *Ergonomics and Health in Modern Offices*, edited by E. Grandjean. Taylor and Francis, pp. 484–493.

Orne MT (1962) On the social psychology of the psychological experiment with particular references to demand characteristics and their implications. *American Psychologist*, 17(Nov.): 776–783.

OSHA 3123 (1993) *Ergonomics Program Management Guidelines for Meatpacking Plants*. US Department of Labor, OSHA.

Ostberg O, Warell B, Nordell L (1984) ComforTable – a generic desk for the automated office. *Behaviour and Information Technology*, 3: 411–416.

Overstall FW, Exton-Smith AN, Imms FJ, Johnson AL (1977) Falls in the elderly related to postural imbalance. *British Medical Journal*, Jan: 261–264.

Owen RD (1994) Carpal tunnel syndrome: a products liability perspective. *Ergonomics*, 37: 449–476.

Owens DA, Wolf-Kelley K (1987) Near work, visual fatigue and variations of oculomotor tonus. *Investigative Ophthalmology and Vision Science*, 28: 743–749.

Oxenburgh M (1991) *Increasing Productivity and Profit Through Health and Safety*. CCH International, Australia.

Oxenburgh M (1994) *Improving Productivity and Profit Through Occupational Health and Safety*. CIH International, Australia.

Paffenbarger RS, Hyde RT, Jung DL, Wing Al (1984) Epidemiology of exercise and coronary heart disease. *Clinics in Sports Medicine*, 3: 297–318.

Palmer KT, Cooper C, Walker-Bone K, Syddall H, Coggon D (2001) Use of keyboards and symptoms in the neck and arm: evidence from a national survey. *Occupational Medicine*, 51: 392–395.

Papageorgiou AC, Croft PR, Ferry S, Jayson MIV, Silman AJ (1995) Estimating the prevalence of low back pain in the general population *Spine*, 20: 1889–1894.

Parcells C, Stommel M, Hubbard RP (1999) Mismatch of classroom furniture and student body dimensions. *Journal of Adolescent Health*, 24: 265–273.

Parenmark G, Engvall B, Malmkvist AK (1988) Ergonomic on the job training of assembly workers. *Applied Ergonomics*, 19: 143–146.

Paris CR, Thomas MH, Gilson RD, Kincaid JP (2000) Linguistic cues and memory for synthetic and natural speech. *Human Factors*, 42, 421–431.

Parkes AM (1991) Drivers' business decision making ability whilst using carphones. In *Contemporary Ergonomics 1991*, edited by EJ Lovesy, Taylor & Francis, pp. 427–432.

Parsons HM (1974) What happened at Hawthorne? *Science*, 183: 922–933.

Parsons K (1990) Human response to thermal environments. In *Principles and Methods in Evaluation of Human Work. A Practical Ergonomics Methodology*, edited by JR Wilson and EN Corlett, Taylor and Francis.

Parsons K (1993) *Human Thermal Environments*. Taylor and Francis.

Parsons KC (1995) International heat stress standards: a review. *Ergonomics*, 38: 6–22.

Patel BCK, Morgan LH (1991) Work-related penetrating eye injuries. *Acta Ophthalmologica*, 69: 377–381.

Patkin M (1989) Hand and arm pain in office workers. *Modern Medicine of South Africa*, March: 53–67.

Pearcy MJ (1993) Twisting mobility of the human back in flexed postures. *Spine*, 18: 114–119.

Peebles L, Norris B (1998) Adult data. Departmant of Trade and Industry, UK DTI/Pub 2917/3k/6/98/NPURN 98/376.

Perrot DR, Sadralodabai T, Saberi K, Strybel TZ (1991) Aurally aided visual search in the central visual field. Effects of visual load and visual enhancement of the target. *Human Factors*, 33: 389–400.

Peterson LR, Peterson MJ (1959) Short term retention of individual verbal items. *Journal of Experimental Psychology*, 58: 193–218.

Peterson S, Hoffer G (1995) Are drivers of air-bag-equipped cars more aggressive: a test of the offsetting behaviour hypothesis. *The Journal of Economics and Law*, 35: 251–264.

Pheasant ST (1982) A technique for estimating anthropometric data from the parameters of the distribution of stature. *Ergonomics*, 25: 981–982.

Pheasant ST (1986) *Bodyspace*. Taylor and Francis.

Pheasant ST (1991) *Ergonomics, Work and Health*. Macmillan Press.

Pheasant ST, O'Neill D (1975) Performance in gripping and turning – a study in hand/handle effectiveness. *Applied Ergonomics*, 6: 205–208.

Pheasant ST, Scriven JG (1983) Sex differences in strength – some implications for the design of hand tools. In *Proceedings of the Ergonomics Society's Conference 1983*, edited by K Coombes: Taylor and Francis, pp. 9–13.

Phoon WH, Lee SH, Chia SE (1993) Tinnitus in noise-exposed workers. *Occupational Medicine*, 43: 35–38.

Pinker S (1994) *The Language Instinct*. Penguin Books.

Pitner M (1990) Pathophysiology of overuse injuries of the hand and wrist. *Hand Injuries in Sports and Performing Arts, Hand Clinics of North America*, 6: 355–363.

Pope MH, Bevins T, David RPT, Wilder G, Frymower JW (1985) The relationship between anthropometric, postural, muscular and mobility characteristics of males aged 18–55. *Spine*, 10: 644–648.

Porritt, The Rt Hon Lord (1985) Slipping, tripping and falling, familiarity breeds contempt. *Ergonomics*, 28: 947–948.

Porter JM, Gyi DE, Robertson J (1992) Evaluation of a tilting computer desk. In *Contemporary Ergonomics 1992*, edited by EJ Lovesy. Taylor and Francis.

Posner M (1980) Orienting of attention. *Quarterly Journal of Experimental Psychology*, 32: 3–25.

Posner MI, Nissen MJ, Klein RM (1976) Visual dominance: an information processing account of its origins and significance. *Psychological Review*, 83: 157–171.

Poulton EC (1976a) Arousing environmental stresses can improve performance, whatever people say. *Aviation, Space and Environmental Medicine*, 47: 1193–1204.

Poulton EC (1976b) Continuous noise interferes with work by masking auditory feedback and inner speech. *Applied Ergonomics*, 7: 79–84.

Poulton EC (1977) Continuous intense noise masks auditory feedback and inner speech. *Psychological Bulletin*, 84: 977–1001.

Preece J (1993) *A Guide to Usability*. The Open University Press, Addison-Wesley.

Priatna BL (1987) Ergonomic practice and improvement of productivity in women workers. *Ergonomics in Developing Countries International Symposium*, ILO, Geneva, pp. 446–448.

Prumper J, Zapf D, Brodbeck FC, Frese M (1992) Some surprising differences between novice and expert errors in computerised office work. *Behaviour and Information Technology*, 11: 319–328.

Punnett L, Fine LJ, Keyserling WM, Herrin GD, Chaffin DB (2000) Shoulder disorders and postural stress in automobile assembly work. *Scandinavian Journal of Work Environment and Health*, 26: 283–291.

Puri V (1997) Smart cards – the smart way for banks to go? *International Journal of Bank Marketing*, 15: 134–139.

Putz-Anderson V (1988) *Cumulative Trauma Disorders: A Manual for Musculoskeletal Diseases of the Upper Limbs*. Taylor and Francis.

Pyykko I, Jantti P, Aalto H (1990) Postural control in elderly subjects. *Age and Ageing*, 19: 215–221.

Quillian MR (1969) The teachable language comprehender: a simulation program and theory of language. *Communications of the Association for Computing Machinery*, 12: 459–476.

Quinter JL (1989) The pain of 'RSI': a historical perspective. *Australian Family Physician*, 18: 1003–1009.

Quinter JL, Elvey RL (1993) Understanding 'RSI': a review of the role of peripheral neural pain and hyperalgesia. *The Journal of Manual and Manipulative Therapy*, 1: 99–105.

Rabinowitz D, Bridger RS, Lambert MI (1998) Lifting technique and abdominal belt usage. *Safety Science*, 28: 155–164.

Radwin RG, Lin ML (1993) An analytical method for characterising repetitive motion and postural stress using spectral analysis. *Ergonomics*, 36: 379–389.

Rafacz W, McGill SM (1996) Wearing an abdominal belt increases diastolic blood pressure. *Journal of Occupational and Environmental Medicine*, 38: 925–927.

Ralston JV, Pisoni DB, Lively SE, Greene BG, Mullenix JW (1991) Comprehension of synthetic speech produced by rule: word monitoring and sentence-by-sentence listening times. *Human Factors*, 33: 471–491.

Ramsey JD (1995) Task performance in the heat: a review. *Ergonomics*, 38: 154–165.

Rasmussen J (1983) Skills, rules and knowledge; signals, signs and symbols and other distinctions in human performance models. *IEEE Transactions on Systems, Man and Cybernetics*, SMC-13: 257–266.

Rasmussen J, Rouse WB (1981) *Human Detection and Diagnosis of System Failures*. Plenum.

Rasmussen J, Pejterse AM, Goodstein LP (1994) *Cognitive Systems Engineering*. Wiley.

Rasyid R, Siswanto A (1986) The effects of illumination on productivity. *Occupational Safety and Health Series*, 58: 32–34. International Labour Office.

Rauterberg M (1992) An empirical comparison of menu-selection (CUI) and desktop (GUI) computer programmes carried out by beginners and experts. *Behaviour and Information Technology*, 11: 227–236.

Ream E, Richardson A (1996) Fatigue, a concept analysis. *International Journal of Nursing Studies*, 33: 519–529.

Reason J (1990) *Human Error*. Cambridge University Press.

Reddell CR, Congleton JJ, Huchingson RD, Montgomery JF (1992) An evaluation of a weight-lifting belt and back injury prevention training class for airline baggage handlers. *Applied Ergonomics*, 23: 319–329.

Reilly T (1991) Assessment of some aspects of physical fitness. *Applied Ergonomics*, 22: 291–294.

Reilly T, Davies S (1995) Effects of a weightlifitng belt on spinal loading during performance of the dead lift. In *Sport, Leisure and Ergonomics*, edited by G Atkinson, T Reilly. E and FN Spon.

Reynolds KL, White JS, Knapik JJ, Witt CE, Amoroso P (1999) Injuries and risk factors in a 100-mile march. *Preventive Medicine*, 28: 163–173.

Rhomert W (1960) Ermittlung von Ehrolungspausen fuer statische arbeit des Menchen. *Internationale Zeitschrift Angewandte Physiologie*, 123–164.

Richardson C, Toppenburg R, Jull G (1990) An initial evaluation of eight abdominal exercises for their ability to provide stabilisation for the lumbar spine. *The Australian Journal of Physiotherapy*, 36: 6–11.

Richardson JK, Chung T, Schultz SJ, Hurvitz E (1997) A familial predisposition to lumbar disc injury. *Spine*, 22(3): 1487–1493.

Ring L (1981) *Facts on Backs*. Institute Press.

Rintamaki H, Korhonen E, Rissanen J, Oksa H, Peinimaki T (2001) Cold problems and upper limb muscular strain in food processing industry. *Proceedings of the Australian Physiological and Pharmacological Society*, 32(2): Supplement 1.

Rivas FJ, Diaz JA, Santos R (1984) Valores maximos de esfuerzo admisibles en los puestos de trabajo. *Prevencion*, 87: 20–25.

Robert JJ, Blide RW, McWhorter RPT, Coursey C (1995) The effects of a work hardening program on cardiovascular fitness and muscular strength. *Spine*, 20: 1187–1193.

Roberts DF (1995) The pervasiveness of plasticity. In: *Human Variability and Plasticity*, edited by CGN Mascie-Taylor, B Bogin. Cambridge University Press.

Robinson W (1963) Lighting and the production engineer – the 1962 Sir Alfred Herbert Paper to the Institution of Production Engineers. *The Production Engineer*, 42: 123–146.

Rock I, Palmer S (1990) The legacy of Gestalt psychology. *Scientific American*, 48–61.

Rogan A, O'Neill DH (1993) Ergonomics aspects of crop production in tropical developing countries: a literature review. *Applied Ergonomics*, 24(6): 371–386.

Rohlmann A, Claes LE, Bergmann G, Graichen F, Neef P, Wilke H-J (2001) Comparison of intradiscal pressures and spinal fixator loads for different body positions and exercises. *Ergonomics*, 44: 781–794.

Rohmert W (1987) Physiological and psychological work load measurement and analysis. In *Handbook of Human Factors*, edited by G Salvendy. Wiley.

Roquelaure Y, Mechali S, Dano C, Fanello S, Benetti F, Bureau D, Mariel M, Martin H, Derriennic F, Penneau-Fontbonne D (1997) Occupational and personal risk factors for carpal tunnel syndrome. *Scandinavian Journal of Work, Environment and Health*, 23: 264–269.

Rosa RR, Bonnet MH (1993) Performance and alertness on 8 h and 12 h rotating shifts at a natural gas utility. *Ergonomics*, 36: 1177–1193.

Roscoe AH (1993) Heart rate as a psychophysiological measure for in-flight workload assessment. *Ergonomics*, 36: 1055–1062.

Rosenberg D (1989) A cost–benefit analysis for corporate user-interface standards. In *Coordinating User-Interfaces for Consistency*, edited by D Rosenberg. Academic Press.

Rothlisberger FJ, Dickson WJ (1939) *Management and the Worker*. Harvard University Press.

Rouse WB (1988) Adaptive aiding for human–computer control. *Human Factors*, 30: 431–443.

Roy SH, DeLuca CJ, Casavant DA (1989) Lumbar muscle fatigue and chronic lower-back pain. *Spine*, 14: 992–1001.

Roy SH, DeLuca CJ, Synder-Mackler L, Emley MS, Crenshaw RL, Lyond JP (1990) Fatigue, recovery and low back pain in varsity rowers. *Medicine and Science in Sports and Exercise*, 22: 515–523.

Rubeck PA (1975) Hawthorne concept – does it affect reading progress? *The Reading Teacher*, 374–379.

Ruff C (1985) Merck case history. *Architecture* 74(11): Supplement 5.

Rushton S, Riddell PM (1999) Developing visual systems and exposure to virtual reality and stereo displays. *Applied Ergonomics*, 30: 69–78.

Rydevik B, Lundborg G, Bagge U (1981) Effects of graded compression on intraneural blood flow. *Journal of Hand Surgery*, 6: 3–12.

Rys M, Konz S (1994) Standing. *Ergonomics*, 37: 677–687.

Sanders AF (1970) Some aspects of the selective process in the functional visual field. *Ergonomics*, 13: 101–117.

Sandmark H, Hogstedt C, Vingard E (2000) Primary osteoarthrosis of the knee in men and women as a result of lifelong physical load from work. *Scandinavian Journal of Work, Environment and Health*, 26: 20–25.

Sanwo S (1996) The role of indigenous people in organisational ergonomics in Africa. *Cyberg Conference*, Curtin University, Australia, September, 1996.

Scannel K (1998) Low-cost methods of noise-control at source increase production and reduce the risk of hearing damage. *Journal of Occupational Health and Safety, Australia and New Zealand*, 14: 493–503.

Scansetti G (1984) Toxic agents emitted from office machines and materials. In *Ergonomics and Health in the Modern Office*, edited by E Grandjean. Taylor and Francis.

Schaafstal A (1993) Knowledge and strategy in diagnostic skill. *Ergonomics*, 11: 1305–1316.

Schaffer LH (1975) Multiple attention in continuous verbal tasks. In *Attention and Performance V*, edited by PMA Rabbitt, S Dornic. Academic Press.

Schierhout G, Myers JE, Brider RS (1992) Musculoskeletal pain, postural stress and ergonomic design factors in factory floor occupations in sectors of manufacturing in South Africa. *Proceedings of the PREMUS '92 Conference*, Stockholm, Sweden.

Schliefer LM, Ley R (1994) End-tidal P_{CO_2} as an index of psychophysiological activity during VDU data entry work and relaxation. *Ergonomics*, 37: 245–254.

Schultz TZ (1969) *The Life of the Primates*. Weidenfeld and Nicholson.

Schutte PC, Kielblock AJ, Van der Walt WH, Celliers CP, Strydom NB (1982) *An Analysis of the Viability of Microclimate Acclimatisation*. Chamber of Mines Research Report No. 28/82, Chamber of Mines, Johannesburg, South Africa.

Schwartz JP, Norman KL (1986) The importance of item distinctiveness on performance using a menu selection system. *Behaviour and Information Technology*, 5: 173–182.

Schwarzer AC, Aprill CN, Bogduk N (1995) The sacroiliac joint in chronic low back pain. *Spine*, 20(1): 31–37.

Sears A, Jacko JA, Brewer B, Robelo LD (1998) An empirical perspective on icon design. *Proceedings of the Human Factors and Ergonomics Society 42nd Annual Meeting*, pp. 448–452.

Sears A, Jacko JA, Chu J, Moro F (2001) The role of visual search in the design of soft keyboards. *Behaviour and Information Technology*, 3: 159–166.

Sell RG (1980) Success and failure in implementing changes in job design. *Ergonomics*, 23: 809–816.

Sell RG (1986) The politics of workplace participation. *Personnel Management*, June.

Sellen AJ (1994) Detection of everyday errors. *Applied Psychology: An International Review*, **43**: 475–498.

Selye H (1956) *The Stress of Life*. McGraw-Hill.

Seminara JL, Smith DL (1983) Remedial human factors engineering – Part 1. *Applied Ergonomics*, **14**: 253–264.

Seminara JL, Smith DL (1984) Remedial human factors engineering – Part 2. *Applied Ergonomics*, **15**: 31–44.

Sen RN, Ganguli AK, Ray GG, Chakrabarti D (1983) Tea-leaf plucking – workloads and environmental studies. *Ergonomics*, **26**: 887–893.

Shah RK (1993) A pilot survey of the traditional use of the patuka round the waist for the prevention of back pain in Nepal. *Applied Ergonomics*, **24**: 337–344.

Shelton JK, Mann-Janosi J (1992) Unhealthy health care costs. *The Journal of Medicine and Philosophy*, **17**: 7–19.

Shepard RN, Cooper LA (1982) *Mental Images and Their Transformations*. Bradford Books, MIT Press.

Shepherd A (1993) An approach to information requirements specification for process control tasks. *Ergonomics*, **36**: 1425–1437.

Shilling R, Anderson N (1986) Occupational epidemiology in developing countries. *Occupational Health and Safety, Australia and New Zealand*, **2**: 468–471.

Shinar D, Meir M, Ben-Shoham I (1998) How automatic is manual gear shifting? *Human Factors*, **42**: 647–653.

Shirazi-Adl A (1992) Finite element simulation of changes in the fluid content of human lumbar discs: mechanical and clinical implications. *Spine*, **17**: 206–212.

Shneiderman B (1988) We can design better user-interfaces: a review of human–computer interaction styles. In *Designing a Better World*, Proceedings of the 10th Congress of the International Ergonomics Association, edited by AS Adams, RR Hall, BJ McPhee, MS Oxenburgh.

Shneiderman B (1991) A taxonomy and rule base for the selection of interaction styles. In *Human Factors for Informatics Usability*, edited by B Shackel, S Richardson. Cambridge University Press.

Shneiderman B (1992) *Designing the User-Interface*. Addison-Wesley.

Shortliffe EH (1984) Reasoning methods in medical consultation systems: artificial intelligence approaches. *Computer Programs in Biomedicine*, **18**: 5–14.

Shortliffe EH, Buchanan BG, Feigenbaum EA (1979) Knowledge engineering for medical decision making: a review of computer-based decision aids. *Proceedings of the IEEE*, **67**: 1209–1224.

Shute SJ, Starr SJ (1984) Effects of adjustable furniture on VDU users. *Human Factors*, **2**: 157–70.

Siekmann H (1990) Recommended maximum temperatures for touchable surfaces. *Applied Ergonomics*, **21**: 69–73.

Simon HA (1979) Information processing models of cognition. *Annual Reviews of Psychology*, **30**: 33–396.

Simpson GC (1988) The economic justification for ergonomics. *International Journal of Industrial Ergonomics*, **2**: 157–163.

Sims M, Gillies G, Drury R (1977) Using a cool spot to improve the thermal comfort of glassmakers. *Applied Ergonomics*, **8**: 2–6.

Singleton WT (1960) An experimental investigation of speed controls for sewing machines. *Ergonomics*, **3**: 365–375.

Singleton WT (1972) *Man–Machine Systems*. Penguin Books.

Skrabenek P, McCormick. (1990) *Follies and Fallacies in Medicine*. Prometheus Books.

Sluiter J, Frings-Dressen M, van der Beek AJ, Meijman T, Heisterkamp SH (2000) Neuro-endocrine reactivity and recovery from work with different physical and mental demands. *Scandinavian Journal of Work, Environment and Health*, **26**: 306–316.

Smith A (1991) A review of the non-auditory effects of noise on health. *Work and Stress*, 5: 49–62.

Smith MJ, Colligan MJ, Tasto DL (1982) Health and safety consequences of shiftwork in the food processing industry. *Ergonomics*, 25: 133–144.

Smith MJ, Karsh BT, Conway FT, Cohen WJ, James CA, Morgan JI, Sanders K, Zehel DJ (1998) Effects of a split keyboard design and wrist rest on performance, posture and comfort. *Human Factors*, 40: 324–336.

Smith MJ, Stammerjohn LW, Cohen GF, Lalich NR (1980) Job stress in video display operations. In *Ergonomics Aspects of Visual Display Terminals*, edited by E Grandjean, E Vigliari. Taylor and Francis.

Smith PA, Wilson JR (1993) Navigation in hypertext through virtual environments. *Applied Ergonomics*, 24: 271–278.

Smythe H (1988) The 'repetitive strain injury syndrome' is referred pain from the neck. *The Journal of Rheumatology*, 15: 1604–1608.

Snijders CJ, Slagter AHE, van Strik R, Vleeming A, Stoeckart R, Stam HJ (1995) Why leg crossing? The influence of common postures on abdominal muscle activity. *Spine*, 20: 1989–1993.

Snook SH (1978) The design of manual handling tasks. *Ergonomics*, 21: 963–985.

Snook SH, Ciriello VM (1974) The effects of heat stress on a manual handling task. *American Industrial Hygiene Association Journal*, November: 681–684.

Snook SH, Ciriello VM (1991) The design of manual handling tasks: revised tables of maximum acceptable weights and forces. *Ergonomics*, 34: 1197–1213.

Snook SH, Campanelli RA, Hart JW (1978) A study of three preventative approaches to low back injury. *Journal of Occupational Medicine*, 20: 479–481.

Snook SH, Webster BS, Mcorry RW, Fogelman MT, McCann KB (1998) The reduction of chronic nonspecific low back pain through the control of early morning lumbar flexion. *Spine*, 23: 2601–1607.

Snyder KM (1991) *A Guide to Software Usability*. IBM Internal Publication.

So RHY, Lo WT, Ho TK (2001) Effects of navigation speed on motion sickness caused by an immersive virtual environment. *Human Factors*, 43: 452–461.

Soderberg GL, Cook TM (1984) Electromyography in biomechanics. *Physical Therapy*, 64: 1813–1820.

Soderberg I, Calissendorff B, Elofsson S, Knave B, Nyman KG (1983) Investigation of the visual strain experienced by microscope operators at an electronics plant. *Applied Ergonomics*, 14: 297–305.

Solomonow M, D'Ambrosia R (1987) Biomechanics of muscle overuse injuries: a theoretical approach. *Clinics in Sports Medicine*, 6: 241–257.

Sommerich CM, McGlothlin JD, Marras WS (1993) Occupational risk factors associated with soft tissue disorders of the shoulder: a review of recent literature. *Ergonomics*, 36: 697–717.

Sorkin RD (1987) Design of auditory and tactile displays. In *Handbook of Human Factors*, edited by G Salvendy. Wiley, pp. 549–576.

Soule RG, Goldman RF (1969) Energy cost of loads carried on the head, hands and feet. *Journal of Applied Physiology*, 27: 687–690.

Sperling G (1960) The information available in brief visual presentations. *Psychological Monographs*, 74.

Spurr GB, Prentice AM, Murgatroyd PR, Goldberg GR, Reina JC, Christman NT (1988) Energy expenditure from minute-by-minute heart-rate recording: a comparison with indirect calorimetry. *American Journal of Clinical Nutrition*, 48: 552–559.

Stagner R (1982) The importance of a historical context. *American Psychologist*, July: 856.

Stamper MT (1987) Good health is not for sale. *Ergonomics*, 30: 199–206.

Stanney KM, Kennedy RS, Drexler JM, Harm DL (1999) Motion sickness and proprioceptive after effects following virtual environment exposre. *Applied Ergonomics*, 30: 27–38.

Stanton N, Edworthy J (1994) Towards a methodology for constructing and evaluating representational auditory alarm displays. In *Contemporary Ergonomics 1994*, edited by SA Robertson. Taylor and Francis, pp. 360–365.

Stanton NA, Taylor RG, Tweedie LA (1992) Maps as navigational aids in hypertext environments. *Journal of Educational Multimedia and Hypermedia*, 1: 431–444.

Starr S (1983) Video display terminals: preliminary guidelines for selection, installation and use. Bell Telephone Laboratories, Short Hills, NJ.

Steeneken HJM (1998) Personal active noise reduction with integrated speech communication devices: development and assessment. *Noise and Health*, 1: 67–75.

Sterling E (1990) Designing a user-friendly tall building. In *Human Factors in Organisational Design and Management – III*, edited by K Noro, O Brown Jr. Elsevier Science (North Holland).

Sterling TD, Sterling E, Dimich-Ward (1983) Building illness in the white collar workplace. *International Journal of Health Services*, 13: 277–287.

Stevenson JM, Weber CL, Smith T, Dumas GA, Albert W (2001) A longitudinal study of the development of low back pain in an industrial population. *Spine*, 26: 1370–1377.

Stone B (2001) Virtual reality for interactive training. *International Journal of Human Computer Studies*, 55: 699–711.

Stone PT (1986) Issues in vision and lighting for users of VDUs. In *Health Hazards of VDUs?*, edited by B Pearce. Wiley.

Stotts DB (1998) The usefulness of icons on the computer interface. *Proceedings of the Human Factors and Ergonomics Society 42nd Annual Meeting*, pp. 453–457.

Stranden E (2000) Dynamic leg volume changes when sitting in a locked and free floating tilt office chair. *Ergonomics*, 43: 421–433.

Stuart-Buttle C, Marras WS, Kim JY (1993) The influence of anti-fatigue mats on back and leg fatigue. *Proceedings of the Human Factors and Ergonomics Society 37th Annual Meeting*. pp. 769–773.

Stubbs DA, Buckle PW, Hudson MP, Rivers PM (1983a) Back pain in the nursing profession I. Epidemiology and pilot methodology. *Ergonomics*, 26: 755–765.

Stubbs DA, Buckle PW, Hudson MP, Rivers PM (1983b) Back pain in the nursing profession II. The effectiveness of training. *Ergonomics*, 26: 767–779.

Stuhlen FB, DeLuca CJ (1982) Muscle fatigue monitor: a non-invasive device for observing localised muscular fatigue. *IEEE Transactions on Biomedical Engineering*, 29: 760–768.

Sturrock F, Kirwan B, Baber C (1997) Using knowledge requirements to define interface designs. *European Conference on Cognitive Science*, 9–11 April, Manchester UK.

Suleck EA (1965) Eye safety, how and how much? *Safety Maintenance*, 129: 11–12.

Sulotto F, Romano C, Dori S, Piolatto G, Chiesa A, Chiacco C, Scansetti G (1994) The prediction of recommended energy expenditure for an 8 hour work-day using an air-purifying respirator. *Ergonomics*, 36: 1479–1487.

Sundstrom ED (1986) *Workplaces*. Cambridge University Press.

Swenson EE, Purswell JL, Schlegel RE, Stanevich RL (1992) Coefficient of friction and subjective assessment of worksurfaces. *Human Factors*, 34: 67–77.

Sykes JM (1988) *Sick Building Syndrome: A Review*. Health and Safety Executive, Specialist Inspector Reports No. 10.

Talbott EO, Gibson LB, Burks A, Engberg R, McHugh P (1999) Evidence for a dose–response relationship between occupational noise exposure and blood pressure. *Occupational Medicine*, 54: 71–76.

Tanebaum D (1996) Dvorak keyboards: the typist's long-lost friend. *Technology Review*, July: 21–23.

Taylor FW (1911) *The Principles of Scientific Management*. Harper and Brothers.

Teel KS (1971) Is human factors worth the investment? *Human Factors*, 13(1): 17–21.

Teniswood CF (1987) Ergonomics in industry. In *Ergonomics in Developing Countries: An International Symposium*. International Labour Office, Geneva, pp. 183–191.

Tewari VK, Datta RK, Murthy ASR (1991) Evaluation of three manually operated weeding devices. *Applied Ergonomics*, 22: 111–116.

Thimbleby H (1991) Can humans think? The Ergonomics Society Lecture 1991. *Ergonomics*, 34: 1269–1287.

Tichauer ER (1978) *The Biomechanical Basis of Ergonomics*. Wiley.

Tile M (1984) *Fractures of the Pelvis and Acetabulum*. Williams and Wilkins.

Tipton M, Eglin C, Genser M, Golden F (1999) Immersion deaths and deterioration in swimming performance in cold water. *The Lancet*, 354: 626–628.

Tittiranonda P, Rempel D, Armstrong T, Burastero S (1999) Effect of four computer keyboards in computer users with upper extremity musculoskeletal disorders. *American Journal of Industrial Medicine*, 35: 647–661.

Tola S, Riihimaki H, Videman T (1988) Neck and shoulder symptoms among men in machine operating, dynamic physical work and sedentary work. *Scandinavian Journal of Work Environment and Health*, 14: 299–305.

Tomei F, Baccolo TP, Palmi S, Rosati MV (1999) Chronic venous disorders and occupation. *American Journal of Industrial Medicine*, 36: 653–665.

Tomlinson RW, Mannenica I (1977) A study of physiological and work study indices of forestry work. *Applied Ergonomics*, 8: 165–172.

Trezies AJ, Lyons AR, Fielding K, Davis TR (1997) Is occupation an aetiological factor in the development of trigger finger? *Journal of Hand Surgery*, 23(4): 539–540.

Trist EL, Bamforth KW (1951) Some social and psychological consequences of the longwall method of coal getting. *Human Relations*, 4: 3–38.

Tsubota K, Nakamori K (1995) Effects of ocular surface area and blink rate on tear dynamics. *Archives of Ophthalmology*, 113: 155–158.

Tulving E (1972) Episodic and semantic memory. In *Organisation of Memory*, edited by E Tulving, W Donaldson. Academic Press.

Tversky A, Kahneman D (1974) Judgement under uncertainty: heuristics and biases. *Science*, 185: 1124–1131.

Tversky A, Kahneman D (1981) The framing of decisions and the psychology of choice. *Science*, 211: 453–458.

Tynan O (1980) *Improving the Quality of Working Life in the 1980s*. WRU Occasional Paper 16, Work Research Unit, Almack House, 26 King St, London SW1Y 6RB, UK.

Udo H, Fujimora M, Yoshinaga F (1999) The effect of a tilting seat on back, lower back and legs during sitting work. *Industrial Health*, 37: 369–381.

Uiterwaal M, Glerum EBC, Busser HJ, van Lummel RC (1998) Ambulatory monitoring of physical activity in working situations – a validation study. *Journal of Medical Engineering and Technology*, 22(4): 168–172.

Ulijaszek SJ (1995) Plasticity, growth and energy balance. In *Human Variability and Plasticity*, edited by CGN Mascie-Taylor, B Bogin. Cambridge University Press.

Ulrich RS (1984) View through a window can improve recovery from surgery. *Science*, 224, (Apr.): 420–421.

Umbers IG (1979) Models of the process operator. *International Journal of Man–Machine Studies*, 11: 263–284.

University of California (1993) Not seeing red. *University of California at Berkeley Wellness Letter*, 9(2 Aug.).

Valle-Sole J, Sole A, Vallderiola F, Munoz E, gonzalez LE, Tolosa ES (1995) Reaction time and acoustic startle in normal human subjects. *Neuroscience Letters*, 195: 95–100.

van Deursen DL, Goosens RHM, Evers JJM, van der Helm FCT, van Deursen LLJM (2000) Length of the spine while sitting on a new concept of office chair. *Applied Ergonomics*, 31: 95–98.

van Deursen LL, Patijn J, Durinck JR, Brouwer R, Erven-Sommers JR, Vortman BJ (1999) Sitting and low back pain: the positive effect of rotary dynamic stimuli during prolonged sitting. *European Spine Journal*, 8: 187–193.

Van Dieen JH, de Looze MP, Hermans V (2001) Effects of dynamic office chairs on trunk kinematics, trunk extensor EMG, spinal shrinkage. *Ergonomics*, 44: 739–750.

Van Dieen JH, Toussant HM, Thissen C, Van den Ven A (1993) Spectral analysis of erector spinae AMG during intermittent isometric fatiguing exercise. *Ergonomics*, 36: 407–414.

Van Nes FL (1986) Space, colour and typography on visual display unit. *Behaviour and Information Technology*, 5: 99–118.

Van Tulder M, Malmivaara A, Esmail R, Koes B (2000) Exercise therapy for low back pain. *Spine*, 21: 2784–2796.

Van Wely PA (1961) Manual lifting of loads. *Ergonomics Research Society Annual Conference.*

Van Wyk RJ (1992) Corporate wide technological literacy. Seminar given at the Graduate School of Business, University of Cape Town, 10 November 1992.

Varghese MA, Saha PN, Atreya N (1995) Aerobic capacity of urban women homemakers in Bombay. *Ergonomics*, 38(9): 1877–1883.

Velmans M (1991) Is human information processing conscious? *Behavioural and Brain Sciences*, 14: 651–726.

Verbeek J (1991) The use of adjustable furniture: evaluation of an instruction programme for office workers. *Applied Ergonomics*, 22: 179–184.

Verhagen P, Bervoets R, Debrandere G, Millet F, Santermans G, Stuyck M, Vandermoere D, Willems G (1975) Direction of movement stereotypes in different cultural groups. In *Ethnic Variables in Human Factors Engineering*, edited by A Chapanis. Johns Hopkins University Press.

Vernon HM (1924) The influence of rest pauses and changes of posture on the capacity for muscular work. *The Medical Research Council Report*, 29: 28–55.

Vernon-Roberts B (1989) Pathology of intervertebral discs and apophyseal joints. In *The Lumbar Spine and Back Pain*, 3rd edn, edited by I Jayson, Malcolm IV. Churchill Livingstone.

Vidulich MA (1988) Speech responses and dual task performance: better time-sharing asymmetric. *Human Factors*, 30: 517–529.

Visick D, Johnson P, Long J (1984) The use of some simple speech recognisers in industrial applications. In *Interact '84*, Proceedings of the First IFIP Conference on Human–Computer Interaction, vol. 1, pp. 99–103.

Volinn E (1997) The epidemiology of low back pain in the rest of the world. *Spine*, 22(15): 1747–1754.

Waddell G (1982) An approach to backache. *British Journal of Hospital Medicine*, 28: 187–219.

Wahl G (1998) Ergonomic intervention has a return of investment of 17 to 1. *Applied Occupational and Environmental Hygiene*, 13(4): 212–215.

Walsh NE, Schwartz MS (1990) The influence of prophylactic orthoses on abdominal strength and low back injury in the workplace. *American Journal of Physical Medicine and Rehabilitation*, 69: 245–250.

Wanner HU (1984) Indoor air quality in offices. In *Ergonomics and Health in Modern Offices*, edited by E Grandjean, Taylor and Francis.

Warktosch W (1994) Ergonomic research in South African forestry. *Suid Afrikaanse Bosboutydskrik*, 171: 53–62.

Warm JS (1984) *Sustained Attention in Human Performance*. Wiley.

Wason PC, Johnson-Laird PN (1972) *Psychology of Reasoning: Structure and Content*. Batsford.

Wason PC, Shapiro D (1971) Natural and contrived experience in a reasoning problem. *Quarterly Journal of Experimental Psychology*, 23: 63–71.

Waters TR, Putz-Anderson V, Garg A, Fine LJ (1993) Revised NIOSH equation for the design and evaluation of manual lifting tasks. *Ergonomics*, 36: 749–776.

Webb JM, Kramer AF (1990) Maps or analogies? A comparison of instructional aids for menu navigation. *Human Factors*, 32: 251–266.

Weber BH (1970) Silencing of hand-held percussive rockdrills for underground operations. *CIM Bulletin*, February.

Webster BS, Snook SH (1990) The cost of compensable low back pain. *Journal of Occupational Medicine*, 32: 13–15.

Webster BS, Snook SH (1994) The cost of compensable upper extremity cumulative trauma disorders. *Journal of Occupational Medicine*, 36: 713–717.

Wedderburn AAI (1992) How fast should the shift rotate? A rejoinder. *Ergonomics*, 35: 1447–1451.

Welch RB, Langley EO, Lamaev O (1971) The measurement of fatigue in hot working conditions. *Ergonomics*, 14: 85–90.

West M, Patterson M (1998) People power. The link between job satisfaction and productivity. *CentrePiece*, Autumn: 2–5.

Westerman SJ, Cribbin T, Wilson R (2001) Virtual information space navigation: evaluating the use of head tracking. *Behaviour and Information Technology*, 20: 419–426.

Wetherall A (1981) The efficacy of some auditory–vocal subsidiary tasks as measures of the mental load on male and female drivers. *Ergonomics*, 24: 197–214.

Whistance RS (1995) Postural adaptations to workbench modifications in standing workers. *Ergonomics*, 38: 2485–2503.

Whistance RS (1996) *An Investigation Into the Ergonomics of Standing*. PhD thesis, University of Cape Town.

Whistance RS, van Geems B, Adams LP, Bridger RS (1995) Postural adaptations to workplace modifications in standing work. *Ergonomics*, 38(12): 2485–2503.

Wickens CD (1980) The structure of attentional resources. In *Attention and Performance VIII*, edited by R Nickerson. Erlbaum.

Wickens CD (1984) *Engineering Psychology and Human Performance*. Charles E Merrill.

Wickens CD (1987) Decision making. In *Human Factors Design Handbook*, edited by G Salvendy. Wiley.

Wickens CD (1992) *Engineering Psychology and Human Performance*, 2nd edn. Pearson Education.

Wickstrom G, Bendix T (2000) The 'Hawthorne effect' – what did the Hawthorne studies actually show? *Scandinavian Journal of Work, Environment and Health*, 26: 363–367.

Wieder DL (1992) Impingement syndrome: a question of mechanics. *Rehab Management*, Feb./Mar.: 87–92.

Wierwille WW (1979) Physiological measures of aircrew mental workload. *Human Factors*, 21: 575–593.

Wikstrom BO, Kjellberg A, Landstrom U (1994) Health effects of long-term occupational exposure to vibration: a review. *International Journal of Industrial Ergonomics*, 14: 273–292.

Wilde GJS (1994) *Target Risk*. PDE Publications.

Wilensky JL, Wilensky HL (1951) Personnel counselling: the Hawthorne case. *The American Journal of Sociology*, 57: 265–280.

Wilkins AJ, Nimmo-Smith I (1989) Fluorescent lighting, headaches and eyestrain. *Lighting Research and Technology*, 21: 11–18.

Wilkinson RT (1992) How fast should the shift rotate? *Ergonomics*, 35: 1425–1446.

Williamson AM, Gower CGI, Clarke BC (1994) Changing the hours of shiftwork: a comparison of 8 and 12 hour shift rosters in a group of computer operators. *Ergonomics*, 37: 287–298.

Williges RC, Wierwille WW (1979) Behavioural measures of aircrew mental workload. *Human Factors*, 21: 549–574.

Wilson J (1996a) It's tough for teachers. *The Ergonomist*, January: 4.

Wilson J (1996b) Effects of participating in virtual environments. A review of current knowledge. *Safety Science*, 23: 39–51.

Withey WR (1982) The provision of energy. In *The Body at Work*, edited by WT Singleton. Cambridge University Press.

Wogalter MS, Young SL (1991) Behavioural compliance to voice and print warnings. *Ergonomics*, 34: 79–89.

Wogalter MS, Begley PB, Scancorelli L, Brelsford JW (1997) Effectiveness of elevator service signs: measurement of perceived understandability, willingness to comply and behaviour. *Applied Ergonomics*, 28: 181–187.

Woods DD, Roth EM (1988) Cognitive engineering: human problem solving with tools. *Human Factors*, 30: 415–430.

Woodson W (1981) *Human Factors Design Handbook*. McGraw-Hill.

World Health Organization (1995) *Physical Status: The Use and Interpretation of Anthropometry*. WHO Technical Report Series 854. World Health Organization, Geneva.

Wright P (1978) Feeding the information eaters: suggestions for integrating pure and applied research on language comprehension. *Instructional Science*, 7: 249–312.

Wright P (1982) Some factors determining when instructions will be read. *Ergonomics*, 25: 225–237.

Wright P, Bartram C, Rogers N, Emslie H, Evans J, Wilson B, Belt S (2000) Text entry on handheld computers by older users. *Ergonomics*, 43: 702–716.

Wyon DP (1974) The effects of moderate heat stress on typewriting performance. *Ergonomics*, 17: 309–318.

Yadav BG, Panigrahi BK, Jena D (1976) Comparative study of hand hoes for operator comfort. *Journal of Agricultural Engineering*, 13: 91–93.

Yang KH, King AI (1984) Mechanism of facet load transmission as a hypothesis for low back pain. *Spine*, 9: 557–565.

Yeow PT, Taylor SP (1991) Effects of long term visual display unit usage on visual functions. *Optometry and Vision Science*, 68: 930–941.

Yoshitake R, Ise N, Yamada S, Tsuchiya K (1997) Ana analysis of users' preference on keyboards through ergonomic comparison among four keyboards. *Applied Human Sciences*, 16: 205–211.

Young L (1973) Human control capabilities. In *Bioastronautics Data Book*, 2nd edn, ed. JF Parker and VR West. NASA SP 3006, pp. 751–805.

Young RM (1981) The machine inside the machine: users' models of pocket calculators. *International Journal of Man–Machine Studies*, 15: 51–85.

Young SL (1990) Comprehension and memory of instruction manual warnings: conspicuous print and pictorial icons. *Human Factors*, 32: 637–649.

Zacharkow D (1988) *Posture: Sitting, Standing, Chair Design and Exercise*. Thomas Springfield.

Zhai S, Milgram P (1997) Anisotropic human performance in six degrees of freedom tracking: an evaluation of three dimensional display and control interfaces. *IEEE Transactions on Systems, Man and Cybernetics*, 27: 518–528.

Zimmerman CM, Bridger RS (2000) Effects of dialogue design on automatic teller machine (ATM) usability. Transaction times and card loss. *Behaviour and Information Technology*, 19: 441–449.

Zipp P, Haider E, Halpern N, Rhomert W (1983) Keyboard design through physiological strain measurements. *Applied Ergonomics*, 14: 117–122.

Zohar D, Fussfeld N (1981) A systems approach to organisational behaviour modification: theoretical considerations and empirical evidence. *International Review of Applied Psychology*, 30: 491–505.

Index

City of Westminster College
Learning Services / Queens Park Centre